Springer Finance

Springer Finance

Risk-Neutral Valuation: Pricing and Hedging of Financial Derivatives, 2nd Edition
N.H. Bingham and Rüdiger Kiesel
ISBN 1-85233-001-5 (2004)

Visual Exploration in Finance with Self-Organizing Maps
Guido Deboeck and Teuvo Kohonen (Editors)
ISBN 3-540-76266-3 (1998)

Mathematics of Financial Markets
Robert J. Elliott and P. Ekkehard Kopp
ISBN 0-387-98553-0 (1999)

Mathematical Models of Financial Derivatives
Y.-K. Kwok
ISBN 981-3083-25-5 (1998)

Stochastic Calculus for Finance I: The Binomial Asset Pricing Model
Steven E. Shreve
ISBN 0-387-40100-8 (2004)

Mathematics of Financial Markets, 2nd Edition
Robert J. Elliott and P. Ekkehard Kopp
ISBN 0-387-98553-0 (2004)

You-lan Zhu, Xiaonan Wu,
and I-Liang Chern

Derivative Securities and Difference Methods

With 92 Illustrations

 Springer

ory.s

You-lan Zhu
Department of Mathematics
University of North Carolina
   at Charlotte
Charlotte, NC 28223
USA

Xiaonan Wu
Department of Mathematics
Hong Kong Baptist University
Kowloon Tong
Hong Kong

I-Liang Chern
Department of Mathematics
National Taiwan University
11704 Taipei
Taiwan

Library of Congress Cataloging-in-Publication Data
Zhu, You-lan.
    Derivative securities and difference methods / You-lan Zhu, Xiaonan Wu, I-Liang Chern.
      p. cm.
    Includes bibliographical references and index.
    ISBN 0-387-20842-9 (alk. paper)
    1. Derivative securities.   2. Difference equations.   I. Wu, Xiaonan.   II. Chern, I-Liang.   III. Title.
    HG6024.A3Z497 2004
    332.64'57—dc22
                                                       2004045549

ISBN 0-387-20842-9   Printed on acid-free paper.

Printed in the United States of America.   (EB)

9 8 7 6 5 4 3 2 1   SPIN 10942215

Springer is a part of Springer Science + Business Media
springeronline.com

To our families

Preface

In the past three decades, great progress has been made in the theory and practice of financial derivative securities. Now huge volumes of financial derivative securities are traded on the market every day. This causes a big demand for experts who know how to price financial derivative securities. This book is designed as a textbook for graduate students in a mathematical finance program and as a reference book for the people who already work in this field. We hope that a person who has studied this book and who knows how to write codes for engineering computation can handle the business of providing efficient derivative-pricing codes. In order for this book to be used by various people, the prerequisites to study the majority of this book are multivariable calculus, linear algebra, and basic probability and statistics.

In this book, the determination of the prices of financial derivative securities is reduced to solving partial differential equation problems, i.e., a PDE approach is adopted in order to find the price of a derivative security. This book is divided into two parts. In the first part, we discuss how to establish the corresponding partial differential equations and find the final and necessary boundary conditions for a specific derivative product. If possible, we derive its explicit solution and describe some properties of the solution. In many cases, no explicit solution has been found so far. In these situations, we have to use numerical methods to determine the value of financial derivative securities. Therefore, the second part is devoted to numerical methods for derivative securities. There are two styles of financial derivatives: European and American. The numerical methods for both styles of derivatives are described. The main numerical method discussed is the finite-difference method. The binomial/trinomial method is also introduced as a version of an explicit finite-difference method, and the pseudo-spectral method is discussed as a high-order finite-difference method. In this part, numerical methods for determining the market price of risk are also studied as numerical methods for inverse problems. From the viewpoint of partial differential equations, solving an inverse problem means to determine a function as a variable coefficient in a partial differential equation, according to certain values of some solutions.

During the past few years, a great number of books on financial derivative securities have been published. For example: Duffie[26], Baxter and Rennie[5], Hull[39], James and Webber[43], Jarrow[44], Kwok[49], Lamberton and Lapeyre[50], Lyuu[54], Musiela and Rutkowski[58], Pelsser[60], Tavella and Randall[69], Wilmott, Dewynne, and Howison[71], Wilmott[72], Wilmott[73], and Yan[76] have published books on this subject. However, each book has its own features and gives emphasis to some aspects of this subject. Relatively speaking, this book is similar to the books by Wilmott, Dewynne, and Howison[71], Kwok[49], and Tavella and Randall[69] because all of them deal with the partial differential equation problems in finance and their numerical methods. However, this book pays more attention to numerical methods. At least the following features of this book are unique:

1. The slopes of the payoff functions for many derivative securities are discontinuous, and American-style derivative securities usually have free boundaries. These features downgrade the efficiency of numerical methods. In this book, we will discuss how to make computation more efficient even though the solutions have such types of weak singularities.

2. Many derivative security problems are defined on an infinite domain. When a numerical method is used to solve such a problem, usually a large finite domain is taken, and some artificial boundary conditions are adopted for implicit methods. This book will discuss how to convert such a problem into a problem defined on a finite domain and without requiring any artificial boundary conditions. Also, conditions guaranteeing that a random variable is defined on a finite domain are derived. When these conditions hold, any derivative security problems will be defined on a finite domain and do not need any artificial boundary conditions in order to solve them numerically.

3. A numerical method for an inverse problem in finance, for determination of the market price of risk on the spot interest rate, has been provided. As soon as having the market price of risk on the spot interest rate, we can use partial differential equations for evaluating interest rate derivatives in practice.

4. A three-factor interest rate model has been provided. All the parameters in the model and the final values of derivatives are determined from the market data. Because of this, it can be expected that the model reflects the real market. The evaluation of interest rate derivatives is reduced to solving a final value problem of a three-dimensional partial differential equation on a finite domain. Because the correctness of the formulation of the problem is proven, the numerical method for such a problem can be designed without difficulties.

The first four chapters are related to partial differential equations in finance. Chapter 1 is an introduction, where basic features of several assets and financial derivative securities are briefly described. Chapter 2 discusses basic options. In this chapter, Itô's lemma and the Black–Scholes equation are introduced, along with the derivation of the Black–Scholes formulae. These topics are followed by a discussion on American options as both linear complementarity and free-boundary problems. Also in Chapter 2, the put–call parity relation for European options as well as the put–call symmetry relations for American options are introduced. Finally, the general equations for derivative securities are derived.

In Chapter 3, exotic options such as barrier, Asian, lookback, and multi-asset are introduced. The equations, final conditions, and necessary boundary conditions for these options are provided. In this chapter, we examine a few cases in which a two-dimensional problem may be reduced to a one-dimensional problem. Explicit solutions for some of these options are provided whenever possible. Also, the formulations as free-boundary problems have been given for several American exotic options.

In Chapter 4, one-factor interest rate models, namely, the Vasicek, Cox–Ingersoll–Ross, Ho-Lee, and Hull–White models, are carefully discussed. Then, we describe how the problem of determining the market price of risk from the market data may be formulated as an inverse problem. After that, the formulations of interest rate derivatives such as bond options and swaptions are given. Then, we discuss multi-factor models and give the details of a three-factor model that can reflect the real market and be used in practice readily. The final topics in Chapter 4 are a discussion on two-factor convertible bonds and the derivation of the equivalent free-boundary problem.

Most of basic materials in these four chapters can be found from many books, for example, from the books listed above. Readers who need to know more about these subjects are referred to those books. Some of the materials are the authors' research results. For more details, see those corresponding papers given in the references.

As is well-known, exact solutions to the vanilla American option problems are not known, and the problems need to be solved numerically. For vanilla European options, if σ depends on S or the dividend is paid discretely, then explicit solutions may not exist. Therefore, in order to evaluate their prices, we often rely on numerical methods. For pricing exotic options and interest rate derivatives, we rely on numerical methods even more due to the complexity of these problems.

The next four chapters are devoted to numerical methods for partial differential equations in finance. In Chapter 5, we provide the basic numerical methods that will be used for solving partial differential equation problems and discuss the basic theory on finite-difference methods – stability, convergence and the extrapolation technique of numerical solutions. Most of these concepts can be found in many books. In the next chapter, Initial-Boundary Value and LC (Linear Complementarity) Problems, we discuss the numerical methods for European-style derivative securities and for American-style derivative securities formulated as an LC problem. In Chapter 7, Free-Boundary Problems, we carefully discuss how to solve one-factor and two-factor American option problems as free-boundary problems by implicit finite-difference methods. We also describe how to solve a two-factor convertible bond problem as a free-boundary problem by the pseudo-spectral method. In this chapter, we provide a comparison among these methods given in this chapter and in Chapter 6 as well. In the last chapter, Interest Rate Modeling, we begin with another formulation of the inverse problem and some numerical examples on the market price of risk. Then, we discuss how to price interest rate derivatives, such as swaptions, using one-factor models with numerical market prices of risk and show some numerical results. Finally, how to use the three-factor model to price interest rate derivatives in practice is discussed. Most of the materials presented in the last three chapters are from research results, especially from the authors' research.

This book can be used as a text book for two courses as a sequence. In the first course, the subject "Partial Differential Equations in Finance" is taught by using the materials in Part I. The second one is a course on "Numerical Methods for Derivative Securities" based on Part II of this book. In order to help students to understand the materials and check whether or not students have understood them, a number of problems are given at the end of each chapter. Also, at the ends of Chapters 5–8, some projects are given in order for students to be trained in evaluating derivative securities. This book is considered as a book between a textbook for graduate students and a monograph. If time is not enough, some portions can be omitted and left to students as reference materials. We have used it as a textbook in our mathematical finance program and almost all the materials can be taught in class. The following materials are basic and more important:

- Sections 1.1–2;
- Sections 2.1–4, Subsections 2.5.1–2, 2.6.1–3, 2.9.1–4, 2.10.1–2;
- Section 3.1, Subsections 3.2.1, 3.3.1–4, 3.4.1–2;
- Sections 4.1–2, 4.6–7;
- Subsections 5.1.1–2, 5.2.1–2, Section 5.3, Subsection 5.4.1, Section 5.5;
- Subsections 6.1.1–3, 6.1.5, 6.2.1–3, 6.3.2–3, 6.3.6;
- Section 7.1, Subsections 7.2.1, 7.2.3, 7.2.5–6, Section 7.3;
- Section 8.3.

These materials can be taught in one semester. Thus, if only one course is offered, this book can also be used.

During the production of this book, we received great help from our colleagues and former and current graduate students. We would like to express our thanks to them, especially, to Bing-mu Chen, Jinliang Li, Yingjun Sun, Xionghua Wu, Chenggao Yang, and Jainqing Zhang, who provided many data and plots for this book. Our gratefulness is also extended to Jeremy Lane for his grammatical editing of the entire text and Doris Huneycutt for her careful and patient word processing. Finally, special thanks go to Achi Dosanjh, the editor of this book, and the reviewers of the book for their many suggestions, which greatly improved the quality of the book.

University of North Carolina at Charlotte,   **You-lan Zhu**
Hong Kong Baptist University,   **Xiaonan Wu**
National Taiwan University,   **I-Liang Chern**

Contents

Part I

Partial Differential Equations in Finance

1

Introduction

1.1 Assets

We first introduce some basic knowledge on stocks, bonds, foreign currencies, commodities, and indices, all of which are called **assets** in this book.

Huge volumes of stocks are traded on the stock market every day, and the price of a stock changes all the time. Such a price is a typical random variable. As examples, the prices of the stocks issued by IBM and GE during the period 1990–2000 are plotted in Figs. 1.1 and 1.2. Stocks are issued by corporations. A corporation like IBM, for example, is a business unit, which gets its capital through issuing stocks. A holder of a share of stock owns a fixed portion of the corporation. For example, if a corporation issues 10 million shares of stock, then the holder of a share of stock owns 10^{-7} portion of the corporation. Stock prices, especially those of high technology stocks, have large volatilities. However, stocks usually have higher returns than bonds, which attracts people to buy them. Many corporations distribute a small amount of cash to its stockholders in proportion to the number of shares of stock held periodically. The amount is not fixed and is determined by the corporation after the stocks have been issued. This payment is commonly known as the dividend. A corporation sometimes splits its stock. When a stock split occurs, the value of the stock changes. If one share splits into two shares, the value of a new share of stock is one half of the value of an old share of stock because the value of the corporation does not change when the stock split occurs.

Bonds and other debt instruments are other types of securities that are traded on the market frequently. Besides issuing stocks, a corporation can also get its capital through issuing bonds. Governments at various levels issue bonds for some special purposes, too. The holder of a bond will get the face value (the par value) at the maturity as long as the issuer has the ability to pay. Therefore, the price of a bond usually goes to the face value as the maturity approaches, which is a feature any price of stock does not have and is called the pull-to-par phenomenon. Periodically, a bondholder will receive a fixed amount of cash, usually a few percent of the face value. This percentage is

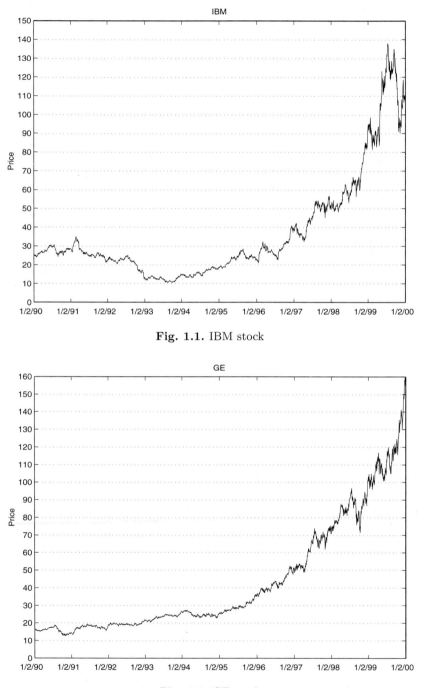

Fig. 1.1. IBM stock

Fig. 1.2. GE stock

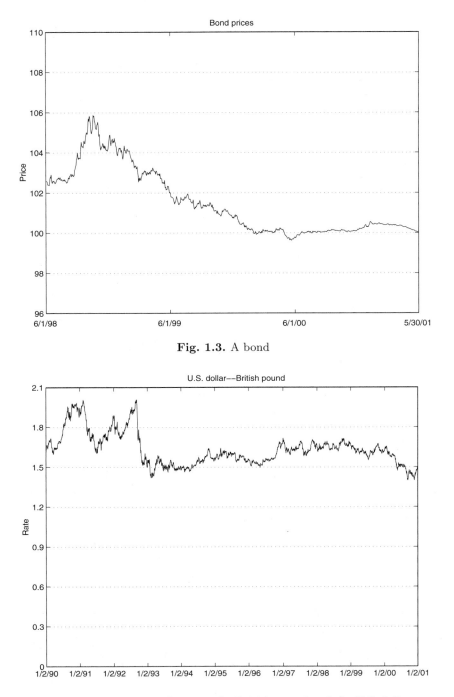

Fig. 1.3. A bond

Fig. 1.4. Exchange rate between the British pound and the U.S. dollar
(The quote is the number of U.S. dollars per British pound)

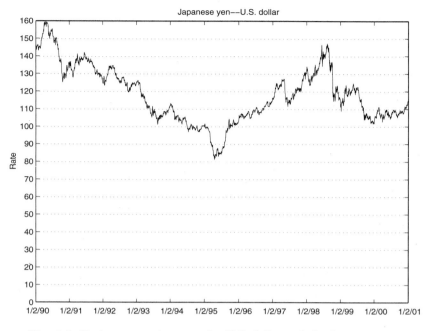

Fig. 1.5. Exchange rate between the U.S. dollar and the Japanese yen
(The quote is the number of Japanese yen per U.S. dollar)

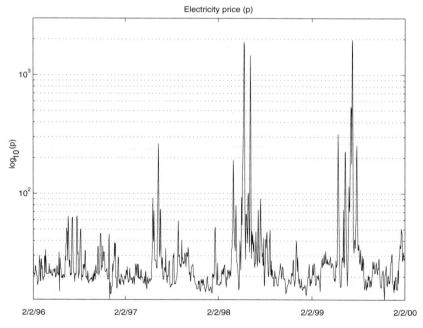

Fig. 1.6. Electricity price

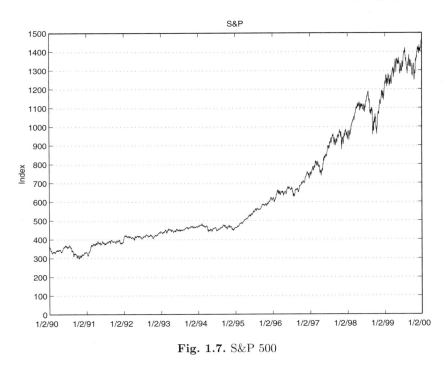

Fig. 1.7. S&P 500

specified when a bond is issued, and this amount of money is called a coupon. In Fig. 1.3, the price of a five-year government bond with a coupon 6.5% and maturing on May 31, 2001, is plotted, and the pull-to-par phenomenon can be seen clearly. Bonds usually have less risk than stocks. The bonds issued by the central government of a developed country have almost no risk.

Foreign currencies can also be sold or bought on the foreign currencies market. The exchange rate of a foreign currency, similar to the price of a stock, is also changing continuously. Fig. 1.4 shows the exchange rate between the British pound and the U.S. dollar during 1990–2000. For the same period, the exchange rate between the Japanese yen and the U.S. dollar is plotted in Fig. 1.5. A person who holds foreign currencies can always deposit them into a bank to earn some interest. The interest is paid every day. Therefore, a foreign currency can be seen as a stock that pays dividends continuously. The interest rate of the foreign currency plays the role of the dividend yield. The price of a foreign currency usually has lower volatility.

Another important financial market is the commodity market. Similar to stocks, bonds, and foreign currencies, commodities are traded on the commodity market. However, a holder of commodities sometimes has to spend money every day in order to store them in a safe place. In this case, commodities pay negative dividend yields. In addition, the prices of some commodities have certain periodicity due to the periodicity in climate. In Fig. 1.6, the electricity price of a company from 1996 to 2000 is given. The unit of the price is dollar/Megawatt. Because the range of the price p is from about \$20 to \$2,000, we use $\log_{10} p$ instead of p as the ordinate in the figure. From this figure, we can see that the price possesses some "periodicity" with a period of one year. In the peak season, late June–late August, the price is higher because the demand is higher. On a few days the price almost reached \$2,000, but usually the price is \$20–\$30. The other markets do not possess such a feature.

We can also sell or buy indices, for example, S&P 500 and S&P 100. An index is a mixture of many stocks in which the percentage for each stock is fixed. Some of the stocks may pay dividends on different days, and the dividend payment of an index can be approximately understood as a continuous payment. Figure 1.7 shows the index level of S&P 500 during 1990–2000.

1.2 Derivative Securities

On markets, not only stocks, bonds, foreign currencies, commodities, and indices can be sold or bought, but also any contracts related to an asset can be traded. A contract is an agreement on something between two parties for a specified period. Those contracts are called **derivative securities** or **contingent claims**. The assets are called the underlying assets because those securities are derived from and their prices are contingent on the assets. Forward contracts, futures contracts, and options are such securities. Moreover, many other types of derivative securities exist. For example, an interest rate derivative is a contract derived from interest rates, rather than an asset. If one party earns a certain amount of money, then the other party loses the same amount of money. When we mention the value of a contract, generally speaking, the party should be specified. For a forward contract, a futures contract, or an option, one party agrees to sell the asset or writes the option and the other party agrees to buy the asset or purchases the option. We often say that the former takes the short position and the latter takes the long position of the contract. The values of the contract for the holders of the long and short positions are the same in magnitude but have the opposite signs. In this book, the value of a forward contract, a futures contract, or an option means the value of the contract for the holder of the long position, i.e., for the buyer. The end of the specified period usually is called the expiry, the expiration date, or the maturity date. The value of the contract at expiry is called the payoff of the contract. If a contract can be exercised at any time during the period, then the derivative is called an American-style derivative; if it can be exercised at a certain time specified in the contract, then we say that it is a European-style derivative. In what follows, we give some details on three types of derivatives: forward and futures contracts, options, and interest rate derivatives.

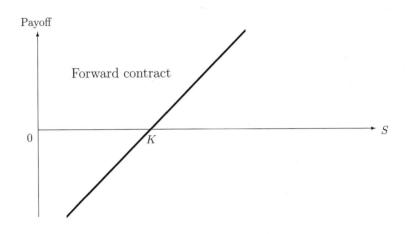

Fig. 1.8. Payoffs from forward contracts

1.2.1 Forward and Futures Contracts

A forward contract is an agreement between two parties according to which one party will buy an asset from another party at the expiry for a specified price in the agreement. The specified price, called the delivery or forward price, is chosen so that its value is zero at the initiation of the contract. As time passes, the value of the contract can become positive or negative, depending on movement in the price of the asset. If the price of the asset rises sharply soon after the contract is entered into, the value of the forward contract becomes positive. If the price drops, it becomes negative. Its payoff is positive if the price of the asset is greater than the delivery price. Otherwise, the payoff is less than or equal to zero. Concretely, the payoff of a forward contract is

$$S - K, \tag{1.1}$$

where S is the price of the asset at expiry, and K is the delivery price. The graph of the payoff as a function of S is given in Fig. 1.8.

Corporations facing foreign exchange exposure frequently enter into forward contracts on foreign currencies with financial institutions in order to avoid potential loss in profits caused by the sharp change in foreign currency exchange rates. Such a contract is usually not traded on an exchange, and we say that it is traded on the over-the-counter market.

Like a forward contract, a futures contract is also an agreement between two parties to buy or sell an asset at a certain time in the future for a specified price called the delivery or futures price. However, futures contracts are usually traded on an exchange. In order to guarantee that the contract will be honored, the exchange requests each party to deposit funds in a margin account. At the end of each day, the difference between the closing futures prices on the day and the previous day is added to or subtracted from the margin account of each party, so the net profit or loss is paid over the lifetime of the contract. Another difference between a forward and a futures contract is that an exact delivery date is sometimes not specified in a futures contract. For commodities, the delivery period is often the entire month. These differences make determining how much its holder owns more complicated than evaluating the value of a forward contract for many situations.

1.2.2 Options

Options on stocks were first traded on an organized exchange in 1973. Now, options are traded on a large number of exchanges throughout the world. Huge volumes of options are also traded in the over-the-counter market by financial institutions. Not only can an option be on assets, but it can also be on another derivative, for example, a futures contract.

An option gives the holder a right, not an obligation, to do something. Hence, the holder does not have to exercise this right. There are two basic types of options: call options and put options. A call option gives the holder a right to buy the underlying asset at or by a certain date for a specified price. A put option gives the holder a right to sell the underlying asset at or by a certain date for a specified price. The price in the contract is known as the exercise or strike price. Let E denote this price. If the price of stock is less than E, then a holder of a call option will not exercise the option because there is no point in buying for E a stock that has a market value less than E. That is, the payoff of a call option is

$$\max(S - E, 0), \tag{1.2}$$

where S is the price of the stock at the end of the option's life. Similarly, the payoff of a put option is

$$\max(E - S, 0). \tag{1.3}$$

The graphs of these two functions are given in Figs. 1.9 and 1.10, respectively. If S is greater than E and if the holder of a call option could immediately exercise the option, then the holder would earn some money. In this case, we say that the call option is "in the money." If S is less than E, it is said that the call option is "out of the money" because the holder would lose money if the option were exercised immediately. If $S = E$, we say that the call option is "at the money" because no cash flow would come in or go out if exercising the option. For a put option, the situation is similar.

Unlike a forward or futures contract, where its value is equal to zero at the initiation of the contract, the holder of an option has to pay a certain amount of money to the writer of the option in order to enter into the option contract because the payoff is always nonnegative. This payment is usually called a premium. Because the holder of an option paid a premium when the option

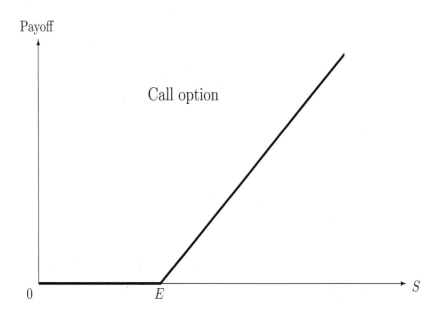

Fig. 1.9. The payoff diagram for a call option

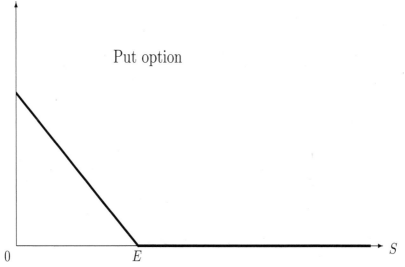

Fig. 1.10. The payoff diagram for a put option

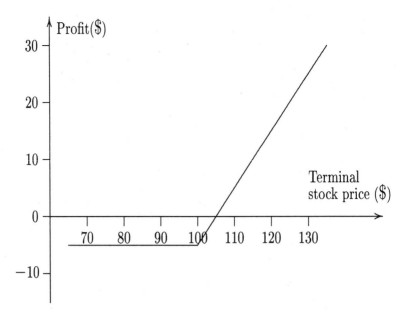

Fig. 1.11. Profit of a call option

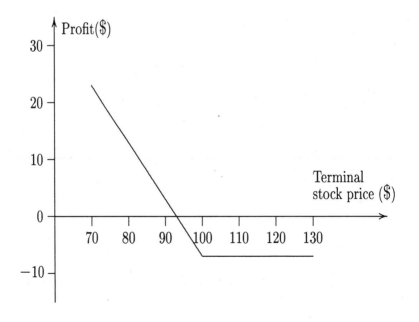

Fig. 1.12. Profit of a put option

was bought, the value of the payoff is not the money earned. The money earned, the profit, is the payoff function minus the value of the premium at the end of the option's life. In Figs. 1.11 and 1.12, the profits are shown for a call option and a put option, respectively.

Many newspapers, such as the *Wall Street Journal* and the *Financial Times*, carry vanilla option quotations, which refer to trading that took place on exchanges on the previous workday. Most of the options that are traded on exchanges are American.

1.2.3 Interest Rate Derivatives

On markets, the interest rate of a loan depends on the term of the loan. For example, one-year loans and five-year loans usually have different interest rates. The interest rate for the shortest possible deposit is commonly called the spot rate. The interest rates are constantly changing, at least on a daily basis, therefore in some cases it is necessary to consider interest rates as random variables. A security dependent on interest rates is called an interest rate derivative.

A typical interest rate derivative is a swap. Such a contract is an agreement between two parties A and B, usually between a bank and a corporation. In the contract, they agree that during the next few years A will pay interest on certain capital to B at a fixed interest rate, and B will pay interest on the same capital to A at a floating interest rate. One needs to determine what the fixed interest rate should be according to the current market and the value of the contract with a given fixed interest rate at a specific time. Other than swaps, there are many other types of interest rate derivatives, such as caps, floors and collars and options on swaps, caps, floors, and collars, which will be discussed in Chapter 4.

1.2.4 Factors Affecting Derivative Prices

It is clear that the value of a derivative depends on underlying random variables. For example, the value of an option on a stock depends on the price of the stock, and the manner of dependence is determined by the feature of the option. For a call option, the profit of the holder of the option will increase when the price of the stock rises. Therefore, the price of a call option is an increasing function of the stock price. The price of a put option is a decreasing function because the holder of the option will receive less when the stock price increases. The value of a swap depends on the floating interest rate. For the party who receives the floating interest, the swap is more valuable when the floating interest rate rises.

The price of a derivative also depends on some parameters. For example, the volatility of the underlying random variable, the time to expiration, the strike price of an option, and the fixed interest rate of a swap are such parameters. If the volatility is large, then the chance that the underlying random variable becomes very large or very small increases. For a call option, no matter whether the stock price is $10 less than the strike price or $100 less than the strike price at expiry, the value of the option is zero. However, if at expiry the stock price changes from $10 higher than the strike price to $100 higher than the strike price, then the value of the call option increases from $10 to $100. Therefore, a holder of a call option benefits from a large volatility. Similarly, a holder of a put option also benefits from a large volatility. That is, both calls and puts become more valuable as the volatility increases. If the time to expiration is longer, then the value of an American-style derivative should increase because the holder of the long-life American derivative has more exercise opportunities than the owner of the short-life American derivative. A rise in the strike price makes a call option less valuable because in order to get one share, the owner of the call option needs to pay more. However, a rise in the strike price makes the price of a put option go up because the owner of an option will get more money from one share. Increasing the fixed interest rate of a swap causes the party who receives the floating interest to pay more money, so the value to this party decreases. Derivatives depend on more parameters, such as the short-term interest rate and the dividend.

1.2.5 Functions of Derivative Securities

Generally speaking, derivatives have two primary uses: speculation and hedging.

Speculation. Suppose the price of a particular stock, for example, IBM, is $120 today. An investor who believes that the stock is going to rise can purchase shares in that company. If he is correct, he makes money; if he is wrong, he loses money. This investor is speculating. If the share price rises from $120 to $150, he makes a profit of $30 per share or

$$\frac{30}{120} = 25\%.$$

If it falls to $90, he takes a loss of $30 or

$$\frac{30}{120} = 25\%.$$

Alternatively, instead of shares, he buys call options with exercise price $120. Assume that such an option costs $15. If the share price rises to $150, the payoff is $30 with a profit of $15; if the share price drops to $90, then the loss is $15 because the payoff is zero and the premium is $15. Therefore, the profit or loss is magnified to

$$\frac{15}{15} = 100\%!$$

Consequently, options can be a cheap way of exposing a portfolio to a large amount of risk.

On the other hand, if the investor thinks that IBM shares are going to fall, he can do one of two things: sell shares or buy puts. If he speculates by selling shares that he does not own,[1] he will profit from a fall in IBM shares. He can also buy puts and will earn money from a fall of the stock.

Hedging. An owner of an asset will lose money when the price of the asset falls. The value of a put option rises when the asset price falls. What happens to the value of a portfolio containing both the asset and the put when the asset price falls? Clearly, the answer depends on the ratio of assets to options in the portfolio. If the ratio is equal to zero, the value rises, whereas if the ratio is infinity, the value falls. Somewhere between these two extremes is a ratio at which a small movement in the asset does not result in any movement in the value of the portfolio. Such a portfolio is risk-free. The reduction of risk by taking advantage of such correlations between the asset and option price movements is called hedging. This is one example explaining how options are used in hedging. Call options can also be used for the purpose of hedging.

Problems

1. What is the difference between taking a long position in a forward contract and in a call option?
2. Suppose the futures price of gold is currently $324 per ounce. An investor takes a short position in a futures contract for the delivery of 1,000 ounces. How much does the investor gain or lose if the price of gold at the end of the contract is (a) $310 per ounce; (b) $340 per ounce?
3. An investor holds a European call option on a stock with an exercise price of $88 and the option costs $3.50. For what value of the stock at maturity will the investor exercise the option, and for what value of the stock at maturity will the investor make a profit?
4. An investor holds a European put option for a stock with an exercise price of $88 and the option costs $3.50. Find the gain or loss to the investor if the stock price at maturity is (a) $93.50; (b) $81.50.
5. A company will receive a certain amount of foreign currency in one year. To reduce the risk of the changes in the exchange rate, what type of contract is appropriate for hedging?
6. Suppose a fund manager holds 10 million shares of IBM stock worth $68 each. What option is suitable for reducing the risk of decline of the stock price in the next three months?
7. A stock price is $67 just before a dividend of $1.50 is paid. What is the stock price immediately after the payment?

---

[1] This is perfectly legal in many markets. This action is called a short selling.

2

Basic Options

2.1 Asset Price Model and Itô's Lemma

2.1.1 A Model for Asset Prices

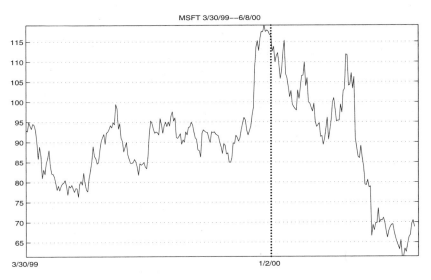

Fig. 2.1. Stock price of Microsoft

As examples, in Figs. 1.1–1.7 we showed how the prices of assets vary with time t. Fig. 2.1 shows the stock price of Microsoft Inc. in the period March 30, 1999, to June 8, 2000. From these figures, we can see the following: the graphs are jagged, and the size of the jags changes all the time. This means that we cannot express S as a smooth function of t, and it is difficult to predict exactly the price at time t from the price before time t. It is natural to think of the price at time t as a random variable. Now let us give a model for such a random variable.

Suppose that at time t the asset price is S. Let us consider a small subsequent time interval dt, during which S changes to $S+dS$. (We use the notation df for the small change in any quantity f over this time interval.) How might we model the corresponding return rate on the asset, dS/S?

Assume that the return rate on the asset can be described by the following stochastic differential equation:

$$\frac{dS}{S} = \mu(S,t)dt + \sigma(S,t)dX, \tag{2.1}$$

where μ and σ are called the **drift** and the **volatility**, respectively, and dX is known as a Wiener process defined by

$$\begin{cases} dX = \phi\sqrt{dt}, \\ \phi \text{ being a standardized normal random variable.} \end{cases}$$

In this model, the first part is an anticipated and deterministic return rate, and the second part is the random return rate of the asset price in response to unexpected events. As we can see, the random increment dS depends solely on today's price. This independence from the past is known as the Markov property. In many situations, it is assumed that μ and σ are constants. Due to its simplicity, this is a popular model for asset prices

For a random variable ψ with a probability density function $f(\psi)$ defined on $(-\infty, \infty)$, the expectation of any function $F(\psi)$, $\mathrm{E}\left[F\left(\psi\right)\right]$, is given by

$$\mathrm{E}\left[F\left(\psi\right)\right] = \int_{-\infty}^{\infty} F(\psi)f(\psi)d\psi.$$

The variance of $F(\psi)$, $\mathrm{Var}\left[F(\psi)\right]$, is defined by

$$\mathrm{Var}\left[F(\psi)\right] = \mathrm{E}\left[(F(\psi) - \mathrm{E}\left[F(\psi)\right])^2\right].$$

According to these definitions, for any constants a, b, c, and random variable W, we have

$$\mathrm{E}\left[aW - b\right] = a\mathrm{E}\left[W\right] - b,$$
$$\mathrm{Var}\left[W\right] = \mathrm{E}\left[(W - \mathrm{E}\left[W\right])^2\right]$$
$$= \mathrm{E}\left[W^2\right] - (\mathrm{E}\left[W\right])^2$$

and

$$\mathrm{Var}\left[\frac{W}{c}\right] = \frac{1}{c^2}\mathrm{Var}\left[W\right].$$

For a standardized normal random variable ϕ, the probability density function is

$$\frac{1}{\sqrt{2\pi}}e^{-\phi^2/2}, \qquad -\infty < \phi < \infty.$$

As a probability density function, this function satisfies[1]

$$\int_{-\infty}^{\infty} \frac{1}{\sqrt{2\pi}} e^{-\phi^2/2} d\phi = 1.$$

Therefore we have

$$\mathrm{E}\,[\phi] = \int_{-\infty}^{\infty} \phi \frac{1}{\sqrt{2\pi}} e^{-\phi^2/2} d\phi = 0$$

and

$$\begin{aligned}
\mathrm{Var}\,[\phi] &= \mathrm{E}\,[\phi^2] \\
&= \int_{-\infty}^{\infty} \phi^2 \frac{1}{\sqrt{2\pi}} e^{-\phi^2/2} d\phi \\
&= -\frac{1}{\sqrt{2\pi}} \int_{-\infty}^{\infty} \phi d\left(e^{-\phi^2/2}\right) \\
&= \frac{1}{\sqrt{2\pi}} \int_{-\infty}^{\infty} e^{-\phi^2/2} d\phi \\
&= 1.
\end{aligned}$$

From these we obtain

$$\mathrm{E}\,[dX] = \mathrm{E}\,[\phi]\sqrt{dt} = 0$$

and

$$\mathrm{Var}\,[dX] = \mathrm{E}\,[dX^2] = \mathrm{E}\,[\phi^2]\,dt = dt.$$

Consequently[2]

$$\mathrm{E}\,[dS] = \mathrm{E}\,[\sigma S\,dX + \mu S\,dt] = \mu S\,dt,$$

and

$$\begin{aligned}
\mathrm{Var}\,[dS] &= \mathrm{E}\,[dS^2] - (\mathrm{E}\,[dS])^2 \\
&= \mathrm{E}\,[\sigma^2 S^2 dX^2 + 2\sigma S^2 \mu\,dt\,dX + \mu^2 S^2 dt^2] - \mu^2 S^2 dt^2 \\
&= \sigma^2 S^2 dt.
\end{aligned}$$

The square root of the variance is known as the standard deviation. Thus, the deviation of the return on the asset is proportional to σ. This means that an asset price with a larger σ would appear more jagged. Typically, for stocks, indices, exchange rates, and bonds, the value of σ is in the range

---

[1] Because $\int_0^\infty e^{-x^2/2}dx \times \int_0^\infty e^{-y^2/2}dy = \int_0^{\pi/2} \int_0^\infty e^{-r^2/2}r\,dr\,d\theta = \pi/2$, we have $\int_0^\infty e^{-\phi^2/2}d\phi = \sqrt{\pi/2}$.

[2] Here, dX is a random variable and S is unchanged. In stochastic calculus, it is called conditional expectation (see [46] and [5]).

0.02 to 0.4. Usually, the volatility of stocks is greater than indices, exchange rates, and bonds, and government bonds have the smallest volatility among these. Among shares, high-tech companies tend to have higher volatility than other companies. For example, assume that the volatility of the price of IBM stock is a constant during 1990–2000, then its value is 0.31. Under the same assumption, for the price of GE stock, $\sigma = 0.23$. For S&P 500, British pound–U.S. dollar exchange rate, Japanese yen–U.S. dollar exchange rate, and a five-year government bond with coupon 6.5% and maturing on May 31, 2001, $\sigma = 0.10, 0.11, 0.12$, and 0.03, respectively. For the bond, we assume that σ depends on the time to maturity. Clearly, at maturity σ is zero. The value 0.03 means that the maximum value of σ is 0.03. In practice, the volatility is often quoted as a percentage so that $\sigma = 0.2$ would be 20% volatility.

If $\sigma = 0$, then

$$\frac{dS}{S} = \mu dt$$

and

$$S(t) = S_0 e^{\mu(t-t_0)},$$

where S_0 is the value of the asset at $t = t_0$.

In this asset price model, μ and σ are two parameters. In general, these parameters depend on the asset price S and time t, i.e., $\mu = \mu(S,t)$, $\sigma = \sigma(S,t)$. According to the historical data, we can determine these parameters (or parameter functions) for the past by statistical analysis. If we assume that μ and σ depend on S only, then the functions $\mu(S)$ and $\sigma(S)$ determined by the historical data can be used for the future.

A Wiener process is also referred to as a Brownian motion. There are many excellent books on the Brownian motion. Readers interested in this subject can read, for example, [46]. A basic and very important feature of the Wiener process is that the sum of two independent Wiener processes is also a Wiener process, and the variance of the sum is the sum of the two original variances. That is, if $dX_1 = \phi_1 \sqrt{dt_1}$ and $dX_2 = \phi_2 \sqrt{dt_2}$ are two Wiener processes and they are independent, namely, $\mathrm{E}[\phi_1 \phi_2] = 0$, then

$$dX_3 = dX_1 + dX_2 = \phi_1 \sqrt{dt_1} + \phi_2 \sqrt{dt_2} = \phi_3 \sqrt{dt_1 + dt_2}, \qquad (2.2)$$

where ϕ_3 is also a standardized normal random variable.

2.1.2 Itô's Lemma

There is a practical lower bound for the basic time-step dt of the random walk of an asset price. Thus, an asset price is a discrete random variable. However, sometimes the lower bound is so small that we consider an asset price as a continuous random variable. Also, because it is much more efficient to solve the resulting differential equations than to evaluate options by direct simulation of the random walk on a practical time scale, we will assume that an asset price is a continuous random variable even if the basic time-step is not very small.

Before coming to Itô's lemma, we need one result, which we do not prove. This result is, with probability one,

$$dX^2 = \phi^2 dt \to dt \quad \text{as} \quad dt \to 0.$$

This can be explained as follows. Because

$$\mathrm{E}\left[dX^2\right] = \mathrm{E}\left[\phi^2\right] dt = dt$$

and

$$\mathrm{Var}\left[dX^2\right] = \mathrm{E}\left[dX^4\right] - (\mathrm{E}\left[dX^2\right])^2 = O(dt^2),$$

the variance of dX^2 is very small and the smaller dt becomes, the closer dX^2 comes to being equal to dt.

Assume

$$dS = a(S,t)dt + b(S,t)dX$$

and suppose $f(S,t)$ is a smooth function of a random variable S and time t. We need to find a stochastic differential equation for f. If we vary S and t by a small amount dS and dt, then f also varies by a small amount. From the Taylor series expansion we can write

$$df = \frac{\partial f}{\partial S}dS + \frac{\partial f}{\partial t}dt + \frac{1}{2}\left(\frac{\partial^2 f}{\partial S^2}dS^2 + 2\frac{\partial^2 f}{\partial t \partial S}dt\, dS + \frac{\partial^2 f}{\partial t^2}dt^2\right) + \cdots.$$

Because

$$dS^2 = [a(S,t)dt + b(S,t)dX]^2 = \left(adt + b\phi\sqrt{dt}\right)^2$$

$$= a^2(dt)^2 + 2ab\phi(dt)^{3/2} + b^2\phi^2 dt \to b^2 dt \quad \text{as} \quad dt \to 0,$$

we have

$$df = \frac{\partial f}{\partial S}dS + \left(\frac{\partial f}{\partial t} + \frac{1}{2}b^2\frac{\partial^2 f}{\partial S^2}\right)dt \quad \text{as} \quad dt \to 0 \qquad (2.3)$$

or in the form of a stochastic differential equation

$$df = b\frac{\partial f}{\partial S}dX + \left(\frac{\partial f}{\partial t} + \frac{1}{2}b^2\frac{\partial^2 f}{\partial S^2} + a\frac{\partial f}{\partial S}\right)dt.$$

This is Itô's lemma. If in the asset price model (2.1), μ and σ are constants, i.e.,

$$dS = \mu S dt + \sigma S dX,$$

then Itô's lemma is in the form:

$$df = \frac{\partial f}{\partial S}dS + \left(\frac{\partial f}{\partial t} + \frac{1}{2}\sigma^2 S^2\frac{\partial^2 f}{\partial S^2}\right)dt$$

$$= \sigma S\frac{\partial f}{\partial S}dX + \left(\frac{\partial f}{\partial t} + \frac{1}{2}\sigma^2 S^2\frac{\partial^2 f}{\partial S^2} + \mu S\frac{\partial f}{\partial S}\right)dt.$$

2.1.3 Expectation and Variance of Lognormal Random Variables

As a simple example, consider the function $f(S) = \ln S$. Differentiation of this function gives

$$\frac{df}{dS} = \frac{1}{S} \quad \text{and} \quad \frac{d^2 f}{dS^2} = -\frac{1}{S^2}.$$

Suppose that S satisfies (2.1) with constant μ and σ, i.e., $dS = \mu S dt + \sigma S dX$. Using Itô's lemma, for $\ln S$ we have

$$d\ln S = \sigma dX + \left(\mu - \frac{\sigma^2}{2}\right) dt = m dt + \sigma dX, \qquad (2.4)$$

where

$$m = \mu - \frac{\sigma^2}{2}. \qquad (2.5)$$

It is clear that

$$\mathrm{E}\left[d\ln S\right] = \mathrm{E}\left[m dt + \sigma dX\right] = m dt$$

and

$$\begin{aligned}
\mathrm{Var}\left[d\ln S\right] &= \mathrm{E}\left[(d\ln S)^2\right] - (\mathrm{E}\left[d\ln S\right])^2 \\
&= \mathrm{E}\left[\sigma^2 dX^2 + 2\sigma m\, dt\, dX + m^2 dt^2\right] - m^2 dt^2 \\
&= \sigma^2 \mathrm{E}\left[\phi^2 dt\right] = \sigma^2 dt.
\end{aligned}$$

From (2.4), the probability density function for $d\ln S$ is[3]

$$\frac{1}{\sigma\sqrt{2\pi dt}} e^{-(d\ln S - m dt)^2/2\sigma^2 dt}.$$

Let $d\ln S = \ln S' - \ln S$. Then for $\ln S'$, the probability density function is

$$G_1\left(\ln S'\right) = \frac{1}{\sigma\sqrt{2\pi dt}} e^{-\left[\ln S' - \ln S - m dt\right]^2/2\sigma^2 dt}.$$

Here, S is the value of the asset at time t and S' is the value of the asset at time $t + dt$ which is a random variable. In Fig. 2.2, the curve of $G_1(\ln S')$ with $\ln S + m dt = 0$ and $\sigma\sqrt{dt} = 0.2$ is shown.

---

[3] • Here $e^{-(d\ln S - m dt)^2/2\sigma^2 dt}$ means $e^{-(d\ln S - m dt)^2/(2\sigma^2 dt)}$. That is, in the expression $(d\ln S - m dt)^2/2\sigma^2 dt$, the division between $(d\ln S - m dt)^2$ and $2\sigma^2 dt$ should be done after $2 \times \sigma^2 \times dt$ is obtained. Throughout the entire book we use such a notation.

• If x is a normal random variable and its mean and variance are a and b^2, then its probability density function is

$$\frac{1}{b\sqrt{2\pi}} e^{-(x-a)^2/2b^2}.$$

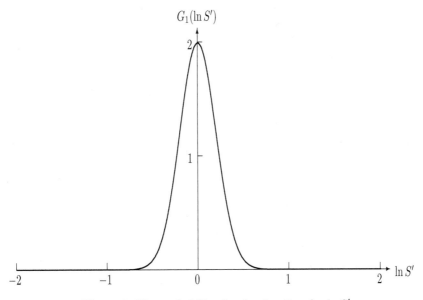

Fig. 2.2. The probability density function for $\ln S'$
with $\ln S + mdt = 0$ and $\sigma\sqrt{dt} = 0.2$

Suppose that for S' the probability density function is $G(S')$. Because[4]

$$G(S')\,dS' = \frac{1}{\sigma\sqrt{2\pi dt}}e^{-(\ln S' - \ln S - mdt)^2/2\sigma^2 dt}\,d\ln S',$$

we have

$$G(S') = \frac{1}{S'\sigma\sqrt{2\pi dt}}e^{-(\ln S' - \ln S - mdt)^2/2\sigma^2 dt}.$$

In Fig. 2.3, the corresponding curve of $G(S')$ is given. This is called a lognormal distribution because the corresponding distribution for $\ln S'$ is normal.

Now the question is what are $\mathrm{E}[S']$ and $\mathrm{Var}[S']$. Because we have the probability density function, let

$$y = \frac{\ln S' - \ln S - mdt}{\sigma\sqrt{dt}}$$

---

[4]If for x the probability density function is $f(x)$, then the probability of $x \in [x, x+dx]$ is $f(x)dx$. If $y = y(x)$ and $y(x)$ is a nondecreasing function, then $x \in [x, x+dx]$ if and only if $y \in [y(x),\ y(x+dx)] \approx \left[y(x),\ y(x) + \dfrac{dy}{dx}dx\right]$. Thus, the probability of the event $y \in \left[y(x), y(x) + \dfrac{dy}{dx}dx\right]$ is also $f(x)dx$. If for y the probability density function is $f_1(y)$, then $f_1(y)dy = f(x)dx$, from which we have $f_1(y) = f(x(y))\dfrac{dx}{dy}$. If $x = \ln S'$ and $y = S'$, then $f_1(S') = f(x(y))\dfrac{dx}{dy} = f(\ln S')\dfrac{1}{S'}$.

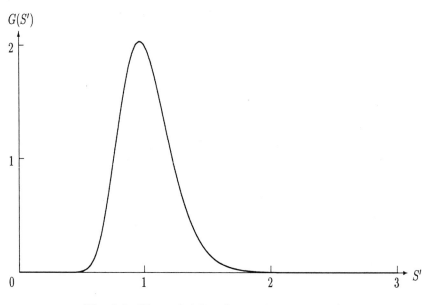

Fig. 2.3. The probability density function for S'
with $\ln S + mdt = 0$ and $\sigma\sqrt{dt} = 0.2$

and we have

$$
\mathrm{E}\left[S'\right] = \int_0^\infty G(S')S'dS'
$$

$$
= \frac{1}{\sigma\sqrt{2\pi dt}} \int_0^\infty e^{-(\ln S' - \ln S - mdt)^2/2\sigma^2 dt} \frac{1}{S'} \times S'dS'
$$

$$
= \frac{1}{\sqrt{2\pi}} \int_{-\infty}^\infty e^{-y^2/2} e^{y\sigma\sqrt{dt} + \ln S + mdt} dy
$$

$$
= \frac{1}{\sqrt{2\pi}} \int_{-\infty}^\infty e^{-\left(y-\sigma\sqrt{dt}\right)^2/2} \times e^{\sigma^2 dt/2 + \ln S + mdt} dy
$$

$$
= e^{\sigma^2 dt/2 + \ln S + mdt} = Se^{\mu dt},
$$

$$
\mathrm{E}\left[S'^2\right] = \int_0^\infty G(S')S'^2 dS'
$$

$$
= \frac{1}{\sigma\sqrt{2\pi dt}} \int_0^\infty e^{-(\ln S' - \ln S - mdt)^2/2\sigma^2 dt} \frac{1}{S'} S'^2 dS'
$$

$$
= \frac{1}{\sqrt{2\pi}} \int_{-\infty}^\infty e^{-y^2/2} \times e^{2(y\sigma\sqrt{dt} + \ln S + mdt)} dy
$$

$$
= \frac{1}{\sqrt{2\pi}} \int_{-\infty}^\infty e^{-\left(y-2\sigma\sqrt{dt}\right)^2/2} e^{2\sigma^2 dt + 2(\ln S + mdt)} dy
$$

$$
= e^{2\sigma^2 dt + \ln S^2 + 2mdt} = S^2 e^{2\mu dt + \sigma^2 dt}
$$

and

$$\mathrm{Var}\left[S'\right] = S^2 e^{2\mu dt + \sigma^2 dt} - S^2 e^{2\mu dt}$$
$$= S^2 e^{2\mu dt}\left(e^{\sigma^2 dt} - 1\right),$$

where we have used (2.5).

If m and σ in (2.4) are constants, then for a large time period $T - t$, we can have

$$\ln S_T - \ln S = \int_t^T d\ln S = m\int_t^T dt + \sigma\int_t^T dX(t) = m(T - t) + \sigma\phi\sqrt{T - t},$$

where S_T is the stock price at time T, S is the stock price at time t, and ϕ is a standardized normal random variable. Here we used the relation $\int_t^T dX(t) = \phi\sqrt{T - t}$, which can be obtained from (2.2). Therefore, in this case, the probability density function for S_T is

$$G(S_T) = \frac{1}{S_T\sigma\sqrt{2\pi(T - t)}} e^{-[\ln S_T - \ln S - m(T - t)]^2/2\sigma^2(T - t)}$$

and

$$\begin{cases} \mathrm{E}\left[S_T\right] = S e^{\mu(T-t)}, \\ \mathrm{Var}\left[S_T\right] = S^2 e^{2\mu(T-t)}\left[e^{\sigma^2(T-t)} - 1\right], \end{cases} \tag{2.6}$$

where μ is given by (2.5):

$$\mu = m + \frac{\sigma^2}{2}.$$

2.2 Derivation of the Black–Scholes Equation

2.2.1 Arbitrage Arguments

In the modern world, financial transactions may be done simultaneously in more than one market. Suppose the price of gold is $324 per ounce in New York and 37,275 Japanese Yen in Tokyo, while the exchange rate is 1 U.S. dollar for 115 Japanese Yen. An arbitrageur, who is always looking for any arbitrage opportunities to make money, could simultaneously buy 1,000 ounces in New York, sell them in Tokyo to obtain a risk-free profit of

$$37{,}275 \times 1{,}000/115 - 324 \times 1{,}000 = \$130.43$$

if the transaction costs can be ignored. Small investors may not profit from such opportunity due to the transaction costs. However, the transaction costs for large investors might be a small portion of the profit, which makes the arbitrage opportunity very attractive.

Arbitrage opportunities usually cannot last long. As arbitrageurs buy the gold in New York, the price of the gold will rise. Similarly, as they sell the gold in Tokyo, the price of the gold will be driven down. Very quickly, the ratio between the two prices will become closer to the current exchange rate. In practice, due to the existence of arbitrageurs, very few arbitrage opportunities can be observed. Therefore, throughout this book we will assume that there are no arbitrage opportunities for financial transactions.

Let us make the following assumptions: both the borrowing interest rate and the lending interest rate are equal to r, short selling is permitted, the assets and options are divisible, and there is no transaction cost. Then, we can conclude that the absence of arbitrage opportunities is equivalent to all risk-free portfolios having the same return rate r.

Let us show this point. Suppose that Π is the value of a portfolio and that during a time step dt the return of the portfolio $d\Pi$ is risk-free. If

$$d\Pi > r\Pi dt,$$

then an arbitrageur could make a risk-free profit $d\Pi - r\Pi dt$ during the time step dt by borrowing an amount Π from a bank to invest in the portfolio. Conversely, if

$$d\Pi < r\Pi dt,$$

then the arbitrageur would short the portfolio and invest Π in a bank and get a net income $r\Pi dt - d\Pi$ during the time step dt without taking any risk. Only when

$$d\Pi = r\Pi dt$$

holds, is it guaranteed that there are no arbitrage opportunities.

In the next subsection, we will derive the equation the prices of derivative securities should satisfy by using the conclusion that all risk-free portfolios have the same return rate r.

2.2.2 The Black–Scholes Equation

Let V denote the value of an option that depends on the value of the underlying asset S and time t, i.e., $V = V(S,t)$. It is not necessary at this stage to specify whether V is a call or a put; indeed, V can even be the value of a whole portfolio of various options. For simplicity, the reader may think of a simple call or put.

Assume that in a time step dt, the underlying asset pays out a dividend $SD_0 dt$, where D_0 is a constant known as the dividend yield.[5] Suppose S satisfies (2.1) with constant μ and σ:

---

[5]This dividend structure is a good model for an index. In this case, many discrete dividend payments are paid at many different times, and the dividend payment can be approximated by a continuous yield without serious error. Also, if the asset is a foreign currency, then the interest rate for the foreign currency plays the role of D_0.

$$dS = \mu S dt + \sigma S dX.$$

According to Itô's lemma (2.3), the random walk followed by V is given by

$$dV = \frac{\partial V}{\partial S} dS + \left(\frac{\partial V}{\partial t} + \frac{1}{2} \sigma^2 S^2 \frac{\partial^2 V}{\partial S^2} \right) dt. \qquad (2.7)$$

Here we require V to have at least one t derivative and two S derivatives.

Now construct a portfolio consisting of one option and a number $-\Delta$ of the underlying asset. This number is as yet unspecified. The value of this portfolio is

$$\Pi = V - \Delta S. \qquad (2.8)$$

Because the portfolio contains one option and a number $-\Delta$ of the underlying asset, and the owner of the portfolio receives $S D_0 dt$ for every asset held, the earnings for the owner of the portfolio during the time step dt is

$$d\Pi = dV - \Delta dS - \Delta S D_0 dt.$$

Using (2.7), we find that Π follows the random walk

$$d\Pi = \left(\frac{\partial V}{\partial S} - \Delta \right) dS + \left(\frac{\partial V}{\partial t} + \frac{1}{2} \sigma^2 S^2 \frac{\partial^2 V}{\partial S^2} - \Delta S D_0 \right) dt.$$

The random component in this random walk can be eliminated by choosing

$$\Delta = \frac{\partial V}{\partial S}. \qquad (2.9)$$

This results in a portfolio whose increment is wholly deterministic:

$$d\Pi = \left(\frac{\partial V}{\partial t} + \frac{1}{2} \sigma^2 S^2 \frac{\partial^2 V}{\partial S^2} - \Delta S D_0 \right) dt. \qquad (2.10)$$

Because the return for any risk-free portfolio should be r, we have

$$r \Pi dt = d\Pi = \left(\frac{\partial V}{\partial t} + \frac{1}{2} \sigma^2 S^2 \frac{\partial^2 V}{\partial S^2} - \Delta S D_0 \right) dt. \qquad (2.11)$$

Substituting (2.8) and (2.9) into (2.11) and dividing by dt, we arrive at

$$\frac{\partial V}{\partial t} + \frac{1}{2} \sigma^2 S^2 \frac{\partial^2 V}{\partial S^2} + (r - D_0) S \frac{\partial V}{\partial S} - rV = 0. \qquad (2.12)$$

When we take different Π for different S and t, we can conclude that (2.12) holds on a domain. In this book, (2.12) is called the Black–Scholes partial differential equation, or the Black–Scholes equation,[6] even though $D_0 = 0$ in the equation originally given by Black and Scholes (see [10]). With its extensions and variants, it plays the major role in the rest of the book.

About the derivation of this equation and the equation itself, we give more explanation here.

---

[6]It is also called Black–Scholes–Merton differential equation (see [39]).

- The key idea of deriving this equation is to eliminate the uncertainty or the risk. $d\Pi$ is not a differential in the usual sense. It is the earning of the holder of the portfolio during the time step dt. Therefore, $\Delta S D_0 dt$ appear. In the derivation, in order to eliminate any small risk, Δ is chosen before an uncertainty appears and does not depend on the coming risk. Therefore, no differential of Δ is needed.

- The linear differential operator given by

$$\frac{\partial}{\partial t} + \frac{1}{2}\sigma^2 S^2 \frac{\partial^2}{\partial S^2} + (r - D_0)S\frac{\partial}{\partial S} - r$$

has a financial interpretation as a measure of the difference between the return on a hedged option portfolio

$$\frac{\partial}{\partial t} + \frac{1}{2}\sigma^2 S^2 \frac{\partial^2}{\partial S^2} - D_0 S\frac{\partial}{\partial S}$$

and the return on a bank deposit

$$r\left(1 - S\frac{\partial}{\partial S}\right).$$

Although the difference between the two returns is identically zero for European options, we will later see that the difference between the two returns may be nonzero for American options.

- From the Black–Scholes equation (2.12), we know that the parameter μ in (2.1) does not affect the option price, i.e., the option price determined by this equation is independent of the average return rate of an asset price per unit time.

- From the derivation procedure of the Black-Scholes equation we know that the Black-Scholes equation still holds if σ, r, and D_0 are functions of S and t.

- If dividends are paid only on certain dates, then the money the owner of the portfolio will get during the time period $[t, t+dt]$ is

$$dV - \Delta dS - \Delta D(S,t)dt,$$

where $D(S,t)$ is a sum of several Dirac delta functions. Suppose that a stock pays dividend $D_1(S)$ at time t_1 and $D_2(S)$ at time t_2 for a share, where $D_1(S) \le S$ and $D_2(S) \le S$. Then

$$D(S,t) = D_1(S)\delta(t - t_1) + D_2(S)\delta(t - t_2),$$

where the Dirac delta function[7] $\delta(t)$ is defined as follows:

---

[7]It is the limit as $\varepsilon \to 0$ of the one-parameter family of functions:

$$\delta_\varepsilon(x) = \begin{cases} \dfrac{1}{2\varepsilon}, & -\varepsilon \le x \le \varepsilon, \\ 0, & |x| > \varepsilon. \end{cases}$$

$$\delta(t) = \begin{cases} 0, & \text{if } t \neq 0, \\ \infty, & \text{if } t = 0 \end{cases} \quad \text{and} \quad \int_{-\infty}^{\infty} \delta(t) = 1.$$

In this case, the modified Black–Scholes equation is in the form

$$\frac{\partial V}{\partial t} + \frac{1}{2}\sigma^2 S^2 \frac{\partial^2 V}{\partial S^2} + (rS - D(S,t))\frac{\partial V}{\partial S} - rV = 0. \qquad (2.13)$$

2.2.3 Final Conditions for the Black–Scholes Equation

From the derivation of the Black–Scholes equation (2.12), we know that this partial differential equation holds for any option (or portfolio of options) whose value depends only on S and t. In order to determine a unique solution of the Black–Scholes equation, the solution at the expiry, $t = T$, needs to be given. This condition is called the final condition for the partial differential equation. Different options satisfy the same partial differential equation, but different final conditions. Therefore, in order to determine the price of an option, we need to know the value of the option at time T. In what follows, we will derive the final conditions for call and put options.

Final condition for call options. Let us examine what a holder of a call option will do just at the moment of expiry. If $S > E$ at expiry, it makes financial sense for the holder to exercise the call option, handing over an amount E for an asset worth S. The money earned by the holder from such a transaction is then $S - E$. On the other hand, if $S < E$ at expiry, the holder should not exercise the option because the holder would lose an amount of $E - S$. In this case, the option expires valueless. Thus, the value of the call option at expiry can be written as

$$C(S,T) = \max(S - E, 0). \qquad (2.14)$$

This function giving the value of a call option at expiry is usually called the payoff function of a call option. In Fig. 1.9, we plot $\max(S-E,0)$ as a function of S, which is usually known as a payoff diagram. A call option with such a payoff is the simplest call option and is known as a vanilla call option.

Final condition for put options. Each option or each portfolio of options has its own payoff at expiry. An argument similar to that given above for the value of a call at expiry leads to the payoff for a put option. At expiry, the put option is worthless if $S > E$ but has the value $E - S$ for $S < E$. Thus, the payoff function of a put option is

$$P(S,T) = \max(E - S, 0). \qquad (2.15)$$

The payoff diagram for a put is shown in Fig. 1.10 where the line shows the payoff function $\max(E-S,0)$. In order to distinguish this put option from other more complicated put options, sometimes it is referred to as the vanilla put option.

2.2.4 Hedging and Greeks

The way to reduce the sensitivity of a portfolio to the movement of something by taking opposite positions in different financial instruments is called hedging. Hedging is a basic concept in finance. When we derived the Black–Scholes equation in Subsection 2.2.2, we chose the delta to be $\dfrac{\partial V}{\partial S}$, so that the portfolio Π became risk-free. This gives an important example on how hedging is applied. Let us see another example of hedging that is similar to what we have used in deriving the Black–Scholes equation.

Consider a call option on a stock. Figure 2.4 shows the relation between the call price and the underlying stock price. When the stock price corresponds to point A, the option price corresponds to point B and the Δ of the call is the slope of the line indicated. As an approximation

$$\Delta = \frac{\delta c}{\delta S},$$

where δS is a small change in the stock price and δc is the corresponding change in the call price.

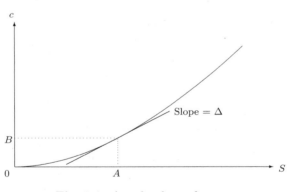

Fig. 2.4. Δ = the slope of a curve

Assume that the delta of the call option is 0.7 and a writer sold 10,000 shares of call options. Then, the writer's position could be hedged by buying $0.7 \times 10{,}000 = 7{,}000$ shares of stocks. If the stock price goes up by $0.50, the writer will earn $3,500 from the 7,000 shares of stocks held. At the same time, the price of call options will go up approximately $0.7 \times 0.5 = \$0.35$, and he will lose $10{,}000 \times \$0.35 = \$3{,}500$ from 10,000 shares of option he sold. Therefore, the net profit or loss is about zero. If the price falls down by a small amount, the situation is similar. Consequently, the writer's position has been hedged quite well as long as the movement of the price is small. This is called delta hedging.

In the example above, we have considered only a call option. Actually, any portfolio can be hedged in this way. If Π denotes the price of option, then the slope is

$$\Delta = \frac{\partial \Pi}{\partial S}.$$

If the movement of the price is not very small, then it might be necessary to use the value of the second derivative of the portfolio with respect to S in order to eliminate most of the risk. The second derivative is known as gamma

$$\Gamma = \frac{\partial^2 \Pi}{\partial S^2}.$$

When hedging in practice, some other values, for example, $\frac{\partial \Pi}{\partial t}$, $\frac{\partial \Pi}{\partial \sigma}$, $\frac{\partial \Pi}{\partial r}$, $\frac{\partial \Pi}{\partial D_0}$, may need to be known. Usually, $\frac{\partial \Pi}{\partial t}$, $\frac{\partial \Pi}{\partial \sigma}$, and $\frac{\partial \Pi}{\partial r}$ are called theta, vega, and rho, respectively; namely, the following notation is used:

$$\Theta = \frac{\partial \Pi}{\partial t}, \quad \mathcal{V} = \frac{\partial \Pi}{\partial \sigma},$$

and

$$\rho = \frac{\partial \Pi}{\partial r}.$$

In currency options, D_0 is the interest rate in the foreign country. Thus, $\frac{\partial \Pi}{\partial D_0}$ is also known as rho. In order to distinguish $\frac{\partial \Pi}{\partial r}$ and $\frac{\partial \Pi}{\partial D_0}$, here we define

$$\rho_d = \frac{\partial \Pi}{\partial D_0}.$$

These quantities are usually referred to as Greeks.

2.3 Two Transformations on the Black–Scholes Equation

In this section, we introduce two transformations. One transformation reduces the Black–Scholes equation to the heat equation. Because Green's function[8] of the heat equation has an analytic expression, we can obtain an analytic expression of Green's function for the Black–Scholes equation using the inverse transformation. Based on this, analytic expressions of European call and put option prices can be derived. These are the famous Black–Scholes formulae. When σ depends on S, or the coefficient of $\frac{\partial V}{\partial S}$ is more complicated, analytic

---

[8]The definitions of Green's functions of the heat equation and the Black–Scholes equation are given in Section 2.4.

expressions of option prices may not exist. In this case, we have to use numerical methods. Also sometimes (for example, for a call option), the solution is unbounded. It is not convenient to solve a problem numerically on an infinite domain with an unbounded solution. Therefore in Subsection 2.3.2, we also provide a transformation under which the Black–Scholes equation on $[0, \infty)$ becomes an equation on $[0, 1)$ with a bounded solution.

2.3.1 Converting the Black–Scholes Equation into a Heat Equation

The price of a European option is a solution of the following problem:

$$\begin{cases} \dfrac{\partial V}{\partial t} + \dfrac{1}{2}\sigma^2 S^2 \dfrac{\partial^2 V}{\partial S^2} + (r - D_0)S\dfrac{\partial V}{\partial S} - rV = 0, \quad 0 \le S, \ t \le T, \\ V(S,T) = V_T(S), \quad 0 \le S. \end{cases} \tag{2.16}$$

The payoff function $V_T(S)$ is determined by the feature of the option. For example, the payoffs of European calls and puts are given by

$$V(S,T) = \max(\pm(S - 1), \ 0), \quad 0 \le S,$$

where $+$ and $-$ in $\pm$ correspond to call and put options, respectively. Here, the exercise price is 1 because we assume that both the price of the stock and the price of option have been divided by the exercise price. We call a problem with such a payoff a standard put/call problem. Let us set

$$\begin{cases} y = \ln S, \\ \tau = T - t, \\ V(S,t) = e^{-r(T-t)}v(y,\tau). \end{cases} \tag{2.17}$$

Because

$$\frac{\partial V}{\partial t} = e^{-r(T-t)}\left(rv - \frac{\partial v}{\partial \tau}\right),$$

$$\frac{\partial V}{\partial S} = e^{-r(T-t)}\frac{\partial v}{\partial y}\frac{dy}{dS} = e^{-r(T-t)}\frac{1}{S}\frac{\partial v}{\partial y},$$

$$\frac{\partial^2 V}{\partial S^2} = \frac{\partial}{\partial S}\left(\frac{\partial V}{\partial S}\right) = \frac{\partial}{\partial S}\left(e^{-r(T-t)}\frac{1}{S}\frac{\partial v}{\partial y}\right)$$

$$= e^{-r(T-t)}\left(-\frac{1}{S^2}\frac{\partial v}{\partial y} + \frac{1}{S^2}\frac{\partial^2 v}{\partial y^2}\right),$$

the Black–Scholes equation is converted into

$$-\frac{\partial v}{\partial \tau} + \frac{1}{2}\sigma^2\left(\frac{\partial^2 v}{\partial y^2} - \frac{\partial v}{\partial y}\right) + (r - D_0)\frac{\partial v}{\partial y} = 0,$$

and the problem above becomes

$$\begin{cases} \dfrac{\partial v}{\partial \tau} = \dfrac{1}{2}\sigma^2 \dfrac{\partial^2 v}{\partial y^2} + \left(r - D_0 - \dfrac{1}{2}\sigma^2\right)\dfrac{\partial v}{\partial y}, & -\infty < y < \infty, \ \ 0 \le \tau, \\ v(y,0) = V_T(e^y), \quad -\infty < y < \infty. \end{cases} \quad (2.18)$$

Furthermore, we let

$$\begin{cases} x = y + \left(r - D_0 - \dfrac{1}{2}\sigma^2\right)\tau, \\ \bar{\tau} = \dfrac{1}{2}\sigma^2\tau, \\ v(y,\tau) = u(x,\bar{\tau}). \end{cases} \quad (2.19)$$

Noticing the relations

$$\frac{\partial v}{\partial \tau} = \frac{1}{2}\sigma^2 \frac{\partial u}{\partial \bar{\tau}} + \left(r - D_0 - \frac{1}{2}\sigma^2\right)\frac{\partial u}{\partial x},$$

$$\frac{\partial v}{\partial y} = \frac{\partial u}{\partial x},$$

$$\frac{\partial^2 v}{\partial y^2} = \frac{\partial^2 u}{\partial x^2},$$

we finally arrive at

$$\begin{cases} \dfrac{\partial u}{\partial \bar{\tau}} = \dfrac{\partial^2 u}{\partial x^2}, & -\infty < x < \infty, \ \ 0 \le \bar{\tau}, \\ u(x,0) = V_T(e^x), \quad -\infty < x < \infty. \end{cases} \quad (2.20)$$

The partial differential equation in this problem is usually called the heat or diffusion equation.

Before we go to the next subsection, we point out the following:

1. From (2.17) and (2.19), we know

$$V(S,t) = e^{-r(T-t)}u(\ln S + (r - D_0 - \sigma^2/2)(T-t), \ \sigma^2(T-t)/2)$$

$$= e^{-r(T-t)}u\left(\ln \frac{Se^{-D_0(T-t)}}{e^{-r(T-t)}} - \sigma^2(T-t)/2, \ \sigma^2(T-t)/2\right).$$

Therefore, besides those parameters in the payoff function $V_T(S)$, $V(S,t)$ depends on only three parameters: $Se^{-D_0(T-t)}$, $e^{-r(T-t)}$, and $\sigma^2(T-t)/2$.

2. Actually, the transformations (2.17) and (2.19) can be combined into one transformation

$$\begin{cases} x = \ln S + \left(r - D_0 - \dfrac{1}{2}\sigma^2\right)(T - t), \\[2mm] \bar{\tau} = \dfrac{1}{2}\sigma^2(T - t), \\[2mm] V(S, t) = e^{-r(T-t)}u(x, \bar{\tau}). \end{cases} \tag{2.21}$$

That is, through the transformation (2.21), the Black–Scholes equation can be directly converted into the heat equation. The reason we complete the transformation through two steps is to see the function of each single transformation. In fact, from the derivation we know the following:

- Through setting $\tau = T - t$, we change a problem with a final condition to a problem with an initial condition and let the initial time be zero.
- The transformation $y = \ln S$ is to reduce an equation with variable coefficients to one with constant coefficients. This is the transformation by which the Euler equation in ordinary differential equations becomes a differential equation with constant coefficients.
- Letting $V = e^{-r(T-t)}v(y, \tau)$, we eliminate the term rV in the equation. This is similar to the fact that an equation $\dfrac{dV}{d\tau} - rV = f$ can be written as $\dfrac{d(e^{-r\tau}V)}{d\tau} = e^{-r\tau}f$ after the equation is multiplied by $e^{-r\tau}$. The factor $e^{-r\tau}$ is called an integrating factor for the ordinary differential equation. If r depends on t, then the integrating factor is $e^{-\int_0^\tau r(T-s)ds} = e^{-\int_t^T r(s)ds}$ and the term rV can be eliminated in the same way.
- The transformation $x = y + (r - D_0 - \sigma^2/2)\tau$ is to eliminate the term $(r - D_0 - \sigma^2/2)\dfrac{\partial v}{\partial y}$. This is similar to reducing the simplest hyperbolic partial differential equation $\dfrac{\partial v}{\partial \tau} - a\dfrac{\partial v}{\partial y} = 0$ to an ordinary differential equation. For this case, the characteristic equation is $\dfrac{dy}{d\tau} = -a$ and its solution is $y = -a\tau + c$ or $y + a\tau = c$. Let $x = y + a\tau$ and $v(y, \tau) = u(x, \tau)$, then the hyperbolic partial differential equation becomes $\dfrac{\partial u(x, \tau)}{\partial \tau} = 0$. If a depends on t, then the solution of the characteristic equation is $y = -\int_0^\tau a(T - s)ds + c = -\int_t^T a(s)ds + c$. Letting $x = y + \int_t^T a(s)ds$ and $v(y, \tau) = u(x, \tau)$, we still have $\dfrac{\partial u(x, \tau)}{\partial \tau} = 0$.
- In order for the coefficient of $\dfrac{\partial^2 u}{\partial x^2}$ to be one, we let $\bar{\tau} = \sigma^2\tau/2$. If σ depends on t, then letting $\bar{\tau} = \frac{1}{2}\int_0^\tau \sigma^2(T - s)ds = \frac{1}{2}\int_t^T \sigma^2(s)ds$ can still make the coefficient of $\dfrac{\partial^2 u}{\partial x^2}$ be one.

3. From the explanation on the function of each single transformation given above, we can see that if r, D_0, and σ are not constant, but depend on t

only, then the Black–Scholes equation can still be converted into a heat equation by the following transformation

$$\begin{cases} x = \ln S + \int_t^T \left[r(s) - D_0(s) - \sigma^2(s)/2 \right] ds, \\[2mm] \bar{\tau} = \dfrac{1}{2} \int_t^T \sigma^2(s) ds, \\[2mm] V(S,t) = \mathrm{e}^{-\int_t^T r(s)ds} u(x,\bar{\tau}) \end{cases} \qquad (2.22)$$

and the solution $V(S,t)$ possesses the following form:

$$\mathrm{e}^{-\int_t^T r(s)ds} u \left(\ln \frac{S \mathrm{e}^{-\int_t^T D_0(s)ds}}{\mathrm{e}^{-\int_t^T r(s)ds}} - \frac{1}{2} \int_t^T \sigma^2(s)ds, \ \frac{1}{2} \int_t^T \sigma^2(s)ds \right),$$

$$(2.23)$$

where $u(x,\bar{\tau})$ is a solution of the heat equation (see [71]). This is left for the reader as an exercise (Problem 11). There, in order to see the function of each part of the transformation, the reader is asked to reduce the Black–Scholes equation with time-dependent parameters to a heat equation through two steps.

4. The transformation to convert the Black–Scholes equation into a heat equation is not unique. In fact, we can let $x = \ln S$, $\bar{\tau} = \dfrac{1}{2}\sigma^2(T - t)$, $V(S,t) = \mathrm{e}^{\alpha x + \beta \bar{\tau}} u(x,\bar{\tau})$, and choose constants α and β such that $u(x,\bar{\tau})$ satisfies the heat equation (see [71]).

2.3.2 Transforming the Black–Scholes Equation into an Equation Defined on a Finite Domain

Let us consider the following option problem:

$$\begin{cases} \dfrac{\partial V}{\partial t} + \dfrac{1}{2}\sigma^2(S)S^2 \dfrac{\partial^2 V}{\partial S^2} + (r - D_0)S \dfrac{\partial V}{\partial S} - rV = 0, \\[3mm] \hspace{4cm} 0 \le S < \infty, \quad t \le T, \\[2mm] V(S,T) = V_T(S), \hspace{1.5cm} 0 \le S < \infty. \end{cases} \qquad (2.24)$$

The transformation to be described in this subsection works even when σ, r, or D_0 depends on S and t. For simplicity, we assume in the derivation that σ depends on S and that r, D_0 are constant. In this case, an analytic expression of the solution $V(S,t)$ may not exist, and numerical methods may be necessary. Also for a call option, the solution $V(S,t)$ is not bounded. Therefore, we introduce new independent variables and dependent variable through the following transformation:

$$\begin{cases} \xi = \dfrac{S}{S + P_m}, \\ \tau = T - t, \\ V(S,t) = (S + P_m)\overline{V}(\xi,\tau). \end{cases} \tag{2.25}$$

From (2.25) we have

$$S = \frac{P_m \xi}{1 - \xi}, \quad S + P_m = \frac{P_m}{1 - \xi}$$

and

$$\frac{d\xi}{dS} = \frac{P_m}{(S + P_m)^2} = \frac{(1 - \xi)^2}{P_m}.$$

Because

$$\frac{\partial V}{\partial t} = \frac{\partial}{\partial t}\left[(S + P_m)\overline{V}(\xi,\tau)\right] = -(S + P_m)\frac{\partial \overline{V}}{\partial \tau} = -\frac{P_m}{1 - \xi}\frac{\partial \overline{V}}{\partial \tau},$$

$$\frac{\partial V}{\partial S} = \frac{\partial}{\partial S}\left[(S + P_m)\overline{V}(\xi,\tau)\right] = (S + P_m)\frac{\partial \overline{V}}{\partial \xi}\frac{d\xi}{dS} + \overline{V} = (1 - \xi)\frac{\partial \overline{V}}{\partial \xi} + \overline{V},$$

$$\frac{\partial^2 V}{\partial S^2} = \frac{\partial}{\partial \xi}\left[(1 - \xi)\frac{\partial \overline{V}}{\partial \xi} + \overline{V}\right]\frac{d\xi}{dS} = \frac{(1 - \xi)^3}{P_m}\frac{\partial^2 \overline{V}}{\partial \xi^2},$$

and let

$$\bar{\sigma}(\xi) = \sigma(S(\xi)) = \sigma\left(\frac{P_m \xi}{1 - \xi}\right),$$

the original equation becomes

$$\frac{P_m}{1 - \xi}\frac{\partial \overline{V}}{\partial \tau} = \frac{\bar{\sigma}^2(\xi)P_m\xi^2(1 - \xi)}{2}\frac{\partial^2 \overline{V}}{\partial \xi^2} + (r - D_0)P_m\xi\frac{\partial \overline{V}}{\partial \xi} + \frac{(r - D_0)\xi - r}{1 - \xi}P_m\overline{V}$$

or

$$\frac{\partial \overline{V}}{\partial \tau} = \frac{\bar{\sigma}^2(\xi)\xi^2(1 - \xi)^2}{2}\frac{\partial^2 \overline{V}}{\partial \xi^2} + (r - D_0)\xi(1 - \xi)\frac{\partial \overline{V}}{\partial \xi} - [r(1 - \xi) + D_0\xi]\overline{V},$$

$$0 \le \xi < 1, \quad 0 \le \tau.$$

Assume that $\overline{V}$ is a smooth function of ξ, then the equation also holds at $\xi = 1$. Because $V(S,T) = (S + P_m)\overline{V}(\xi,0) = \overline{V}(\xi,0)\dfrac{P_m}{1 - \xi}$, the condition $V(S,T) = V_T(S)$ can be rewritten as $\overline{V}(\xi,0) = V_T\left(\dfrac{P_m \xi}{1 - \xi}\right)\dfrac{1 - \xi}{P_m}$. Consequently, the problem (2.24) becomes

$$\begin{cases} \dfrac{\partial \overline{V}}{\partial \tau} = \dfrac{1}{2}\bar{\sigma}^2(\xi)\xi^2(1-\xi)^2\dfrac{\partial^2 \overline{V}}{\partial \xi^2} + (r-D_0)\xi(1-\xi)\dfrac{\partial \overline{V}}{\partial \xi} - [r(1-\xi)+D_0\xi]\overline{V}, \\[2mm] \qquad\qquad\qquad\qquad\qquad\qquad\qquad 0 \le \xi \le 1, \quad 0 \le \tau, \\[2mm] \overline{V}(\xi,0) = \dfrac{1-\xi}{P_m}V_T\left(\dfrac{P_m\xi}{1-\xi}\right), \qquad\qquad\qquad 0 \le \xi \le 1. \end{cases}$$

$$(2.26)$$

Thus, the transformation (2.25) converts a problem on an infinite domain into a problem on a finite domain. For a parabolic equation defined on a finite domain to have a unique solution, besides an initial condition, boundary conditions are usually needed. However, in this equation the coefficients of $\dfrac{\partial^2 \overline{V}}{\partial \xi^2}$ and $\dfrac{\partial \overline{V}}{\partial \xi}$ at $\xi = 0$ and at $\xi = 1$ are equal to zero, i.e., the equation degenerates to ordinary differential equations at the boundaries. Actually, at $\xi = 0$ the equation becomes

$$\frac{\partial \overline{V}(0,\tau)}{\partial \tau} = -r\overline{V}(0,\tau)$$

and the solution is

$$\overline{V}(0,\tau) = \overline{V}(0,0)e^{-r\tau}. \qquad\qquad (2.27)$$

Similarly, at $\xi = 1$ the equation reduces to

$$\frac{\partial \overline{V}(1,\tau)}{\partial \tau} = -D_0\overline{V}(1,\tau),$$

from which we have

$$\overline{V}(1,\tau) = \overline{V}(1,0)e^{-D_0\tau}. \qquad\qquad (2.28)$$

Therefore for this equation, the two solutions of the ordinary differential equations provide the values at the boundaries, and no boundary conditions are needed in order for the problem (2.26) to have a unique solution.

Consequently, in order to price an option, we need to solve a problem on a finite domain if this formulation is adopted. From the point view of numerical methods, such a formulation is better. This is its first advantage. Actually, the uniqueness of solution for problem (2.26) can easily be proven (see Section 2.9). Indeed, not only the uniqueness can be proven, but the stability of the solution with respect to the initial value can also be shown easily. That is, this formulation makes proof of some theoretical problems easy. This is its other advantage.

For a call option, the payoff is

$$V(S,T) = \max(S-E,0),$$

so the initial condition in (2.26) for a call is

$$\overline{V}(\xi, 0) = \max(S - E, 0)(1 - \xi)/P_m$$

$$= \max\left(\frac{P_m\xi}{1 - \xi} - E, 0\right)(1 - \xi)/P_m$$

$$= \max\left(\xi - \frac{E}{P_m}(1 - \xi), 0\right).$$

For a put option

$$V(S, T) = \max(E - S, 0).$$

Therefore

$$\overline{V}(\xi, 0) = \max\left(\frac{E}{P_m}(1 - \xi) - \xi, 0\right).$$

Let $P_m = E$, the two initial conditions become

$$\overline{V}(\xi, 0) = \max(2\xi - 1, 0)$$

and

$$\overline{V}(\xi, 0) = \max(1 - 2\xi, 0)$$

respectively. Therefore, a European call option is the solution of the following problem:

$$\begin{cases} \dfrac{\partial \overline{V}}{\partial \tau} = \dfrac{1}{2}\bar{\sigma}^2(\xi)\xi^2(1 - \xi)^2\dfrac{\partial^2 \overline{V}}{\partial \xi^2} + (r - D_0)\xi(1 - \xi)\dfrac{\partial \overline{V}}{\partial \xi} - [r(1 - \xi) + D_0\xi]\overline{V}, \\ \\ \qquad\qquad\qquad\qquad\qquad\qquad\qquad\qquad 0 \le \xi \le 1, \quad 0 \le \tau, \\ \\ \overline{V}(\xi, 0) = \max(2\xi - 1, 0), \qquad\qquad\qquad 0 \le \xi \le 1 \end{cases}$$

$$(2.29)$$

and the solution of the problem

$$\begin{cases} \dfrac{\partial \overline{V}}{\partial \tau} = \dfrac{1}{2}\bar{\sigma}^2(\xi)\xi^2(1 - \xi)^2\dfrac{\partial^2 \overline{V}}{\partial \xi^2} + (r - D_0)\xi(1 - \xi)\dfrac{\partial \overline{V}}{\partial \xi} - [r(1 - \xi) + D_0\xi]\overline{V}, \\ \\ \qquad\qquad\qquad\qquad\qquad\qquad\qquad\qquad 0 \le \xi \le 1, \quad 0 \le \tau, \\ \\ \overline{V}(\xi, 0) = \max(1 - 2\xi, 0), \qquad\qquad\qquad 0 \le \xi \le 1 \end{cases}$$

$$(2.30)$$

gives the price of a European put option. In (2.29) the initial condition is bounded, so $\overline{V}(\xi, \tau)$, as a solution of a linear parabolic equation, is also bounded. Therefore in this case, the solution that needs to be found numerically is bounded.

So far, we assumed that σ depends only on S and that r and D_0 are constant. However, the result will be the same if σ depends on both S and t, and r and D_0 also depend on S and t.

Finally, we would like to point out that from (2.28) we can have an asymptotic expression of the solution of the Black–Scholes equation at infinity. Because at $\xi = 1$ there is an analytic solution (2.28), noticing

$$V(S,t) = (S + P_m)\overline{V}(\xi, \tau),$$

for $S \approx \infty$ we have

$$\begin{aligned}
V(S,t) &= (S + P_m)\overline{V}(\xi, \tau) \approx (S + P_m)\overline{V}(1, \tau) \\
&= (S + P_m)\overline{V}(1, 0)e^{-D_0\tau} \\
&\approx V(S,T)e^{-D_0\tau} = V(S,T)e^{-D_0(T-t)}.
\end{aligned} \tag{2.31}$$

This is an asymptotic expression of the solution of the Black–Scholes equation at infinity.

2.4 Solutions of European Options

A linear partial differential equation

$$A\frac{\partial^2 u}{\partial t^2} + 2B\frac{\partial^2 u}{\partial t \partial x} + C\frac{\partial^2 u}{\partial x^2} = F\left(x, t, u, \frac{\partial u}{\partial t}, \frac{\partial u}{\partial x}\right)$$

is called a parabolic partial differential equation if $AC - B^2 = 0$, where $A, B,$ and C are not all equal to zero. The diffusion equation is the simplest parabolic equation. The Black–Scholes equation is another parabolic equation. In Section 2.3 we reduced the Black–Scholes equation to a diffusion equation. Here we will first find out the analytic expression of the solution of the diffusion equation, then that of the Black–Scholes equation, and finally the Black–Scholes formulae for European options are derived.

2.4.1 The Solutions of Parabolic Equations

In order for a parabolic differential equation to have a unique solution, one has to specify some conditions. For example, the initial value problem for a heat equation

$$\frac{\partial u}{\partial \bar{\tau}} = \frac{\partial^2 u}{\partial x^2}, \quad -\infty < x < \infty, \quad \bar{\tau} \geq 0 \tag{2.32}$$

with

$$u(x, 0) = u_0(x) \tag{2.33}$$

has a unique solution under certain conditions that usually hold for cases considered in this book.

Let us find the solution of the equation (2.32) with initial condition (2.33). The way to find the solution is not unique. Here, we use the following method (see [47]). We first try to find a special solution of (2.32) in the form

$$u(x, \bar{\tau}) = \bar{\tau}^{-1/2}U(\eta),$$

where

$$\eta = \frac{x - \xi}{\sqrt{\bar{\tau}}}, \qquad \xi \text{ being a parameter.}$$

Because

$$\frac{\partial u}{\partial \bar{\tau}} = -\frac{\bar{\tau}^{-3/2}}{2}\left(U + \eta\frac{dU}{d\eta}\right) = -\frac{\bar{\tau}^{-3/2}}{2}\frac{d}{d\eta}[\eta U(\eta)],$$

$$\frac{\partial u}{\partial x} = \bar{\tau}^{-1/2}\frac{dU}{d\eta}\frac{1}{\sqrt{\bar{\tau}}} = \bar{\tau}^{-1}\frac{dU}{d\eta},$$

$$\frac{\partial^2 u}{\partial x^2} = \bar{\tau}^{-3/2}\frac{d^2 U}{d\eta^2},$$

from (2.32) we have

$$-\frac{\bar{\tau}^{-3/2}}{2}\frac{d}{d\eta}(\eta U) = \bar{\tau}^{-3/2}\frac{d^2 U}{d\eta^2},$$

that is,

$$\frac{d^2 U}{d\eta^2} + \frac{1}{2}\frac{d}{d\eta}(\eta U) = 0.$$

Integrating this equation, we have

$$\frac{dU}{d\eta} + \frac{\eta}{2}U = c_1,$$

where c_1 is a constant. Let us choose $c_1 = 0$, so now we have a linear homogeneous equation. The solution of this equation is

$$U(\eta) = ce^{-\eta^2/4},$$

where c is a constant. Thus, for the diffusion equation we have a special solution in the form

$$c\bar{\tau}^{-1/2}e^{-(x-\xi)^2/4\bar{\tau}}.$$

If we further require

$$\int_{-\infty}^{\infty} c\bar{\tau}^{-1/2}e^{-(x-\xi)^2/4\bar{\tau}}d\xi = 1,$$

then

$$c = \frac{1}{\displaystyle\int_{-\infty}^{\infty}\bar{\tau}^{-1/2}e^{-(x-\xi)^2/4\bar{\tau}}d\xi} = \frac{1}{\sqrt{2}\displaystyle\int_{-\infty}^{\infty}e^{-\eta^2/2}d\eta} = \frac{1}{2\sqrt{\pi}}$$

and the special solution is

$$\frac{1}{2\sqrt{\pi\bar{\tau}}}e^{-(x-\xi)^2/4\bar{\tau}}.$$

This solution is called the fundamental solution, or Green's function, for the heat equation (2.32). Let $g(\xi; x, \bar{\tau})$ represent this class of functions with ξ as parameters, i.e., the relation

$$\frac{\partial g(\xi; x, \bar{\tau})}{\partial \bar{\tau}} = \frac{\partial^2 g(\xi; x, \bar{\tau})}{\partial x^2}$$

holds for any ξ. Thus, for any $u_0(\xi)$ we have

$$\int_{-\infty}^{\infty} u_0(\xi) \frac{\partial g(\xi; x, \bar{\tau})}{\partial \bar{\tau}} d\xi = \int_{-\infty}^{\infty} u_0(\xi) \frac{\partial^2 g(\xi; x, \bar{\tau})}{\partial x^2} d\xi,$$

that is,

$$\frac{\partial \left[\int_{-\infty}^{\infty} u_0(\xi) g(\xi; x, \bar{\tau}) d\xi \right]}{\partial \bar{\tau}} = \frac{\partial^2 \left[\int_{-\infty}^{\infty} u_0(\xi) g(\xi; x, \bar{\tau}) d\xi \right]}{\partial x^2}.$$

Consequently,

$$u(x, \bar{\tau}) = \int_{-\infty}^{\infty} u_0(\xi) \times \frac{1}{2\sqrt{\pi\bar{\tau}}} e^{-(x-\xi)^2/4\bar{\tau}} d\xi \qquad (2.34)$$

is also a solution of (2.32). Because

$$\lim_{\bar{\tau} \to 0} \frac{1}{2\sqrt{\pi\bar{\tau}}} e^{-(x-\xi)^2/4\bar{\tau}} = \begin{cases} 0, & x - \xi \neq 0, \\ \infty, & x - \xi = 0 \end{cases}$$

and

$$\int_{-\infty}^{\infty} \frac{1}{2\sqrt{\pi\bar{\tau}}} e^{-(x-\xi)^2/4\bar{\tau}} d\xi = 1$$

is true for any $\bar{\tau}$, we have

$$\lim_{\bar{\tau} \to 0} \frac{1}{2\sqrt{\pi\bar{\tau}}} e^{-(x-\xi)^2/4\bar{\tau}} = \delta(x - \xi)$$

and

$$\lim_{\bar{\tau} \to 0} \int_{-\infty}^{\infty} u_0(\xi) \times \frac{1}{2\sqrt{\pi\bar{\tau}}} e^{-(x-\xi)^2/4\bar{\tau}} d\xi = u_0(x).$$

Consequently, (2.34) is the solution of the initial-value problem

$$\begin{cases} \dfrac{\partial u}{\partial \bar{\tau}} = \dfrac{\partial^2 u}{\partial x^2}, & -\infty < x < \infty, \quad \bar{\tau} > 0, \\ u(x, 0) = u_0(x), & -\infty < x < \infty. \end{cases}$$

2.4.2 Solutions of the Black–Scholes Equation

From (2.21), (2.20), and (2.34), we know that the solution of the final value problem

$$\begin{cases} \dfrac{\partial V}{\partial t} + \dfrac{1}{2}\sigma^2 S^2 \dfrac{\partial^2 V}{\partial S^2} + (r - D_0)S\dfrac{\partial V}{\partial S} - rV = 0, & 0 \le S, \quad 0 \le t \le T, \\ V(S,T) = V_T(S), & 0 \le S \end{cases}$$

is

$$V(S,t) = e^{-r(T-t)} \int_{-\infty}^{\infty} u_0(\xi) \frac{1}{2\sqrt{\pi\bar{\tau}}} e^{-(x-\xi)^2/4\bar{\tau}} d\xi$$

$$= e^{-r(T-t)} \int_{-\infty}^{\infty} V_T\left(e^\xi\right) \frac{1}{2\sqrt{\pi\bar{\tau}}} e^{-(\xi-x)^2/4\bar{\tau}} d\xi$$

$$= e^{-r(T-t)} \frac{1}{\sigma\sqrt{2\pi(T-t)}}$$

$$\times \int_0^\infty V_T(S')e^{-\left\{\ln S' - \left[\ln S + (r-D_0-\sigma^2/2)(T-t)\right]\right\}^2/2\sigma^2(T-t)} \frac{dS'}{S'}.$$

This result can be written as

$$V(S,t) = e^{-r(T-t)} \int_0^\infty V_T(S')G(S',T;S,t)dS', \tag{2.35}$$

where

$$G(S',T;S,t)$$
$$= \frac{1}{\sigma\sqrt{2\pi(T-t)}S'} e^{-\left\{\ln S' - \left[\ln S + (r-D_0-\sigma^2/2)(T-t)\right]\right\}^2/2\sigma^2(T-t)}. \tag{2.36}$$

This function is usually referred to as Green's function of the Black-Scholes equation.

From Subsection 2.1.3, this is the probability density function for a lognormal distribution with $m = r - D_0 - \dfrac{1}{2}\sigma^2$, and according to (2.6) the expectation of S' is

$$E[S'] = Se^{(r-D_0)(T-t)}. \tag{2.37}$$

This function can also be written as

$$G(S',T;S,t) = \frac{1}{\sqrt{2\pi}bS'} e^{-\left[\ln(S'/a)+b^2/2\right]^2/2b^2},$$

where

$$a = Se^{(r-D_0)(T-t)} \quad \text{and} \quad b = \sigma\sqrt{T-t}.$$

For this function, there are the following useful formulae:

$$\int_c^\infty G(S',T;S,t)dS' = N\left(\frac{\ln(a/c) - b^2/2}{b}\right) \tag{2.38}$$

and

$$\int_c^\infty S'G(S',T;S,t)dS' = aN\left(\frac{\ln(a/c) + b^2/2}{b}\right), \tag{2.39}$$

where $N(z)$ is the cumulative distribution function for the standardized normal variable defined by[9]

$$N(z) = \frac{1}{\sqrt{2\pi}}\int_{-\infty}^z e^{-\xi^2/2}d\xi. \tag{2.40}$$

The proof of the two formulae is straightforward. Let

$$\eta(S') = \frac{\ln(S'/a) + b^2/2}{b},$$

that is,

$$S' = ae^{b\eta - b^2/2}.$$

Thus

$$dS' = ae^{b\eta - b^2/2}bd\eta = S'bd\eta.$$

Consequently, we have

$$\int_c^\infty \frac{1}{\sqrt{2\pi}bS'}e^{-[\ln(S'/a)+b^2/2]^2/2b^2}dS'$$
$$= \int_{\eta(c)}^\infty \frac{1}{\sqrt{2\pi}bS'}e^{-\eta^2/2}S'bd\eta$$
$$= N(-\eta(c))$$
$$= N\left(-\frac{\ln(c/a) + b^2/2}{b}\right)$$
$$= N\left(\frac{\ln(a/c) - b^2/2}{b}\right)$$

and

---

[9] The value of this function has to be obtained by numerical methods. If $z \le 0$, this function can be approximated by

$$N(z) = 0.5t\exp(-x^2 - 1.26551223 + t(1.00002368 + t(0.37409196 + t(0.09678418$$
$$+t(-0.18628806 + t(0.27886807 + t(-1.13520398 + t(1.48851587$$
$$+t(-0.82215223 + t \times 0.17087277))))))))),$$

where $x = -z \times 0.707106781186550$ and $t = 1.0/(1.0 + 0.5x)$. If $z > 0$, then $N(z) = 1 - N(-z)$. The fractional error is less than 0.6×10^{-7} everywhere. See NUMERICAL RECIPES IN C: The Art of Scientific Computing. Cambridge University Press, Cambridge (1988–1992).

$$\int_{c}^{\infty} S' \frac{1}{\sqrt{2\pi}bS'} e^{-\left[\ln(S'/a)+b^2/2\right]^2/2b^2} dS'$$

$$= \int_{\eta(c)}^{\infty} \frac{1}{\sqrt{2\pi}b} e^{-\eta^2/2} ae^{b\eta-b^2/2} bd\eta$$

$$= \frac{a}{\sqrt{2\pi}} \int_{\eta(c)}^{\infty} e^{-(\eta-b)^2/2} d\eta$$

$$= \frac{a}{\sqrt{2\pi}} \int_{\eta(c)-b}^{\infty} e^{-\xi^2/2} d\xi$$

$$= aN\left(-\frac{\ln(c/a)+b^2/2}{b}+b\right)$$

$$= aN\left(\frac{\ln(a/c)+b^2/2}{b}\right).$$

2.4.3 Prices of Forward Contracts and Delivery Prices

From Subsection 1.2.1, we know that the payoff function for a forward contract is

$$V(S,T) = S - K.$$

Therefore, according to (2.35) and using (2.37), we see that its price is

$$V(S,t) = e^{-r(T-t)} \int_{0}^{\infty} (S' - K)G(S',T;S,t)dS'$$

$$= e^{-r(T-t)}(Se^{(r-D_0)(T-t)} - K)$$

$$= Se^{-D_0(T-t)} - Ke^{-r(T-t)}.$$

Because for a forward contract the buyer does not need to pay any premium at $t = 0$, we have

$$V(S,0) = Se^{-D_0 T} - Ke^{-rT} = 0.$$

Consequently, the delivery price should be

$$K = e^{(r-D_0)T} S_0,$$

where in order to make it clear, we use S_0, instead of S, to denote the price of the underlying asset at the initiation of the contract.

2.4.4 Derivation of the Black–Scholes Formulae

At $t = T$, the value of a call option is

$$c(S,T) = \max(S - E, 0).$$

According to (2.35), (2.38), and (2.39), the value of a European call is

$$c(S,t) = e^{-r(T-t)} \int_0^\infty \max(S' - E, 0) G(S', T; S, t) dS'$$

$$= e^{-r(T-t)} \int_E^\infty (S' - E) G(S', T; S, t) dS'$$

$$= e^{-r(T-t)} \left[\int_E^\infty S' G(S', T; S, t) dS' - \int_E^\infty E G(S', T; S, t) dS' \right]$$

$$= e^{-r(T-t)} \left[S e^{(r-D_0)(T-t)} N(d_1) - E N(d_2) \right]$$

$$= S e^{-D_0(T-t)} N(d_1) - E e^{-r(T-t)} N(d_2), \tag{2.41}$$

where

$$d_1 = \left[\ln \frac{S e^{(r-D_0)(T-t)}}{E} + \frac{1}{2}\sigma^2(T-t) \right] \bigg/ \left(\sigma\sqrt{T-t} \right)$$

$$= \left[\ln \frac{S e^{-D_0(T-t)}}{E e^{-r(T-t)}} + \frac{1}{2}\sigma^2(T-t) \right] \bigg/ \left(\sigma\sqrt{T-t} \right),$$

$$d_2 = \left[\ln \frac{S e^{(r-D_0)(T-t)}}{E} - \frac{1}{2}\sigma^2(T-t) \right] \bigg/ \left(\sigma\sqrt{T-t} \right)$$

$$= \left[\ln \frac{S e^{-D_0(T-t)}}{E e^{-r(T-t)}} - \frac{1}{2}\sigma^2(T-t) \right] \bigg/ \left(\sigma\sqrt{T-t} \right)$$

$$= d_1 - \sigma\sqrt{T-t}.$$

For a put, the final value is

$$p(S, T) = \max(E - S, 0).$$

Thus, the value of a European put is

$$p(S,t) = e^{-r(T-t)} \int_0^\infty \max(E - S', 0) G(S', T; S, t) dS'$$

$$= e^{-r(T-t)} \int_0^E (E - S') G(S', T; S, t) dS'$$

$$= e^{-r(T-t)} \left[E \int_0^E G(S', T; S, t) dS' - \int_0^E S' G(S', T; S, t) dS' \right]$$

$$= e^{-r(T-t)} \left\{ E[1 - N(d_2)] - S e^{(r-D_0)(T-t)} [1 - N(d_1)] \right\}$$

$$= E e^{-r(T-t)} N(-d_2) - S e^{-D_0(T-t)} N(-d_1). \tag{2.42}$$

It is interesting that the values of European call and put options can be expressed in terms of the cumulative distribution function for the standardized normal random variable, $N(z)$. Expressions (2.41) and (2.42) give closed-form solutions for European vanilla options and are usually referred to as the Black–Scholes formulae.

When hedging is involved, we not only seek the value of options, but also the value of the first and second derivatives with respect to S, Δ, and Γ. For a European call, $\Delta = \dfrac{\partial c}{\partial S}$ is

$$
\begin{aligned}
\frac{\partial c}{\partial S} &= e^{-D_0(T-t)} N(d_1) + Se^{-D_0(T-t)} \frac{1}{\sqrt{2\pi}} e^{-d_1^2/2} \frac{\partial d_1}{\partial S} \\
&\quad - Ee^{-r(T-t)} \frac{1}{\sqrt{2\pi}} e^{-d_2^2/2} \frac{\partial d_2}{\partial S} \\
&= e^{-D_0(T-t)} N(d_1) + \frac{1}{\sqrt{2\pi}} \frac{\partial d_1}{\partial S} \left(Se^{-D_0(T-t)-d_1^2/2} - Ee^{-r(T-t)-d_2^2/2} \right).
\end{aligned}
$$

Noticing

$$
\begin{aligned}
&-r(T-t) - d_2^2/2 \\
={}& -r(T-t) - \left[d_1^2 - 2d_1\sigma\sqrt{T-t} + \sigma^2(T-t) \right]/2 \\
={}& -r(T-t) - \left[d_1^2 - 2\ln(S/E) - 2\left(r - D_0 + \sigma^2/2\right)(T-t) + \sigma^2(T-t) \right]/2 \\
={}& -d_1^2/2 - D_0(T-t) + \ln(S/E),
\end{aligned}
$$

that is,

$$
Se^{-D_0(T-t)-d_1^2/2} = Ee^{-r(T-t)-d_2^2/2},
$$

we have

$$
\frac{\partial c}{\partial S} = e^{-D_0(T-t)} N(d_1).
$$

Taking the derivative with respect to S again yields

$$
\frac{\partial^2 c}{\partial S^2} = \frac{1}{S\sigma\sqrt{2\pi(T-t)}} e^{-D_0(T-t)-d_1^2/2}.
$$

Similarly, for put options

$$
\frac{\partial p}{\partial S} = -e^{-D_0(T-t)} N(-d_1) \quad \text{and} \quad \frac{\partial^2 p}{\partial S^2} = \frac{\partial^2 c}{\partial S^2}.
$$

We need to point out that if the value of an option and the price of the underlying asset are divided by E, then the dimensionless option value V/E and the derivatives of V/E can still be obtained by the same formulae. The only change is to let $E = 1$ and S should have dimensionless value.

What do the functions $c(S, t)$ and $p(S, t)$ look like? The prices of the European call and put options for the case $r = 0.1$, $D_0 = 0.05$, $\sigma = 0.2$ are shown in Figs. 2.5 and 2.6. Clearly, the curves should approach the payoff functions as $t \to T$, which can be seen from the two figures. From Fig. 2.6, we can

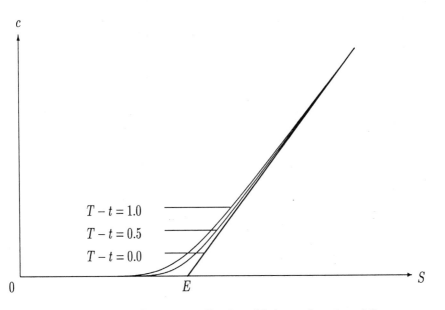

Fig. 2.5. The European call value $c(S,t)$ as a function of S
($r = 0.1$, $D_0 = 0.05$, $\sigma = 0.2$, and $T - t = 0$, 0.5, and 1.0)

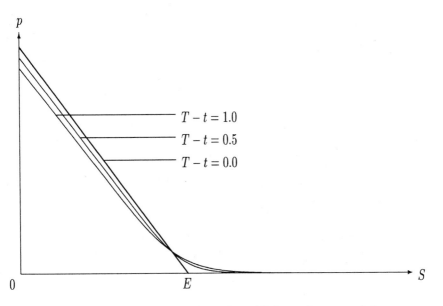

Fig. 2.6. The European put value $p(S,t)$ as a function of S
($r = 0.1$, $D_0 = 0.05$, $\sigma = 0.2$, and $T - t = 0$, 0.5, and 1.0)

also see that when S is close to zero, the curves approach the payoff from the bottom and when S is large, the curves tend to the payoff from the top. That is, $p(S,t)$ is less than the payoff for small S and greater than the payoff for large S. In Sections 2.5 and 2.6, we will see that for American options, the price should always be at least the payoff. Because of this, the Black–Scholes equation cannot be used to determine the price of American options in some situations.

When σ, r, and D_0 depend on t, closed-form solutions can still be obtained (see [57] and [71]). Actually, through the transformation (2.22), the Black–Scholes equation

$$\frac{\partial V}{\partial t} + \frac{1}{2}\sigma^2(t)S^2\frac{\partial^2 V}{\partial S^2} + [r(t) - D_0(t)]S\frac{\partial V}{\partial S} - r(t)V = 0$$

can still be reduced to a diffusion equation. Let

$$\begin{cases} \alpha(t) = \dfrac{1}{2}\int_t^T \sigma^2(s)ds, \\[2mm] \delta(t) = \int_t^T D_0(s)ds, \\[2mm] \gamma(t) = \int_t^T r(s)ds, \end{cases}$$

then the solution of the Black–Scholes equation in this case is

$$V(S,t) = e^{-\gamma(t)}\int_{-\infty}^{\infty} V_T\left(e^{\xi}\right)\frac{1}{2\sqrt{\pi\bar{\tau}}}e^{-(\xi-x)^2/4\bar{\tau}}d\xi,$$

where $x = \ln S + \gamma(t) - \delta(t) - \alpha(t)$ and $\bar{\tau} = \alpha(t)$. Therefore, for a call with coefficients $r(t)$, $D_0(t)$, and $\sigma(t)$, the solution should be

$$c(S,t) = Se^{-\delta(t)}N(\bar{d}_1) - Ee^{-\gamma(t)}N(\bar{d}_2),$$

where

$$\bar{d}_1 = \left[\ln\frac{Se^{-\delta(t)}}{Ee^{-\gamma(t)}} + \alpha(t)\right]\Big/[2\alpha(t)]^{1/2},$$

$$\bar{d}_2 = \left[\ln\frac{Se^{-\delta(t)}}{Ee^{-\gamma(t)}} - \alpha(t)\right]\Big/[2\alpha(t)]^{1/2}.$$

2.4.5 Put–Call Parity Relation

Although call and put options are superficially different, they can be combined in such a way that they are perfectly correlated. In fact, there is the following relation:

$$c(S,t) - p(S,t) = Se^{-D_0(T-t)} - Ee^{-r(T-t)}, \tag{2.43}$$

which is usually called the put–call parity relation. It can be obtained in different ways. From the Black–Scholes formulae, (2.41) and (2.42), we can have

$$c(S,t) - p(S,t) = Se^{-D_0(T-t)}N(d_1) - Ee^{-r(T-t)}N(d_2)$$
$$- Ee^{-r(T-t)}N(-d_2) + Se^{-D_0(T-t)}N(-d_1)$$
$$= Se^{-D_0(T-t)} - Ee^{-r(T-t)}.$$

This is one way to get it.

We can also find this relation without finding the concrete expressions of $c(S,t)$ and $p(S,t)$. Let us examine the portfolio

$$\Pi = S + p - c - E.$$

The payoff for this portfolio at expiry is

$$\Pi(S,T) = S + \max(E - S, 0) - \max(S - E, 0) - E = 0.$$

According to (2.35),

$$\Pi(S,t)$$
$$= e^{-r(T-t)} \int_0^\infty [S' + \max(E - S', 0) - \max(S' - E, 0) - E]G(S', T; S, t)dS'$$
$$= Se^{-D_0(T-t)} + p(S,t) - c(S,t) - Ee^{-r(T-t)}$$
$$= 0.$$

Here, we are actually using the superposition principle of homogeneous linear partial differential equations. From this relation, we immediately have the put–call parity. In Section 2.11, we will derive this relation again without using a partial differential equation. Here, we need to point out that the put–call parity relation is true only for European options. For American options, the equality becomes an inequality, which will be discussed in Section 2.11.

2.4.6 An Explanation in Terms of Probability

The function $G(S', T; S, t)$ given by (2.36) represents a probability density function of a random variable S', and S' can be interpreted as the random price of a stock at time T. Then, we can understand S as the price of the stock at time t because $G(S', T; S, t)$ goes to a Dirac delta function $\delta(S' - S)$ as $T \to t$. $V_T(S')$ is the value of an option at time T if the price is S'. Therefore

$$\int_0^\infty V_T(S')G(S', T; S, t)dS'$$

is the expectation of the value of the option at time T if the price is S at time t, and

$$e^{-r(T-t)} \int_0^\infty V_T(S')G(S', T; S, t)dS'$$

is the present (or discounted) value of the expectation at time T. That is, the price of an option at time t given by (2.35) is the present value of the expectation of the option value at time T. This is the explanation of the solution given by (2.35) in terms of probability.

Suppose that S and S' are the prices of a stock at time $T - \Delta t$ and time T, respectively, and that S' has the probability density function $G(S', T; S, T - \Delta t)$. According to (2.6) we have

$$E[S'] = Se^{(r-D_0)\Delta t}$$

and

$$\text{Var}[S'] = S^2 e^{2(r-D_0)\Delta t}\left(e^{\sigma^2 \Delta t} - 1\right) \approx S^2 \sigma^2 \Delta t.$$

Therefore[10]

$$E\left[\frac{S' - S}{S}\right] = \frac{Se^{(r-D_0)\Delta t} - S}{S} \approx (r - D_0)\Delta t$$

and

$$\text{Var}\left[\frac{S' - S}{S}\right] \approx \sigma^2 \Delta t.$$

Consequently

$$\frac{dS}{S} = (r - D_0)dt + \sigma dX.$$

However, in the real world

$$\frac{dS}{S} = \mu dt + \sigma dX.$$

Therefore, the random variable in the expression of the solution is a different random variable from that in the real world. Usually, we say that the random variable in the expression of the solution is in a "risk-neutral" world. In this case, the expected return rate per unit time on any asset is the difference between the riskless interest rate r and the dividend yield D_0.

It is clear that if we let

$$\overline{V}(S,t) = e^{r(T-t)}V(S,t),$$

then $\overline{V}$ is the solution of the problem

$$\begin{cases} \dfrac{\partial \overline{V}}{\partial t} + \dfrac{1}{2}\sigma^2 S^2 \dfrac{\partial^2 \overline{V}}{\partial S^2} + (r - D_0)S\dfrac{\partial \overline{V}}{\partial S} = 0, 0 \le S, \quad t \le T, \\ \overline{V}(S,T) = \overline{V}_T(S), \qquad\qquad\qquad\qquad\quad 0 \le S \end{cases}$$

and

$$\overline{V}(S,t) = \int_0^\infty \overline{V}_T(S')G(S', T; S, t)dS' = E\left[\overline{V}_T(S')\right].$$

In probability theory, when this relation holds, it is said that $\overline{V}(S,t)$ is a martingale under the probability density function $G(S', T; S, t)$.

---

[10] Here we take a conditional expectation, i.e., S' is a random variable and S is fixed.

2.5 American Option Problems as Linear Complementarity Problems

In the next two sections, we will discuss how to determine the value of an American option. In this section, we first derive the additional constraints on American call and put options. Then, we formulate the American call and put problems as linear complementarity (LC) problems. In the next section, we will discuss how to formulate an American option problem as a free-boundary problem (FBP).

2.5.1 Constraints on American Options

Let $C(S,t)$ and $P(S,t)$ denote the prices of American call and put options, respectively. As we know from Section 1.2, an American option has the additional feature that it may be exercised at any time during the life of the option. What does this additional feature mean in mathematics? It means that the value of an American call option must satisfy

$$C(S,t) \geq \max(S - E, 0), \tag{2.44}$$

and that the value of an American put option must fulfill

$$P(S,t) \geq \max(E - S, 0). \tag{2.45}$$

Usually, $\max(S - E, 0)$ and $\max(E - S, 0)$ are called the intrinsic values of call and put options, respectively. Thus, satisfying the two inequalities above means that the value of an option must be at least equal to its intrinsic value. Because of this fact, $C(S,t) - \max(S - E, 0)$ and $P(S,t) - \max(E - S, 0)$ must be nonnegative and are usually called the time value of the American call and put options, respectively. Conditions (2.44) and (2.45) are usually referred to as the constraints on American vanilla options. These conclusions can be proven by arbitrage arguments.

First, let us consider an American call option. For $S \leq E$, the condition (2.44) means $C(S,t) \geq 0$. This is always true because a solution of the Black–Scholes equation with a nonnegative payoff function as a final condition is always nonnegative. From the financial point of view, it is also clear that the option price should not be negative because a holder of an option has only rights, no obligation. Thus, (2.44) always holds for any $S \in [0, E]$. Suppose that for a price $S > E$, the condition (2.44) is not fulfilled, i.e., $C(S,T) < S - E$. Then, an obvious arbitrage opportunity arises: by short selling the asset on the market for S, purchasing the option for C, and exercising the call option, a risk-free profit of $S - E - C$ is made. Of course, such an opportunity would not last long before the value of the option was pushed up by the demand of arbitrageurs. We conclude that for a value of an American call, we must impose the constraint (2.44). For an American put option the situation is similar. For any $S \geq E$, the condition (2.45) holds naturally. Suppose the option price satisfies $P(S,t) < E - S$ for a price $S < E$. Then, by purchasing the option for P, purchasing the asset from the market for S, and exercising the put option, an immediate risk-free profit of $E - P - S$ is made, and the demand will push the option price up so that condition (2.45) holds.

Bermudan options are similar to American options but can be exercised only at several predetermined dates, instead of the entire period $[0,T]$. This means that for a Bermudan option, condition (2.44) or condition (2.45) should be required at several predetermined dates but not on the entire period $[0,T]$, which is the only difference between American and Bermudan options.

Because an American option can be exercised at any time by expiry, a holder of an American option has more rights than does a holder of a European option. Thus, the holder of an American option needs to pay at least as much premium as does the holder of a European option with the same parameters. A Bermudan option can be exercised at several predetermined dates including the expiration date, its holder has less rights than does the holder of an American option and more rights than does the holder of a European option. Thus, its premium should be between the premiums of the American and European options with the same parameters.

2.5.2 Formulation of the Linear Complementarity Problem in (S,t)-Plane

As we easily see, at $S = 0$ the Black–Scholes equation degenerates to an ordinary differential equation

$$\frac{\partial V(0,t)}{\partial t} - rV(0,t) = 0$$

and its solution is

$$V(0,t) = V(0,T)e^{-r(T-t)}.$$

For a put, $V(0,T) = E$. Therefore, the price of a European put option at $S = 0$ is

$$p(0,t) = Ee^{-r(T-t)} < E$$

for any $t < T$ if $r > 0$. Consequently, the price of a European put option will not satisfy the constraint (2.45). Therefore, in order to price an American put option, we must modify the method for determining the price of an option in the following way. If the Black–Scholes equation gives a price satisfying the constraint (2.45), then it is the price of the American put option; if not, the value needs to be replaced by $\max(E - S, 0)$. Thus, the price of an American option usually is not a solution of (2.16) but a solution of a so-called linear complementarity (LC) problem. In such a problem, usually in some regions the solution satisfies the partial differential equation (PDE) and in other regions it is not determined by the PDE. Those boundaries between the two types of regions are called free boundaries.

Before we formulate the American option problems as linear complementarity problems, we would like to show the following theorem.

Theorem 2.1 *Let* $\mathbf{L_{s,t}}$ *be an operator in an option problem in the form:*

$$\mathbf{L_{s,t}} = a(S,t)\frac{\partial^2}{\partial S^2} + b(S,t)\frac{\partial}{\partial S} + c(S,t)$$

and $G(S,t)$ *be the constraint function for an American option. Furthermore, we assume that* $\dfrac{\partial G}{\partial t} + \mathbf{L_{s,t}}G$ *exists. Suppose* $V(S,t^*) = G(S,t^*)$ *on an open interval* (A_1, B_1) *on the S-axis. Let* $t = t^* - \Delta t$, *where* Δt *is a sufficiently small positive number and let* (A, B) *be an open interval in* (A_1, B_1). *Show the following conclusions: If for any* $S \in (A, B)$,

$$\frac{\partial G}{\partial t}(S,t^*) + \mathbf{L_{s,t^*}}G(S,t^*) \geq 0,$$

then the value $V(S,t)$ *determined by the equation*

$$\frac{\partial V}{\partial t}(S,t) + \mathbf{L_{s,t}}V(S,t) = 0$$

satisfies the condition $V(S,t) - G(S,t) \geq 0$ *on* (A, B); *and if for any* $S \in (A, B)$,

$$\frac{\partial G}{\partial t}(S,t^*) + \mathbf{L_{s,t^*}}G(S,t^*) < 0,$$

then the equation

$$\frac{\partial V}{\partial t}(S,t) + \mathbf{L_{s,t}}V(S,t) = 0$$

cannot give a solution satisfying the condition $V(S,t) - G(S,t) \geq 0$ *for any* $S \in (A, B)$.

Proof. Because $V(S,t^*) = G(S,t^*)$, the fact that $V(S,t) - G(S,t) \geq 0$ holds for any $t = t^* - \Delta t$, Δt being a sufficiently small positive number, is equivalent to the fact that at time t^*, $V(S,t) - G(S,t)$ is a nonincreasing function with respect to t, that is,

$$\frac{\partial V}{\partial t}(S,t^*) - \frac{\partial G}{\partial t}(S,t^*) \leq 0.$$

If

$$\frac{\partial G}{\partial t}(S,t^*) + \mathbf{L_{s,t^*}}G(S,t^*) \geq 0$$

and

$$\frac{\partial V}{\partial t}(S,t^*) + \mathbf{L_{s,t^*}}V(S,t^*) = \frac{\partial V}{\partial t}(S,t^*) + \mathbf{L_{s,t^*}}G(S,t^*) = 0,$$

then

$$\frac{\partial G}{\partial t}(S,t^*) \geq -\mathbf{L_{s,t^*}}G(S,t^*) = \frac{\partial V}{\partial t}(S,t^*)$$

or

$$\frac{\partial V}{\partial t}(S,t^*) - \frac{\partial G}{\partial t}(S,t^*) \leq 0.$$

Therefore, in this case we can use the equation

$$\frac{\partial V}{\partial t}(S,t) + \mathbf{L}_{\mathbf{s},\mathbf{t}} V(S,t) = 0$$

to get a solution satisfying the condition $V(S,t) - G(S,t) \geq 0$.

If

$$\frac{\partial G}{\partial t}(S,t^*) + \mathbf{L}_{\mathbf{s},\mathbf{t}^*} G(S,t^*) < 0$$

and

$$\frac{\partial V}{\partial t}(S,t^*) + \mathbf{L}_{\mathbf{s},\mathbf{t}^*} V(S,t^*) = \frac{\partial V}{\partial t}(S,t^*) + \mathbf{L}_{\mathbf{s},\mathbf{t}^*} G(S,t^*) = 0,$$

then

$$\frac{\partial G}{\partial t}(S,t^*) < -\mathbf{L}_{\mathbf{s},\mathbf{t}^*} G(S,t^*) = \frac{\partial V}{\partial t}(S,t^*)$$

or

$$\frac{\partial V}{\partial t}(S,t^*) - \frac{\partial G}{\partial t}(S,t^*) > 0,$$

which will cause $V(S,t) - G(S,t) < 0$ for any $t = t^* - \Delta t$. Therefore, we cannot get the solution by using the equation

$$\frac{\partial V}{\partial t}(S,t) + \mathbf{L}_{\mathbf{s},\mathbf{t}} V(S,t) = 0.$$

Instead we have to let $V(S,t) - G(S,t) = 0$ in order to get a solution satisfying the condition $V(S,t) - G(S,t) \geq 0$. ∎

About this theorem, we would like to make the following remarks.

- Let $\mathbb{D}_{ge}$ denote the open domain where $\frac{\partial G}{\partial t}(S,t) + \mathbf{L}_{\mathbf{s},\mathbf{t}} G(S,t) \geq 0$ and $\mathbb{D}_l$ the open domain where $\frac{\partial G}{\partial t}(S,t) + \mathbf{L}_{\mathbf{s},\mathbf{t}} G(S,t) < 0$. From this theorem, we see that there are two possibilities for a point to be on a free boundary. One possibility is: in a neighborhood of the point, $V(S,t) = G(S,t)$ and some portion of the neighborhood belongs to $\mathbb{D}_{ge}$ and another portion of the neighborhood belongs to $\mathbb{D}_l$. The other possibility is: in some portion of a neighborhood of the point, $V(S,t) > G(S,t)$ and in another portion of the neighborhood, $V(S,t) = G(S,t)$ and this portion belongs to $\mathbb{D}_l$. Therefore, a free boundary cannot appear in the open domain $\mathbb{D}_{ge}$. A free boundary will appear only in the open domain $\mathbb{D}_l$ and on the boundary of $\mathbb{D}_l$. If $V(S,T) = G(S,T)$, then a free boundary will start at a point between the open domains $\mathbb{D}_{ge}$ and $\mathbb{D}_l$. If $V(S,T) > G(S,T)$ on some portion of the interval $[0,\infty)$ and $V(S,T) = G(S,T)$ on another portion, then a free boundary will start at a boundary between an open interval belonging to $\mathbb{D}_l$ and an open interval where $V(S,T) > G(S,T)$. Later, a free boundary may move but never move into the open domain $\mathbb{D}_{ge}$.

● Let time be τ instead of t and we want to have the solution at $\tau = \tau^* + \Delta\tau$ from the solution at $\tau = \tau^*$, where $\Delta\tau > 0$. Then the theorem and the remark above are still true if the condition

$$\frac{\partial G}{\partial t}(S, t) + \mathbf{L}_{\mathrm{s},t} G(S, t) \geq 0$$

is changed into

$$\frac{\partial G}{\partial \tau}(S, \tau) + \mathbf{L}_{\mathrm{s},\tau} G(S, \tau) \leq 0$$

and the condition

$$\frac{\partial G}{\partial t}(S, t) + \mathbf{L}_{\mathrm{s},t} G(S, t) < 0$$

is changed into

$$\frac{\partial G}{\partial \tau}(S, \tau) + \mathbf{L}_{\mathrm{s},\tau} G(S, \tau) > 0.$$

Now let us formulate the problem the price of the American option should satisfy. Let us assume that at time t we have obtained $P(S, t)$ satisfying (2.45) and we need to determine $P(S, t - \Delta t)$ satisfying (2.45), where Δt is a sufficiently small positive number. Define $G_p(S, t) = \max(E - S, 0)$. For simplicity, we assume that the entire interval consists of three open intervals plus their boundaries. On the first open interval, $P(S, t) > G_p(S, t)$. For any point in this interval, $P(S, t - \Delta t)$ must be still greater than $G_p(S, t)$ if Δt is small enough. Therefore, at any point in this open interval

$$\begin{cases} \dfrac{\partial P}{\partial t} + \dfrac{1}{2}\sigma^2 S^2 \dfrac{\partial^2 P}{\partial S^2} + (r - D_0)S\dfrac{\partial P}{\partial S} - rP = 0, \\ P(S, t) > G_p(S, t). \end{cases}$$

On the second open interval $P(S, t) = G_p(S, t)$ and

$$\left[\frac{\partial}{\partial t} + \frac{1}{2}\sigma^2 S^2 \frac{\partial^2}{\partial S^2} + (r - D_0)S\frac{\partial}{\partial S} - r \right] G_p(S, t) \geq 0$$

and on the third open interval $P(S, t) = G_p(S, t)$ and

$$\left[\frac{\partial}{\partial t} + \frac{1}{2}\sigma^2 S^2 \frac{\partial^2}{\partial S^2} + (r - D_0)S\frac{\partial}{\partial S} - r \right] G_p(S, t) < 0.$$

According to Theorem 2.1, for a point (S, t) in the second open interval the Black–Scholes can be used to determine $P(S, t - \Delta t)$ and the following is true:

$$\begin{cases} \dfrac{\partial P}{\partial t} + \dfrac{1}{2}\sigma^2 S^2 \dfrac{\partial^2 P}{\partial S^2} + (r - D_0)S\dfrac{\partial P}{\partial S} - rP = 0, \\ P(S, t) = G_p(S, t) \end{cases}$$

On the third interval, the Black–Scholes equation cannot be used to determine $P(S, t-\Delta t)$. Instead, $P(S, t-\Delta t)$ should equal $G_p(S, t-\Delta t)$. In this situation

$$\frac{P(S,t) - P(S,t-\Delta t)}{\Delta t} = \frac{G_p(S,t) - G_p(S,t-\Delta t)}{\Delta t} \to \frac{\partial G_p(S,t)}{\partial t}$$

as $\Delta t \to 0$ and we have

$$\begin{cases} \left[\dfrac{\partial}{\partial t} + \dfrac{1}{2}\sigma^2 S^2 \dfrac{\partial^2}{\partial S^2} + (r - D_0)S\dfrac{\partial}{\partial S} - r \right] P \\ = \left[\dfrac{\partial}{\partial t} + \dfrac{1}{2}\sigma^2 S^2 \dfrac{\partial^2}{\partial S^2} + (r - D_0)S\dfrac{\partial}{\partial S} - r \right] G_p(S,t) < 0, \\ P(S,t) = G_p(S,t). \end{cases}$$

Because $P(S,T) = G_p(S,T)$, we can use this argument from T to 0. Putting all the cases together, for $S \in [0, \infty)$ and $t \leq T$ we have

$$\begin{cases} \left[\dfrac{\partial P}{\partial t} + \dfrac{1}{2}\sigma^2 S^2 \dfrac{\partial^2 P}{\partial S^2} + (r - D_0)S\dfrac{\partial P}{\partial S} - rP \right](P - G_p) = 0, \\ \dfrac{\partial P}{\partial t} + \dfrac{1}{2}\sigma^2 S^2 \dfrac{\partial^2 P}{\partial S^2} + (r - D_0)S\dfrac{\partial P}{\partial S} - rP \leq 0, \\ P(S,t) - G_p(S,t) \geq 0, \\ P(S,T) = G_p(S,T), \end{cases} \qquad (2.46)$$

where $G_p(S,t) = \max(E - S, 0)$. Here, we use the fact that (2.46) is also true in some sense at the boundary points of these open intervals. This problem is called the linear complementarity problem for an American put option.

Similarly, for an American call option, the corresponding linear complementarity problem is

$$\begin{cases} \left[\dfrac{\partial C}{\partial t} + \dfrac{1}{2}\sigma^2 S^2 \dfrac{\partial^2 C}{\partial S^2} + (r - D_0)S\dfrac{\partial C}{\partial S} - rC \right](C - G_c) = 0, \\ \dfrac{\partial C}{\partial t} + \dfrac{1}{2}\sigma^2 S^2 \dfrac{\partial^2 C}{\partial S^2} + (r - D_0)S\dfrac{\partial C}{\partial S} - rC \leq 0, \\ C(S,t) - G_c(S,t) \geq 0, \\ C(S,T) = G_c(S,T), \end{cases} \qquad (2.47)$$

where $0 \leq S$, $t \leq T$, and $G_c(S,t) = \max(S - E, 0)$. From the derivation of (2.46), when σ, r, D_0 depend on S and t, the formulations are still correct.

In the last subsection, we concluded that the price of an American option is at least as much as the price of a European option with the same parameters by using the financial reason. Here, we assume that σ, r, and D_0 are constants and explain this conclusion by using mathematical tools. Let $V(S,t)$, $v(S,t)$

denote the prices of the American and European options, respectively, and let $G_v(S,t)$ be the constraint for the American option satisfying the condition $G_v(S,t^*) \geq G_v(S,t^{**})$ if $t^* \leq t^{**}$. Set $\Delta t = T/N$, where N is a positive integer. Let $\tilde{V}(S,T) = G_v(S,T)$ and for $t_n = n\Delta t$, $n = N-1, N-2, \cdots, 0$, define

$$\tilde{V}(S,t_n) = \max\left(e^{-r\Delta t}\int_0^\infty \tilde{V}(S',t_{n+1})G(S',t_{n+1};S,t_n)dS', G_v(S,t_n)\right),$$

where $G(S',t_{n+1};S,t_n)$ is given by (2.36). Suppose $\tilde{V}(S,t_{n+1}) \geq v(S,t_{n+1})$. From (2.35) we know

$$v(S,t_n) = e^{-r\Delta t}\int_0^\infty v(S',t_{n+1})G(S',t_{n+1};S,t_n)dS'$$

$$\leq e^{-r\Delta t}\int_0^\infty \tilde{V}(S',t_{n+1})G(S',t_{n+1};S,t_n)dS'$$

$$\leq \tilde{V}(S,t_n).$$

At $t = t_N = T$, the condition $\tilde{V}(S,t_N) = \tilde{V}(S,T) \geq v(S,T) = v(S,t_N)$ holds. Therefore, using the induction method, we can prove $\tilde{V}(S,t_n) \geq v(S,t_n)$ for $n = N-1, N-2, \cdots, 0$ successively. Letting $N \to \infty$ and noticing that $\tilde{V}(S,t)$ generates $V(S,t)$ as $N \to \infty$, we can have the conclusion we need.

The price of an American option has another property: $V(S,t^*) \geq V(S,t^{**})$ if $t^* \leq t^{**}$. Let us explain this fact by using mathematical tools. Suppose $\tilde{V}(S,t_n) \geq \tilde{V}(S,t_{n+1})$. According to the definition of $\tilde{V}(S,t_n)$, we have

$$\tilde{V}(S,t_n) = \max\left(e^{-r\Delta t}\int_0^\infty \tilde{V}(S',t_{n+1})G(S',t_{n+1};S,t_n)dS', G_v(S,t_n)\right)$$

$$\leq \max\left(e^{-r\Delta t}\int_0^\infty \tilde{V}(S',t_n)G(S',t_n;S,t_{n-1})dS', G_v(S,t_{n-1})\right)$$

$$= \tilde{V}(S,t_{n-1}).$$

Here we have used the facts

$$G(S',t_{n+1};S,t_n) = G(S',t_n;S,t_{n-1}) \quad \text{and} \quad G_v(S,t_n) \leq G_v(S,t_{n-1}).$$

Because

$$\tilde{V}(S,t_{N-1}) = \max\left(e^{-r\Delta t}\int_0^\infty \tilde{V}(S',t_N)G(S',t_N;S,t_{N-1})dS', G_v(S,t_{N-1})\right)$$

$$\geq G_v(S,t_{N-1}) \geq G_v(S,t_N) = \tilde{V}(S,t_N),$$

we can prove

$$\tilde{V}(S,t_n) \geq \tilde{V}(S,t_{n+1}) \quad \text{for} \quad n = N-2, N-3, \cdots, 0$$

successively. This means

$$\tilde{V}(S, t_n) \geq \tilde{V}(S, t_m) \quad \text{for} \quad n \leq m \leq N.$$

Letting $N \to \infty$ and noticing that $\tilde{V}(S, t)$ generates $V(S, t)$ as $N \to \infty$, we arrive at the conclusion

$$V(S, t^*) \geq V(S, t^{**}) \quad \text{if} \quad t^* \leq t^{**}.$$

2.5.3 Formulation of the Linear Complementarity Problem in $(x, \bar{\tau})$-Plane

As we know from Subsection 2.3.1, if we set

$$\begin{cases} x = \ln S + \left(r - D_0 - \dfrac{1}{2}\sigma^2\right)(T - t), \\[2mm] \bar{\tau} = \dfrac{1}{2}\sigma^2(T - t), \\[2mm] V(S, t) = e^{-r(T-t)}u(x, \bar{\tau}), \end{cases}$$

then

$$\frac{\partial V}{\partial t} + \frac{1}{2}\sigma^2 S^2 \frac{\partial^2 V}{\partial S^2} + (r - D_0)S\frac{\partial V}{\partial S} - rV$$

becomes

$$-\frac{1}{2}\sigma^2 e^{-r(T-t)}\left(\frac{\partial u}{\partial \bar{\tau}} - \frac{\partial^2 u}{\partial x^2}\right).$$

Thus,

$$\frac{\partial P}{\partial t} + \frac{1}{2}\sigma^2 S^2 \frac{\partial P}{\partial S^2} + (r - D_0)S\frac{\partial P}{\partial S} - rP < 0$$

is equivalent to

$$\frac{\partial u}{\partial \bar{\tau}} - \frac{\partial^2 u}{\partial x^2} > 0$$

and the Black–Scholes equation holds if and only if

$$\frac{\partial u}{\partial \bar{\tau}} - \frac{\partial^2 u}{\partial x^2} = 0.$$

Let us define

$$g_p(x, \bar{\tau}) = \max\left(e^{2r\bar{\tau}/\sigma^2} - e^{x+(2D_0/\sigma^2+1)\bar{\tau}}, 0\right),$$

then

$$\begin{aligned} P - G_p &= P(S, t) - \max(1 - S, 0) \\ &= e^{-r(T-t)}u(x, \bar{\tau}) - \max\left(1 - e^{x-(r-D_0-\sigma^2/2)(T-t)}, 0\right) \\ &= e^{-r(T-t)}\left[u(x, \bar{\tau}) - \max\left(e^{r(T-t)} - e^{x+(D_0+\sigma^2/2)(T-t)}, 0\right)\right] \\ &= e^{-r(T-t)}[u(x, \bar{\tau}) - g_p(x, \bar{\tau})], \end{aligned}$$

where we suppose $E = 1$ for simplicity. Thus, $P - G_p > 0$ is equivalent to

$$u(x, \bar{\tau}) - g_p(x, \bar{\tau}) > 0$$

and $P - G_p = 0$ if and only if

$$u(x, \bar{\tau}) - g_p(x, \bar{\tau}) = 0.$$

Therefore, the American put option is the solution of the following problem:

$$\begin{cases} \left(\dfrac{\partial u}{\partial \bar{\tau}} - \dfrac{\partial^2 u}{\partial x^2} \right) [u(x, \bar{\tau}) - g_p(x, \bar{\tau})] = 0, \\[2mm] \dfrac{\partial u}{\partial \bar{\tau}} - \dfrac{\partial^2 u}{\partial x^2} \geq 0, \\[2mm] u(x, \bar{\tau}) - g_p(x, \bar{\tau}) \geq 0, \\[2mm] u(x, 0) = g_p(x, 0), \end{cases} \tag{2.48}$$

where $x \in (-\infty, \infty)$ and $0 \leq \bar{\tau}$. Similarly, for American call options we have

$$\begin{cases} \left(\dfrac{\partial u}{\partial \bar{\tau}} - \dfrac{\partial^2 u}{\partial x^2} \right) [u(x, \bar{\tau}) - g_c(x, \bar{\tau})] = 0, \\[2mm] \dfrac{\partial u}{\partial \bar{\tau}} - \dfrac{\partial^2 u}{\partial x^2} \geq 0, \\[2mm] u(x, \bar{\tau}) - g_c(x, \bar{\tau}) \geq 0, \\[2mm] u(x, 0) = g_c(x, 0), \end{cases} \tag{2.49}$$

where $x \in (-\infty, \infty)$, $0 \leq \bar{\tau}$ and

$$g_c(x, \bar{\tau}) = \max \left(e^{x + (2D_0/\sigma^2 + 1)\bar{\tau}} - e^{2r\bar{\tau}/\sigma^2}, 0 \right).$$

The derivation of (2.49) is almost identical to the American put. The only difference is that instead of using $P - G_p = e^{-r(T-t)} [u(x, \bar{\tau}) - g_p(x, \bar{\tau})]$, we need to use the relation

$$\begin{aligned} C - G_c &= C(S, t) - \max(S - 1, 0) \\ &= e^{-r(T-t)} u(x, \bar{\tau}) - \max \left(e^{x - (r - D_0 - \sigma^2/2)(T-t)} - 1, 0 \right) \\ &= e^{-r(T-t)} \left[u(x, \bar{\tau}) - \max \left(e^{x + (D_0 + \sigma^2/2)(T-t)} - e^{r(T-t)}, 0 \right) \right] \\ &= e^{-r(T-t)} [u(x, \bar{\tau}) - g_c(x, \bar{\tau})], \end{aligned}$$

where we also assume $E = 1$.

It is clear that if r, D_0, and σ depend on t, then similar results hold. However, if σ depends on S, then we may not be able to convert the problems (2.46) and (2.47) into (2.48) and (2.49) by a simple transformation.

2.5.4 Formulation of the Linear Complementarity Problem on a Finite Domain

Generally speaking, r, D_0, and σ are not constants. For simplicity, we assume that σ depends on S in this subsection even though the derivation is almost the same when r, D_0, and σ all depend on S and t.

From Subsection (2.3.2), we know that through the transformation

$$\begin{cases} \xi = \dfrac{S}{S+E}, \\ \tau = T - t, \\ V(S,t) = (S+E)\overline{V}(\xi,\tau) = \dfrac{E}{1-\xi}\overline{V}(\xi,\tau), \end{cases}$$

the operator

$$\frac{\partial}{\partial t} + \frac{1}{2}\sigma^2(S)S^2\frac{\partial^2}{\partial S^2} + (r-D_0)S\frac{\partial}{\partial S} - r$$

is converted into

$$\frac{-E}{1-\xi}\left\{ \frac{\partial}{\partial \tau} - \frac{1}{2}\bar{\sigma}^2(\xi)\xi^2(1-\xi)^2\frac{\partial^2}{\partial \xi^2} - (r-D_0)\xi(1-\xi)\frac{\partial}{\partial \xi} + [r(1-\xi)+D_0\xi] \right\},$$

where $\bar{\sigma}(\xi) = \sigma(E\xi/(1-\xi))$, and the function $\max(\pm(S-E),0)$ becomes

$$\frac{E}{1-\xi}\max(\pm(2\xi-1),0).$$

Therefore, the problem (2.46) can be rewritten as

$$\begin{cases} \left(\dfrac{\partial \overline{V}}{\partial \tau} - \mathbf{L}_\xi \overline{V}\right)\left[\overline{V}(\xi,\tau) - \max(1-2\xi,0)\right] = 0, \\ \\ \dfrac{\partial \overline{V}}{\partial \tau} - \mathbf{L}_\xi \overline{V} \geq 0, \\ \\ \overline{V}(\xi,\tau) - \max(1-2\xi,0) \geq 0, \\ \\ \overline{V}(\xi,0) = \max(1-2\xi,0), \end{cases} \qquad (2.50)$$

where $0 \leq \xi \leq 1$, $0 \leq \tau$ and

$$\mathbf{L}_\xi = \frac{1}{2}\bar{\sigma}^2(\xi)\xi^2(1-\xi)^2\frac{\partial^2}{\partial \xi^2} + (r-D_0)\xi(1-\xi)\frac{\partial}{\partial \xi} - [r(1-\xi)+D_0\xi].$$

This is the American put option problem reformulated as a linear complementarity problem on a finite domain. Similarly, from (2.47) we know that the American call option problem can be reformulated as the following linear complementarity problem:

$$\begin{cases} \left(\dfrac{\partial \overline{V}}{\partial \tau} - \mathbf{L}_\xi \overline{V} \right) \left[\overline{V}(\xi, \tau) - \max(2\xi - 1, 0) \right] = 0, \\[2mm] \dfrac{\partial \overline{V}}{\partial \tau} - \mathbf{L}_\xi \overline{V} \ge 0, \\[2mm] \overline{V}(\xi, \tau) - \max(2\xi - 1, 0) \ge 0, \\[2mm] \overline{V}(\xi, 0) = \max(2\xi - 1, 0), \end{cases} \qquad (2.51)$$

where

$$0 \le \xi \le 1 \quad \text{and} \quad 0 \le \tau.$$

In this section, an American option is reduced to a linear complementarity problem. Such a problem usually needs to be solved numerically. Here, we need to point out that the version given in Subsection 2.5.3 can be applied only if σ does not depend on S and that the other two versions can be applied for any case. However, the version given in Subsection 2.5.3 has the simplest equation. Also, if an implicit scheme is used, then for the versions given in Subsections 2.5.2 and 2.5.3, artificial boundary conditions are needed at the boundaries because numerical methods have to be performed on a finite domain. However, the version given in this subsection does not have such a problem.

2.5.5 More General Form of the Linear Complementarity Problems

From the three previous subsections, we see that a linear complementarity problem could be in the form:

$$\begin{cases} \left[\dfrac{\partial V(S, t)}{\partial t} + \mathbf{L}_{\mathbf{s}, t} V(S, t) \right] \left[V(S, t) - G(S, t) \right] = 0, \\[2mm] \dfrac{\partial V(S, t)}{\partial t} + \mathbf{L}_{\mathbf{s}, t} V(S, t) \le 0, \\[2mm] V(S, t) - G(S, t) \ge 0, \\[2mm] V(S, T) = G(S, T), \end{cases}$$

where[11]

$$S_l \le S \le S_u, \quad t \le T$$

and

$$\mathbf{L}_{\mathbf{s}, t} = a(S, t) \frac{\partial^2}{\partial S^2} + b(S, t) \frac{\partial}{\partial S} + c(S, t).$$

However, a linear complementarity problem could have a more general form such as

---

[11]If $S_l = -\infty$, then the first "$\le$" needs to be changed into "$<$," and if $S_u = \infty$, then the second "$\le$" needs to be changed into "$<$."

$$\begin{cases} \left[\dfrac{\partial V(S,t)}{\partial t} + \mathbf{L}_{\mathbf{s},t}V(S,t) + d(S,t)\right][V(S,t) - G(S,t)] = 0, \\[2mm] \dfrac{\partial V(S,t)}{\partial t} + \mathbf{L}_{\mathbf{s},t}V(S,t) + d(S,t) \leq 0, \\[2mm] V(S,t) - G(S,t) \geq 0, \\[2mm] V(S,T) = G_1(S) \geq G(S,T), \end{cases} \qquad (2.52)$$

where[12] $S_l \leq S \leq S_u$, $t \leq T$. For example, the linear complementarity problem for one-factor convertible bonds has such a form. For two-factor convertible bonds, the form of the linear complementarity problem is similar, but the operator $\mathbf{L}_{\mathbf{s},t}$ is two-dimensional (see Chapter 4).

2.6 American Option Problems as Free-Boundary Problems

2.6.1 Free Boundaries

From the past section, we discovered that there are some regions where the Black–Scholes equation cannot be used. Therefore, there exist two different types of regions: one where the Black–Scholes equation is valid, and the other where the Black–Scholes equation cannot be used and the solution is equal to the constraint. Because we do not know *a priori* the location of the boundaries between the two types of different regions, these boundaries are called free boundaries. Because in some regions the solution is known, we only need to determine the price in other regions and the locations of these free boundaries. In order to do that, we reformulate the American option problems as so-called free-boundary problems (FBPs).

Let us first discuss how to find the locations of the free boundaries at time T. Using Theorem 2.1, we can easily determine the locations of free boundaries at time T, namely, the starting points of free boundaries. We will show that for an American put option with $r > 0$, there is a free boundary starting from the point $(\min(E, rE/D_0), T)$ on the (S,t)-plane. If $r = 0$, then there is no free boundary. This implies that the Black–Scholes equation is valid everywhere and that the prices of the American and European put options are the same. For an American call option, the situation is similar. If $D_0 > 0$, then there is a free boundary starting from the point $(\max(E, rE/D_0), T)$ on the (S,t)-plane. If $D_0 = 0$, then there is no free boundary, implying that an American call option is the same as a European call option.

First, let us consider an American put option and let $P(S,t)$ denote its value as we did in Subsection 2.5.1. In this case

---

[12]The last footnote can also be applied in this case.

$$G(S,t) = \max(E - S, 0) = \begin{cases} E - S, & \text{for } S < E, \\ 0, & \text{for } S \geq E \end{cases}$$

and the operator $\mathbf{L_{s,t}}$ in this case does not depend on t and is equal to

$$\mathbf{L_s} = \frac{1}{2}\sigma^2 S^2 \frac{\partial^2}{\partial S^2} + (r - D_0)S\frac{\partial}{\partial S} - r.$$

For $S < \min(E, rE/D_0)$, $G(S,t) = E - S$ and

$$\frac{\partial G}{\partial t}(S,T) + \mathbf{L_s}G(S,T)$$

$$= \frac{1}{2}\sigma^2 S^2 \frac{\partial^2 G}{\partial S^2} + (r - D_0)S\frac{\partial G}{\partial S} - rG$$

$$= \frac{1}{2}\sigma^2 S^2 \frac{\partial^2}{\partial S^2}(E - S) + (r - D_0)S\frac{\partial}{\partial S}(E - S) - r(E - S)$$

$$= D_0 S - rE < 0$$

because $S < rE/D_0$. Consequently, for $S < \min(E, rE/D_0)$, we cannot get the solution by using the Black–Scholes equation at time T. Instead, the price will be pushed up to satisfy $P(S,T) - G(S,T) = 0$, so $P(S,T) = E - S$ and

$$\frac{\partial P}{\partial t} + \frac{1}{2}\sigma^2 S^2 \frac{\partial^2 P}{\partial S^2} + (r - D_0)S\frac{\partial P}{\partial S} - rP < 0.$$

For $S > E$, $G(S,t) = 0$ and

$$\frac{\partial G}{\partial t}(S,T) + \mathbf{L_s}G(S,T) = 0.$$

If $rE/D_0 < E$, then we have an interval $(rE/D_0, E)$ and for $S \in (rE/D_0, E)$, $G(S,t) = E - S$ and

$$\frac{\partial G}{\partial t}(S,T) + \mathbf{L_s}G(S,T) = D_0 S - rE > 0$$

because $S > rE/D_0$. Thus, for $S > \min(E, rE/D_0)$, the Black–Scholes equation gives a solution satisfying

$$P(S, T - \Delta t) \geq G(S, T - \Delta t) = \max(E - S, 0)$$

for any sufficiently small positive Δt.

Therefore, at $t = T$, if $r > 0$, then the S-axis is divided into two parts: $[0, \min(E, rE/D_0))$ where the Black–Scholes equation cannot be used and $(\min(E, rE/D_0), \infty)$ where the Black–Scholes equation gives the price of the American put option. Thus, at time T there is a free boundary at $S = \min(E, rE/D_0)$. If $r = 0$, then $\min(E, rE/D_0) = 0$, so in the entire interval $(0, \infty)$, the Black–Scholes equation can be used, and there is no free boundary.

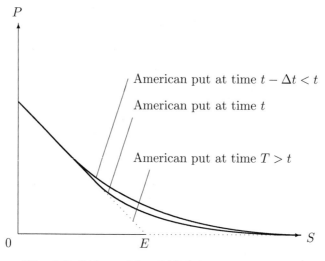

Fig. 2.7. $P(S, t - \Delta t) \geq P(S, t)$ for any positive Δt

Now let us explain that in the case $r > 0$, no new free boundary can appear at any time $t < T$, so the free boundary starting from the point $(\min(E, rE/D_0), T)$ is the only free boundary in this problem. Let $S_f(t)$ denote this free boundary. Because it starts from the point $(\min(E, rE/D_0), T)$, we have

$$S_f(T) = \min\left(E, \frac{rE}{D_0} \right). \tag{2.53}$$

A holder of a long-life American put option has more exercise opportunities than a holder of a short-life American put option, so we have $P(S, t - \Delta t) \geq P(S, t)$ for any positive Δt. Therefore, if $P(S, t) > \max(E - S, 0)$ for time t, then $P(S, t - \Delta t) \geq P(S, t) > \max(E - S, 0)$ for any $\Delta t > 0$ (see Fig. 2.7). Consequently, at time t there is no chance for a new free boundary to appear in the interval $(S_f(t), \infty)$ where $P(S, t) > \max(E - S, 0)$ and $S_f(t)$ must be less than $S_f(T) = \min(E, rE/D_0)$ for any t. Because of $S_f(t) \leq \min(E, rE/D_0)$, we have $P(S, t) = E - S$ at any point (S, t) with $S \in (0, S_f(t))$. From Theorem 2.1, we know that it is impossible for a new free boundary to appear at such a point because at a neighborhood of the point

$$\frac{\partial G}{\partial t} + \frac{1}{2}\sigma^2 S^2 \frac{\partial^2 G}{\partial S^2} + (r - D_0)S\frac{\partial G}{\partial S} - rG < 0.$$

Therefore, no new free boundary can appear anywhere. However, if a point is close enough to the point $(S_f(t), t)$, there is a chance for the point to be on the free boundary at some time $t - \Delta t$. Thus, the free boundary starting from the point $(\min(E, rE/D_0), T)$ is movable.

Consequently, if $r > 0$, then there is a unique free boundary, and the entire domain is divided into two regions by the free boundary (see Fig. 2.8): one

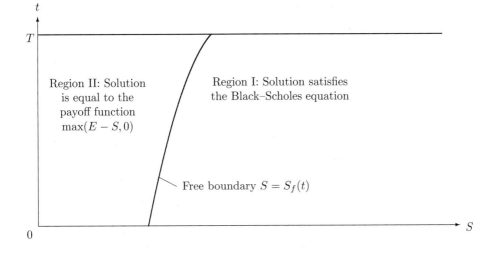

Fig. 2.8. Structure of solution to American put options $(r > 0)$

region is $[0, S_f(t)) \times [0, T]$, where

$$
\begin{cases}
P = E - S = \max(E - S, 0), \\
\dfrac{\partial P}{\partial t} + \dfrac{1}{2}\sigma^2 S^2 \dfrac{\partial^2 P}{\partial S^2} + (r - D_0)S\dfrac{\partial P}{\partial S} - rP < 0
\end{cases}
$$

and the other is $(S_f(t), \infty) \times [0, T]$, where

$$
\begin{cases}
P > \max(E - S, 0), \\
\dfrac{\partial P}{\partial t} + \dfrac{1}{2}\sigma^2 S^2 \dfrac{\partial^2 P}{\partial S^2} + (r - D_0)S\dfrac{\partial P}{\partial S} - rP = 0
\end{cases}
$$

if $t < T$.

Before going further, it is necessary to explain the meaning of the inequality

$$
\frac{\partial P}{\partial t} + \frac{1}{2}\sigma^2 S^2 \frac{\partial^2 P}{\partial S^2} + (r - D_0)S\frac{\partial P}{\partial S} - rP < 0.
$$

As pointed out in Subsection 2.2.2, this means that the difference between the return from the portfolio and the return from an equivalent bank deposit is negative. Hence, it is optimal to exercise the portfolio. If $P(S, t) > \max(E - S, 0)$, one should hold the option, as one should not give up a higher value (the option) for a lower value (the payoff). Therefore, the free boundary is the optimal exercise price that divides the exercise region and the non-exercise region.

As pointed out above, if $P(S, t) > \max(E - S, 0)$, then $P(S, t - \Delta t) > P(S, t)$ for any positive Δt. From this fact, we can further see that the following inequality holds (see Fig. 2.7):

$$S_f(t) > S_f(t - \Delta t), \quad \Delta t > 0,$$

implying that $S_f(t)$ is an increasing function of t (see Fig. 2.8).

Now let us consider an American call option. As we know, at very large S, the solution of the Black–Scholes equation with final condition $V(S, t) = \max(S - E, 0)$ has the following asymptotic expression

$$V(S, t) \approx V(S, T)e^{-D_0(T-t)} = \max(S - E, 0)e^{-D_0(T-t)},$$

so if $D_0 > 0$, then $V(S, t) < \max(S - E, 0)$ for any $t < T$. Therefore, if $D_0 > 0$, the American call problem is a free-boundary problem. Now let us show that the free-boundary problem has only one free boundary, which is also denoted by $S_f(t)$ in what follows, and determine the location of the free boundary at $t = T$ from the constraint condition $C(S, t) \geq G(S, t)$.

In the case of an American call option,

$$G(S, t) = \max(S - E, 0) = \begin{cases} S - E, & S > E, \\ 0, & S \leq E. \end{cases}$$

Let $S > \max(E, rE/D_0)$. In this case

$$G(S, t) = S - E$$

and

$$\begin{aligned}
&\frac{\partial G}{\partial t}(S, T) + \mathbf{L_s}G(S, T) \\
&= \frac{1}{2}\sigma^2 S^2 \frac{\partial^2 G}{\partial S^2} + (r - D_0)S\frac{\partial G}{\partial S} - rG \\
&= \frac{1}{2}\sigma^2 S^2 \frac{\partial^2 (S - E)}{\partial S^2} + (r - D_0)S\frac{\partial(S - E)}{\partial S} - r(S - E) \\
&= rS - D_0 S - rS + rE = -D_0 S + rE < 0
\end{aligned}$$

because $S > rE/D_0$. Therefore, the Black–Scholes equation cannot hold in this case, and $C(S, T - \Delta t)$ should be equal to $S - E$ for $S > \max(E, rE/D_0)$. For $S < \max(E, rE/D_0)$, the Black–Scholes equation can hold. Thus, a free boundary starts at $S = \max(E, rE/D_0)$, i.e.,

$$S_f(T) = \max\left(E, \frac{rE}{D_0}\right). \tag{2.54}$$

Using the same argument we have used for an American put option, we can show that the free boundary starting from the point $(\max(E, rE/D_0), T)$ is the only free boundary because no new free boundary can appear at time $t < T$. Just like the put case, the entire domain is divided into two parts by the free boundary. However, the situation is a little different from the American

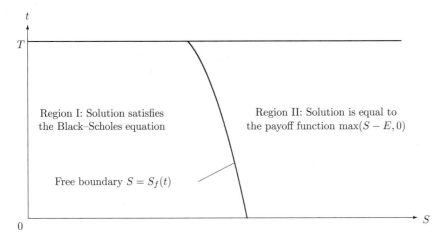

Fig. 2.9. Structure of solution to American call options ($D_0 > 0$)

put. Here in the region $[0, S_f(t)) \times [0, T]$, the Black–Scholes equation holds, whereas in the region $(S_f(t), \infty) \times [0, T]$, the Black–Scholes equation cannot be used. In other words, for $S \in [0, S_f(t))$ and $t < T$,

$$\begin{cases} C(S,t) \geq \max(S - E, 0), \\ \dfrac{\partial C}{\partial t} + \dfrac{1}{2}\sigma^2 S^2 \dfrac{\partial^2 C}{\partial S^2} + (r - D_0)S\dfrac{\partial C}{\partial S} - rC = 0, \end{cases}$$

where the equal sign in $C(S,t) \geq \max(S - E, 0)$ holds only at $S = 0$; whereas for $S \in (S_f(t), \infty)$,

$$\begin{cases} C(S,t) = S - E = \max(S - E, 0), \\ \dfrac{\partial C}{\partial t} + \dfrac{1}{2}\sigma^2 S^2 \dfrac{\partial^2 C}{\partial S^2} + (r - D_0)S\dfrac{\partial C}{\partial S} - rC < 0 \end{cases}$$

and the option should be exercised (see Fig. 2.9). It can also be shown that for an American call option, the free boundary $S_f(t)$ is a decreasing function of t, as graphed in Fig. 2.9, and that the price of an American call option is the same as a European call if $D_0 = 0$.

2.6.2 Free-Boundary Problems

In this subsection, we will describe the formulation of American option problems as free-boundary problems. In order to give a complete formulation, we need to give the conditions on the free boundary. For an initial-boundary value problem of a parabolic equation on a finite interval, if the locations of the boundaries are given and if the coefficient of the second derivative at the boundaries is not equal to zero, one boundary condition at each boundary is needed in order for the problem to have a unique solution. However, the location of the free boundary is unknown, so two conditions are needed at the free boundary in order for the problem to have a unique solution. One boundary condition determines the option value on the free boundary and the other boundary condition determines the location of the free boundary. Now the question is what the two conditions should be. For some other linear complementarity problems, it has been proved that on the free boundary the value and the first derivative are continuous (see [28]). For this problem, from the proof given by Badea and Wang (see [3] and [4]), we know that the situation is still the same. Therefore, the two conditions on the free boundary are: both the value and the derivative with respect to S are continuous.

For an American put option, in the region $[0, \ S_f(t))$,

$$P(S,t) = E - S \quad \text{and} \quad \frac{\partial P}{\partial S} = -1.$$

Therefore, the boundary conditions on the free boundary $S_f(t)$ are

$$P(S_f(t), t) = E - S_f(t) \tag{2.55}$$

and

$$\frac{\partial P}{\partial S}(S_f(t), t) = -1. \tag{2.56}$$

It is clear that when (2.56) holds, the gradient $\dfrac{\partial P}{\partial S}$ must be continuous at $S = S_f$ (see Fig. 2.10).

Now we can formulate the American put option problem. On the domain $[0, S_f(t)) \times [0, T]$, $P(S, t) = E - S$, while on the domain $[S_f(t), \infty) \times [0, T]$, $P(S, t)$ is the solution of the free-boundary problem for American put options

$$
\begin{cases}
\dfrac{\partial P}{\partial t} + \dfrac{1}{2}\sigma^2 S^2 \dfrac{\partial^2 P}{\partial S^2} + (r - D_0)S\dfrac{\partial P}{\partial S} - rP = 0, \\[2mm]
\qquad\qquad\qquad\qquad\qquad S_f(t) \le S, \quad 0 \le t \le T, \\[2mm]
P(S,T) = \max(E - S, 0), \quad S_f(T) \le S, \\[2mm]
P(S_f(t), t) = E - S_f(t), \qquad t \le T, \\[2mm]
\dfrac{\partial P(S_f(t), t)}{\partial S} = -1, \qquad\quad t \le T, \\[2mm]
S_f(T) = \min\left(E, \dfrac{rE}{D_0}\right).
\end{cases}
\tag{2.57}
$$

Similarly, for call options we need two boundary conditions on the free boundary. One is

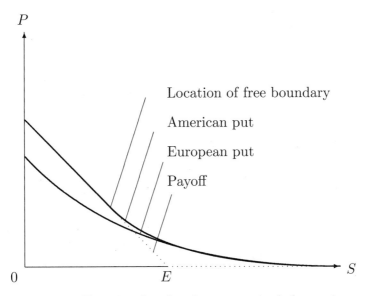

Fig. 2.10. The price of an American put option before expiry

$$C(S,t) = S_f(t) - E \tag{2.58}$$

and the other still can be obtained by requiring the continuity of the slope of the solution at $S = S_f(t)$. In this case, the condition is

$$\frac{\partial C(S_f(t),t)}{\partial S} = 1. \tag{2.59}$$

Therefore for the American call option, the formulation is as follows. On the domain $[0, S_f(t)] \times [0, T]$, $C(S,t)$ is the solution of the free-boundary problem for American call options

$$
\begin{cases}
\dfrac{\partial C}{\partial t} + \dfrac{1}{2}\sigma^2 S^2 \dfrac{\partial^2 C}{\partial S^2} + (r - D_0)S\dfrac{\partial C}{\partial S} - rC = 0, \\
\qquad\qquad\qquad\qquad\qquad 0 \le S \le S_f(t), \quad 0 \le t \le T, \\
C(S,T) = \max(S - E, 0), \quad 0 \le S \le S_f(T), \\
C(S_f(t),t) = S_f(t) - E, \qquad 0 \le t \le T, \\
\dfrac{\partial C}{\partial S}(S_f(t),t) = 1, \qquad\qquad 0 \le t \le T, \\
S_f(T) = \max\left(E, \dfrac{rE}{D_0}\right);
\end{cases} \tag{2.60}
$$

whereas on the domain $(S_f(t), \infty) \times [0, T]$, $C(S,t) = S - E$. In Fig. 2.11, the value of an American call option is plotted, from which we can see how the

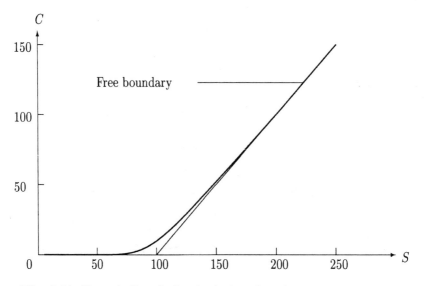

Fig. 2.11. Numerically calculated solution of an American call problem
with $E = 100$, $r = 0.1$, $D_0 = 0.05$, $\sigma = 0.2$, and $T = 1$ year

two parts of solution are connected smoothly. The parameters of the problem
are $E = 100$, $r = 0.1$, $D_0 = 0.05$, $\sigma = 0.2$, and $T = 1$ year.

Finally, we need to point out that $S_f(T)$ is determined by the partial
differential operator and the final condition. Therefore, in a free-boundary
problem, the starting location of the free boundary is not arbitrary and should
be consistent with the partial differential operator and the final condition.

2.6.3 Put–Call Symmetry Relations

From Subsection 2.3.2, we know that under the transformation

$$
\begin{cases}
\xi = \dfrac{S}{S + E}, \\
\tau = T - t, \\
c(S, t) = (S + E)\bar{c}(\xi, \tau),
\end{cases}
\tag{2.61}
$$

a call option problem with a constant σ is converted into the following problem:

$$
\begin{cases}
\dfrac{\partial \bar{c}}{\partial \tau} = \dfrac{1}{2}\sigma^2 \xi^2 (1 - \xi)^2 \dfrac{\partial^2 \bar{c}}{\partial \xi^2} + (r - D_0)\xi(1 - \xi)\dfrac{\partial \bar{c}}{\partial \xi} - [r(1 - \xi) + D_0\xi]\bar{c}, \\
\qquad\qquad\qquad\qquad\qquad\qquad\qquad\qquad 0 \le \xi \le 1, \quad 0 \le \tau, \\
\bar{c}(\xi, 0) = \max(2\xi - 1, 0), \qquad\qquad\qquad\qquad 0 \le \xi \le 1.
\end{cases}
$$

If we introduce the transformation

$$\begin{cases} \xi = 1 - \dfrac{S}{S+E} = \dfrac{E}{S+E}, \\[2mm] \tau = T - t, \\[2mm] p(S,t) = (S+E)\tilde{p}(\xi,\tau), \end{cases} \tag{2.62}$$

the put option with a constant σ is reduced to

$$\begin{cases} \dfrac{\partial \tilde{p}}{\partial \tau} = \dfrac{1}{2}\sigma^2 \xi^2 (1-\xi)^2 \dfrac{\partial^2 \tilde{p}}{\partial \xi^2} + (D_0 - r)\xi(1-\xi)\dfrac{\partial \tilde{p}}{\partial \xi} - [D_0(1-\xi) + r\xi]\tilde{p}, \\[3mm] \hspace{5cm} 0 \le \xi \le 1, \quad 0 \le \tau, \\[2mm] \tilde{p}(\xi,0) = \max(2\xi - 1, 0), \hspace{1.5cm} 0 \le \xi \le 1. \end{cases}$$

Comparing the two problems above, we know that if $r = a$ and $D_0 = b$ in the first problem and $r = b$ and $D_0 = a$ in the second problem, then the two solutions should be equal:

$$\bar{c}(\xi,\tau;a,b) = \tilde{p}(\xi,\tau;b,a).$$

Here, the first and second arguments after the semicolon in $\bar{c}, \tilde{p}$ are the values of the interest rate and dividend yield, respectively, and in this subsection for $c, p, C, P, S_{pf}, S_{cf}$, and so forth, the same notation will be used. Because

$$\bar{c}(\xi,\tau;a,b) = \dfrac{c\left(\dfrac{E\xi}{1-\xi}, t;a,b\right)(1-\xi)}{E}$$

and

$$\tilde{p}(\xi,\tau;b,a) = \dfrac{p\left(\dfrac{E(1-\xi)}{\xi}, t;b,a\right)\xi}{E},$$

we can rewrite the above relation between $\bar{c}$ and $\tilde{p}$ as

$$\dfrac{c\left(\dfrac{E\xi}{1-\xi}, t;a,b\right)(1-\xi)}{E} = \dfrac{p\left(\dfrac{E(1-\xi)}{\xi}, t;b,a\right)\xi}{E}$$

and further have

$$\begin{cases} c(S,t;a,b) = p\left(\dfrac{E^2}{S}, t;b,a\right)S/E, \quad \text{or} \\[3mm] p(S,t;b,a) = c\left(\dfrac{E^2}{S}, t;a,b\right)S/E \end{cases} \tag{2.63}$$

and its special case

$$p(E, t; b, a) = c(E, t; a, b).$$

Unlike the put–call parity, the relation (2.63) still holds for American options. By using the transformation (2.61), an American call problem can indeed be written as

$$
\begin{cases}
\left\{ \dfrac{\partial \bar{C}}{\partial \tau} - \dfrac{1}{2}\sigma^2 \xi^2 (1-\xi)^2 \dfrac{\partial^2 \bar{C}}{\partial \xi^2} - (r - D_0)\xi(1-\xi)\dfrac{\partial \bar{C}}{\partial \xi} + [r(1-\xi) + D_0\xi]\bar{C} \right\} \\[2mm]
\times \left[\bar{C}(\xi,\tau) - \max(2\xi - 1, 0) \right] = 0, \\[2mm]
\dfrac{\partial \bar{C}}{\partial \tau} - \dfrac{1}{2}\sigma^2 \xi^2 (1-\xi)^2 \dfrac{\partial^2 \bar{C}}{\partial \xi^2} - (r - D_0)\xi(1-\xi)\dfrac{\partial \bar{C}}{\partial \xi} + [r(1-\xi) + D_0\xi]\bar{C} \geq 0, \\[2mm]
\bar{C}(\xi,\tau) - \max(2\xi - 1, 0) \geq 0, \\[2mm]
\bar{C}(\xi,0) = \max(2\xi - 1, 0),
\end{cases}
$$

where $\xi \in [0, 1]$ and $0 \leq \tau$. Under the transformation (2.62), an American put option can be converted into

$$
\begin{cases}
\left\{ \dfrac{\partial \tilde{P}}{\partial \tau} - \dfrac{1}{2}\sigma^2 \xi^2 (1-\xi)^2 \dfrac{\partial^2 \tilde{P}}{\partial \xi^2} - (D_0 - r)\xi(1-\xi)\dfrac{\partial \tilde{P}}{\partial \xi} + [D_0(1-\xi) + r\xi]\tilde{P} \right\} \\[2mm]
\times \left[\tilde{P}(\xi,\tau) - \max(2\xi - 1, 0) \right] = 0, \\[2mm]
\dfrac{\partial \tilde{P}}{\partial \tau} - \dfrac{1}{2}\sigma^2 \xi^2 (1-\xi)^2 \dfrac{\partial^2 \tilde{P}}{\partial \xi^2} - (D_0 - r)\xi(1-\xi)\dfrac{\partial \tilde{P}}{\partial \xi} + [D_0(1-\xi) + r\xi]\tilde{P} \geq 0, \\[2mm]
\tilde{P}(\xi,\tau) - \max(2\xi - 1, 0) \geq 0, \\[2mm]
\tilde{P}(\xi,0) = \max(2\xi - 1, 0),
\end{cases}
$$

where $\xi \in [0, 1]$ and $0 \leq \tau$. Just as in the European case, if $r = a, D_0 = b$ in the former problem and $r = b, D_0 = a$ in the latter problem, then the two solutions should be the same and

$$
\begin{cases}
C(S, t; a, b) = P\left(\dfrac{E^2}{S}, t; b, a \right) S/E, \quad \text{or} \\[4mm]
P(S, t; b, a) = C\left(\dfrac{E^2}{S}, t; a, b \right) S/E
\end{cases}
\tag{2.64}
$$

holds. Also, the location of free boundary in the former problem, $\xi_{cf}(\tau; a, b)$, must be the same as the location of free boundary of the latter problem, $\xi_{\tilde{p}f}(\tau; b, a)$, i.e.,

$$\xi_{cf}(\tau; a, b) = \xi_{\tilde{p}f}(\tau; b, a)$$

or

$$\frac{S_{cf}(t; a, b)}{E + S_{cf}(t; a, b)} = \frac{E}{E + S_{pf}(t; b, a)}.$$

Therefore

$$E\, S_{cf}(t; a, b) + S_{cf}(t; a, b)\, S_{pf}(t; b, a) = E^2 + E S_{cf}(t; a, b),$$

which can be reduced to

$$S_{cf}(t; a, b) \times S_{pf}(t; b, a) = E^2. \tag{2.65}$$

The relations (2.63)–(2.65) are called the put–call symmetry relations.

They can also be derived in another way, which gives the financial meaning of these relations. Suppose that S is the price of a British pound in U.S. dollars and that ξ is the price of a U.S. dollar in British pounds. Then $\xi = 1/S$. Let P be the value of a put option on S in U.S. dollars with exercise price E, and let $\tilde{C}$ be the value of a call option on ξ in British pounds with an exercise price $\dfrac{1}{E}$. The holder of the put option has the right to sell one pound for E U.S. dollars even if $S \leq E$. The holder of E units of the call option has the right to buy E dollars by paying one British pound even if $\xi \geq \dfrac{1}{E}$. The condition $S \leq E$ is equivalent to $\xi \geq \dfrac{1}{E}$. Thus, both the holder of one unit of the put option and the holder of E units of the call options have the right to exchange one British pound for E U.S. dollars even if $S < E$, i.e., even if the price of a British pound on the market is less than E U.S. dollars. The two holders have the same rights, so the value of one unit of the put option and the value of E units of the call option in U.S. dollars should be equal, i.e.,

$$P = SE\tilde{C},$$

Here, we need to notice that P and $\tilde{C}$ have different but related volatilities, interest rates, dividend yields, and exercise prices. According to Itô's lemma, if

$$dS = \mu S dt + \sigma S dX,$$

then

$$d\xi = (-\mu + \sigma^2)\xi dt - \sigma \xi dX.$$

Hence, the volatilities of S and $\xi = 1/S$ are the same in this case. Consequently, the volatilities of the two problems are the same if the volatilities are constants.

Suppose that $\sigma, r,$ and D_0 are constant and that the interest rates of the British pound and the U.S. dollar are a and b, respectively. Then for the call, $r = a$ and $D_0 = b$ and for the put $r = b$ and $D_0 = a$, and the volatilities are the same. In this case, the relation above can be written as

$$P(S, t; b, a, E) = SE\tilde{C}\left(\frac{1}{S}, t; a, b, 1/E\right),$$

where the third argument after the semicolon in P and $\tilde{C}$ represents the exercise price. Let $\eta = E^2 \xi$ and $C(\eta, t) = E^2 \tilde{C}$. It is clear that for C, the corresponding partial differential equation is

$$\frac{\partial C}{\partial t} + \frac{1}{2}\sigma^2\eta^2\frac{\partial^2 C}{\partial\eta^2} + (a-b)\eta\frac{\partial C}{\partial\eta} - aC = 0$$

and the final condition and the constraint for C are

$$C(\eta, T) = E^2\max(\xi - 1/E, 0) = \max(\eta - E,\ 0),$$

and

$$C(\eta, t) = E^2\tilde{C}(\xi, t) \geq E^2\max(\xi - 1/E, 0) = \max(\eta - E,\ 0).$$

C represents the value of a call option with an exercise price E. From $P = SE\tilde{C}$ and $C = E^2\tilde{C}$, we have $P = CS/E$, i.e.,

$$P(S, t; b,\ a) = C\left(\frac{E^2}{S}, t; a,\ b\right)S/E.$$

Here, both the put option P and the call option C have the same exercise price E, so the dependence on E is suppressed. Let $S_{pf}(t; b, a, E)$ be the position of the free boundary corresponding to $P(S, t; b, a, E)$, and let $\xi_{\tilde{c}f}(t; a, b, 1/E)$ be the position of the free boundary corresponding to $\tilde{C}(1/S, t; a, b, 1/E)$. From the above, $\xi_{\tilde{c}f}(t; a, b, 1/E)$ is equal to $1/S_{pf}(t; b, a, E)$. Suppose that corresponding to $C(E^2/S, t; a, b, E)$, the position of the free boundary is $\eta_{cf}(t; a, b, E)$. Because $\eta_{cf}(t; a, b, E)$ is equal to $E^2\xi_{\tilde{c}f}(t; a, b, 1/E)$, we have

$$S_{pf}(t; b, a, E) \times \eta_{cf}(t; a, b, E) = E^2.$$

Here, the exercise prices for both C and P are E, and the dependence on the exercise price can also be suppressed. Moreover, if we use S, instead of η, as the state variable for the call option C, then the relation above can be written as

$$S_{pf}(t; b,\ a) \times S_{cf}(t; a,\ b) = E^2.$$

Besides the two ways above, the put–call symmetry relations can be derived by a third way. Since

$$\max(E - S,\ 0) = \frac{S}{E}\max\left(\frac{E^2}{S} - E,\ 0\right),$$

when we take, instead of S and P, E^2/S and PE/S as the new variables, a vanilla put option problem is converted into a call option problem even if the parameters depend on S and t. Thus such relations exist for more complicated cases. If σ depends upon S, then the following relations hold:

$$\begin{cases} C(S, t; a, b, \sigma(S)) = P\left(\dfrac{E^2}{S}, t; b, a, \sigma(S)\right)S/E, \quad \text{or} \\[3mm] P(S, t; b, a, \sigma(S)) = C\left(\dfrac{E^2}{S}, t; a, b, \sigma(S)\right)S/E \end{cases}$$

and
$$S_{cf}(t; a, b, \sigma(S)) \times S_{pf}(t; b, a, \sigma(E^2/S)) = E^2.$$

Here, the third argument after the semicolon is the function for the volatility. The proof is left for the reader as an exercise (Problem 34).

Since there exist such relations, if one has a code to price all the vanilla call option problems, then one can also evaluate any vanilla put option problem by using the code. The converse is also true. If one already has a code that can deal with both American call and put options, then the symmetry relations can be used for checking the accuracy of the numerical results. Because the numerical results have errors, they will not exactly satisfy the symmetry relation and can be used as indicators to show how accurate the numerical results are if the values of a call and the corresponding put option have been obtained. For details, see the paper [86] by Zhu, Ren, and Xu. For more about symmetry relations and similar results, see [49], [56], [48], and [23].

2.7 Equations for Some Greeks

Here, for American options we would like to derive the equations and boundary conditions that $\mathcal{V} = \dfrac{\partial \Pi}{\partial \sigma}$, $\rho = \dfrac{\partial \Pi}{\partial r}$, and $\rho_d = \dfrac{\partial \Pi}{\partial D_0}$ should satisfy. Let us first consider American call options and write the dependence of C and S_f on r, D_0, and σ explicitly, that is, instead of $C(S, t)$ and $S_f(t)$, we use $C(S, t; r, D_0, \sigma)$ and $S_f(t; r, D_0, \sigma)$ to denote the price of American call options and the free boundary in what follows. Differentiating the partial differential equation in (2.60) with respect to r, D_0, or σ yields the equations for $\dfrac{\partial C}{\partial r}$, $\dfrac{\partial C}{\partial D_0}$ or $\dfrac{\partial C}{\partial \sigma}$.

For example, for $\dfrac{\partial C}{\partial \sigma}$ we have

$$\frac{\partial C_\sigma}{\partial t} + \frac{1}{2}\sigma^2 S^2 \frac{\partial^2 C_\sigma}{\partial S^2} + (r - D_0)S\frac{\partial C_\sigma}{\partial S} - rC_\sigma + \sigma S^2 \frac{\partial^2 C}{\partial S^2} = 0,$$

where C_σ denotes the partial derivative of the call option with respect to σ. The final condition for the price of American call options is

$$C(S, T; r, D_0, \sigma) = \max(S - E, 0).$$

Therefore $\dfrac{\partial C}{\partial \sigma} = 0$ at $t = T$. The boundary conditions on the free boundary are

$$C(S_f(t; r, D_0, \sigma), t; r, D_0, \sigma) = S_f(t; r, D_0, \sigma) - E \tag{2.66}$$

and

$$\frac{\partial C(S_f(t; r, D_0, \sigma), t; r, D_0, \sigma)}{\partial S} = 1. \tag{2.67}$$

From (2.66) we have

$$\frac{\partial C}{\partial S}\frac{\partial S_f}{\partial \sigma} + \frac{\partial C}{\partial \sigma} = \frac{\partial S_f}{\partial \sigma}$$

on the free boundary. Noticing (2.67), we have $\dfrac{\partial C}{\partial \sigma} = 0$ at the free boundary. Consequently, $\dfrac{\partial C}{\partial \sigma}$ is the solution of the following final-boundary value problem

$$\begin{cases} \dfrac{\partial C_\sigma}{\partial t} + \dfrac{1}{2}\sigma^2 S^2 \dfrac{\partial^2 C_\sigma}{\partial S^2} + (r - D_0)S\dfrac{\partial C_\sigma}{\partial S} - rC_\sigma + \sigma S^2 \dfrac{\partial^2 C}{\partial S^2} = 0, \\ \qquad\qquad\qquad\qquad\qquad 0 \le S \le S_f(t), \quad 0 \le t \le T, \\ C_\sigma(S,T) = 0, \qquad\qquad\quad 0 \le S \le S_f(T), \\ C_\sigma(S_f(t),t) = 0, \qquad\quad 0 \le t \le T, \end{cases} \qquad (2.68)$$

where $\dfrac{\partial^2 C}{\partial S^2}$ and $S_f(t)$ are known functions obtained from the solution of problem (2.60).

For $\dfrac{\partial C}{\partial r}$ and $\dfrac{\partial C}{\partial D_0}$, we can derive the same final and boundary conditions as $\dfrac{\partial C}{\partial \sigma}$, namely,

$$\frac{\partial C}{\partial r} = \frac{\partial C}{\partial D_0} = 0 \qquad (2.69)$$

at $t = T$ and

$$\frac{\partial C}{\partial r} = \frac{\partial C}{\partial D_0} = 0 \qquad (2.70)$$

at the free boundary. The only difference is the equation. Differentiating the partial differential equation in (2.60) with respect to r and D_0 yields

$$\frac{\partial C_r}{\partial t} + \frac{1}{2}\sigma^2 S^2 \frac{\partial^2 C_r}{\partial S^2} + (r - D_0)S\frac{\partial C_r}{\partial S} - rC_r + S\frac{\partial C}{\partial S} - C = 0 \qquad (2.71)$$

and

$$\frac{\partial C_{D_0}}{\partial t} + \frac{1}{2}\sigma^2 S^2 \frac{\partial^2 C_{D_0}}{\partial S^2} + (r - D_0)S\frac{\partial C_{D_0}}{\partial S} - rC_{D_0} - S\frac{\partial C}{\partial S} = 0 \qquad (2.72)$$

respectively, where C_r stands for $\dfrac{\partial C}{\partial r}$ and C_{D_0} for $\dfrac{\partial C}{\partial D_0}$.

For American put options, the Greeks are solutions of similar problems. This is left for the reader to show as Problem 36 of this chapter.

2.8 Perpetual Options

If an option does not have an expiry date but rather an infinite time horizon, then the option is called a perpetual option. Let $C(S, 0; T)$ be the today's price of an American call option with expiry T, and let $C_\infty(S)$ be the price of the corresponding perpetual American call option. Between them, there is the following relation:

$$C_\infty(S) = \lim_{T \to \infty} C(S, 0; T).$$

Since $\left. \dfrac{\partial C(S, t; T)}{\partial t} \right|_{t=0} = 0$ as $T \to \infty$, we know from (2.60) that $C_\infty(S)$ is the solution of the following problem

$$
\begin{cases}
\dfrac{1}{2}\sigma^2 S^2 \dfrac{d^2 C_\infty}{dS^2} + (r - D_0)S \dfrac{dC_\infty}{dS} - rC_\infty = 0, \quad 0 \le S \le S_f, \\[2mm]
C_\infty(S_f) = S_f - E, \\[2mm]
\dfrac{dC_\infty(S_f)}{dS} = 1.
\end{cases}
\tag{2.73}
$$

Let

$$C_\infty(S) = S^\alpha,$$

then

$$\frac{dC_\infty}{dS} = \alpha S^{\alpha-1}$$

and

$$\frac{d^2 C_\infty}{dS^2} = \alpha(\alpha - 1)S^{\alpha-2}.$$

Substituting these into the ordinary differential equation in (2.73), we get

$$\frac{1}{2}\sigma^2\alpha^2 + \left(r - D_0 - \frac{1}{2}\sigma^2\right)\alpha - r = 0.$$

The two roots of this equation are

$$\alpha_\pm = \frac{1}{\sigma^2}\left[-\left(r - D_0 - \frac{1}{2}\sigma^2\right) \pm \sqrt{\left(r - D_0 - \frac{1}{2}\sigma^2\right)^2 + 2\sigma^2 r}\,\right].$$

Thus

$$C_\infty(S) = C_+ S^{\alpha_+} + C_- S^{\alpha_-}.$$

It is clear that $\alpha_+ > 0$ and $\alpha_- < 0$. In order to guarantee the solution to be bounded at $S = 0$, C_- should equal zero. Consequently, we arrive at

$$C_\infty(S) = C_+ S^{\alpha_+}.$$

From the free-boundary conditions in (2.73) we obtain

$$C_+ S_f^{\alpha_+} = S_f - E,$$
$$C_+ \alpha_+ S_f^{\alpha_+ - 1} = 1.$$

Solving these two equations we get

$$S_f = \frac{E}{1 - 1/\alpha_+} \quad \text{and} \quad C_+ = \frac{1}{\alpha_+ S_f^{\alpha_+ - 1}}.$$

Thus, the solution of problem (2.73) is

$$C_\infty(S) = \frac{S_f}{\alpha_+} \left(\frac{S}{S_f} \right)^{\alpha_+}. \tag{2.74}$$

For an American put option, as $T \to \infty$, the free-boundary problem (2.57) becomes

$$\begin{cases} \frac{1}{2}\sigma^2 S^2 \frac{d^2 P_\infty}{dS^2} + (r - D_0)S\frac{dP_\infty}{dS} - rP_\infty = 0, & S_f \le S, \\ P_\infty(S_f) = E - S_f, \\ \frac{dP_\infty(S_f)}{dS} = -1. \end{cases}$$

Similar to the call option, the price of a perpetual American put option is

$$P_\infty(S) = \frac{-S_f}{\alpha_-} \left(\frac{S}{S_f} \right)^{\alpha_-}, \tag{2.75}$$

where

$$S_f = \frac{E}{1 - 1/\alpha_-}.$$

2.9 General Equations for Derivatives

Generally speaking, a financial derivative could depend on several random variables, and a random variable may not represent a price of an asset that can be traded on the market. For example, a derivative could depend on prices of several assets and interest rates, whereas volatilities may need to be treated as random variables. As we know, both interest rates and volatilities are not prices of assets. In this section, we will derive the general partial differential equations satisfied by derivatives, where there exist several state variables, and a state variable may not be a price of an asset traded on the market. The derivation of general equations can be found from other books, for example, the books by Hull[39], and Wilmott, Dewynne, and Howison[71].

2.9.1 Models for Random Variables

Suppose a financial derivative depends on time t and n random state variables, namely, $S_1, S_2, \cdots, S_n$. Each of them satisfies a stochastic differential equation

$$dS_i = a_i dt + b_i dX_i, \quad i = 1, 2, \cdots, n, \tag{2.76}$$

where a_i, b_i are functions of $S_1, S_2, \cdots, S_n$ and t, and $dX_i = \phi_i \sqrt{dt}$ are Wiener processes. In addition, the random variables ϕ_i and ϕ_j could be correlated. Let us suppose

$$E[\phi_i \phi_j] = \rho_{ij}, \tag{2.77}$$

where $-1 \le \rho_{ij} \le 1$. If $\rho_{ij} = 0$, then ϕ_i and ϕ_j are not correlated. If $\rho_{ij} = \pm 1$, then ϕ_i and ϕ_j are completely correlated. It is clear that $\rho_{ii} = 1$.

In practice, a random variable always has a lower bound and an upper bound. How do we model a random variable with such a property? For simplicity, we consider problems with only one random variable S. Suppose that we want a random variable S to have a lower boundary S_l, i.e., if $S \ge S_l$ at time t, then we want to guarantee that S is still greater than or equal to S_l after time t even though the movement of S possesses some uncertainty. In this case, we need to require that $a(S,t)$ and $b(S,t)$ at $S = S_l$ satisfy either the condition:

$$\begin{cases} a(S_l, t) - b(S_l, t) \dfrac{\partial}{\partial S} b(S_l, t) \ge 0, & 0 \le t \le T, \\ b(S_l, t) = 0, & 0 \le t \le T \end{cases} \tag{2.78}$$

or when $b(S,t)$ is differentiable, the condition

$$\begin{cases} a(S_l, t) \ge 0, & 0 \le t \le T, \\ b(S_l, t) = 0, & 0 \le t \le T. \end{cases}$$

In Subsection 2.9.5, we will see that if (2.78) holds at $S = S_l$, then a unique solution of the corresponding partial differential equation can be determined by a final condition on $[S_l, \infty)$ without any boundary conditions at $S = S_l$. Therefore, what happens at $S = S_l$ will not affect the solution at $t = 0$ for any S. This fact can be interpreted as follows. If (2.78) holds for any $t \in [t_0, T]$, then for any such time t, S will be greater than or equal to S_l if $S > S_l$ at $t = t_0$. That is, S is either reflected into the region or is absorbed by the boundary in the event S hits the lower bound S_l at some time $t \in [t_0, T]$. This is because if there are paths that pass through a point (S_l, t) and go to the outside of $[S_l, \infty)$, then the solution at the point $(S, 0)$ should depend on the value of the option at the point (S_l, t). For example, in the popular model

$$dS = \mu S dt + \sigma S dX,$$

we have $a = \mu S$ and $b = \sigma S$. Therefore, the condition (2.78) holds at $S = 0$, and S is always greater than or equal to zero. In the Cox–Ingersoll–Ross interest rate model (see [22])

$$dr = (\bar{\mu} - \bar{\gamma} r) dt + \sqrt{\alpha r} dX, \quad \bar{\mu}, \bar{\gamma}, \alpha > 0,$$

which will be discussed in Chapter 4, $a = \bar{\mu} - \bar{\gamma}r$, $b = \sqrt{\alpha r}$, and the condition
(2.78) is reduced to $\bar{\mu} - \alpha/2 \geq 0$ if the lower bound is zero. This means
that if $\bar{\mu} - \alpha/2 \geq 0$, then at $r = 0$, no boundary condition is needed. In
fact, if $\bar{\mu} - \alpha/2 \geq 0$, the upward drift is sufficiently large to make the origin
inaccessible (see [22]). Therefore, no boundary condition at $r = 0$ is related
to inaccessibility to the origin.

Actually, S_l may not be zero, and a similar condition

$$\begin{cases} a\left(S_u, t\right) - b(S_u, t)\dfrac{\partial}{\partial S}b(S_u, t) \leq 0, & 0 \leq t \leq T, \\ b\left(S_u, t\right) = 0, & 0 \leq t \leq T \end{cases} \tag{2.79}$$

can also be required at $S = S_u > S_l$ so that S will always be in $[S_l, S_u]$. If
$a\left(S_l, t\right) \geq 0$ and $a\left(S_u, t\right) \leq 0$, then it is usually said that the model has a mean
reversion property. However, if $b\left(S_l, t\right) \neq 0$ or $b\left(S_u, t\right) \neq 0$, then there is still a
chance for S to become less than S_l or greater than S_u. If the conditions (2.78)
and (2.79) hold, then we say that the model really has a reversion property
because S will always be in $[S_l, S_u]$. In this book, the conditions (2.78) and
(2.79) will be referred to as the reversion conditions.

The two random variables given above as examples are defined on $[0, \infty)$.
In what follows, we will show that they can be converted into new random
variables whose domains can be naturally extended to $[0, 1]$, and for them the
reversion conditions hold at both the lower and upper boundaries.

Let us introduce a new random variable

$$\xi = \frac{S}{S + P_m},$$

where P_m is a positive parameter. From this relation, we can have

$$S = \frac{P_m\xi}{1 - \xi}, \qquad S + P_m = \frac{P_m}{1 - \xi},$$

and

$$\frac{d\xi}{dS} = \frac{P_m}{(S + P_m)^2} = \frac{(1 - \xi)^2}{P_m}, \qquad \frac{d^2\xi}{dS^2} = \frac{-2P_m}{(S + P_m)^3} = \frac{-2(1 - \xi)^3}{P_m^2}.$$

According to Itô's lemma, if S satisfies $dS = \mu S dt + \sigma S dX$, then for ξ the
stochastic differential equation is

$$\begin{aligned} d\xi &= \frac{(1 - \xi)^2}{P_m}dS - \frac{(1 - \xi)^3}{P_m^2}\sigma^2 S^2 dt \\ &= \left[\mu\xi(1 - \xi) - \sigma^2\xi^2(1 - \xi)\right]dt + \sigma\xi(1 - \xi)dX. \end{aligned}$$

Consequently for ξ, the conditions (2.78) and (2.79) are fulfilled at $\xi = 0$ and
$\xi = 1$, respectively.

Similarly for the Cox–Ingersoll–Ross interest rate model, let

$$\xi = \frac{r}{r + P_m},$$

then we get

$$d\xi = \left[\frac{(1 - \xi)^2}{P_m} \left(\bar{\mu} - \frac{\bar{\gamma} P_m \xi}{1 - \xi} \right) - \frac{\alpha \xi (1 - \xi)^2}{P_m} \right] dt + \frac{\sqrt{\alpha} \xi^{1/2} (1 - \xi)^{3/2}}{P_m^{1/2}} dX.$$

Thus, both (2.78) and (2.79) hold. In what follows, we always assume that the reversion conditions hold. If the domain of the random variable is infinite, then after converting it into another random variable defined on a finite domain, the reversion conditions are fulfilled.

2.9.2 Generalization of Itô's Lemma

Let $V = V(S_1, S_2, \cdots, S_n, t)$. According to the Taylor expansion, we have

$$dV = V(S_1 + dS_1, S_2 + dS_2, \cdots, S_n + dS_n, t + dt) - V(S_1, S_2, \cdots, S_n, t)$$

$$= \sum_{i=1}^{n} \frac{\partial V}{\partial S_i} dS_i + \frac{\partial V}{\partial t} dt + \frac{1}{2} \sum_{i=1}^{n} \sum_{j=1}^{n} \frac{\partial^2 V}{\partial S_i \partial S_j} dS_i dS_j$$

$$+ \frac{1}{2} \sum_{i=1}^{n} \frac{\partial^2 V}{\partial S_i \partial t} dS_i dt + \cdots.$$

Because

$$\lim_{dt \to 0} dS_i dS_j / dt = b_i b_j \rho_{ij}$$

and $dS_i dt$ is a quantity of order $(dt)^{3/2}$, the relation above as $dt \to 0$ becomes

$$dV = f dt + \sum_{i=1}^{n} \frac{\partial V}{\partial S_i} dS_i, \tag{2.80}$$

where

$$f = \frac{\partial V}{\partial t} + \frac{1}{2} \sum_{i=1}^{n} \sum_{j=1}^{n} \frac{\partial^2 V}{\partial S_i \partial S_j} b_i b_j \rho_{ij}.$$

This is called the generalized Itô's lemma.

2.9.3 Derivation of Equations for Financial Derivatives

Suppose that there are $n+1$ distinct financial derivatives dependent on S_1, S_2, $\cdots$, S_n and t. They could have different expiries or different exercise prices.

Even some of the derivatives may depend on only some of the random variables S_i. Let V_k stand for the value of the k-th derivative. According to the generalized Itô's lemma in Subsection 2.9.2, we have

$$dV_k = f_k dt + \sum_{i=1}^{n} \nu_{i,k} dS_i,$$

where

$$f_k = \frac{\partial V_k}{\partial t} + \frac{1}{2} \sum_{i=1}^{n} \sum_{j=1}^{n} \frac{\partial^2 V_k}{\partial S_i \partial S_j} b_i b_j \rho_{ij}$$

and

$$\nu_{i,k} = \frac{\partial V_k}{\partial S_i}.$$

Furthermore, we suppose that the holder of a derivative might receive some cash payment, such as a coupon, during the life of the derivative. Let the cash payment for the k-th derivative during the time interval $[t, t+dt]$ be $K_k dt$, K_k being a known function that may depend on $S_1, S_2, \cdots, S_n$ and t. Consider a portfolio consisting of the $n+1$ derivatives:

$$\Pi = \sum_{k=1}^{n+1} \Delta_k V_k,$$

where Δ_k is the amount of the k-th derivative in the portfolio. During the time interval $[t, t+dt]$, the holder of this portfolio will earn

$$\sum_{k=1}^{n+1} \Delta_k \left(dV_k + K_k dt \right)$$

$$= \sum_{k=1}^{n+1} \Delta_k \left(f_k dt + \sum_{i=1}^{n} \nu_{i,k} dS_i + K_k dt \right)$$

$$= \sum_{k=1}^{n+1} \Delta_k \left(f_k + K_k \right) dt + \sum_{i=1}^{n} \left(\sum_{k=1}^{n+1} \Delta_k \nu_{i,k} \right) dS_i.$$

Let us choose Δ_k such that

$$\sum_{k=1}^{n+1} \Delta_k \nu_{i,k} = 0, \quad i = 1, 2, \cdots, n.$$

In this case the portfolio is risk-free, so its return rate is r, i.e.,

$$\sum_{k=1}^{n+1} \Delta_k \left(f_k + K_k \right) dt = r \Pi dt = r \sum_{k=1}^{n+1} \Delta_k V_k dt$$

or

$$\sum_{k=1}^{n+1} \Delta_k \left(f_k + K_k - rV_k \right) = 0.$$

This relation and the relations that the chosen Δ_k satisfy can be written together in a matrix form as follows:

$$\begin{bmatrix} \nu_{1,1} & \nu_{1,2} & \cdots & \nu_{1,n+1} \\ \vdots & \vdots & \vdots & \vdots \\ \nu_{n,1} & \nu_{n,2} & \vdots & \nu_{n,n+1} \\ g_1 & g_2 & \cdots & g_{n+1} \end{bmatrix} \begin{bmatrix} \Delta_1 \\ \Delta_2 \\ \vdots \\ \Delta_{n+1} \end{bmatrix} = 0,$$

where $g_k = f_k + K_k - rV_k$, $k = 1, 2, \cdots, n+1$.

In order for the system to have a non-trivial solution, the determinant of the matrix must be zero, or the $n+1$ row vectors of the matrix must be linearly dependent. Therefore, we let the last row be expressed as a linear combination of the other rows with coefficients $\tilde{\lambda}_1, \tilde{\lambda}_2, \cdots, \tilde{\lambda}_n$,

$$\begin{bmatrix} f_1 + K_1 - rV_1 \\ f_2 + K_2 - rV_2 \\ \vdots \\ f_{n+1} + K_{n+1} - rV_{n+1} \end{bmatrix} = \tilde{\lambda}_1 \begin{bmatrix} \nu_{1,1} \\ \nu_{1,2} \\ \vdots \\ \nu_{1,n+1} \end{bmatrix}$$

$$+ \tilde{\lambda}_2 \begin{bmatrix} \nu_{2,1} \\ \nu_{2,2} \\ \vdots \\ \nu_{2,n+1} \end{bmatrix} + \cdots + \tilde{\lambda}_n \begin{bmatrix} \nu_{n,1} \\ \nu_{n,2} \\ \vdots \\ \nu_{n,n+1} \end{bmatrix}.$$

This relation can be rewritten as

$$f_k + K_k - rV_k - \sum_{i=1}^{n} \tilde{\lambda}_i \nu_{i,k} = 0, \qquad k = 1, 2, \cdots, n+1,$$

which means that any derivative satisfies an equation of the form

$$f + K - rV - \sum_{i=1}^{n} \tilde{\lambda}_i \nu_i = 0.$$

Usually, $\tilde{\lambda}_i$ is written in the form:

$$\tilde{\lambda}_i = \lambda_i b_i - a_i$$

so the equation above can be written as

$$\frac{\partial V}{\partial t} + \frac{1}{2} \sum_{i=1}^{n} \sum_{j=1}^{n} b_i b_j \rho_{ij} \frac{\partial^2 V}{\partial S_i \partial S_j} + \sum_{i=1}^{n} (a_i - \lambda_i b_i) \frac{\partial V}{\partial S_i} - rV + K = 0. \quad (2.81)$$

This equation is called the general partial differential equation for derivatives. In (2.81), a_i and b_i are given functions in the model of dS_i, ρ_{ij} is the correlation coefficient between dS_i and dS_j, and K depends on the individual derivative security. For each i, λ_i is an unknown function, which could depend on $S_1, S_2, \cdots, S_n$ and t, and is called the market price of risk for S_i. The reason is as follows. According to Itô's lemma and using the equation (2.81), we have

$$dV = \left(\frac{\partial V}{\partial t} + \frac{1}{2} \sum_{i=1}^{n} \sum_{j=1}^{n} \frac{\partial^2 V}{\partial S_i \partial S_j} b_i b_j \rho_{ij} \right) dt + \sum_{i=1}^{n} \frac{\partial V}{\partial S_i} dS_i$$

$$= \left[\sum_{i=1}^{n} (\lambda_i b_i - a_i) \frac{\partial V}{\partial S_i} + rV - K \right] dt + \sum_{i=1}^{n} \frac{\partial V}{\partial S_i} (a_i dt + b_i dX_i)$$

or

$$dV + K dt - rV dt = \sum_{i=1}^{n} \frac{\partial V}{\partial S_i} b_i (dX_i + \lambda_i dt).$$

Here, $dV + K dt$ is the return for the derivative including the cash payment, such as a coupon, and $rV dt$ is the return if the investment is risk-free. Therefore, $dV + K dt - rV dt$ is the excess return above the risk-free rate during the time interval $[t, t + dt]$. This equals the right-hand side of the equation. Its expectation is $\sum_{i=1}^{n} \frac{\partial V}{\partial S_i} b_i \lambda_i dt$ because $E[dX_i] = 0$, $i = 1, 2, \cdots, n$. Therefore, the term $\frac{\partial V}{\partial S_i} b_i \lambda_i dt$ may be interpreted as an excess return above the risk-free return for taking the risk dX_i. Consequently, λ_i is a price of risk for S_i that is associated with dX_i and is often called the market price of risk for S_i.

2.9.4 Three Types of State Variables

There are three types of state variables, for which the term $a_i - \lambda_i b_i$ in (2.81) will be determined in different ways.

Suppose S_i is an asset price that can be traded on the market. For example, S_i is a stock price per share. In this case, the stock itself can be considered as a derivative security. Suppose the stock pays a dividend continuously with dividend yield D_{0i}. In this case, the price of this derivative security should be $S_i e^{-D_{0i}(T-t)}$ (see Problem 9), i.e., $S_i e^{-D_{0i}(T-t)}$ should be a solution of the equation (2.81). Substituting $V = S_i e^{-D_{0i}(T-t)}$ into (2.81) yields $e^{-D_0(T-t)}(D_{0i}S_i + a_i - \lambda_i b_i - rS_i) = 0$. Therefore, for this case

$$a_i - \lambda_i b_i = (r - D_{0i})S_i. \tag{2.82}$$

We obtain the same result as we had when the Black–Scholes equation for continuous dividend-paying assets was derived. If the dividend is paid discretely, the situation is similar:

$$a_i - \lambda_i b_i = rS_i - D_i(S_i, t) \tag{2.83}$$

because if V depends only on S_i and t, then (2.81) should become (2.13). Here, $D_i(S_i, t)dt$ is the dividend paid during the time period $[t, t + dt]$.

A state variable S_i with $b_i = 0$ in (2.76) is another type of state variable. From $b_i = 0$, we have

$$a_i - \lambda_i b_i = a_i, \tag{2.84}$$

so λ_i disappears in the equation (2.81). As we will see from Chapter 3, if S_i is the maximum, minimum, or average price of the stock during a time period, then $dS_i = a_i dt$.

If S_i is the spot interest rate, then in order to determine λ_i, we have to solve an inverse problem. We will discuss this problem in detail in Chapter 4. This is an example of the third type of state variable. Besides the interest rate, the random volatility also falls into this type of state variable.

2.9.5 Uniqueness of Solutions

The equation (2.81) is a parabolic equation. If $b_i = 0$ at $S_i = S_{i,l}$ and $S_i = S_{i,u}$, $i = 1, 2, \cdots, n$, then we say that the equation is a degenerate parabolic partial differential equation. In this subsection, we are going to discuss when a degenerate equation has a unique solution. The conclusion expected is that if for any i,

$$a_i(S_{i,l}, t) - b_i(S_{i,l}, t)\frac{\partial}{\partial S_i} b_i(S_{i,l}, t) \geq 0 \tag{2.85}$$

and

$$a_i(S_{i,u}, t) - b_i(S_{i,u}, t)\frac{\partial}{\partial S_i} b_i(S_{i,u}, t) \leq 0 \tag{2.86}$$

hold,[13] the solution of the degenerate parabolic equation on a rectangular domain with a final condition at $t = T$ is unique.[14] If

$$a_i(S_{i,l}, t) - b_i(S_{i,l}, t)\frac{\partial}{\partial S_i} b_i(S_{i,l}, t) < 0$$

or

$$a_i(S_{i,u}, t) - b_i(S_{i,u}, t)\frac{\partial}{\partial S_i} b_i(S_{i,u}, t) > 0,$$

then a boundary condition at $S_i = S_{i,l}$ or $S_i = S_{i,u}$ needs to be imposed besides the final condition in order to have a unique solution. We now prove this conclusion for the one-dimensional case.

---

[13] a_i and b_i could also depend on $S_1, \cdots, S_{i-1}, S_{i+1}, \cdots, S_n$. Here, the dependence of a_i and b_i on them is suppressed, and the two relations hold for $S_j \in [S_{j,l}, S_{j,u}]$, $j = 1, \cdots, i-1, i+1, \cdots, n$.

[14] For a parabolic equation defined on a general domain, the conditions for a parabolic partial differential equation to be degenerate and the conditions for the solution of its initial-value problem to be unique, see the paper [81] by Zhu.

In the case $n = 1$, (2.81) simplifies to

$$\frac{\partial V}{\partial t} + \frac{1}{2}b^2\frac{\partial^2 V}{\partial S^2} + (a - \lambda b)\frac{\partial V}{\partial S} - rV + K = 0.$$

Here, the sign of the coefficient of the second derivative is opposite of the coefficient of the second derivative in the heat equation. We say that such a parabolic equation has an "anti-directional" time. For a heat equation, an initial condition is given at $t = 0$, and the solution for $t \geq 0$ needs to be determined. Therefore, for the equation with an "anti-directional" time, a final condition should be given at $t = T$, and the solution for $t \leq T$ is needed to be determined. Consequently, we consider the following problem:

$$\begin{cases} \dfrac{\partial V}{\partial t} + \dfrac{1}{2}b^2\dfrac{\partial^2 V}{\partial S^2} + (a - \lambda b)\dfrac{\partial V}{\partial S} - rV + K = 0, \\[2mm] \qquad\qquad\qquad 0 \leq t \leq T, \quad S_l \leq S \leq S_u, \\[2mm] V(S,T) = f(S), \qquad S_l \leq S \leq S_u, \\[2mm] V(S_l, t) \begin{cases} \text{needs not to be given if (2.85) holds,} \\ = f_l(t) \text{ if (2.85) does not hold,} \end{cases} \\[4mm] V(S_u, t) \begin{cases} \text{needs not to be given if (2.86) holds,} \\ = f_u(t) \text{ if (2.86) does not hold.} \end{cases} \end{cases} \qquad (2.87)$$

It is not difficult to convert (2.87) into a problem in the form:

$$\begin{cases} \dfrac{\partial u}{\partial \tau} = f_1(x,\tau)\dfrac{\partial^2 u}{\partial x^2} + f_2(x,\tau)\dfrac{\partial u}{\partial x} + f_3(x,\tau)u + g(x,\tau), \\[2mm] \qquad\qquad\qquad 0 \leq x \leq 1, \quad 0 \leq \tau \leq T, \\[2mm] u(x,0) = f(x), \qquad 0 \leq x \leq 1, \\[2mm] u(0,\tau) \begin{cases} \text{needs not to be given if } f_2(0,\tau) - \dfrac{\partial f_1(0,\tau)}{\partial x} \geq 0, \\[2mm] = f_l(\tau) \text{ if } f_2(0,\tau) - \dfrac{\partial f_1(0,\tau)}{\partial x} < 0, \end{cases} \\[5mm] u(1,\tau) \begin{cases} \text{needs not to be given if } f_2(1,\tau) - \dfrac{\partial f_1(1,\tau)}{\partial x} \leq 0, \\[2mm] = f_u(\tau) \text{ if } f_2(1,\tau) - \dfrac{\partial f_1(1,\tau)}{\partial x} > 0, \end{cases} \end{cases} \qquad (2.88)$$

where $f_1(0,\tau) = f_1(1,\tau) = 0$ and $f_1(x,\tau) \geq 0$. Thus, if we can prove the uniqueness of the solution of (2.88), then we have the uniqueness of the solution of (2.87). The third and fourth relations in (2.88) are the boundary conditions for degenerate parabolic equations. For parabolic equations, there is always a boundary condition at any boundary, that is, the number of boundary conditions for parabolic equations is always one. However, for degenerate parabolic equations, sometimes there is a boundary condition and sometimes there is not, depending on the value of $f_2(x,\tau) - \dfrac{\partial f_1(x,\tau)}{\partial x}$ at the boundary. For (2.88), we have the following theorem (see [68]).

Theorem 2.2 *Suppose that the solution of (2.88) exists and is bounded[15] and that there exist a constant c_1 and two bounded functions $c_2(\tau)$ and $c_3(\tau)$ such that*

$$1 + \max_{0 \le x \le 1,\, 0 \le \tau \le T} \left(\left| \frac{\partial^2 f_1(x,\tau)}{\partial x^2} - \frac{\partial f_2(x,\tau)}{\partial x} + 2 f_3(x,\tau) \right| \right) \le c_1,$$

$$- \min\left(0, f_2(0,\tau) - \frac{\partial f_1(0,\tau)}{\partial x} \right) \le c_2(\tau)$$

and

$$\max\left(0, f_2(1,\tau) - \frac{\partial f_1(1,\tau)}{\partial x} \right) \le c_3(\tau).$$

In this case, its solution is unique and stable with respect to the initial value $f(x)$, inhomogeneous term $g(x,\tau)$, and the boundary values $f_l(\tau), f_u(\tau)$ if there are any.

Proof.   Because the partial differential equation in (2.88) can be rewritten as

$$\frac{\partial u}{\partial \tau} = \frac{\partial}{\partial x}\left[f_1(x,\tau) \frac{\partial u}{\partial x} \right] + \left[f_2(x,\tau) - \frac{\partial f_1(x,\tau)}{\partial x} \right] \frac{\partial u}{\partial x} + f_3(x,\tau) u + g(x,\tau),$$

multiplying that equation by $2u$, we have

$$\frac{\partial(u^2)}{\partial \tau} = 2\frac{\partial}{\partial x}\left(f_1 u \frac{\partial u}{\partial x} \right) + \left(f_2 - \frac{\partial f_1}{\partial x} \right) \frac{\partial(u^2)}{\partial x} - 2 f_1 \left(\frac{\partial u}{\partial x} \right)^2 + 2 f_3 u^2 + 2 g u$$

$$= 2\frac{\partial}{\partial x}\left(f_1 u \frac{\partial u}{\partial x} \right) + \frac{\partial}{\partial x}\left[\left(f_2 - \frac{\partial f_1}{\partial x} \right) u^2 \right] - 2 f_1 \left(\frac{\partial u}{\partial x} \right)^2$$

$$+ \left(\frac{\partial^2 f_1}{\partial x^2} - \frac{\partial f_2}{\partial x} + 2 f_3 \right) u^2 + 2 g u.$$

Integrating this equality with respect to x on the interval $[0,1]$, we obtain the second equality

$$\frac{d}{d\tau} \int_0^1 u^2(x,\tau)\, dx$$

$$= 2\left(f_1 u \frac{\partial u}{\partial x} \right)\Big|_{x=0}^{1} + \left[\left(f_2 - \frac{\partial f_1}{\partial x} \right) u^2 \right]\Big|_{x=0}^{1} - 2\int_0^1 f_1 \left(\frac{\partial u}{\partial x} \right)^2 dx$$

$$+ \int_0^1 \left(\frac{\partial^2 f_1}{\partial x^2} - \frac{\partial f_2}{\partial x} + 2 f_3 \right) u^2 dx + 2\int_0^1 g u\, dx.$$

---

[15]This is proven in the paper [6] by Behboudi.

Because

$$\left[\left(f_2 - \frac{\partial f_1}{\partial x}\right)u^2\right]\Big|_{x=0}^{1}$$

$$= \left[f_2(1,\tau) - \frac{\partial f_1(1,\tau)}{\partial x}\right]u^2(1,\tau) - \left[f_2(0,\tau) - \frac{\partial f_1(0,\tau)}{\partial x}\right]u^2(0,\tau)$$

$$\leq \max\left(0, f_2(1,\tau) - \frac{\partial f_1(1,\tau)}{\partial x}\right)f_u^2(\tau) - \min\left(0, f_2(0,\tau) - \frac{\partial f_1(0,\tau)}{\partial x}\right)f_l^2(\tau),$$

from the equality above and the relations $f_1(0,\tau) = f_1(1,\tau) = 0$ and $f_1(x,\tau) \geq 0$, we have

$$\frac{d}{d\tau}\int_0^1 u^2(x,\tau)dx$$

$$\leq c_1\int_0^1 u^2(x,\tau)dx + \int_0^1 g^2(x,\tau)dx + c_2(\tau)f_l^2(\tau) + c_3(\tau)f_u^2(\tau).$$

Based on this inequality and by the Gronwall inequality, we arrive at

$$\int_0^1 u^2(x,\tau)dx$$

$$\leq e^{c_1\tau}\left\{\int_0^1 f^2(x)dx + \int_0^\tau\left[\int_0^1 g^2(x,s)dx + c_2(s)f_l^2(s) + c_3(s)f_u^2(s)\right]ds\right\},$$

$$t \in [0,T].$$

From the last inequality, we know that the solution is stable with respect to $f(x)$ and $g(x,\tau)$. Also if

$$f_2(0,\tau) - \frac{\partial f_1(0,\tau)}{\partial x} \geq 0 \quad \text{and} \quad f_2(1,\tau) - \frac{\partial f_1(1,\tau)}{\partial x} \leq 0$$

hold and

$$f(x) \equiv 0, \quad g(x,\tau) \equiv 0,$$

then the solution of (2.88) must be zero. Hence, the functions $f(x)$ and $g(x,\tau)$ determine the solution uniquely. If

$$f_2(0,\tau) - \frac{\partial f_1(0,\tau)}{\partial x} < 0 \quad \text{and} \quad f_2(1,\tau) - \frac{\partial f_1(1,\tau)}{\partial x} \leq 0$$

hold, then the solution is determined by

$$f(x), \quad g(x,\tau), \quad \text{and} \quad f_l(\tau)$$

uniquely. The situation for other cases are similar. Therefore, we may conclude that if the solution of (2.88) exists, then it is unique and stable with respect to the initial value $f(x)$, the inhomogeneous term $g(x,\tau)$, and the boundary values $f_l(\tau), f_u(\tau)$ if there are any. This completes the proof and gives an explanation on when a boundary condition is necessary. ∎

Here we give some remarks.

- From the probabilistic point of view, a boundary condition on a boundary is needed if and only if there are paths reaching the boundary from a point $x \in (0,1)$ and $t = 0$. Therefore, on whether or not a random variable can reach a boundary from the interior, there are similar conclusions (see [30]).
- For the case $n = 2$ and on a rectangular finite domain, if $b_1(S_1, S_2, t)$ and $b_2(S_1, S_2, t)$ are analytic, the uniqueness is also proved (see [85]) and the idea can be generalized to the case with $n > 2$. On a general finite three-dimensional domain, a similar result is also obtained (see [81]). Therefore, a degenerate parabolic equation at boundaries is similar to a hyperbolic equation.[16] Due to this fact, roughly speaking, we might say that the parabolic equation degenerates into a hyperbolic-parabolic equation (a hyperbolic equation for one-dimensional case) at the boundaries. When conditions (2.85) and (2.86) hold, incoming information is not needed at boundaries, that is, the value of V at the boundaries at $t = t^*$ is determined by the value V on the region: $S_{i,l} \leq S_i \leq S_{i,u}$, $i = 1, 2, \cdots, n$ and $t^* \leq t \leq T$. Therefore, in this case, in order for a degenerate parabolic equation to have a unique solution, only the final condition is needed.[17] If b_i is not analytic at boundaries, this conclusion has not been proved for $n > 1$. However, it is expected that the conclusion is still true.
- When the domain of S_i is not finite, a final condition is still enough for such an equation to have a unique solution if S_i can be converted into a random variable for which the reversion conditions hold. The reason is that a final condition can determine a unique solution if the new random variable is used. However, a transformation will not change the nature of the problem. If the problem has a unique solution as a function of a random variable, the problem will also have a unique solution as a function of another random variable associated by a transformation. Applying this theorem to problem (2.26), we know that its solution is unique and stable with respect to the initial value. Problem (2.26) is obtained through a transformation from the European option problem (2.24). Therefore, the European option problem (2.24) also has a unique solution.

2.10 Jump Conditions

2.10.1 Hyperbolic Equations with a Dirac Delta Function

Consider the following linear hyperbolic partial differential equation

$$\frac{\partial u}{\partial t} + f_1(x_1, x_2, \cdots, x_K, t)\frac{\partial u}{\partial x_1} + \cdots + f_K(x_1, x_2, \cdots, x_K, t)\frac{\partial u}{\partial x_K} = 0.$$

---

[16]When $f_1(x, t) \equiv 0$, the partial differential equation in (2.88) is called a hyperbolic equation.

[17]Oleĭnik and Radkevič in their book [59] discussed the uniqueness of solutions of this type of partial differential equations under different conditions.

Let C be a curve defined by the system of ordinary differential equations

$$\frac{dx_1(t)}{dt} = f_1(x_1, x_2, \cdots, x_K, t),$$

$$\vdots$$

$$\frac{dx_K(t)}{dt} = f_K(x_1, x_2, \cdots, x_K, t)$$

with initial conditions

$$x_1(0) = \xi_1, \ x_2(0) = \xi_2, \ \cdots, \ x_K(0) = \xi_K.$$

Along the curve we have

$$\frac{du}{dt} = \frac{\partial u}{\partial t} + \frac{\partial u}{\partial x_1}\frac{dx_1}{dt} + \cdots + \frac{\partial u}{\partial x_K}\frac{dx_K}{dt} = 0.$$

Therefore, u is a constant along the curve:

$$u\left(x_1(t^*), x_2(t^*), \cdots, x_K(t^*), t^*\right) = u\left(x_1(t^{**}), x_2(t^{**}), \cdots, x_K(t^{**}), t^{**}\right),$$

where t^* and t^{**} are any two times. If

$$f_k(x_1, x_2, \cdots, x_K, t) = F_k(x_1, x_2, \cdots, x_K, t)\delta(t - t_i),$$

where $\delta(t - t_i)$ is the Dirac delta function, then

$$x_k(t_i^+) - x_k(t_i^-) = \int_{t_i^-}^{t_i^+} F_k(x_1, x_2, \cdots, x_K, t)\delta(t - t_i)dt$$

$$= F_k\left(x_1(t_i^-), x_2(t_i^-), \cdots, x_K(t_i^-), t_i^-\right)$$

and

$$u\left(x_1(t_i^-), x_2(t_i^-), \cdots, x_K(t_i^-), t_i^-\right)$$
$$= u\left(x_1(t_i^+), x_2(t_i^+), \cdots, x_K(t_i^+), t_i^+\right)$$
$$= u\left(x_1(t_i^-) + F_{1i}^-, x_2(t_i^-) + F_{2i}^-, \cdots, x_K(t_i^-) + F_{Ki}^-, t_i^+\right), \quad (2.89)$$

where t_i^- and t_i^+ denote the time just before and after t_i, respectively, and

$$F_{ki}^- \equiv F_k\left(x_1(t_i^-), x_2(t_i^-), \cdots, x_K(t_i^-), t_i^-\right).$$

For such a jump condition, a similar derivation is given in the book [71] by Wilmott, Dewynne, and Howison.

2.10.2 Jump Conditions for Options with Discrete Dividends and Discrete Sampling

From (2.89), jump conditions of various options can be derived. Here, we give two examples. One is simple and the other is quite complicated. Jump conditions for other options will be given when they are discussed.

Suppose $V(S,t)$ is the value of an option on a stock, which pays a dividend D_i at time t_i, $i = 1, 2, \cdots, I$. Here, we assume that $t_i \leq T$, T being expiry. From Section 2.2, we know that $V(S,t)$ satisfies (2.13):

$$\frac{\partial V}{\partial t} + \frac{1}{2}\sigma^2 S^2 \frac{\partial^2 V}{\partial S^2} + [rS - D(S,t)]\frac{\partial V}{\partial S} - rV = 0,$$

where

$$D(S,t) = \sum_{i=1}^{I} D_i(S)\delta(t - t_i), \quad \text{with} \quad D_i(S) \leq S.$$

This means that at $t \neq t_i, i = 1, 2, \cdots, I$, V satisfies

$$\frac{\partial V}{\partial t} + \frac{1}{2}\sigma^2 S^2 \frac{\partial^2 V}{\partial S^2} + rS\frac{\partial V}{\partial S} - rV = 0$$

and at $t = t_i, i = 1, 2, \cdots$, or I, the equation

$$\frac{\partial V}{\partial t} - D_i(S)\delta(t - t_i)\frac{\partial V}{\partial S} = 0$$

holds. According to (2.89), at $t = t_i$ we have

$$V(S, t_i^-) = V(S - D_i(S), t_i^+). \qquad (2.90)$$

This is the jump condition for options on stocks with discrete dividends. We now explain the financial meaning of this relation. At $t = t_i$, the stock pays a dividend D_i, so the stock price will drop by D_i. If the price is S at t_i^-, then the price is $S - D_i$ at t_i^+. However, the price of the option is unchanged at time t_i because the holder of the option does not receive any money at time t_i.

The second example involves several independent variables. Suppose the stock price is measured discretely and let $S_1, S_2, \cdots, S_N$ be the first N largest sampled stock prices until time t and $S_1 \geq S_2 \geq \cdots \geq S_N$. Assume that the value of option V depends on $S, S_1, \cdots, S_N, t$. From Subsection 3.4.6, we will see that if sampling occurs at $t = t_i$, then

$$\frac{dS_n}{dt} = \begin{cases} \left[\max(S, S_1(t_i^-)) - S_1(t_i^-)\right]\delta(t - t_i), & \text{if } n = 1, \\ \left[\max(\min(S, S_{n-1}(t_i^-)), S_n(t_i^-)) \right. \\ \left. -S_n(t_i^-)\right]\delta(t - t_i), & \text{if } n = 2, 3, \cdots, N; \end{cases}$$

otherwise

$$\frac{dS_n}{dt} = 0.$$

According to Section 2.9, in this case, the option price is the solution of

$$\frac{\partial V}{\partial t} + \frac{\partial V}{\partial S_1}\frac{dS_1}{dt} + \frac{\partial V}{\partial S_2}\frac{dS_2}{dt} + \cdots + \frac{\partial V}{\partial S_N}\frac{dS_N}{dt}$$
$$+ \frac{1}{2}\sigma^2 S^2 \frac{\partial^2 V}{\partial S^2} + (r - D_0)S\frac{\partial V}{\partial S} - rV = 0.$$

Consequently, at $t = t_i$, V satisfies

$$\frac{\partial V}{\partial t} + \frac{\partial V}{\partial S_1}\frac{dS_1}{dt} + \frac{\partial V}{\partial S_2}\frac{dS_2}{dt} + \cdots + \frac{\partial V}{\partial S_N}\frac{dS_N}{dt} = 0.$$

From (2.89) we know when $t = t_i$, the jump condition

$$V(S, S_1^-, S_2^-, \cdots, S_N^-, t_i^-) = V(S, \max(S, S_1^-), \max(\min(S, S_1^-), S_2^-),$$
$$\cdots, \max(\min(S, S_{N-1}^-), S_N^-), t_i^+) \qquad (2.91)$$

holds, where S_n^- denotes $S_n(t_i^-)$ for brevity.

It is clear how to use such a jump condition when a European-style deriva-tive is evaluated. When the price of an American-style derivative needs to be calculated, such a condition should be used on the solution obtained by the PDE. After that, taking the maximum between the new solution and the constraint yields the solution for the American derivative.

2.11 More Arbitrage Theory

In Section 2.2, we derived the Black–Scholes equation by using arbitrage argu-ments. Here, we will further use arbitrage arguments to obtain some properties of option prices. Similar materials can be found in the book [39] by Hull.

2.11.1 Three Conclusions and Some Portfolios

Consider two portfolios $\mathbf{X}$ and $\mathbf{Y}$, whose values depend on a stock price S and time t. Let $\mathbf{X}(S,t)$ and $\mathbf{Y}(S,t)$ denote the values of portfolios $\mathbf{X}$ and $\mathbf{Y}$, respectively. $\mathbf{X}$ and $\mathbf{Y}$ could involve options, and all their expires are T. By using arbitrage arguments, we can have three conclusions, which are written in the form of theorems.

Theorem 2.3 *If only European options are involved and* $\mathbf{X}(S,T) \geq \mathbf{Y}(S,T)$ *for any* S, *then for any* $t \leq T$, $\mathbf{X}(S,t)$ *must be greater than or equal to* $\mathbf{Y}(S,t)$. *This result also implies that if* $\mathbf{X}(S,T) \leq \mathbf{Y}(S,T)$, *then for any* $t \leq T$, $\mathbf{X}(S,t) \leq \mathbf{Y}(S,t)$.

Proof.   Suppose that at time $\bar{t}$ the value of portfolio $\mathbf{X}$ is less than the value of portfolio $\mathbf{Y}$ and that the latter is higher than the former by an amount of $Z(\bar{t})$. In this case, an arbitrageur can earn at least $Z(\bar{t})e^{r(T-\bar{t})}$ at time T by doing the following: sell $\mathbf{Y}$, buy $\mathbf{X}$, and invest $Z(\bar{t})$ into a bank at an interest rate r at time $\bar{t}$, and get $\mathbf{X}(S,T)$ from portfolio $\mathbf{X}$, pay $\mathbf{Y}(S,T)$ for portfolio $\mathbf{Y}$, and obtain $Z(\bar{t})e^{r(T-\bar{t})}$ from the risk-free investment at time T. Because $\mathbf{X}(S,T) \geq \mathbf{Y}(S,T)$ for any S, the arbitrageur will always earn at least $Z(\bar{t})e^{r(T-\bar{t})}$ at the time T, which means that the earning is risk-free. Thus, everyone will do such a thing. Because so many people sell $\mathbf{Y}$ and buy $\mathbf{X}$, the price of $\mathbf{Y}$ will drop and the price of $\mathbf{X}$ will rise and will be immediately equal to or greater than the price of $\mathbf{Y}$. Therefore, Theorem 2.3 holds. ∎

From this result, assuming $\mathbf{X}(S,T) \leq \mathbf{Y}(S,T)$, we can immediately get that for any time $t \leq T$, $\mathbf{X}(S,t) \leq \mathbf{Y}(S,t)$ and furthermore we can have

Theorem 2.4 *If $\mathbf{X}(S,T) = \mathbf{Y}(S,T)$ for any S, then for any $t \leq T$, $\mathbf{X}(S,t)$ must be equal to $\mathbf{Y}(S,t)$ for any S.*

Proof.   Because $\mathbf{X}(S,T) = \mathbf{Y}(S,T)$ means $\mathbf{X}(S,T) \geq \mathbf{Y}(S,T)$ and $\mathbf{X}(S,T) \leq \mathbf{Y}(S,T)$, from the conclusion above we have for any t

$$\mathbf{X}(S,t) \geq \mathbf{Y}(S,t) \quad \text{and} \quad \mathbf{X}(S,t) \leq \mathbf{Y}(S,t),$$

which means

$$\mathbf{X}(S,t) = \mathbf{Y}(S,t).$$

Thus we have Theorem 2.4. ∎

We can also have the following conclusion.

Theorem 2.5 *Suppose that portfolio $\mathbf{Y}$ involves only one American option and no European option and that portfolio $\mathbf{X}$ involves only European options. If $\mathbf{X}(S,T) \geq \mathbf{Y}(S,T)$ at time T and if the amount of cash and the number of stocks in $\mathbf{X}$ is greater than or equal to the amount of cash and the number of stocks the holder of $\mathbf{Y}$ has when the American option is exercised at time $\bar{t} < T$, then $\mathbf{X}(S,t) \geq \mathbf{Y}(S,t)$ for any time t.*

Proof.   The argument is similar to the argument for proving Theorem 2.3. Suppose $\mathbf{X}(S,t) < \mathbf{Y}(S,t)$ at time $t < T$. Then, an arbitrageur can purchase $\mathbf{X}$, sell $\mathbf{Y}$, and earn some money. Later, when the American option is exercised early at time $\bar{t} < T$, the arbitrageur will never lose money because the amount of cash and the number of stocks in $\mathbf{X}$ are greater than or equal to the amount of cash and the number of stocks the holder of $\mathbf{Y}$ has. When the American option is not exercised before time T, the arbitrageur will also never lose any money because the value of $\mathbf{X}$ is greater than or equal to the value of $\mathbf{Y}$ at time T. Therefore, the earning is risk-free, which means $\mathbf{X}(S,t)$ should not be less than $\mathbf{Y}(S,t)$ at any time. ∎

Before applying these conclusions, we define some portfolios and find their values at time T along with what their holders will have if American options are exercised at time $\bar{t} < T$.

Portfolio A: An amount of cash equal to $Ee^{-r(T-t)}$ invested at an interest rate r. It is clear that its value at time T is E.

Portfolio B: $e^{-D_0(T-t)}$ shares of a stock with dividends being reinvested in the stock if the stock pays the dividend continuously or one share of a stock plus a loan $D(t)$ if the stock pays cash dividends discretely. Here, $D(t)$ is equal to the present value of these dividends to be paid from time t to time T, and the money will be returned to the loaner as soon as the stock pays a dividend. Obviously, its value at time T is the price of the stock S.

Portfolio C: One European call option plus portfolio **A**. The value of this portfolio at time T is $\max(S-E,0)+E=\max(S,E)$.

Portfolio D: One European put option plus portfolio **B**. Its value at time T is $\max(E-S,0)+S=\max(S,E)$.

Portfolio E: One American call option plus portfolio **A**. If the American call option is not exercised before time T, its value at time T is $\max(S-E,0)+E=\max(S,E)$. If at some time $\bar{t}<T$, the stock price S is greater than E and the American option is exercised, then the holder of the portfolio has one share plus a loan of $(1-e^{-r(T-\bar{t})})E$.

Portfolio F: One American put option plus portfolio **B**. $\max(S,E)$ is its value at time T if the put option is not exercised before time T; while its holder has an amount of cash E minus $(1-e^{-D_0(T-\bar{t})})$ shares or an amount of cash $E-D(\bar{t})$ if the stock price S is less than E and the put option is exercised at some time $\bar{t}<T$.

Portfolio G: One European call option plus E. Its value at time T is equal to $\max(S,E)$.

Portfolio H: One European put option plus one share. Its value is equal to $\max(S,E)$ at expiry.

2.11.2 Bounds of Option Prices

Consider a European call option and portfolio **B**. At time T, $c(S,T)=\max(S-E,0)\le \mathbf{B}(S,T)=S$. From Theorem 2.3, we have

$$c(S,t)\le Se^{-D_0(T-t)}$$

or

$$c(S,t)\le S-D(t).$$

Now let us compare portfolio **C** with portfolio **B**. Because at time T

$$\mathbf{C}(S,T)=\max(S,E)\ge \mathbf{B}(S,T)=S,$$

we have

$$c(S,t)+Ee^{-r(T-t)}\ge Se^{-D_0(T-t)}$$

or

$$c(S,t)+Ee^{-r(T-t)}\ge S-D(t).$$

Clearly, $c(S,t) \geq 0$ for any case. Therefore, for a European call option we have

$$\max\left(Se^{-D_0(T-t)} - Ee^{-r(T-t)}, 0\right) \leq c(S,t) \leq Se^{-D_0(T-t)} \qquad (2.92)$$

or

$$\max\left(S - D(t) - Ee^{-r(T-t)}, 0\right) \leq c(S,t) \leq S - D(t). \qquad (2.93)$$

Consequently, the lower bound of $c(S,t)$ is $\max\left(Se^{-D_0(T-t)} - Ee^{-r(T-t)}, 0\right)$ or $\max\left(S - D(t) - Ee^{-r(T-t)}, 0\right)$ and the upper bound is $Se^{-D_0(T-t)}$ or $S - D(t)$. Here, we assume that $S - D(t)$ is always greater than zero. If $S < D(t)$ at time t, then any person will buy one share of the stock by finding a loan of amount S at time t and returning the loan as soon as the stock pays a dividend. In this way, the person gets one share and some cash free at time T. Therefore, the price must rise until $S \geq D(t)$.

Because $C(S,t) \geq c(S,t)$, we require that $C(S,t)$ is greater than or equal to the lower bound of $c(S,t)$. Also, $C(S,t)$ needs to be greater than or equal to the constraint $\max(S - E, 0)$. Thus

$$\max\left(Se^{-D_0(T-t)} - Ee^{-r(T-t)}, S - E, 0\right)$$

or

$$\max\left(S - D(t) - Ee^{-r(T-t)}, S - E, 0\right)$$

is a lower bound. Clearly, S is an upper bound for an American call option. Consequently, for the price of an American call option, we have

$$\max\left(Se^{-D_0(T-t)} - Ee^{-r(T-t)}, S - E, 0\right) \leq C(S,t) \leq S \qquad (2.94)$$

or

$$\max\left(S - D(t) - Ee^{-r(T-t)}, S - E, 0\right) \leq C(S,t) \leq S. \qquad (2.95)$$

Now let us compare a European put option with portfolio $\mathbf{A}$. At time T,

$$p(S,T) = \max(E - S, 0) \leq \mathbf{A}(S,T) = E.$$

Thus

$$p(S,t) \leq Ee^{-r(T-t)}.$$

In order to get a lower bound of $p(S,t)$, let us look at portfolios $\mathbf{D}$ and $\mathbf{A}$. Because at time T, $\mathbf{D}(S,T) = \max(S, E) \geq \mathbf{A}(S,T) = E$, we arrive at

$$p(S,t) + Se^{-D_0(T-t)} \geq Ee^{-r(T-t)}$$

or

$$p(S,t) + S - D(t) \geq Ee^{-r(T-t)}.$$

Also, $p(S,t)$ must be nonnegative. Therefore, we have

$$\max\left(Ee^{-r(T-t)} - Se^{-D_0(T-t)}, 0\right) \leq p(S,t) \leq Ee^{-r(T-t)} \tag{2.96}$$

or

$$\max\left(Ee^{-r(T-t)} - S + D(t), 0\right) \leq p(S,t) \leq Ee^{-r(T-t)}. \tag{2.97}$$

These give the lower and upper bounds of European put options.

For an American put option, we can also get the lower and upper bounds. Because $P(S,t) \geq p(S,t)$, we have

$$P(S,t) \geq \max\left(Ee^{-r(T-t)} - Se^{-D_0(T-t)}, 0\right)$$

or

$$P(S,t) \geq \max\left(Ee^{-r(T-t)} - S + D(t), 0\right).$$

Also, $P(S,t)$ must be greater than or equal to $\max(E - S, 0)$. Therefore, we further obtain

$$P(S,t) \geq \max\left(Ee^{-r(T-t)} - Se^{-D_0(T-t)}, E - S, 0\right)$$

or

$$P(S,t) \geq \max\left(Ee^{-r(T-t)} - S + D(t), E - S, 0\right).$$

E is a upper bound of $P(S,t)$, consequently we have

$$\max\left(Ee^{-r(T-t)} - Se^{-D_0(T-t)}, E - S, 0\right) \leq P(S,t) \leq E \tag{2.98}$$

or

$$\max\left(Ee^{-r(T-t)} - S + D(t), E - S, 0\right) \leq P(S,t) \leq E. \tag{2.99}$$

From the proofs we know that if one of these relations is not true, then we can find an arbitrage opportunity to earn some money. This means that the lower bound is the greatest lower bound and that the upper bound is the least upper bound. From Subsection 1.2.4, we know that the price of an option is an increasing function of the volatility. Therefore, if the lower bound is the greatest lower bound, then as the volatility approaches zero, the limit of option should be the lower bound. Similarly, if the upper bound is the least upper bound, then as the volatility approaches infinity, the limit of the option should be the upper bound. When r, D_0, and σ are constant, the European option price is given by the Black–Scholes formulae in Subsection 2.4.4:

$$c(S,t) = Se^{-D_0(T-t)}N(d_1) - Ee^{-r(T-t)}N(d_2)$$

and

$$p(S,t) = Ee^{-r(T-t)}N(-d_2) - Se^{-D_0(T-t)}N(-d_1),$$

where

$$d_1 = \left[\ln\frac{Se^{-D_0(T-t)}}{Ee^{-r(T-t)}} + \frac{1}{2}\sigma^2(T-t)\right] \bigg/ \left(\sigma\sqrt{T-t}\right)$$

and

$$d_2 = \left[\ln \frac{Se^{-D_0(T-t)}}{Ee^{-r(T-t)}} - \frac{1}{2}\sigma^2(T-t)\right] \Big/ \left(\sigma\sqrt{T-t}\right).$$

Therefore we have

$$
\begin{cases}
\lim_{\sigma \to 0} c(S,t) = \begin{cases} 0, & \text{if } Se^{-D_0(T-t)} < Ee^{-r(T-t)}, \\ Se^{-D_0(T-t)} - Ee^{-r(T-t)}, & \text{if } Se^{-D_0(T-t)} > Ee^{-r(T-t)}, \end{cases} \\
\lim_{\sigma \to \infty} c(S,t) = Se^{-D_0(T-t)}, \\
\lim_{\sigma \to 0} p(S,t) = \begin{cases} 0, & \text{if } Ee^{-r(T-t)} < Se^{-D_0(T-t)}, \\ Ee^{-r(T-t)} - Se^{-D_0(T-t)}, & \text{if } Ee^{-r(T-t)} > Se^{-D_0(T-t)}, \end{cases} \\
\lim_{\sigma \to \infty} p(S,t) = Ee^{-r(T-t)}.
\end{cases}
$$

That is, (2.92) and (2.96) truly provide the least upper and greatest lower bounds of European options, respectively.

Here, we give an example to show that if the price of an option does not satisfy a related condition, then there exists an arbitrage opportunity. More examples are given as problems for the reader to study.

Example 1. Consider a European call option on a dividend-paying stock. Suppose the following: $S = \$102, E = \$100, c = \$8.50, r = 0.1$, the time to maturity is 9 months, and the present value of the dividend $D(t)$ is \$0.50. Is there any arbitrage opportunity?

Solution: As we know, the price of a call option has to satisfy the condition (2.93):

$$\max\left(S - D(t) - Ee^{-r(T-t)}, 0\right) \leq c(S,t) \leq S - D(t).$$

In this case

$$\max\left(S - D(t) - Ee^{-r(T-t)}, 0\right) = \max\left(102 - 0.5 - 100e^{-0.9/12}, 0\right) = 8.73.$$

Therefore, the price of the call option is less than the lower bound. In this case, if we own one share of the stock or if you can borrow one share of the stock for the period $[t, T]$, then we should take a long position in a portfolio **C** and a short position in a portfolio **B**. In other words, buy one call option, sell one share, and deposit $Ee^{-r(T-t)} + D(t)$ in a bank at time t. In this case we will get $-8.5 + 102 - 100e^{-0.9/12} - 0.5 = \$0.23$ at time t, and this is a risk-free earning. This is because we can get the money from the bank to pay the dividends on the stock during the time interval $[t, T]$ and get E from the bank at time T. If $S \geq E$ at time T, we can exercise the call option and get one share. If $S < E$, we can have one share of the stock that is bought from the market and an amount of cash $E - S$. In any case, we have one share plus at least \$0.23. That is, we can get one share back or return one share to the borrower and earn at least \$0.23 free at time T.

2.11.3 Relations Between Call and Put Prices

Let us look at portfolios $\mathbf{C}$ and $\mathbf{D}$. Because $\mathbf{C}(S,T) = \mathbf{D}(S,T)$, we have

$$c(S,t) + Ee^{-r(T-t)} = p(S,t) + Se^{-D_0(T-t)} \tag{2.100}$$

or

$$c(S,t) + Ee^{-r(T-t)} = p(S,t) + S - D(t) \tag{2.101}$$

according to Theorem 2.4. These are called put–call parities of European options. For stocks with continuous dividends, we obtained such a relation through a very long procedure in Section 2.4. However, the derivation here is so simple. This shows that arbitrage theory is a very powerful tool.

The put–call parity relations hold only for European options. For American options they are not true, but the following inequalities on the difference between the American call and put option prices are fulfilled

$$Se^{-D_0(T-t)} - E \le C(S,t) - P(S,t) \le S - Ee^{-r(T-t)} \tag{2.102}$$

or

$$S - D(t) - E \le C(S,t) - P(S,t) \le S - Ee^{-r(T-t)}. \tag{2.103}$$

The two inequalities can also be written as

$$
\begin{cases}
Se^{-D_0(T-t)} - E + P(S,t) \le C(S,t) \le S - Ee^{-r(T-t)} + P(S,t), \\
C(S,t) - S + Ee^{-r(T-t)} \le P(S,t) \le C(S,t) - Se^{-D_0(T-t)} + E
\end{cases}
$$

or

$$
\begin{cases}
S - D(t) - E + P(S,t) \le C(S,t) \le S - Ee^{-r(T-t)} + P(S,t), \\
C(S,t) - S + Ee^{-r(T-t)} \le P(S,t) \le C(S,t) - S + D(t) + E,
\end{cases}
$$

which gives the lower and upper bounds of an American call (put) option if the price of the corresponding American put (call) option is known.

First, let us prove the left portions of the inequalities (2.102) and (2.103). Consider portfolios $\mathbf{G}$ and $\mathbf{F}$. Because $\mathbf{G}$ contains European options only and $\mathbf{F}$ contains only one American option, it is possible to use Theorem 2.5. According to Theorem 2.5, the value of $\mathbf{G}$ is always greater than or equal to the value of $\mathbf{F}$ if we can prove two things:

1. The value of $\mathbf{G}$ is greater than or equal to the value of $\mathbf{F}$ at time T;
2. The amount of cash and the number of stocks in $\mathbf{G}$ is greater than or equal to the amount of cash and the number of stocks in $\mathbf{F}$ when the American option is exercised at time $\bar{t} < T$.

At time T, the value of $\mathbf{G}$ is equal to the value of $\mathbf{F}$. At any time $\bar{t} < T$, there is an amount of cash E and no stock in $\mathbf{G}$. If the American put option in $\mathbf{F}$ is exercised before time T, $\mathbf{F}$ contains an amount of cash E and $-(1 - e^{-D_0(T-\bar{t})})$

shares or an amount of cash $E - D(t)$. Therefore, both the amount of cash and the number of stocks in $\mathbf{G}$ is greater than or equal to those in $\mathbf{F}$ if the American option in $\mathbf{F}$ is exercised at some time $\bar{t} < T$. Consequently, according to Theorem 2.5, the value of $\mathbf{G}$ is greater than or equal to the value of $\mathbf{F}$ for any case, that is,

$$P(S,t) + Se^{-D_0(T-t)} \le c(S,t) + E$$

or

$$P(S,t) + S - D(t) \le c(S,t) + E.$$

Because $C(S,t) \ge c(S,t)$, we further have

$$Se^{-D_0(T-t)} - E \le C(S,t) - P(S,t)$$

or

$$S - D(t) - E \le C(S,t) - P(S,t).$$

In order to prove the right portions of the relations, we need to look at portfolios $\mathbf{H}$ and $\mathbf{E}$. In $\mathbf{H}$ there is only one European option and in $\mathbf{E}$ the American option is the only option, so we can use Theorem 2.5 again. When the American call option in $\mathbf{E}$ is exercised before time T, the amount of cash and the number of stocks in $\mathbf{H}$ is greater than or equal to those in $\mathbf{E}$. When it is not exercised before expiry, the value of $\mathbf{H}$ is equal to the value of $\mathbf{E}$ at time T. Therefore

$$C(S,t) + Ee^{-r(T-t)} \le p(S,t) + S.$$

Noticing $P(S,t) \ge p(S,t)$, we have

$$C(S,t) - P(S,t) \le S - Ee^{-r(T-t)}.$$

This completes our proof.

Example 2. Suppose there are an American call option and an American put option on the same stock. The stock pays dividends continuously, and $D_0 = 0.05$. For both options, $E = \$100$ and $T = 1$ month. At present, $r = 0.1, S = \$103$, and $C = \$5.50$. Find the upper and lower bounds for the price of the American put option by using (2.102). How do we take the arbitrage opportunity if the price of the American put option is greater than the calculated upper bound?

Solution: According to (2.102), the lower bound of $P(S,t)$ is

$$C(S,t) - S + Ee^{-r(T-t)} = 5.5 - 103 + 100e^{-0.1/12} = 1.67$$

and the upper bound is

$$C(S,t) - Se^{-D_0(T-t)} + E = 5.5 - 103e^{-0.05/12} + 100 = 2.93.$$

Suppose that on the market $P(S,t) = \$3.50$ and that we have $e^{-0.05/12}$ shares in hand or can borrow $e^{-0.05/12}$ shares for a period $[t, t+1/12]$. Now we describe how to take advantage of the arbitrage opportunity. At time t, we can sell the American put option and $e^{-0.05/12}$ shares, purchase one European call option that is less than or equal to \$5.50, and hold at least an amount of cash $3.5+103e^{-0.05/12} -5.5 = \$100.57$. If we want, it can be deposited into a bank. At any time $\bar{t} \in [t, T)$, the holder of the American option wants to exercise the option, we pay \$100 and get one share. In this case, we have one share of stock and at least an amount of cash equal to \$0.57. If the holder of the American option does not exercise the option before time T, we will always have at least \$0.57 in cash plus one share of stock at time T. The reason is that we can exercise the European call option and get one share if $S > E$, whereas we can purchase one share from the market if $S \leq E$. Because we have $e^{-0.05/12}$ shares of stocks at time t, we should have one share of stock in order to have no loss on stocks or we need to return one share to the borrower at time T. Therefore, the risk-free earning in this case is at least \$0.57.

Problems

1. a) Suppose that S_1 and S_2 are two independent normal random variables. The mean and variance of S_1 are μ_1 and σ_1^2 and for S_2 they are μ_2 and σ_2^2. Show that $S_1 + S_2$ is a normal random variable with mean $\mu_1 + \mu_2$ and variance $\sigma_1^2 + \sigma_2^2$.
    b) Suppose $\Delta t = t/n$ and ϕ_i, $i = 1, 2, \cdots, n$ are independent standardized normal random variables. Show

    $$X(t) = \lim_{n \to \infty} \left(\phi_1\sqrt{\Delta t} + \phi_2\sqrt{\Delta t} + \cdots + \phi_n\sqrt{\Delta t} \right)$$

    is a normal random variable with mean zero and variance t.
    c) Define $dX = X(t + dt) - X(t)$. Show that it is a normal random variable with mean zero and variance dt.
    d) Suppose $S(t) = e^{\mu t + \sigma X(t)}$. Show $d\ln S(t) \equiv \ln S(t + dt) - \ln S(t) = \mu dt + \sigma dX$.
2. $*^{18}$ Suppose

    $$dS = a(S, t)dt + b(S, t)dX,$$

    where dX is a Wiener process. Let f be a function of S and t. Show that

    $$df = \frac{\partial f}{\partial S}dS + \left(\frac{\partial f}{\partial t} + \frac{1}{2}b^2\frac{\partial^2 f}{\partial S^2} \right) dt$$

    $$= b\frac{\partial f}{\partial S}dX + \left(\frac{\partial f}{\partial t} + \frac{1}{2}b^2\frac{\partial^2 f}{\partial S^2} + a\frac{\partial f}{\partial S} \right) dt.$$

---

[18] A problem with * in this book means that you can find the answer in this book. It is suggested that a student should first read and understand the corresponding material and then do the problem without looking at the book.

This result is usually referred to as Itô's lemma.

3. Suppose that a random variable satisfies

$$dS = \mu S dt + \sigma S dX,$$

where dX is a Wiener process. Find the stochastic equation for $\xi = \dfrac{1}{S}$ by using Itô's lemma and determine the mean and variance of $\dfrac{d\xi}{\xi}$.

4. Suppose that S satisfies

$$dS = \mu S dt + \sigma S dX, \quad 0 \le S < \infty,$$

where μ, σ are positive constants and dX is a Wiener process. Let

$$\xi = \frac{S}{S + P_m},$$

where P_m is a positive constant. The range of ξ is $[0, 1)$. The stochastic differential equation for ξ is in the form:

$$d\xi = a(\xi)dt + b(\xi)dX.$$

Find the concrete expressions for $a(\xi)$ and $b(\xi)$ by Itô's lemma and show

$$\begin{cases} a(0) = 0, \\ b(0) = 0, \end{cases} \quad \text{and} \quad \begin{cases} a(1) = 0, \\ b(1) = 0. \end{cases}$$

5. Consider a random variable r satisfying the stochastic differential equation

$$dr = (\mu - \gamma r)dt + w dX, \quad -\infty < r < \infty,$$

where μ, γ, w are positive constants and dX is a Wiener process. Define

$$\xi = \frac{r}{|r| + P_m}, \quad P_m > 0,$$

which transforms the domain $(-\infty, \infty)$ for r into $(-1, 1)$ for ξ. Suppose the stochastic equation for the new random variable ξ is

$$d\xi = a(\xi)dt + b(\xi)dX.$$

Find the concrete expressions of $a(\xi)$ and $b(\xi)$ and show that $a(\xi)$ and $b(\xi)$ fulfill the conditions

$$\begin{cases} a(-1) = 0, \\ b(-1) = 0, \end{cases} \quad \text{and} \quad \begin{cases} a(1) = 0, \\ b(1) = 0. \end{cases}$$

6. Suppose that S has the probability density function

$$G(S) = \frac{1}{\sqrt{2\pi b S}} e^{-\left[\ln(S/a) + b^2/2\right]^2/2b^2}.$$

Let $\xi = \dfrac{1}{S}$. Find the probability density function for ξ, $\mathrm{E}\,[\xi]$ and $\mathrm{Var}\,[\xi]$.

7. a) *Show that if an investment is risk-free, then theoretically its return rate must be the spot interest rate.
   b) *Using this fact and Itô's lemma, derive the Black–Scholes equation.

8. Find the solution of the form
   a) $V(S,t) = V(S)$,
   b) $V(S,t) = A(t)B(S)$
   for the Black–Scholes equation.

9. Show by substitution that
   a) $V(S,t) = Se^{-D_0(T-t)}$,
   b) $V(S,t) = Ee^{-r(T-t)}$
   are solutions of the Black–Scholes equation. What do these solutions represent?

10. *Suppose $V(S,t)$ is the solution of the problem

$$\begin{cases} \dfrac{\partial V}{\partial t} + \dfrac{1}{2}\sigma^2 S^2 \dfrac{\partial^2 V}{\partial S^2} + (r - D_0)S\dfrac{\partial V}{\partial S} - rV = 0, & 0 \le S, \quad t \le T, \\ V(S,T) = V_T(S), & 0 \le S. \end{cases}$$

Let

$$x = \ln S + (r - D_0 - \sigma^2/2)(T - t), \quad \bar{\tau} = \sigma^2(T - t)/2$$

and

$$V(S,t) = e^{-r(T-t)}u(x,\bar{\tau}).$$

Show that $u(x,\bar{\tau})$ is the solution of the problem

$$\begin{cases} \dfrac{\partial u}{\partial \bar{\tau}} = \dfrac{\partial^2 u}{\partial x^2}, & -\infty < x < \infty, \quad 0 \le \bar{\tau}, \\ u(x,0) = V_T(e^x), & -\infty < x < \infty. \end{cases}$$

11. Consider the problem **A**

$$\begin{cases} \dfrac{\partial V}{\partial t} + a(t)S^2\dfrac{\partial^2 V}{\partial S^2} + b(t)S\dfrac{\partial V}{\partial S} - r(t)V = 0, & 0 \le S, \quad t \le T, \\ V(S,T) = V_T(S), & 0 \le S \end{cases}$$

and define

$$\alpha(t) = \int_t^T a(s)ds,$$

$$\beta(t) = \int_t^T b(s)ds$$

and

$$\gamma(t) = \int_t^T r(s)ds.$$

Show that

a) Let $x = \ln S + \beta(t) - \alpha(t)$, $\bar{\tau} = \alpha(t)$ and $V(S,t) = e^{-\gamma(t)}u(x,\bar{\tau})$, then $u(x,\bar{\tau})$ is the solution of the problem:

$$\begin{cases} \dfrac{\partial u}{\partial \bar{\tau}} = \dfrac{\partial^2 u}{\partial x^2}, & -\infty < x < \infty, \quad 0 \le \bar{\tau}, \\ u(x,0) = V_T(e^x), & -\infty < x < \infty. \end{cases}$$

b) $V(S,t)$ must be in a form

$$V(S,t) = e^{-\gamma(t)}u\left(\ln S + \beta(t) - \alpha(t), \alpha(t)\right)$$

or

$$V(S,t) = e^{-\gamma(t)}\bar{u}\left(Se^{\beta(t)}, \alpha(t)\right).$$

c) If

$$V(S,t) = e^{-r(T-t)}\bar{u}\left(Se^{b(T-t)}, a(T-t)\right)$$

is the solution of the problem **A** when $a(t) = a$, $b(t) = b$, and $r(t) = r$, then

$$V(S,t) = e^{-\gamma(t)}\bar{u}\left(Se^{\beta(t)}, \alpha(t)\right).$$

is the solution of problem **A** with time-dependent coefficients.

12. *Suppose $V(S,t)$ is the solution of the problem

$$\begin{cases} \dfrac{\partial V}{\partial t} + \dfrac{1}{2}\sigma^2(S)S^2\dfrac{\partial^2 V}{\partial S^2} + (r - D_0)S\dfrac{\partial V}{\partial S} - rV = 0, & 0 \le S, \ t \le T, \\ V(S,T) = V_T(S), \ 0 \le S. \end{cases}$$

Let $\xi = \dfrac{S}{S + P_m}$, $\tau = T - t$, and $V(S,t) = (S + P_m)\overline{V}(\xi,\tau)$, where P_m is a positive constant.

a) Show that $\overline{V}(\xi,\tau)$ is the solution of the problem

$$\begin{cases} \dfrac{\partial \overline{V}}{\partial \tau} = \dfrac{1}{2}\bar{\sigma}^2(\xi)\xi^2(1 - \xi)^2\dfrac{\partial^2 \overline{V}}{\partial \xi^2} + (r - D_0)\xi(1 - \xi)\dfrac{\partial \overline{V}}{\partial \xi} \\ \qquad\qquad - [r(1 - \xi) + D_0\xi]\overline{V}, & 0 \le \xi \le 1, \quad 0 \le \tau, \\ \overline{V}(\xi,0) = \dfrac{1 - \xi}{P_m}V_T\left(\dfrac{P_m\xi}{1 - \xi}\right), & 0 \le \xi \le 1, \end{cases}$$

where

$$\bar{\sigma}(\xi) = \sigma\left(\frac{P_m\xi}{1-\xi}\right).$$

b) What are the advantages of reformulating the problem on a finite domain?

13. *Find an integral expression of the solution of the following problem

$$\begin{cases} \dfrac{\partial u}{\partial \bar{\tau}} = \dfrac{\partial^2 u}{\partial x^2}, & -\infty < x < \infty, \quad 0 \le \bar{\tau}, \\ u(x,0) = u_0(x), & -\infty < x < \infty. \end{cases}$$

14. Suppose that S is a random variable that is defined on $[0, \infty)$ and whose probability density function is

$$G(S) = \frac{1}{\sqrt{2\pi}bS}e^{-\left[\ln(S/a)+b^2/2\right]^2/2b^2},$$

a and b being positive numbers. Show that

a)

$$\int_0^c G(S)dS = N\left(\frac{\ln(c/a)+b^2/2}{b}\right);$$

b)

$$\int_0^c SG(S)dS = aN\left(\frac{\ln(c/a)-b^2/2}{b}\right);$$

c) for any real number n

$$\int_0^c S^n G(S)dS = a^n e^{(n^2-n)b^2/2}N\left(\frac{\ln(c/a)+b^2/2}{b}-nb\right);$$

d) for any real number n

$$E[S^n] = a^n e^{(n^2-n)b^2/2};$$

e) for any real number n

$$\int_c^\infty S^n G(S)dS = a^n e^{(n^2-n)b^2/2}N\left(-\frac{\ln(c/a)+b^2/2}{b}+nb\right),$$

where

$$N(z) = \frac{1}{\sqrt{2\pi}}\int_{-\infty}^z e^{-\xi^2/2}d\xi.$$

15. Using the results given in Problems 10, 13, and 14, derive the Black–Scholes formula for a European put option.

16. Verify that the Black–Scholes formula for a put option is the solution of the following problem:

$$\begin{cases} \dfrac{\partial p}{\partial t} + \dfrac{1}{2}\sigma^2 S^2 \dfrac{\partial^2 p}{\partial S^2} + (r - D_0)S\dfrac{\partial p}{\partial S} - rp = 0, & 0 \le S, \quad 0 \le t \le T, \\ p(S,T) = \max(E - S,\, 0), & 0 \le S. \end{cases}$$

(Hint: Show the identity $Ee^{-r(T-t)-d_2^2/2} = Se^{-D_0(T-t)-d_1^2/2}$ first.)

17. Using the Black–Scholes formula for a put option and the results in Problem 11, find the formula for the price of a put option with time-dependent parameters.

18. Consider the following problem

$$\begin{cases} \dfrac{\partial c_b}{\partial t} + \dfrac{1}{2}\sigma^2 S^2 \dfrac{\partial^2 c_b}{\partial S^2} + (r - D_0)S\dfrac{\partial c_b}{\partial S} - rc_b = 0, \\ 0 \le S < \infty, \quad 0 < t < T, \\ c_b(S,T) = \begin{cases} 0, & \text{if } 0 \le S < S^{**}, \\ f(S), & \text{if } S^{**} \le S < S^*, \\ S - E, & \text{if } S^* \le S < \infty, \end{cases} \end{cases}$$

where

$$f(S) = a_0 + a_1 S + \cdots + a_J S^J.$$

Show that it has a solution in the following closed form:

$$c_b(S,t) = \sum_{n=0}^{J} \left\{ a_n S^n e^{[(n-1)r - nD_0 + (n-1)n\sigma^2/2](T-t)} \right.$$
$$\left. \times \left[N\left(d^* - n\sigma\sqrt{T-t}\right) - N\left(d^{**} - n\sigma\sqrt{T-t}\right) \right] \right\}$$
$$+ Se^{-D_0(T-t)}\left[1 - N\left(d^* - \sigma\sqrt{T-t}\right) \right] - Ee^{-r(T-t)}[1 - N(d^*)],$$

where

$$d^* = \left[\ln(S^*/S) - \left(r - D_0 - \frac{1}{2}\sigma^2\right)(T-t) \right] \Big/ \left(\sigma\sqrt{T-t}\right),$$

$$d^{**} = \left[\ln(S^{**}/S) - \left(r - D_0 - \frac{1}{2}\sigma^2\right)(T-t) \right] \Big/ \left(\sigma\sqrt{T-t}\right).$$

19. Consider a European call option on a non–dividend-paying stock. Use the Black–Scholes formula to find the option price when the stock price is $63, the strike price is $60, the risk-free interest rate is 5% per annum, the volatility is 35% per annum, and the time to maturity is six months.

20. Consider a European put option on a dividend-paying stock. Use the Black–Scholes formula to find the option price when the stock price is $55, the strike price is $60, the risk-free interest rate is 5% per annum, the volatility is 35% per annum, the dividend yield is 3% per annum, and the time to maturity is six months.

21. Consider a European call option on a non–dividend-paying stock. The option price is $4.5, the stock price is $86, the exercise price is $92, the risk-free interest rate is 5% per annum, and the time to maturity is three months. Use the Black–Scholes formula for a call option to find what the corresponding volatility should be. (This volatility is usually referred to as the implied volatility associated with the given option price.)

22. Consider a European option on a non–dividend-paying stock. The stock price is $37, the exercise price is $34, the risk-free interest rate is 5% per annum, the volatility is 30% per annum, and the time to maturity is six months. Find the put and call option prices by using the Black–Scholes formulae and verify that the put–call parity holds.

23. *Suppose that $c(S,t)$ and $p(S,t)$ are the prices of European call and put options with the same parameters, respectively. Show the put–call parity

$$c(S,t) - p(S,t) = Se^{-D_0(T-t)} - Ee^{-r(T-t)}$$

without using the Black–Scholes formulae.

24. By using the put–call parity relation of European options

$$c(S,t) - p(S,t) = Se^{-D_0(T-t)} - Ee^{-r(T-t)},$$

show that the following relations hold:

$$\frac{\partial p}{\partial S} = \frac{\partial c}{\partial S} - e^{-D_0(T-t)}, \quad \frac{\partial^2 p}{\partial S^2} = \frac{\partial^2 c}{\partial S^2}$$

and

$$\frac{\partial^2 p}{\partial S \partial \sigma} = \frac{\partial^2 c}{\partial S \partial \sigma}, \quad \frac{\partial p}{\partial \sigma} = \frac{\partial c}{\partial \sigma}, \quad \frac{\partial^2 p}{\partial \sigma^2} = \frac{\partial^2 c}{\partial \sigma^2}.$$

25. a) Show that the solution of the problem

$$\begin{cases} \dfrac{\partial V}{\partial t} + \dfrac{1}{2}\sigma^2 S^2 \dfrac{\partial^2 V}{\partial S^2} + (r - D_0)S\dfrac{\partial V}{\partial S} - rV = 0, & 0 \le S, \quad t \le T, \\ V(S,T) = \max(S,E), & 0 \le S \end{cases}$$

is

$$V(S,t) = Ee^{-r(T-t)}N(-d_2) + Se^{-D_0(T-t)}N(d_1),$$

where

$$d_1 = \left[\ln\frac{Se^{-D_0(T-t)}}{Ee^{-r(T-t)}} + \frac{1}{2}\sigma^2(T-t)\right] \bigg/ \left(\sigma\sqrt{T-t}\right),$$

$$d_2 = \left[\ln\frac{Se^{-D_0(T-t)}}{Ee^{-r(T-t)}} - \frac{1}{2}\sigma^2(T-t)\right] \bigg/ \left(\sigma\sqrt{T-t}\right).$$

b) Let $\overline{S} = E^2/S$ and $\overline{V}(\overline{S}, t) = EV(S, t)/S$. Show that $\overline{V}(\overline{S}, t)$ is the solution of the following problem:

$$
\begin{cases}
\dfrac{\partial \overline{V}}{\partial t} + \tfrac{1}{2}\sigma^2 \overline{S}^2 \dfrac{\partial^2 \overline{V}}{\partial \overline{S}^2} + (D_0 - r)\overline{S}\dfrac{\partial \overline{V}}{\partial \overline{S}} - D_0 \overline{V} = 0, & 0 \le \overline{S}, \quad t \le T, \\[2mm]
\overline{V}(\overline{S}, T) = \max(\overline{S}, E), & 0 \le \overline{S}
\end{cases}
$$

and find the expression of $\overline{V}(\overline{S}, t)$.

c) Let

$$
G(S', T; S, t, a_1) = \frac{1}{\sigma\sqrt{2\pi(T-t)}S'}\, e^{-\left[\ln(S'/a)+\sigma^2(T-t)/2\right]^2/2\sigma^2(T-t)},
$$

where $a = Se^{a_1(T-t)}$. Show that $N(d_1)$ and $N(-d_1)$ are the probabilities of the events $\overline{S}' \le E$ ($S' \ge E$) and $\overline{S}' \ge E$ ($S' \le E$), respectively, in the world where the probability density function of $\overline{S}'$ is

$$
G(\overline{S}', T; \overline{S}, t, D_0 - r),
$$

and that $N(d_2)$ and $N(-d_2)$ are the probabilities of the events $S' \ge E$ and $S' \le E$, respectively, in the world where the probability density function of S' is

$$
G(S', T; S, t, r - D_0).
$$

d) Replacing $Se^{-D_0(T-t)}$ by $Ee^{-r(T-t)}$ and $Ee^{-r(T-t)}$ by $Se^{-D_0(T-t)}$, we can obtain the second (first) term in the solution in (a) from the first (second) term, and the entire solution is unchanged. Why does the solution have this symmetry?

26. *Explain why an American option is always worth at least as much as a European option on the same asset with the same strike price and exercise date by financial reasoning and by mathematical tools.

27. *Explain why an American option is always worth at least as much as its intrinsic value. What does the time value of an American option represent?

28. *Let $\mathbf{L_{s,t}}$ be an operator in an option problem in the form:

$$
\mathbf{L_{s,t}} = a(S, t)\frac{\partial^2}{\partial S^2} + b(S, t)\frac{\partial}{\partial S} + c(S, t)
$$

and $G(S, t)$ be the constraint function for an American option. Furthermore, we assume that $\dfrac{\partial G}{\partial t} + \mathbf{L_{s,t}}G$ exists. Suppose $V(S, t^*) = G(S, t^*)$ on an open interval (A_1, B_1) on the S-axis. Let $t = t^* - \Delta t$, where Δt is a sufficiently small positive number and let (A, B) be an open interval in (A_1, B_1). Show the following conclusions: If for any $S \in (A, B)$,

$$
\frac{\partial G}{\partial t}(S, t^*) + \mathbf{L_{s,t^*}}G(S, t^*) \ge 0,
$$

then the value $V(S,t)$ determined by the equation

$$\frac{\partial V}{\partial t}(S,t) + \mathbf{L_{s,t}}V(S,t) = 0$$

satisfies the condition $V(S,t) - G(S,t) \geq 0$ on (A,B); and if for any $S \in (A,B)$,

$$\frac{\partial G}{\partial t}(S,t^*) + \mathbf{L_{s,t^*}}G(S,t^*) < 0,$$

then the equation

$$\frac{\partial V}{\partial t}(S,t) + \mathbf{L_{s,t}}V(S,t) = 0$$

cannot give a solution satisfying the condition $V(S,t) - G(S,t) \geq 0$ for any $S \in (A,B)$.

29. A European option is the solution of the problem

$$\begin{cases} \dfrac{\partial V}{\partial t} + \mathbf{L_s}V = 0, & 0 \leq S, \quad t \leq T, \\ V(S,T) = V_T(S), & 0 \leq S, \end{cases}$$

where

$$\mathbf{L_s} = \frac{1}{2}\sigma^2 S^2 \frac{\partial^2}{\partial S^2} + (r - D_0)S\frac{\partial}{\partial S} - r.$$

For an American option, the constraint is that the inequality

$$V(S,t) \geq G(S,t)$$

holds for any S and t, where $G(S,T) = V_T(S)$. Derive the linear complementarity problem for the American option.

30. *Let $V(S,t)$ be the price of a vanilla American option. Explain why $V(S,t - \Delta t) \geq V(S,t)$ is always true by financial reasoning and by mathematical tools, where $\Delta t > 0$.

31. The American call option is the solution of the following linear complementarity problem on a finite domain:

$$\begin{cases} \left(\dfrac{\partial \overline{V}}{\partial \tau} - \mathbf{L_\xi}\overline{V}\right)\left[\overline{V}(\xi,0) - \max(2\xi - 1, 0)\right] = 0, \\[2mm] \dfrac{\partial \overline{V}}{\partial \tau} - \mathbf{L_\xi}\overline{V} \geq 0, \\[2mm] \overline{V}(\xi,\tau) - \max(2\xi - 1, 0) \geq 0, \\[2mm] \overline{V}(\xi,0) = \max(2\xi - 1, 0), \end{cases}$$

where

$$0 \leq \xi \leq 1, \quad 0 \leq \tau$$

and

$$\mathbf{L}_\xi = \frac{1}{2}\bar{\sigma}^2(\xi)\xi^2(1-\xi)^2\frac{\partial^2}{\partial\xi^2} + (r-D_0)\xi(1-\xi)\frac{\partial}{\partial\xi} - [r(1-\xi)+D_0\xi].$$

Reformulate this problem as a free-boundary problem if $D_0 > 0$.

32. The American put option is the solution of the following linear complementarity problem:

$$\begin{cases} \left(\dfrac{\partial u}{\partial\bar{\tau}} - \dfrac{\partial^2 u}{\partial x^2}\right)[u(x,\bar{\tau}) - g_p(x,\bar{\tau})] = 0, \\[2mm] \dfrac{\partial u}{\partial\bar{\tau}} - \dfrac{\partial^2 u}{\partial x^2} \geq 0, \\[2mm] u(x,\bar{\tau}) - g_p(x,\bar{\tau}) \geq 0, \\[2mm] u(x,0) = g_p(x,0), \end{cases}$$

where

$$-\infty < x < \infty, \quad 0 \leq \bar{\tau}$$

and

$$g_p(x,\bar{\tau}) = \max\left(e^{2r\bar{\tau}/\sigma^2} - e^{x+(2D_0/\sigma^2+1)\bar{\tau}}, 0\right).$$

Find the domain where a free boundary may appear and the domain where it is impossible for a free boundary to appear, show that there is only one free boundary at $\bar{\tau} = 0$, and give the starting location of this free boundary.

33. The price of a one-factor convertible bond is the solution of the linear complementarity problem

$$\begin{cases} \left[\dfrac{\partial V}{\partial t} + \dfrac{1}{2}\sigma^2 S^2\dfrac{\partial^2 V}{\partial S^2} + (r-D_0)S\dfrac{\partial V}{\partial S} - rV\right][V(S,t) - nS] = 0, \\[2mm] \dfrac{\partial V}{\partial t} + \dfrac{1}{2}\sigma^2 S^2\dfrac{\partial^2 V}{\partial S^2} + (r-D_0)S\dfrac{\partial V}{\partial S} - rV \leq 0, \\[2mm] V(S,t) - nS \geq 0, \\[2mm] V(S,T) = \max(Z,nS) \geq nS, \end{cases}$$

where $0 \leq S$, $0 \leq t \leq T$, σ, r, and D_0 are constants. Show that if $D_0 > 0$, then the solution of a one-factor convertible bond must involve a free boundary and its location at $t = T$ is $S = Z/n$. Also, derive the corresponding free-boundary problem.

34. a) Suppose $\sigma = \sigma(S,t)$, $r = r(t)$, and $D_0 = D_0(S,t)$. Show that the problem of pricing a put option can always be converted into a problem of pricing a call option. Also explain how to use this conclusion when we write codes if a code for put options is quite a different from a code for call options.

b) Let the exercise price be E. Suppose that r, D_0 are constants and $\sigma = \sigma(S)$. Show

$$P(S,t;b,\ a,\ \sigma(S)) = C\left(E^2/S, t; a,\ b,\ \sigma(S)\right) S/E,$$
$$C\left(S, t; a,\ b,\ \sigma(S)\right) = P\left(E^2/S, t; b,\ a,\ \sigma(S)\right) S/E$$

and

$$S_{cf}(t; a,\ b,\ \sigma(S)) \times S_{pf}(t; b,\ a,\ \sigma(E^2/S)) = E^2.$$

Here, the first, second, and third parameters after the semicolon in P, C, S_{pf}, and S_{cf} are the interest rate, the dividend yield and the volatility function, respectively.

35. Suppose that σ, r, D_0 are constants. In this case, we have the following symmetry relation for European options

$$p(S, t; b,\ a) = c\left(\frac{E^2}{S}, t; a,\ b\right) S/E,$$

where the first and second arguments after the semicolon in p and c are the values of the interest rate and the dividend yield, respectively. For a European call option, the price is

$$c(S,t) = Se^{-D_0(T-t)}N(d_1) - Ee^{-r(T-t)}N(d_2),$$

where

$$d_1 = \frac{\ln\dfrac{Se^{-D_0(T-t)}}{Ee^{-r(T-t)}} + \dfrac{1}{2}\sigma^2(T-t)}{\sigma\sqrt{T-t}},$$

$$d_2 = \frac{\ln\dfrac{Se^{-D_0(T-t)}}{Ee^{-r(T-t)}} - \dfrac{1}{2}\sigma^2(T-t)}{\sigma\sqrt{T-t}}.$$

Find the price of a European put option by using the symmetry relation.

36. Derive the formulation of the problem for $\dfrac{\partial P}{\partial r}$ and write down the formulation of the problems for $\dfrac{\partial P}{\partial \sigma}$ and $\dfrac{\partial P}{\partial D_0}$, where P is the price of an American put option.

37. Find an expression for the value of a perpetual American put option.

38. *Describe the reversion conditions of a stochastic process, and give the intuitive meaning of the conditions.

39. *Describe and derive the generalized Itô's lemma.

40. *Describe and derive the general equations for derivative securities.

41. Suppose $f_1(r,t) \geq 0$ and $f_1(r_l, t) = \dfrac{\partial f_1(r_l, t)}{\partial r} = f_1(r_u, t) = \dfrac{\partial f_1(r_u, t)}{\partial r} = 0$, and $f_2(r_l, t) < 0$, $f_2(r_u, t) > 0$. Explain why problem **A**

$$\begin{cases} \dfrac{\partial V}{\partial t} = f_1 \dfrac{\partial^2 V}{\partial r^2} + f_2 \dfrac{\partial V}{\partial r} + f_3 V, & r_l \le r \le r_u, \quad 0 \le t, \\ V(r,0) = V_0(r), & r_l \le r \le r_u, \\ V(r_l,t) = f_l(t), & 0 \le t, \\ V(r_u,t) = f_u(t), & 0 \le t \end{cases}$$

and problem **B**

$$\begin{cases} \dfrac{\partial V}{\partial t} = -f_1 \dfrac{\partial^2 V}{\partial r^2} + f_2 \dfrac{\partial V}{\partial r} + f_3 V, & r_l \le r \le r_u, \quad t \le T, \\ V(r,T) = V_T(r), & r_l \le r \le r_u \end{cases}$$

have unique solutions.

42. Suppose that S is the price of a dividend-paying stock and satisfies

$$dS = \mu(S,t)S dt + \sigma S dX_1, \quad 0 \le S < \infty,$$

where dX_1 is a Wiener process and σ is another random variable. Let the dividend paid during the time period $[t, t+dt]$ be $D(S,t)dt$. Assume that for σ, the stochastic equation

$$d\sigma = p(\sigma,t)dt + q(\sigma,t)dX_2, \quad \sigma_l \le \sigma \le \sigma_u$$

holds. Here, $p(\sigma,t)$ and $q(\sigma,t)$ are differentiable functions and satisfy the reversion conditions. dX_2 is another Wiener process correlated with dX_1, and the correlation coefficient between them is ρdt. For options on such a stock, derive directly the partial differential equation that contains only the unknown market price of risk for the volatility. Here "Directly" means "without using the general PDE for derivatives." (Hint: Take a portfolio in the form $\Pi = \Delta_1 V_1 + \Delta_2 V_2 + S$, where V_1 and V_2 are two different options.)

43. Explain the financial meaning of the jump conditions for option values.

44. *Use arbitrage arguments to show the put–call parity of European options for the following two cases.

   a) When the dividend is paid continuously, the put–call parity is

$$c(S,t) - p(S,t) = Se^{-D_0(T-t)} - Ee^{-r(T-t)};$$

   b) when the dividend is paid discretely, the put–call parity is

$$c(S,t) - p(S,t) = S - D(t) - Ee^{-r(T-t)},$$

   where $D(t)$ is the value of dividend at time t.

45. *Use arbitrage arguments to show the inequalities of American options for the following two cases.

a) When the dividend is paid continuously, there is the inequality

$$Se^{-D_0(T-t)} - E \le C(S,t) - P(S,t) \le S - Ee^{-r(T-t)}$$

between American put option $P(S,t)$ and American call option $C(S,t)$ with the same parameters.

b) When the dividend is paid discretely, there is the inequality

$$S - D(t) - E \le C(S,t) - P(S,t) \le S - Ee^{-r(T-t)}$$

between American put option $P(S,t)$ and American call option $C(S,t)$ with the same parameters.

46. Consider a European call option with $T = 6$ months and $E = \$80$ on a dividend-paying stock. The dividend is paid continuously with a dividend yield $D_0 = 0.05$. Today, $t = 0$, $r = 0.1$ and $S = \$82$.

a) Find the lower bound of the call option.

b) What are the least profits we could make at time T by arbitrage if the call option price today is $\$0.10$ less than the lower bound and why?

47. Consider a European put option with $T = 3$ months and $E = \$60$ on a dividend-paying stock. Today $t = 0$, $r = 0.05$, and $S = \$55$. The dividends are paid discretely, and the total present value of them is $D(0) = \$0.30$.

a) Find the lower bound of the put option.

b) What are the least profits we could make at time T by arbitrage if the put option price today is $\$0.20$ less than the lower bound and why?

48. Suppose that there are an American call option and an American put option on the same stock that pays dividends discretely. For both of them, $E = \$90$ and $T = 3$ months. At time $t = 0$, the stock price is $\$93$ and the present value of dividend payments during the period $[0,T]$ is $D(0) = \$0.50$. Assume that $r = 0.1$ and $P(S,0) = \$2.50$.

a) Find the upper and lower bounds of the price of the American call option.

b) What are the risk-free profits we could make today by arbitrage if the price of the call option today is $\$0.10$ greater than the calculated upper bound and why?

49. Suppose that $c_1(S,t)$ and $c_2(S,t)$ are the prices of European call options with strikes E_1 and E_2, respectively, where $E_1 < E_2$. Also assume that the two options have the same maturity T and that the interest rate r is a constant. Show

$$0 \le c_1(S,t) - c_2(S,t) \le (E_2 - E_1)e^{-r(T-t)}.$$

50. Suppose that p_1, p_2, and p_3 are the prices of European put options with strike prices E_1, E_2, and E_3 respectively, where $E_2 = \frac{1}{2}(E_1 + E_3)$. All the options have the same maturity. Show

$$p_2 \le \frac{1}{2}(p_1 + p_3).$$

3

Exotic Options

3.1 Introduction

In order to meet a variety of demands, modern financial institutions issue many exotic options besides the vanilla options we have introduced in Chapter 2. An exotic option is an option that is not a vanilla put or call. It usually is traded between companies and banks and not quoted on an exchange. In this case, we usually say that it is traded in the over-the-counter market. Most exotic options are quite complicated, and their final values depend not only on the asset price at expiry but also on the asset price at previous times. They are determined by a part or the whole of the path of the asset price during the life of option. These options are called path-dependent exotic options. Barrier options, Asian options, and lookback options are important examples of path-dependent exotic options.

A barrier option is a derivative product that either becomes worthless, must be exercised, or comes into existence if the underlying asset price reaches a certain level during a certain period of time. For example, a down-and-out call has similar features to a vanilla call option, except that it becomes nullified when the asset price falls below a knock-out level. Because the holder of the option loses some of the rights, the price of such an option is lower than a vanilla call option. However, if the asset price is always higher than the knock-out level (which is expected by any holder of a call option), then the two options are actually the same. Therefore, such a call option is more attractive than a vanilla call option for people who expect the price to rise.

An Asian option is an option whose payoff depends on some form of the average of the underlying asset price over a part or the whole of the life of the option. Consider a call option and let the price of underlying asset in its payoff be replaced by the average of the asset price over a period. Suppose that a manufacturer expects to make a series of crude oil purchases for his factory during some fixed time period. If the price of crude oil drops, then he will be happy because the cost of his product declines; if the price of crude oil rises, then he might lose money because the cost rises. In this case, such an option can be a hedging instrument for him. He can avoid the risk caused by the rise in price and keep the advantage due to the drop in price by holding an average call option on crude oil for that period.

A lookback option is a contract whose payoff depends on the maximum or minimum stock price reached during the life of the option. For example, a lookback put option has a payoff that is the difference between the maximum realized price and the price at expiry. Therefore, the holder of such an option can sell the asset at the highest price.

We have described three examples through which we explain how those exotic options are designed by financial institutions to meet the requirements of their clients. Besides the examples mentioned above, there are many other types of barrier, Asian, lookback options, and other exotic options. Multi-asset options, binary options, forward start options, compound options, and chooser options are all examples of exotic options. In the following sections, we will give some details on these options.

3.2 Barrier Options

3.2.1 Knock-Out and Knock-In Options

As pointed out in Section 3.1, a barrier option is a derivative product that either becomes worthless, must be exercised, or comes into existence if the underlying asset price reaches a certain level during a certain period of time.

A knock-out option is an option that either becomes worthless or must be exercised if the underlying asset value reaches the knock-out level, which is called a barrier. The simplest knock-out options are the down-and-out call and the up-and-out put. An option is called a down-and-out call if it is actually a call when S is always greater than the barrier during the life of the option, and it becomes worthless when S reaches the barrier from above at some time before or at expiry. We call such a barrier a lower barrier B_l, and in this section we mainly consider the case that such a barrier is below the exercise price E. A down-and-out call could be a European-style or an American-style option just like a vanilla option. An up-and-out put is similar to a down-and-out call. However, instead of a lower barrier, it has an upper barrier B_u, which we assume is greater than E. It is a put if S is never above B_u and becomes worthless when S crosses the barrier B_u from below at some time prior to expiry. More complicated knock-out options have two barriers B_l and B_u that might be given as functions of time, and such an option becomes worthless if S enters $[0, B_l]$ or $[B_u, \infty)$ from (B_l, B_u) at some time during the life of the option. Sometimes the holder of the option receives a specified amount of money as a rebate if a barrier is reached. For example, an option is called a knock-out call with a rebate if it has the following three properties. It is actually a call when S is always in (B_l, B_u) during the life of the option, it becomes worthless when S enters $[0, B_l]$ at any time, and it must be exercised for a rebate $B_u - E$ when S crosses the upper barrier B_u from below at any time.

A knock-in option is a contract that comes into existence if the asset price crosses a barrier. For example, a down-and-in call with a lower barrier B_l expires worthless unless the asset price reaches the lower barrier from above prior to or at expiry. If it crosses the lower barrier from above at some time before expiry, then the option becomes a vanilla option. An up-and-in put is similar to a down-and-in call, but the barrier is an upper one and the put option is activated when S crosses the upper barrier from below.

3.2.2 Closed-Form Solutions of Some European Barrier Options

For some European barrier options, closed-form solutions can be obtained. As examples, we first derive such a solution for a European down-and-out call by the method of images and then obtain the solution for a European down-and-in call from the solution for the European down-and-out call.

Now let us look at a European down-and-out call option. Let $c_o(S,t)$ denote the value of this option. If S is always greater than B_l, then it is a call option. Therefore, $c_o(S,t)$ satisfies the Black–Scholes equation for $S > B_l$ and the final condition is

$$c_o(S,T) = \max(S - E, 0), \text{ for } S > B_l.$$

The option becomes worthless if S ever reaches B_l, which means that the boundary condition at $S = B_l$ should be

$$c_o(B_l, t) = 0.$$

Therefore, the fair value of such an option should be the solution of the problem

$$
\begin{cases}
\dfrac{\partial c_o}{\partial t} + \dfrac{1}{2}\sigma^2 S^2 \dfrac{\partial c_o}{\partial S^2} + (r - D_0) S \dfrac{\partial c_o}{\partial S} - r c_o = 0, & S \geq B_l, \quad t \leq T, \\[2mm]
c_o(S,T) = \max(S - E, 0), & S \geq B_l, \\[2mm]
c_o(B_l, t) = 0, & t \leq T.
\end{cases}
$$

Let us consider a more general problem

$$
\begin{cases}
\dfrac{\partial V}{\partial t} + \dfrac{1}{2}\sigma^2 S^2 \dfrac{\partial^2 V}{\partial S^2} + (r - D_0) S \dfrac{\partial V}{\partial S} - r V = 0, & S \geq B_l, \quad t \leq T, \\[2mm]
V(S,T) = V_T(S), & S \geq B_l, \\[2mm]
V(B_l, t) = 0, & t \leq T.
\end{cases}
\tag{3.1}
$$

Set

$$S = e^x, \quad t = T - \frac{2\bar{\tau}}{\sigma^2} \quad \text{and} \quad V = e^{\alpha x + \beta \bar{\tau}} u(x, \bar{\tau}),$$

where α and β are constants. Because

$$\frac{\partial V}{\partial t} = e^{\alpha x + \beta \bar{\tau}} \beta \frac{d\bar{\tau}}{dt} u + e^{\alpha x + \beta \bar{\tau}} \frac{\partial u}{\partial \bar{\tau}} \frac{d\bar{\tau}}{dt}$$

$$= -\frac{\beta \sigma^2}{2} e^{\alpha x + \beta \bar{\tau}} u - \frac{\sigma^2}{2} e^{\alpha x + \beta \bar{\tau}} \frac{\partial u}{\partial \bar{\tau}},$$

$$\frac{\partial V}{\partial S} = e^{\alpha x + \beta \bar{\tau}} \alpha \frac{dx}{dS} u(x, \bar{\tau}) + e^{\alpha x + \beta \bar{\tau}} \frac{\partial u}{\partial x} \frac{dx}{dS}$$

$$= \frac{\alpha}{S} e^{\alpha x + \beta \bar{\tau}} u(x, \bar{\tau}) + \frac{1}{S} e^{\alpha x + \beta \bar{\tau}} \frac{\partial u}{\partial x},$$

$$\frac{\partial^2 V}{\partial S^2} = -\frac{\alpha}{S^2} e^{\alpha x + \beta \bar{\tau}} u(x, \bar{\tau}) + \frac{\alpha^2}{S^2} e^{\alpha x + \beta \bar{\tau}} u(x, \bar{\tau}) + \frac{\alpha}{S} e^{\alpha x + \beta \bar{\tau}} \frac{\partial u}{\partial x} \frac{1}{S}$$

$$- \frac{1}{S^2} e^{\alpha x + \beta \bar{\tau}} \frac{\partial u}{\partial x} + \frac{\alpha}{S^2} e^{\alpha x + \beta \bar{\tau}} \frac{\partial u}{\partial x} + \frac{1}{S} e^{\alpha x + \beta \bar{\tau}} \frac{\partial^2 u}{\partial x^2} \frac{1}{S}$$

$$= \frac{\alpha^2 - \alpha}{S^2} e^{\alpha x + \beta \bar{\tau}} u(x, \bar{\tau}) + \frac{2\alpha - 1}{S^2} e^{\alpha x + \beta \bar{\tau}} \frac{\partial u}{\partial x} + \frac{1}{S^2} e^{\alpha x + \beta \bar{\tau}} \frac{\partial^2 u}{\partial x^2},$$

for $u(x, \bar{\tau})$ we have

$$\frac{\sigma^2}{2} \frac{\partial u}{\partial \bar{\tau}} = \frac{\sigma^2}{2} \frac{\partial^2 u}{\partial x^2} + \left[(r - D_0) + \frac{\sigma^2 (2\alpha - 1)}{2} \right] \frac{\partial u}{\partial x}$$

$$+ \left[-\frac{\beta \sigma^2}{2} + \frac{\sigma^2 (\alpha^2 - \alpha)}{2} + (r - D_0)\alpha - r \right] u(x, \bar{\tau}).$$

Therefore, if we choose

$$\alpha = -\frac{r - D_0 - \sigma^2 / 2}{\sigma^2}$$

and

$$\beta = -\alpha^2 - \frac{2r}{\sigma^2},$$

then under these new variables, the problem above becomes

$$\begin{cases} \dfrac{\partial u}{\partial \bar{\tau}} = \dfrac{\partial^2 u}{\partial x^2}, & \ln B_l \le x, \quad 0 \le \bar{\tau}, \\[2mm] u(x, 0) = e^{-\alpha x} V_T(e^x), & \ln B_l \le x, \\[2mm] u(\ln B_l, \bar{\tau}) = 0, & 0 \le \bar{\tau}. \end{cases} \tag{3.2}$$

As we know, the solution of an initial-value problem of the heat equation is

$$u(x, \bar{\tau}) = \frac{1}{2\sqrt{\pi \bar{\tau}}} \int_{-\infty}^{\infty} u_0(\xi) e^{-(x-\xi)^2 / 4\bar{\tau}} d\xi,$$

where $u_0(\xi)$ is the initial value of u at $\bar{\tau} = 0$. Suppose that the initial condition satisfies the relation

$$u_0(x) = -u_0(2a - x) \quad \text{for} \quad x < a.$$

In this case, the solution can be rewritten as

$$
\begin{aligned}
u\left(x,\bar{\tau}\right) &= \frac{1}{2\sqrt{\pi\bar{\tau}}}\left[\int_{-\infty}^{a} u_0\left(\xi\right)e^{-(x-\xi)^2/4\bar{\tau}}d\xi + \int_{a}^{\infty} u_0\left(\xi\right)e^{-(x-\xi)^2/4\bar{\tau}}d\xi\right] \\
&= \frac{1}{2\sqrt{\pi\bar{\tau}}}\left[-\int_{-\infty}^{a} u_0\left(2a-\xi\right)e^{-(x-\xi)^2/4\bar{\tau}}d\xi\right. \\
&\qquad\qquad\left. + \int_{a}^{\infty} u_0\left(\xi\right)e^{-(x-\xi)^2/4\bar{\tau}}d\xi\right] \\
&= \frac{1}{2\sqrt{\pi\bar{\tau}}}\int_{a}^{\infty} u_0\left(\eta\right)\left(e^{-(x-\eta)^2/4\bar{\tau}} - e^{-(x-2a+\eta)^2/4\bar{\tau}}\right)d\eta.
\end{aligned}
$$

From here we know $u\left(a,\bar{\tau}\right)=0$. Therefore, for $x \geq \ln B_l$, the solution of the problem

$$
\begin{cases}
\dfrac{\partial u}{\partial \tau} = \dfrac{\partial^2 u}{\partial x^2}, & -\infty < x < \infty, \ 0 \leq \bar{\tau}, \\[2mm]
u\left(x,0\right) = \begin{cases} e^{-\alpha x}V_T\left(e^x\right), & \ln B_l \leq x, \\ -e^{-\alpha(2\ln B_l - x)}V_T\left(e^{2\ln B_l - x}\right), & x < \ln B_l \end{cases}
\end{cases}
$$

is the same as that of the problem (3.2). Consequently, the solution of the problem (3.1) is

$$
\begin{aligned}
&V(S,t) \\
&= e^{\alpha x + \beta\bar{\tau}}\frac{1}{2\sqrt{\pi\bar{\tau}}}\int_{\ln B_l}^{\infty} e^{-\alpha\eta}V_T\left(e^\eta\right)\left(e^{-(x-\eta)^2/4\bar{\tau}} - e^{-(x-2\ln B_l+\eta)^2/4\bar{\tau}}\right)d\eta \\
&= e^{-2r\bar{\tau}/\sigma^2}\frac{1}{2\sqrt{\pi\bar{\tau}}}\int_{\ln B_l}^{\infty} V_T\left(e^\eta\right)e^{\alpha(x-\eta)-\alpha^2\bar{\tau}} \\
&\qquad\qquad\qquad \times \left(e^{-(x-\eta)^2/4\bar{\tau}} - e^{-(x-2\ln B_l+\eta)^2/4\bar{\tau}}\right)d\eta \\
&= e^{-r(T-t)}\frac{1}{2\sqrt{\pi\bar{\tau}}}\int_{B_l}^{\infty} V_T\left(S'\right)\left(e^{-\left[(\ln(S'/S))^2+4\alpha\bar{\tau}\ln(S'/S)+4\alpha^2\bar{\tau}^2\right]/4\bar{\tau}}\right. \\
&\qquad\qquad\left. - e^{-\left[(\ln(S'/S)-2\ln(B_l/S))^2+4\alpha\bar{\tau}\ln(S'/S)+4\alpha^2\bar{\tau}^2\right]/4\bar{\tau}}\right)\frac{dS'}{S'} \\
&= e^{-r(T-t)}\frac{1}{2\sqrt{\pi\bar{\tau}}}\int_{B_l}^{\infty} V_T\left(S'\right)\left(e^{-\left(\ln(S'/S)+2\alpha\bar{\tau}\right)^2/4\bar{\tau}}\right. \\
&\qquad\qquad\left. - e^{-\left[(\ln(S'/S)-2\ln(B_l/S)+2\alpha\bar{\tau})^2+8\alpha\bar{\tau}\ln(B_l/S)\right]/4\bar{\tau}}\right)\frac{dS'}{S'} \\
&= e^{-r(T-t)}\frac{1}{\sigma\sqrt{2\pi(T-t)}}\int_{B_l}^{\infty} V_T\left(S'\right)\left(e^{-\left[\ln(S'/S)-(r-D_0-\sigma^2/2)(T-t)\right]^2/2\sigma^2(T-t)}\right. \\
&\qquad\left. - (B_l/S)^{2(r-D_0-\sigma^2/2)/\sigma^2}\,e^{-\left[\ln(S'S/B_l^2)-(r-D_0-\sigma^2/2)(T-t)\right]^2/2\sigma^2(T-t)}\right)\frac{dS'}{S'}.
\end{aligned}
$$

Let us define

$$G_1\left(S',T;S,t,B_l\right)$$

$$= \frac{1}{S'\sigma\sqrt{2\pi\left(T-t\right)}}\left[e^{-\left[\ln(S'/S)-\left(r-D_0-\sigma^2/2\right)(T-t)\right]^2/2\sigma^2(T-t)}\right.$$

$$\left. -\left(B_l/S\right)^{2\left(r-D_0-\sigma^2/2\right)/\sigma^2}e^{-\left[\ln(S'S/B_l^2)-\left(r-D_0-\sigma^2/2\right)(T-t)\right]^2/2\sigma^2(T-t)}\right].$$

Then

$$V\left(S,t\right) = e^{-r(T-t)}\int_{B_l}^{\infty}V_T\left(S'\right)G_1\left(S',T;S,t,B_l\right)dS'.$$

G_1 is usually called Green's function[1] of the problem (3.1), which represents a down-and-out option.

Noticing that Green's function for a vanilla option is

$$G(S',T;S,t) = \frac{1}{S'\sigma\sqrt{2\pi\left(T-t\right)}}e^{-\left[\ln(S'/S)-\left(r-D_0-\sigma^2/2\right)(T-t)\right]^2/2\sigma^2(T-t)},$$

we see

$$G_1\left(S',T;S,t,B_l\right)$$

$$= G\left(S',T;S,t\right) - \left(B_l/S\right)^{2\left(r-D_0-\sigma^2/2\right)/\sigma^2}G\left(S',T;B_l^2/S,t\right).$$

For a down-and-out call option, $V_T\left(S'\right) = \max\left(S'-E,0\right)$. Therefore, for $E \geq B_l$ and $S \geq B_l$,

$$c_o\left(S,t\right) = e^{-r(T-t)}\int_{B_l}^{\infty}\max\left(S'-E,0\right)G_1\left(S',T;S,t,B_l\right)dS'$$

$$= e^{-r(T-t)}\int_0^{\infty}\max\left(S'-E,0\right)G_1\left(S',T;S,t,B_l\right)dS'$$

$$= c\left(S,t\right) - \left(\frac{B_l}{S}\right)^{2\left(r-D_0-\sigma^2/2\right)/\sigma^2}c\left(\frac{B_l^2}{S},t\right). \tag{3.3}$$

Here, we have used the following fact. The function $\max\left(S'-E,0\right)$ is equal to zero for any $S' \in [0,B_l]$ because of $B_l \leq E$. Therefore, the value of integral

---

[1] Actually, $G_1\left(S',T;S,t,B_l\right)dS'$ is the probability of the price at time T being in $[S',S'+dS']$ with the lowest price during the time period $[t,T]$ being greater than B_l. Let us explain this fact. Consider all the paths of the price during the time period $[t,T]$ that start from S at time t. For any path that hits the lower barrier, the contribution to the option value is zero because the option dies. Only those paths that never hit the lower barrier have contribution to the option value. A path that never hits the lower barrier $S = B_l$ during the time period $[t,T]$ is a path whose lowest price during the time period $[t,T]$ is greater than B_l. From the expression for $V(S,t)$, we see that the value of a down-and-out option is equal to the discounting factor times an integral of the product of the payoff function and $G_1\left(S',T;S,t,B_l\right)$ on $[B_l,\infty)$. Consequently, $G_1\left(S',T;S,t,B_l\right)dS'$ actually is the probability of the price at time T being in $[S',S'+dS']$ with the lowest price during the time period $[t,T]$ being greater than B_l. This fact will be used when we derive closed-form solutions for lookback options in Subsection 3.4.3.

will not change if replacing the lower limit B_l by zero. The formula (3.3) is a closed-form solution for a down-and-out call option. From this formula, we know that the price of a down-and-out call option is cheaper than the price of a vanilla call option. From the financial point of view, it is clear that a holder of a down-and-out call option has less rights than a holder of a vanilla call option and should pay less premium. However, if the price is always greater than B_l (which is what a holder of a call expects), then it is the same as a call. This is why a down-and-out call option is so attractive for many people. The method used to derive this formula is called the method of images because the initial condition for $x \geq \ln B_l$ is extended, and because the artificial initial condition for $x < \ln B_l$ is an image of the condition for $x > \ln B_l$.

Let us now consider a down-and-in European call option and let $c_i(S,t)$ stand for its value. The option value $c_i(S,t)$ satisfies the Black–Scholes equation for $S > B_l$, and all we need to do is to determine the correct final and boundary conditions. A down-and-in option expires worthless unless the asset price reaches the lower barrier B_l by expiry, i.e., if S has been greater than B_l right up to time T, then the option is not activated. Thus for $S > B_l$, the final condition is

$$c_i(S,T) = 0.$$

If the asset price S reaches B_l by expiry, then the option immediately turns into a vanilla call and must have the identical value as the vanilla call. The boundary condition is

$$c_i(B_l,t) = c(B_l,t).$$

Therefore, the fair value of a down-and-in option is the solution of the following final-boundary value problem:

$$\begin{cases} \dfrac{\partial c_i}{\partial t} + \dfrac{1}{2}\sigma^2 S^2 \dfrac{\partial^2 c_i}{\partial S^2} + (r - D_0)S\dfrac{\partial c_i}{\partial S} - rc_i = 0, & S \geq B_l, \quad t \leq T, \\ c_i(S,T) = 0, & S \geq B_l, \\ c_i(B_l,t) = c(B_l,t), & t \leq T. \end{cases}$$

Let

$$\bar{c}(S,t) = c(S,t) - c_i(S,t).$$

Because both $c(S,t)$ and $c_i(S,t)$ satisfy the Black–Scholes equation, $\bar{c}(S,t)$ also satisfies the same equation. The final and boundary conditions for $\bar{c}(S,t)$ is

$$\bar{c}(S,T) = c(S,T) - c_i(S,T) = c(S,T) = \max(S - E, 0),$$

and

$$\bar{c}(B_l,t) = c(B_l,t) - c_i(B_l,t) = 0.$$

Therefore, $\bar{c}(S,t)$ actually is $c_o(S,t)$. In other words, we have the identity:

$$c(S,t) = c_o(S,t) + c_i(S,t) \text{ for } S \geq B_l.$$

According to this identity and using the expression for the fair value of a down-and-out call, we have the fair value of a down-and-in option as follows:

$$c_i(S,t) = \left(\frac{B_l}{S}\right)^{2(r-D_0-\sigma^2/2)/\sigma^2} c\left(\frac{B_l^2}{S},t\right) \text{ for } S \geq B_l. \tag{3.4}$$

Obviously, $c_o(S,t) = 0$ and $c_i(S,t) = c(S,t)$ for $S < B_l$. Therefore the identity

$$c(S,t) = c_o(S,t) + c_i(S,t)$$

still holds for $S < B_l$.

For a European up-and-out put option with $B_u > E$, the solution is similar to the formula (3.3). It can also be shown that the sum of a European up-and-out put option and a European up-and-in put option equals a European vanilla put option. These problems are left for the reader as Problems 3 and 4.

Solutions in closed form can still be obtained for more complicated cases. For example, if $B_l = bEe^{-\alpha(T-t)}$, where b and α are constants and $b \in [0,1]$ and $\alpha \geq 0$, then such a solution for a down-and-out call with $D_0 = 0$ is given in the paper [57] by Merton and in the book [49] by Kwok. The solution for $D_0 \neq 0$ can still be obtained (see Problem 5).

If $B_l \geq E$, the closed-form solution for down-and-out call option can still be obtained and it is

$$\begin{aligned}
c_o = &Se^{-D_0(T-t)}N\left(\tilde{d}_1(B_l)\right) - Ee^{-r(T-t)}N\left(\tilde{d}_1(B_l) - \sigma\sqrt{T-t}\right) \\
&- (B_l/S)^{2(r-D_0-\sigma^2/2)/\sigma^2} \\
&\times \left[Se^{-D_0(T-t)}N\left(\bar{d}_1(B_l)\right) - Ee^{-r(T-t)}N\left(\bar{d}_1(B_l) - \sigma\sqrt{T-t}\right)\right],
\end{aligned}$$

where

$$\tilde{d}_1(B_l) = \left[\ln\frac{Se^{(r-D_0)(T-t)}}{B_l} + \frac{1}{2}\sigma^2(T-t)\right] \bigg/ \left(\sigma\sqrt{T-t}\right),$$

$$\bar{d}_1(B_l) = \left[\ln\frac{B_l e^{(r-D_0)(T-t)}}{S} + \frac{1}{2}\sigma^2(T-t)\right] \bigg/ \left(\sigma\sqrt{T-t}\right).$$

For an up-and-out put option with $B_u \leq E$, the closed-form solution can also be obtained. The reader is asked to derive the closed-form formulae for both cases as an exercise. If $B_l \geq E$ or $B_u \leq E$, the total price of the knock-out and knock-in options is still equal to the price of a vanilla option. Therefore, from these two formulae, we can easily obtain the closed-form solutions for a down-and-in call option with $B_l \geq E$ and for a up-and-in put option with $B_u \leq E$.

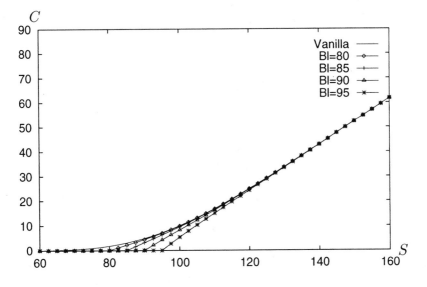

Fig. 3.1. Values of American down-and-out call options
with $B_l = 0.8$, 0.85, 0.9, 0.95, and an American vanilla call option
$(r = 0.1, D_0 = 0.05, \sigma = 0.2, T = 1$ year, and $E = 100)$

3.2.3 Formulation of American Barrier Options

A barrier option could be an American one. As an example, here we give
the formulation of an American down-and-out call option. Let $C_o(S,t)$ be its
price. If $D_0 \neq 0$, then the American down-and-out call option problem with
$B_l < E$ has a free boundary $S_f(t)$. The solution between the free boundary
$S = S_f(t)$ and $S = B_l$ is determined by the following problem:

$$
\begin{cases}
\dfrac{\partial C_o}{\partial t} + \dfrac{1}{2}\sigma^2 S^2 \dfrac{\partial^2 C_o}{\partial S^2} + (r - D_0)S\dfrac{\partial C_o}{\partial S} - rC_o = 0, \\[2ex]
\hspace{5cm} B_l \le S \le S_f(t), \quad t \le T, \\[1ex]
C_o(S,T) = \max(S - E, 0), \quad B_l \le S \le S_f(T), \\[1ex]
C_o(B_l, t) = 0, \hspace{3cm} t \le T, \\[1ex]
C_o(S_f(t), t) = S_f(t) - E, \hspace{1.5cm} t \le T, \\[1ex]
\dfrac{\partial C_o}{\partial S}(S_f(t), t) = 1, \hspace{2.2cm} t \le T, \\[2ex]
S_f(T) = \max\left(E, \dfrac{rE}{D_0}\right).
\end{cases}
\tag{3.5}
$$

The only difference between the formulations of a vanilla call and a down-and-out call is that a boundary condition at $S = B_l$ is imposed on the solution of the down-and-out call option. The problem (3.5) will be referred to as the free-boundary problem for American down-and-out call options.

For American and some other complicated European options, numerical methods might become necessary. Some details on numerical methods for these cases can be found in Chapters 6 and 7. In Fig. 3.1, the prices of American down-and-out call options with $B_l = 0.8$, 0.85, 0.9, and 0.95 and the American vanilla call option obtained by numerical methods are given. As we have pointed out in the case of European barrier call options, the price of a barrier option is cheaper than a vanilla option because the holder of a barrier option has less rights than a holder of a vanilla option. Clearly, this should still be true for the American case, and the greater B_l is, the lower the option price. This fact has been confirmed by the figure. (Obviously, the vanilla call option actually is a down-and-out call option with $B_l = 0$.)

3.2.4 Parisian Options

A Parisian option is a barrier option with the feature that a knock-in or knock-out event only occurs when the price of the underlying asset has been above or below the barrier price for a prescribed continuous length of time if sampling is done continuously or for a prescribed number of contiguous samples if sampling is done discretely (see [19], [34], [69]). As we know, if the knock-in or knock-out event can be activated by one touch of the barrier, such an event can be triggered by manipulating the price of the underlying asset for a short time. Such a thing does not happen with a Parisian option.

Let us consider European Parisian up-and-out options with an upper barrier $B_u > E$. Sampling is done continuously, and the prescribed length of time is T_d. Let t_d be the length of the time period of the stock price being continuously greater than or equal to B_u. It is clear that the option price depends on S, t_d, and t. Let $V(S, t_d, t)$ be the option price, and let time $t = 0$ represent today. We need to find the option price for $t \in [0, T]$. If $S < B_u$ at time t, then t_d must be zero for that time. Therefore, $V(S, t_d, t)$ should have a value only for $t_d = 0$. Suppose that we have the stock price for $t \in [-T_d, 0]$, so t_d can be defined on $[0, T_d]$ for any $t \in [0, T]$. For $t \in [0, T]$ if $S \geq B_u$ and $t_d \geq T_d$, then $V(S, t_d, t)$ must be zero. Therefore, for any $t \in [0, T]$, we only need to find $V(S, t_d, t)$ on the interval $[0, B_u) \times [0, 0]$ and on the domain $[B_u, \infty) \times [0, T_d]$ (see Fig. 3.2). For any $S \in [0, B_u)$ and $t_d = 0$, $V(S, 0, t)$ satisfies the Black–Scholes equation:

$$\frac{\partial V}{\partial t} + \frac{1}{2}\sigma^2 S^2 \frac{\partial^2 V}{\partial S^2} + (r - D_0) S \frac{\partial V}{\partial S} - rV = 0, \quad 0 \leq S < B_u, \quad 0 \leq t \leq T.$$

If $S \in [B_u, \infty)$, then

$$t_d(t + dt) = t_d(t) + dt,$$

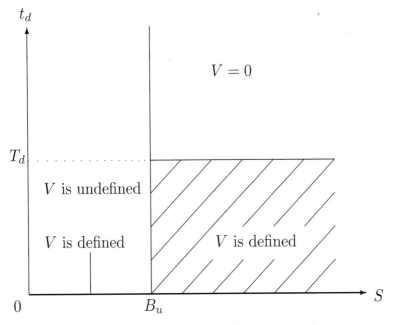

Fig. 3.2. The area of computation for Parisian up-and-out options

and

$$\frac{dt_d}{dt} = 1.$$

Therefore, according to the results in Subsection 2.9.3, $V(S, t_d, t)$ should satisfy

$$\frac{\partial V}{\partial t} + \frac{1}{2}\sigma^2 S^2 \frac{\partial^2 V}{\partial S^2} + (r - D_0) S \frac{\partial V}{\partial S} + \frac{\partial V}{\partial t_d} - rV = 0,$$

$$B_u \le S, \quad 0 \le t_d \le T_d, \quad 0 \le t \le T.$$

Putting these two cases together, we have for $t \in [0, T]$

$$\frac{\partial V}{\partial t} + \frac{1}{2}\sigma^2 S^2 \frac{\partial^2 V}{\partial S^2} + (r - D_0) S \frac{\partial V}{\partial S} + H(S - B_u)\frac{\partial V}{\partial t_d} - rV = 0,$$

$$0 \le S, \quad t_d = 0 \quad \text{and} \quad B_u \le S, \quad 0 < t_d \le T_d,$$

where

$$H(x) = \begin{cases} 0, \ x < 0, \\ 1, \ 0 \le x. \end{cases}$$

Now we consider the case of discrete sampling. Let sampling be done at $T_k = kT/K, k = 1, 2, \cdots, K$, and let the prescribed number of contiguous samples be $N < K$. In this case, for $S \ge B_u$

$$\frac{dt_d}{dt} = \frac{T}{K}\sum_{k=1}^{K}\delta(t - T_k).$$

Therefore, $V(S, t_d, t)$ should satisfy

$$\frac{\partial V}{\partial t} + \frac{1}{2}\sigma^2 S^2\frac{\partial^2 V}{\partial S^2} + (r - D_0)\,S\frac{\partial V}{\partial S} + H(S - B_u)\frac{T}{K}\sum_{k=1}^{K}\delta(t - T_k)\frac{\partial V}{\partial t_d} - rV = 0,$$

$$0 \le S, \quad t_d = 0 \quad \text{and} \quad B_u \le S, \quad 0 < t_d \le T_d, \quad 0 \le t \le T,$$

where $T_d = NT/K$. This partial differential equation can also be rewritten as follows. For any $t \ne T_k, k = 1, 2, \cdots, K$,

$$\frac{\partial V}{\partial t} + \frac{1}{2}\sigma^2 S^2\frac{\partial^2 V}{\partial S^2} + (r - D_0)\,S\frac{\partial V}{\partial S} - rV = 0,$$

$$0 \le S, \quad t_d = 0 \quad \text{and} \quad B_u \le S, \quad 0 < t_d \le T_d$$

and at $t = T_k, k = 1, 2, \cdots$ or K,

$$\begin{cases} \dfrac{\partial V}{\partial t} + \dfrac{1}{2}\sigma^2 S^2\dfrac{\partial^2 V}{\partial S^2} + (r - D_0)\,S\dfrac{\partial V}{\partial S} - rV = 0, 0 \le S < B_u, \quad t_d = 0, \\ V(S, t_d, T_k^-) = V(S, t_d + T/K, T_k^+), \qquad\qquad B_u \le S, \quad 0 \le t_d \le T_d. \end{cases}$$

Here, we have used the jump condition given in Section 2.10, and in order to have the value $V(S, T_d + T/k, T_k^+)$ with $S \ge B_u$, we need to use the fact $V(S, t_d, t) = 0$ for $S \ge B_u$ and $t_d \ge T_d$.

Suppose that at time t the price is B_u and t_d is not equal to zero and could be very close to T_d. If at time $t + dt > t$, S becomes less than B_u, then t_d becomes 0. Thus, the situation $S = B_u$ and $t_d \in [0, T_d]$ can easily become the situation $S = B_u^-$ and $t_d = 0$. Consequently, we require

$$V(B_u, t_d, t) = V(B_u, 0, t), \quad t_d \in (0, T_d).$$

In order to determine the value of an option, we also need to give the value of the option at time T. This is related to the type of the option. As an example, we consider a European Parisian up-and-out call option. Let $c_p(S, t_d, t)$ be its price. For this option, the payoff function is

$$c_p(S, t_d, T) = \begin{cases} \max(S - E, 0), & 0 \le S < B_u, \quad t_d = 0, \\ \max(S - E, 0) = S - E, & B_u \le S, \quad 0 \le t_d < T_d, \\ 0, & B_u \le S, \quad t_d = T_d. \end{cases}$$

Consequently, when sampling is done continuously, $c_p(S, t_d, t)$ is the solution of the problem

$$\begin{cases} \dfrac{\partial c_p}{\partial t} + \dfrac{1}{2}\sigma^2 S^2 \dfrac{\partial^2 c_p}{\partial S^2} + (r - D_0) S \dfrac{\partial c_p}{\partial S} + H(S - B_u) \dfrac{\partial c_p}{\partial t_d} - r c_p = 0, \\[2mm] \quad 0 \le S, \quad t_d = 0 \quad \text{and} \quad B_u \le S, \quad 0 < t_d \le T_d, \quad 0 \le t \le T, \\[2mm] c_p(S, t_d, T) = \begin{cases} \max(S - E, 0), & 0 \le S < B_u, \quad t_d = 0, \\ S - E, & B_u \le S, \quad 0 \le t_d < T_d, \\ 0, & B_u \le S, \quad t_d = T_d, \end{cases} \\[2mm] c_p(B_u, t_d, t) = c_p(B_u, 0, t), \quad t_d \in (0, T_d), \quad 0 \le t \le T, \\[2mm] c_p(S, T_d, t) = 0, \qquad\qquad B_u \le S, \quad 0 \le t \le T; \end{cases} \tag{3.6}$$

and when sampling is done discretely, $c_p(S, t_d, t)$ satisfies

$$\begin{cases} \dfrac{\partial c_p}{\partial t} + \dfrac{1}{2}\sigma^2 S^2 \dfrac{\partial^2 c_p}{\partial S^2} + (r - D_0) S \dfrac{\partial c_p}{\partial S} \\[2mm] \qquad\qquad + H(S - B_u) \dfrac{T}{K} \sum_{k=1}^{K} \delta(t - T_k) \dfrac{\partial c_p}{\partial t_d} - r c_p = 0, \\[2mm] \quad 0 \le S, \quad t_d = 0 \quad \text{and} \quad B_u \le S, \quad 0 < t_d \le T_d, \quad 0 \le t \le T, \\[2mm] c_p(S, t_d, T) = \begin{cases} \max(S - E, 0), & 0 \le S < B_u, \quad t_d = 0, \\ S - E, & B_u \le S, \quad 0 \le t_d < T_d, \\ 0, & B_u \le S, \quad t_d = T_d, \end{cases} \\[2mm] c_p(B_u, t_d, t) = c_p(B_u, 0, t), \quad t_d \in (0, T_d), \quad 0 \le t \le T, \\[2mm] c_p(S, T_d, t) = 0, \qquad\qquad B_u \le S, \quad 0 \le t \le T. \end{cases} \tag{3.7}$$

Table 3.1. Parisian up-and-out call option prices

($r = 0.1$, $D_0 = 0.05$, $\sigma = 0.25$, $E = 100$, $T = 0.5$, and $B_u = 150$)

$T_d \backslash S$	100	120	150
0	6.8669	12.1036	0
0.004	7.1806	13.9689	2.5134
0.02	7.3930	15.1938	4.8226
0.04	7.5759	16.6243	7.3770
0.08	7.7836	18.3608	11.3442

Solving (3.6) or (3.7), we can have the price of Parisian up-and-out call options. In Table 3.1 (see [53]), the prices of some Parisian up-and-out call options are given. From there, we see that the larger the parameter T_d, the higher the Parisian up-and-out call option price. The financial reason is clear.

The larger the parameter T_d, the less the chance of the event of "up-and-out," so the higher the option price.

3.3 Asian Options

3.3.1 Average Strike, Average Price, and Double Average Options

The Asian options are another type of popular path-dependent options. Such a call option can be used by a company to reduce its risk in purchasing raw materials as pointed out in Section 3.1. It can also reduce the risk in selling foreign currency through buying a put whose payoff depends on the difference between the exercise price and the average exchange rate. If the exchange rate drops, the company can get some compensation from the option for the loss in selling foreign currency. In practice, the asset price can be manipulated by some groups so that the asset price can be at a certain level desired by the groups for a short period. An Asian option may also protect option holders from the kind of asset price manipulation that occurs, especially near the end of the option's life.

Asian options can be divided into three types: average strike, average price, and double average options, where average price options are also called average rate options if the underlying asset is an exchange rate. In an average strike option, a payoff function depends on $\alpha S - A$ instead of $S - E$, where $\alpha \approx 1$. Here, A is some type of average price, which depends on the path of the price. Thus, it is also called a floating strike option. As usual, an option with a payoff

$$\max\left(\alpha S - A, 0\right) \tag{3.8}$$

is called an average strike call option and an option with a payoff

$$\max\left(A - \alpha S, 0\right) \tag{3.9}$$

is called an average strike put option.

The payoff of an average price option is a function of $A - E$, i.e., A is in the position of S. Sometimes, it is also called an average value option. Here, the strike price is fixed. Therefore, it is called a fixed strike Asian option as well. For an average price call option, its payoff is

$$\max\left(A - E, 0\right), \tag{3.10}$$

whereas for an average price put option, the payoff is

$$\max\left(E - A, 0\right). \tag{3.11}$$

A double average option has a payoff function of $A - A_1$, where A is an average over one period $[T_s, T_e]$ and A_1 is an average over another period

$[T_{s_1}, T_{e_1}]$. Here, we assume $T_e \leq T_{s_1}$. In what follows, a double average call option is referred to as an option with a payoff

$$\max(A_1 - A, 0) \tag{3.12}$$

and a double average put option is an option with a payoff

$$\max(A - A_1, 0). \tag{3.13}$$

3.3.2 Continuously and Discretely Sampled Arithmetic Averages

Sampling for an arithmetic average may be done either continuously or discretely. Suppose $S(t)$ is the asset price at time t. A continuously sampled average over $[T_s, t] \subset [T_s, T_e]$ is given by

$$A = \frac{1}{t - T_s} \int_{T_s}^{t} S(\tau) \, d\tau.$$

A discretely sampled average over $t_1, t_2, \cdots, t_k$ is

$$A = \frac{1}{k} \sum_{i=1}^{k} S(t_i).$$

In the following, we assume that over $[T_s, T_e]$ the price is sampled K times at $t_1, t_2, \cdots, t_K$, where $T_s = t_1 < t_2 < \cdots < t_K = T_e$. Assume $t_k \leq t < t_{k+1}$. Then, the discretely sampled average over $t_1, t_2, \cdots, t_k$ can be rewritten as an average over $[T_s, t] \subset [T_s, T_e]$ with a weight function

$$A = \frac{1}{\int_{T_s}^{t} \sum_{i=1}^{k} \delta(\tau - t_i) \, d\tau} \int_{T_s}^{t} S(\tau) \sum_{i=1}^{k} \delta(\tau - t_i) \, d\tau,$$

$$= \frac{1}{\int_{T_s}^{t} \sum_{i=1}^{K} \delta(\tau - t_i) \, d\tau} \int_{T_s}^{t} S(\tau) \sum_{i=1}^{K} \delta(\tau - t_i) \, d\tau,$$

where $\delta(t)$ is the Dirac delta function defined in Subsection 2.2.2. Therefore, an average price over $[T_s, t] \subset [T_s, T_e]$ can be written as

$$A = \frac{1}{\int_{T_s}^{t} f(\tau) \, d\tau} \int_{T_s}^{t} S(\tau) f(\tau) \, d\tau,$$

where

$$f(\tau) = \begin{cases} 1, & \text{if sampled continuously} \\ \sum_{i=1}^{K} \delta(\tau - t_i), & \text{if sampled at } t_1, t_2, \cdots, t_K. \end{cases}$$

Here, A and f are defined on $[T_s, T_e]$. We can extend the domain of A from $[T_s, T_e]$ to $[0, T] \supset [T_s, T_e]$ by defining

$$A = \frac{1}{\int_0^t f(\tau)\, d\tau} \int_0^t S(\tau) f(\tau)\, d\tau, \qquad (3.14)$$

where

$$f(\tau) = \begin{cases} 0, & \text{if } \tau \notin [T_s, T_e], \\ 1, & \text{if } \tau \in [T_s, T_e] \text{ and sampled continuously}, \\ \sum_{i=1}^{K} \delta(\tau - t_i), & \text{if sampled at } t_1, t_2, \cdots, t_K. \end{cases}$$

If we give another interval $[T_{s_1}, T_{e_1}]$ or K_1 specified times satisfying $T_{s_1} = t_{11} < t_{12} < \cdots < t_{1K_1} = T_{e_1}$, then we can define another function whose domain is $[0, T]$:

$$A_1 = \frac{1}{\int_0^t f_1(\tau)\, d\tau} \int_0^t S(\tau) f_1(\tau)\, d\tau, \qquad (3.15)$$

where

$$f_1(\tau) = \begin{cases} 0, & \text{if } \tau \notin [T_{s_1}, T_{e_1}], \\ 1, & \text{if } \tau \in [T_{s_1}, T_{e_1}] \text{ and sampled continuously}, \\ \sum_{i=1}^{K_1} \delta(\tau - t_{1i}), & \text{if sampled at } t_{11}, t_{12}, \cdots, t_{1K_1}. \end{cases}$$

Let us define I and I_1 as follows:

$$I = \frac{1}{\int_0^T f(\tau)\, d\tau} \int_0^t S(\tau) f(\tau)\, d\tau \qquad (3.16)$$

and

$$I_1 = \frac{1}{\int_0^T f_1(\tau)\, d\tau} \int_0^t S(\tau) f_1(\tau)\, d\tau. \qquad (3.17)$$

Because

$$A = I\frac{\int_0^T f(\tau)\,d\tau}{\int_0^t f(\tau)\,d\tau},$$

$$A_1 = I_1\frac{\int_0^T f_1(\tau)\,d\tau}{\int_0^t f_1(\tau)\,d\tau}$$

and

$$\frac{\int_0^T f(\tau)\,d\tau}{\int_0^t f(\tau)\,d\tau} \quad \text{and} \quad \frac{\int_0^T f_1(\tau)\,d\tau}{\int_0^t f_1(\tau)\,d\tau}$$

are given functions of t, we can replace A by I or A_1 by I_1 as an independent variable. Furthermore, we will discover in the next two subsections that for some cases, it is more convenient to use I or $D_I = I - I_1$ as an independent variable.

3.3.3 Derivation of Equations

Consider an option whose value V depends on three independent variables, say, S, Y, t. Suppose S satisfies

$$dS = \mu S\,dt + \sigma S\,dX \tag{3.18}$$

and for Y we have

$$dY = g(S,Y,t)\,dt. \tag{3.19}$$

From Subsection 2.9.3, we know that $V(S,Y,t)$ satisfies

$$\frac{\partial V}{\partial t} + \frac{1}{2}\sigma^2 S^2\frac{\partial^2 V}{\partial S^2} + (r - D_0)S\frac{\partial V}{\partial S} + g\frac{\partial V}{\partial Y} - rV = 0. \tag{3.20}$$

If we take

$$Y = A, \quad I \quad \text{or} \quad D_I = I - I_1,$$

from (3.14), (3.16), and (3.17) we have

$$g = \begin{cases} \dfrac{S - A}{\int_0^t f(\tau)\,d\tau}f(t), & \text{if} \quad Y = A, \\[2ex] \dfrac{1}{\int_0^T f(\tau)\,d\tau}Sf(t), & \text{if} \quad Y = I, \\[2ex] \dfrac{1}{\int_0^T f(\tau)\,d\tau}Sf(t) - \dfrac{1}{\int_0^T f_1(\tau)\,d\tau}Sf_1(t), & \text{if} \quad Y = D_I. \end{cases}$$

Therefore for these cases, we can have a partial differential equation that involves, besides S and t, A, I or D_I respectively. For example, if the independent variables are S, A, t, and the average is measured continuously, then $V(S, A, t)$ satisfies

$$
\begin{cases}
\dfrac{\partial V}{\partial t} + \dfrac{1}{2}\sigma^2 S^2 \dfrac{\partial^2 V}{\partial S^2} + (r - D_0) S \dfrac{\partial V}{\partial S} + \dfrac{S - A}{t - T_s} \dfrac{\partial V}{\partial A} - rV = 0, \\
\qquad\qquad\qquad\qquad\text{if}\quad t \in [T_s, T_e], \\[2mm]
\dfrac{\partial V}{\partial t} + \dfrac{1}{2}\sigma^2 S^2 \dfrac{\partial^2 V}{\partial S^2} + (r - D_0) S \dfrac{\partial V}{\partial S} - rV = 0, \\
\qquad\qquad\qquad\qquad\text{if}\quad t \notin [T_s, T_e];
\end{cases}
\tag{3.21}
$$

If the independent variables are S, I, t, and the average is measured discretely, then $V(S, I, t)$ satisfies

$$
\frac{\partial V}{\partial t} + \frac{1}{2}\sigma^2 S^2 \frac{\partial^2 V}{\partial S^2} + (r - D_0) S \frac{\partial V}{\partial S} + \frac{S}{K}\sum_{i=1}^{K} \delta(t - t_i)\frac{\partial V}{\partial I} - rV = 0; \tag{3.22}
$$

and if the independent variables are S, D_I, t, and the average is measured discretely, then $V(S, D_I, t)$ satisfies

$$
\frac{\partial V}{\partial t} + \frac{1}{2}\sigma^2 S^2 \frac{\partial^2 V}{\partial S^2} + (r - D_0) S \frac{\partial V}{\partial S}
$$
$$
+ S\left[\frac{1}{K}\sum_{i=1}^{K}\delta(t - t_i) - \frac{1}{K_1}\sum_{i=1}^{K_1}\delta(t - t_{1i})\right]\frac{\partial V}{\partial D_I} - rV = 0. \tag{3.23}
$$

For average strike and average price options, payoff functions at time T are in the form of $f(S, A)$ or $f(S, I)$, so for a European option we can use (3.21) or (3.22) and the payoff function to determine the value of the option. For double average options, payoff functions at time T are in the form of $f(A - A_1) = f(I - I_1) = f(D_I)$, so for a European option we can use (3.23) and the payoff to compute the option price.

3.3.4 Reducing to One-Dimensional Problems

In the problems mentioned above, there are three independent variables. Usually they are called two-dimensional problems as besides t, the values depend on two independent variables. In many cases, such an Asian option problem can be reduced to a one-dimensional problem, i.e., for a fixed t, the solution depends only on one independent variable. If a problem can be reduced to a one-dimensional problem, the amount of computation needed to obtain numerical solutions will be greatly decreased.

Let us consider an average strike call option. For this case, the payoff function is $\max(\alpha S - A, 0)$. Let

$$\eta = \frac{A}{S} \quad \text{and} \quad W = \frac{V}{S},$$

where the function W actually is the option value in units of the stock price, and consider the case of continuous sampling. Suppose W is a function of η, t. Because

$$\begin{cases} \dfrac{\partial V}{\partial t} = S\dfrac{\partial W}{\partial t}, \\[2mm] \dfrac{\partial V}{\partial S} = S\dfrac{\partial W}{\partial \eta}\dfrac{\partial \eta}{\partial S} + W = W - \eta\dfrac{\partial W}{\partial \eta}, \\[2mm] \dfrac{\partial^2 V}{\partial S^2} = \dfrac{\eta^2}{S}\dfrac{\partial^2 W}{\partial \eta^2}, \\[2mm] \dfrac{\partial V}{\partial A} = \dfrac{\partial W}{\partial \eta}, \end{cases} \tag{3.24}$$

from (3.21) we know that W satisfies

$$\begin{cases} \dfrac{\partial W}{\partial t} + \dfrac{1}{2}\sigma^2\eta^2\dfrac{\partial^2 W}{\partial \eta^2} + \left[(D_0 - r)\eta + \dfrac{1-\eta}{t-T_s}\right]\dfrac{\partial W}{\partial \eta} - D_0 W = 0, \\[3mm] \hspace{5cm} \text{if } t \in [T_s, T_e], \\[3mm] \dfrac{\partial W}{\partial t} + \dfrac{1}{2}\sigma^2\eta^2\dfrac{\partial^2 W}{\partial \eta^2} + (D_0 - r)\eta\dfrac{\partial W}{\partial \eta} - D_0 W = 0, \\[3mm] \hspace{5cm} \text{if } t \notin [T_s, T_e] \end{cases} \tag{3.25}$$

and from the payoff we have

$$W(\eta, T) = \max(\alpha - \eta, 0). \tag{3.26}$$

Equation (3.25) with the final condition (3.26) has a solution, so for a European average strike call option, we only need to solve a one-dimensional problem. For a European average strike put option, the only difference is the payoff.

In the case of an average being measured discretely and $T_s = 0$, the procedure above can still be used to reduce an original average strike option to a one-dimensional problem (see Problem 7).

For an average price call option, the payoff is

$$\max(A - E, 0).$$

We cannot use the transformation mentioned above to reduce the problem into a one-dimensional problem. However, if we consider the fact that the value of the option is a function of S, I, t, and use the transformations

$$\eta = \frac{I - E}{S}, \quad W = \frac{V}{S},$$

then for such a European option, we still only need to solve a one-dimensional problem in order to get its price. Here, taking a discretely sampled average price call option as an example, we explain the situation. Let $V = V(S, I, t)$, then V is the solution of the problem

$$
\begin{cases}
\dfrac{\partial V}{\partial t} + \dfrac{1}{2}\sigma^2 S^2 \dfrac{\partial^2 V}{\partial S^2} + (r - D_0)\, S \dfrac{\partial V}{\partial S} + \dfrac{S}{K}\sum_{i=1}^{K} \delta\,(t - t_i) \dfrac{\partial V}{\partial I} - rV = 0, \\[2mm]
\qquad 0 \le S < \infty, \quad 0 \le I < \infty, \quad t \le T, \\[2mm]
V\,(S, I, T) = \max\,(A - E, 0) = \max\,(I - E, 0), \\[2mm]
\qquad 0 \le S < \infty, \quad 0 \le I < \infty.
\end{cases}
$$

Let $\eta = \dfrac{I - E}{S}$, $W = \dfrac{V}{S}$. In this case, the first three relations in (3.24) are still true and

$$
\frac{\partial V}{\partial I} = \frac{\partial W}{\partial \eta}.
$$

Furthermore, we have

$$
W\,(\eta, T) = \max\,(\eta, 0)\,.
$$

Therefore, $W\,(\eta, T)$ satisfies

$$
\begin{cases}
\dfrac{\partial W}{\partial t} + \dfrac{1}{2}\sigma^2 \eta^2 \dfrac{\partial^2 W}{\partial \eta^2} + \left[(D_0 - r)\,\eta + \dfrac{1}{K}\sum_{i=1}^{K} \delta\,(t - t_i) \right] \dfrac{\partial W}{\partial \eta} \\[2mm]
- D_0 W = 0, \qquad\qquad -\infty < \eta < \infty, \quad t \le T, \\[2mm]
W\,(\eta, T) = \max\,(\eta, 0), \quad -\infty < \eta < \infty.
\end{cases}
\tag{3.27}
$$

That is, this problem can be reduced to a one-dimensional problem. For such a put option or for such a continuously sampled option, the situation is similar.

For a European double average option, we can assume that the value of such an option is a function of S, D_I, t. In this case, we use the transformations

$$
\eta = \frac{D_I}{S} \quad \text{and} \quad W = \frac{V}{S},
$$

then determining the option price can be reduced to solving a one-dimensional problem, and W satisfies

$$
\begin{cases}
\dfrac{\partial W}{\partial t} + \dfrac{1}{2}\sigma^2 \eta^2 \dfrac{\partial^2 W}{\partial \eta^2} + \left[(D_0 - r)\,\eta + \dfrac{1}{K}\sum_{i=1}^{K} \delta\,(t - t_i) \right. \\[4mm]
\qquad\qquad \left. - \dfrac{1}{K_1}\sum_{i=1}^{K_1} \delta\,(t - t_{1i}) \right] \dfrac{\partial W}{\partial \eta} - D_0 W = 0, \\[4mm]
\qquad\qquad\qquad -\infty < \eta < \infty, \quad t \le T, \\[2mm]
W\,(\eta, T) = \max\,(\eta, 0) \ \text{or} \ \max\,(-\eta, 0), \quad -\infty < \eta < \infty.
\end{cases}
\tag{3.28}
$$

Reducing an Asian option problem to a one-dimensional problem can be done in other ways. For example, see [42], [71], [61], [49], [2], and [69].

3.3.5 Jump Conditions

As we can see from Subsection 3.3.4, when the average is measured discretely, at the time a sample is taken, i.e., at $t = t_i \in \mathcal{T} \equiv \{t_1, t_2, \cdots, t_K\}$ or $t = t_{1i} \in \mathcal{T}_1 \equiv \{t_{11}, t_{12}, \cdots, t_{1K_1}\}$, W satisfies

$$\frac{\partial W}{\partial t} + \frac{1}{K}\delta\left(t - t_i\right)\frac{\partial W}{\partial \eta} = 0$$

or

$$\frac{\partial W}{\partial t} - \frac{1}{K_1}\delta\left(t - t_{i1}\right)\frac{\partial W}{\partial \eta} = 0$$

respectively, as the other terms in the equations can be neglected in this case. From Subsection 2.10.2, we know that W fulfills the relation

$$W\left(\eta, t_i^-\right) = W\left(\eta + \frac{1}{K}, t_i^+\right) \tag{3.29}$$

for $t_i \in \mathcal{T}$ and

$$W\left(\eta, t_{1i}^-\right) = W\left(\eta - \frac{1}{K_1}, t_{1i}^+\right) \tag{3.30}$$

for $t_{1i} \in \mathcal{T}_1$. These relations will be referred to as the jump conditions for Asian options with a discrete average and be used when W is determined if a discretely sampled arithmetic average is adopted.

3.3.6 American Asian Options

Some Asian options could be American style. Let us consider an American average strike call option with continuous sampling. Suppose $[T_s, T_e] = [0, T]$. In this case, $V(S, A, t)$ needs to satisfy the following constraint on American average strike call options:

$$V(S, A, t) \geq \max\left(\alpha S - A, 0\right) \quad \text{for} \quad t \in [0, T],$$

which is equivalent to

$$W(\eta, t) \geq \max\left(\alpha - \eta, 0\right) \quad \text{for} \quad t \in [0, T].$$

Thus, for an American average strike call option, we still only need to solve a one-dimensional problem. From (3.25), when we define

$$\mathbf{L}_{a,t} = \frac{1}{2}\sigma^2\eta^2\frac{\partial^2}{\partial \eta^2} + \left[(D_0 - r)\eta + \frac{1 - \eta}{t}\right]\frac{\partial}{\partial \eta} - D_0,$$

this American option should be the solution of the following linear complementarity problem:

$$
\begin{cases}
\left(\dfrac{\partial W}{\partial t} + \mathbf{L}_{a,t}W\right)\left[W\left(\eta, t\right) - \max\left(\alpha - \eta, 0\right)\right] = 0, \\[2ex]
\dfrac{\partial W}{\partial t} + \mathbf{L}_{a,t}W \leq 0, \\[2ex]
W\left(\eta, t\right) - \max\left(\alpha - \eta, 0\right) \geq 0, \\[2ex]
W\left(\eta, T\right) = \max\left(\alpha - \eta, 0\right)
\end{cases}
$$

for $0 \leq \eta$ and $t \leq T$.

Now let us reformulate this problem as a free-boundary problem. In order to do this, we need to find out how many free boundaries it has and where they are located at time T. Theorem 2.1 in Section 2.5 tells us that these locations are the boundaries between the regions where

$$
\left(\frac{\partial}{\partial t} + \mathbf{L}_{a,T}\right) \max\left(\alpha - \eta, 0\right) \geq 0
$$

and the regions where

$$
\left(\frac{\partial}{\partial t} + \mathbf{L}_{a,T}\right) \max\left(\alpha - \eta, 0\right) < 0.
$$

First, we consider the case $\eta > \alpha$. Because $\max\left(\alpha - \eta, 0\right) = 0$, we have

$$
\left(\frac{\partial}{\partial t} + \mathbf{L}_{a,T}\right) \max\left(\alpha - \eta, 0\right) = 0.
$$

Now let us look at the case $\eta < \alpha$. In this case,

$$
\left(\frac{\partial}{\partial t} + \mathbf{L}_{a,T}\right) \max\left(\alpha - \eta, 0\right) = -\left[\left(D_0 - r\right)\eta + \frac{1-\eta}{T}\right] - D_0\left(\alpha - \eta\right)
$$

$$
= -\alpha D_0 + r\eta - \frac{1-\eta}{T}.
$$

The inequality

$$
-\alpha D_0 + r\eta - \frac{1-\eta}{T} > 0
$$

is equivalent to

$$
\eta > \frac{1 + \alpha D_0 T}{1 + rT}.
$$

Thus, when $\dfrac{1 + \alpha D_0 T}{1 + rT} < \alpha$, there exists an interval $\left(\dfrac{1 + \alpha D_0 T}{1 + rT}, \alpha\right)$ where $\left(\dfrac{\partial}{\partial t} + \mathbf{L}_{a,T}\right) \max\left(\alpha - \eta, 0\right) > 0$. Therefore, if $\eta > \min\left(\alpha, \dfrac{1 + \alpha D_0 T}{1 + rT}\right)$, then

$$\left(\frac{\partial}{\partial t} + \mathbf{L}_{a,T}\right)\max\left(\alpha - \eta, 0\right) \geq 0$$

and we can determine the solution by the partial differential equation; and if $\eta < \min\left(\alpha, \dfrac{1 + \alpha D_0 T}{1 + rT}\right)$, then

$$\left(\frac{\partial}{\partial t} + \mathbf{L}_{a,T}\right)\max\left(\alpha - \eta, 0\right) < 0$$

and the solution should be equal to $\alpha - \eta$. Consequently, there is only one free boundary at time T, the location of the free boundary is $\min\left(\alpha, \dfrac{1 + \alpha D_0 T}{1 + rT}\right)$, and in the region $\eta \geq \min\left(\alpha, \dfrac{1 + \alpha D_0 T}{1 + rT}\right)$, we need to solve the partial differential equation. Let $\eta_f(t)$ be the location function of the free boundary. Then

$$\eta_f(T) = \min\left(\alpha, \frac{1 + \alpha D_0 T}{1 + rT}\right).$$

At the free boundary, the solution of the option and its derivative should be continuous, i.e.,

$$\begin{cases} W(\eta_f, t) = \alpha - \eta_f, \\ \dfrac{\partial}{\partial \eta} W(\eta_f, t) = -1 \end{cases}$$

should hold. Therefore, if

$$\eta \geq \eta_f(t),$$

then $W(\eta, t)$ should be the solution of the free-boundary problem

$$\begin{cases} \dfrac{\partial W}{\partial t} + \dfrac{1}{2}\sigma^2\eta^2\dfrac{\partial^2 W}{\partial \eta^2} + \left[(D_0 - r)\eta + \dfrac{1-\eta}{t}\right]\dfrac{\partial W}{\partial \eta} - D_0 W = 0, \\ \qquad\qquad\qquad\qquad\qquad \eta_f(t) \leq \eta, \quad t \leq T, \\ W(\eta, T) = \max\left(\alpha - \eta, 0\right), \quad \eta_f(T) \leq \eta, \\ W(\eta_f, t) = \alpha - \eta_f, \qquad\qquad t \leq T, \\ \dfrac{\partial W}{\partial \eta}(\eta_f, t) = -1, \qquad\qquad t \leq T, \\ \eta_f(T) = \min\left(\alpha, \dfrac{1 + \alpha D_0 T}{1 + rT}\right); \end{cases}$$

(3.31)

while if

$$0 \leq \eta < \eta_f(t),$$

then

$$W(\eta, t) = \alpha - \eta.$$

This problem will be referred to as the free-boundary problem for American average strike call options. For an American average strike put option, the two-dimensional problem can also be reduced into a one-dimensional problem. Furthermore the one-dimensional linear complementarity problem can be converted into a free-boundary problem. We leave this as Problem 8 for the reader to derive.

For an American average price call option, $V(S, I, t)$ satisfies

$$V(S, I, t) \geq \max(A - E, 0)$$

or

$$\frac{V(S, I, t)}{S} \geq \frac{\max(A - E, 0)}{S}.$$

The right-hand side cannot be a function of $\eta = (I - E)/S$ and t. Therefore, for an American average price call option, we cannot use the method described in Subsection 3.3.4 for reducing the problem to a one-dimensional problem. For an American average rate put option, the situation is the same. This means that it is necessary to solve a two-dimensional problem in these cases.

3.3.7 Some Examples

According to the equations, final conditions, boundary conditions, and jump conditions given in Subsections 3.3.4–3.3.6, and using the numerical methods described in Chapters 6–7, we can obtain $W(\eta, t)$ numerically and furthermore find the option price by $V = SW$. In what follows, we give some results in the forms of tables and figures.

Table 3.2. Average strike put option prices for various α

$(r = 0.05, D_0 = 0, \sigma = 0.2, S = 100, T = 1, T_s = 0.1, T_e = 1.0, K = 10,$
and the payoff $= \max(A - \alpha S, 0))$

Parameter α	"Exact" solution
0.900	8.981655
0.925	7.175189
0.950	5.599918
0.975	4.267895
1.000	3.176202
1.025	2.308797
1.050	1.640145
1.075	1.139517
1.100	0.774976

Table 3.3. Average price call option prices for various strike prices

$(r = 0.05,\ D_0 = 0,\ \sigma = 0.2,\ S = 100,\ T = 1,\ T_s = 0.1,\ T_e = 1.0,\ K = 10,$
and the payoff $= \max(A - E, 0))$

Strike price E	"Exact" solution
90.0	12.985323
92.5	11.050426
95.0	9.269009
97.5	7.659745
100.0	6.234515
102.5	4.997539
105.0	3.945496
107.5	3.068492
110.0	2.351591

Table 3.4. Double average call option prices with various sampling intervals Δt

$(r = 0.05,\ D_0 = 0,\ \sigma = 0.2,\ S = 100,\ T = 1,$
$T_s = \Delta t,\ T_e = 0.5,\ T_{s_1} = 0.5 + \Delta t,\ T_{e_1} = 1,$
and the payoff $= \max(A_1 - A, 0))$

Sampling interval Δt (in years)	"Exact" solution
0.1	5.872133
0.05	5.831998
0.025	5.820122
0.0125	5.816244
0.00625	5.814820
0.003125	5.814237
0.0015625	5.813978

In Table 3.2, we list the prices of European average strike put options with various α. The other parameters are $r = 0.05$, $D_0 = 0$, $\sigma = 0.2$, $S = 100$, and $T = 1$. The number of discrete samplings K is 10 and $t_k = k/K$, $k = 1, 2, \cdots, K$. From Table 3.2, we see that the price decreases as α increases. The reason for this fact is clear: when α increases, the money the holder of the option gets at expiry, the payoff $\max(A - \alpha S)$, decreases or does not change.

Table 3.3 shows the results of European average price call options with various strike prices. There, the parameters are the same as those in Table 3.2. From Table 3.3, we see that the option price is a decreasing function of the strike price E. The reason is as follows. When E increases, the money its holder gets at expiry, the payoff $\max(A - E, 0)$, decreases or does not change.

European double average call option prices for various sampling intervals Δt are listed in Table 3.4. For a given $\Delta t = 0.1/2^{n-1}$, n being a positive integer, $t_k = k\Delta t$, $k = 1, 2, \cdots, 10 \times 2^{n-2}$, and $t_{1k_1} = 0.5 + k_1\Delta t$, $k_1 = 1, 2, \cdots, 10 \times 2^{n-2}$. The data show that when the interval goes to zero, the price tends to $5.813\cdots$. As we know, when the interval goes to zero, discrete

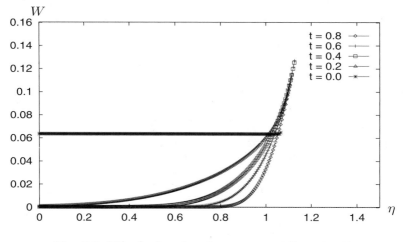

Fig. 3.3. $W(\eta, t)$ of an American average strike put option
($r = 0.1$, $D_0 = 0.1$, $\sigma = 0.2$, and $\alpha = 1$)

sampling becomes continuous sampling. Therefore, the limit should be the price of the option with continuous sampling. From the data, we also see that if the length of interval is less than 0.025 (6.25 business days), then the difference of the prices between an option with discrete sampling and the option with continuous sampling is 10^{-2}. Therefore, if such an error is acceptable, we can use a continuous model to replace a discrete sampling with an interval less than 0.025. Usually, the CPU time needed for continuous sampling is less than that needed for a discrete model with a small Δt, so such a replacement can save CPU time needed. Finally, we would like to point out that in these tables, "exact" solution means that the error of these values is about 10^{-6}.

In Fig. 3.3, for an American average strike put option, the value of $W(\eta, t)$ as a function of η is given. The price of the option is $V(S, A, t) = SW(A/S, t)$. Thus if $t > 0$, then S and A need to be given, whereas if $t = 0$, then $A = S$ and only S needs to be given in order to find the price from the figure.

3.4 Lookback Options

3.4.1 Equations for Lookback Options

Sometimes the payoff of a derivative product depends on the maximum or minimum realized asset price over the life of the option. Such an option is called a lookback option. If the strike price in the payoff depends on the maximum or minimum, then the lookback option is called a lookback strike option or a lookback option with a floating strike; whereas if the stock price in the payoff is replaced by the maximum or minimum, then it is called a lookback price option or a lookback option with fixed strike. When the underlying asset is an exchange rate, a lookback price option is also called a lookback rate option. If the payoff is

$$\max\left(\alpha S - L, 0\right),$$

where L is the minimum realized price and α is a constant satisfying

$$0 < \alpha \le 1,$$

then the option is called a lookback strike call option; whereas if the payoff is

$$\max\left(H - \beta S, 0\right),$$

where H is the maximum realized price and β is a constant satisfying

$$1 \le \beta,$$

then the option is called a lookback strike put option. Similarly, lookback price call and put options have the payoffs

$$\max\left(H - E, 0\right)$$

and

$$\max\left(E - L, 0\right)$$

respectively. A person who holds a European lookback strike call option with $\alpha = 1$ can buy the underlying asset at the lowest realized price at expiration time, and an investor who holds a European lookback strike put option with $\beta = 1$ can sell the asset at the highest realized price. These attractive features are the reason why there exist such options on the market.

As for Asian options, the maximum or minimum realized asset price may be measured continuously, or more commonly, discretely. If it is measured continuously, then

$$L\left(t\right) = \min_{0 \le \tau \le t} S\left(\tau\right)$$

and

$$H\left(t\right) = \max_{0 \le \tau \le t} S\left(\tau\right).$$

If the sampling is done discretely and the sampling times are $t_1, t_2, \cdots, t_K$, where

$$0 \le t_1 < t_2 < \cdots < t_K \le T,$$

then

$$L\left(t\right) = \min\left(S\left(t_1\right), \cdots, S\left(t_{i^*(t)}\right)\right)$$

and

$$H\left(t\right) = \max\left(S\left(t_1\right), \cdots, S\left(t_{i^*(t)}\right)\right),$$

where $i^*\left(t\right)$ is the number of samplings before time t.

We see that the value V of such an option depends on not only S and t but also L or H, i.e., $V = V(S, L, t)$ or $V(S, H, t)$. In what follows, let us derive the equation for the price of a lookback strike put option or a lookback price call option. For such an option, $V = V(S, H, t)$. First, suppose discrete sampling is taken. In this case

$$dH(t) = \begin{cases} 0, & \text{if } t \neq t_i, \ i = 1, 2, \cdots, K, \\ \max\left(S(t_i), H(t_i^-)\right) - H(t_i^-), & \text{if } t = t_i, \ i = 1, 2, \cdots, \text{ or } K, \end{cases}$$

and

$$\frac{dH(t)}{dt} = \sum_{i=1}^{K} \left[\max\left(S(t), H(t^-)\right) - H(t^-)\right] \delta(t - t_i),$$

where

$$H(t^-) = \lim_{\varepsilon \to 0} H(t - \varepsilon)$$

with $\varepsilon > 0$.

Suppose dS satisfies (3.18). According to the results given in Subsection 2.9.3, $V(S, H, t)$ satisfies the equation (3.20) with $g = \dfrac{dH}{dt}$, i.e.,

$$\frac{\partial V}{\partial t} + \frac{1}{2}\sigma^2 S^2 \frac{\partial^2 V}{\partial S^2} + (r - D_0) S \frac{\partial V}{\partial S}$$

$$+ \sum_{i=1}^{K} \left[\max\left(S(t), H(t^-)\right) - H(t^-)\right] \delta(t - t_i) \frac{\partial V}{\partial H} - rV = 0, \quad 0 \leq S, \ 0 \leq H.$$

This means that at $t \neq t_i$, $i = 1, 2, \cdots, K$, V fulfills

$$\frac{\partial V}{\partial t} + \frac{1}{2}\sigma^2 S^2 \frac{\partial^2 V}{\partial S^2} + (r - D_0) S \frac{\partial V}{\partial S} - rV = 0, \ 0 \leq S, \ 0 \leq H \qquad (3.32)$$

and at $t = t_i$, $i = 1, 2, \cdots,$ or K, the equation

$$\frac{\partial V}{\partial t} + \sum_{i=1}^{K} \left[\max\left(S(t), H(t^-)\right) - H(t^-)\right] \delta(t - t_i) \frac{\partial V}{\partial H} = 0, \quad 0 \leq S, \ 0 \leq H$$

holds. It is a hyperbolic equation, and the characteristic relation is

$$\frac{dH}{dt} = \sum_{i=1}^{K} \left[\max\left(S(t), H(t^-)\right) - H(t^-)\right] \delta(t - t_i).$$

According to the results given in Section 2.10, the solution of the characteristic equation is

$$H(t_i^+) - H(t_i^-) = \max\left(S(t_i), H(t_i^-)\right) - H(t_i^-)$$

or
$$H\left(t_i^+\right) = \max\left(S\left(t_i\right), H\left(t_i^-\right)\right).$$

From this relation and the results given in Section 2.10, the solution of the hyperbolic equation above is

$$V\left(S\left(t_i\right), H\left(t_i^-\right), t_i^-\right) = V\left(S\left(t_i\right), H\left(t_i^+\right), t_i^+\right)$$
$$= V\left(S\left(t_i\right), \max\left(S\left(t_i\right), H\left(t_i^-\right)\right), t_i^+\right)$$

or
$$V\left(S, H, t_i^-\right) = V\left(S, \max\left(S, H\right), t_i^+\right), \quad 0 \le S, \quad 0 \le H. \tag{3.33}$$

For simplicity, H stands for $H\left(t_i^-\right)$ here. This is a jump condition for lookback options with a discrete maximum.

Therefore, if the maximum realized asset price is measured discretely, the price of a lookback strike put option or a lookback price call option satisfies (3.32) and (3.33). Consequently, the price of a European lookback strike put option is the solution of the following problem

$$\begin{cases} \dfrac{\partial V}{\partial t} + \dfrac{1}{2}\sigma^2 S^2 \dfrac{\partial^2 V}{\partial S^2} + (r - D_0) S \dfrac{\partial V}{\partial S} - rV = 0, \\ \qquad\qquad \text{if } T \ge t \ne t_i,\ i = 1, 2, \cdots, K, \\ V\left(S, H, t_i^-\right) = V\left(S, \max\left(S, H\right), t_i^+\right), \\ \qquad\qquad \text{if } t = t_i,\ i = 1, 2, \cdots,\ \text{or } K, \\ V\left(S, H, T\right) = \max\left(H - \beta S, 0\right), \end{cases} \tag{3.34}$$

where
$$0 \le H, \quad 0 \le S, \quad \text{and} \quad \beta \ge 1.$$

For an American one, the solution must fulfill the constraint on American lookback strike put options

$$V\left(S, H, t\right) \ge \max\left(H - \beta S, 0\right). \tag{3.35}$$

Therefore, if the first equation in (3.34) gives a value of V that is less than the constraint $\max\left(H - \beta S, 0\right)$ at some point, the value should be replaced by $\max\left(H - \beta S, 0\right)$. In this case, the formulation of problem can be written as a linear complementarity problem.

For a lookback price call option, the final condition and the constraint are

$$V\left(S, H, T\right) = \max\left(H - E, 0\right) \tag{3.36}$$

and
$$V\left(S, H, t\right) \ge \max\left(H - E, 0\right) \tag{3.37}$$

respectively, and the equations for $V(S, H, t)$ still are (3.32) and (3.33).

Now let us derive the equation if continuous sampling is done. The equation (3.33) can be rewritten as

$$V\left(S,H,t_i^-\right) = \begin{cases} V\left(S,S,t_i^+\right), & \text{if } 0 \leq S,\ 0 \leq H < S, \\ V\left(S,H,t_i^+\right), & \text{if } 0 \leq S,\ S \leq H. \end{cases}$$

Therefore, from t_i^+ to t_i^-, V is unchanged if $S \leq H$ and $V\left(S,H,t_i^-\right)$ does not depend on H or $\dfrac{\partial V}{\partial H}\left(S,H,t_i^-\right) = 0$ if $0 \leq H < S$. Consequently, if $S < H$, then $V\left(S,H,t\right)$ always satisfies the equation (3.32) and if $0 \leq H < S$ and $\max\limits_{0 \leq i \leq K}\left(t_{i+1} - t_i\right)$ is very small, where we let $t_0 = 0$ and $t_{K+1} = T$, then $\dfrac{\partial V}{\partial H}$ should be very close to zero because the condition $V\left(S,H,t_i^-\right) = V\left(S,S,t_i^+\right)$ for any $H < S$ is used very frequently. The solution for $t \neq t_i$ should be smooth, thus $\dfrac{\partial V}{\partial H}\left(S,S,t\right)$ should be close to zero and becomes closer to zero when $\max\limits_{0 \leq i \leq K}\left(t_{i+1} - t_i\right)$ goes to zero, i.e., when the measure becomes continuous. Therefore, if sampling is done continuously, then the price of a European lookback strike put option is determined by the problem

$$\begin{cases} \dfrac{\partial V}{\partial t} + \dfrac{1}{2}\sigma^2 S^2 \dfrac{\partial^2 V}{\partial S^2} + \left(r - D_0\right) S \dfrac{\partial V}{\partial S} - rV = 0, \\ \qquad\qquad\qquad\qquad\qquad\qquad 0 \leq S \leq H,\ t \leq T, \\ V\left(S,H,T\right) = \max\left(H - \beta S, 0\right), \qquad 0 \leq S \leq H, \\ \dfrac{\partial V}{\partial H}\left(S,S,t\right) = 0, \qquad\qquad\qquad 0 \leq S. \end{cases} \tag{3.38}$$

For a European lookback price call option, the equation and the boundary condition are the same. The final condition in (3.38) should be replaced by (3.36). The constraint for an American option will be unchanged if the measure is changed from discrete to continuous. Thus the constraints for an American lookback strike put and price call options still are (3.35) and (3.37) respectively.

Similarly, if the minimum realized asset price is measured discretely, then the price of a European lookback strike call option satisfies

$$\begin{cases} \dfrac{\partial V}{\partial t} + \dfrac{1}{2}\sigma^2 S^2 \dfrac{\partial^2 V}{\partial S^2} + \left(r - D_0\right) S \dfrac{\partial V}{\partial S} - rV = 0, \\ \qquad\qquad\qquad\qquad \text{if } T \geq t \neq t_i, i = 1, 2, \cdots, K, \\ V\left(S,L,t_i^-\right) = V\left(S,\min\left(S,L\right),t_i^+\right), \quad \text{if } t = t_i,\ i = 1, 2, \cdots, \text{ or } K, \\ V\left(S,L,T\right) = \max\left(\alpha S - L, 0\right), \end{cases}$$

where

$$0 \leq L, \quad 0 \leq S, \quad \text{and} \quad \alpha \leq 1.$$

If the asset price is measured continuously, then the price is the solution of the problem

$$
\begin{cases}
\dfrac{\partial V}{\partial t} + \dfrac{1}{2}\sigma^2 S^2 \dfrac{\partial^2 V}{\partial S^2} + (r - D_0)\, S \dfrac{\partial V}{\partial S} - rV = 0,\ 0 \leq S, \quad 0 \leq L \leq S, \\[2mm]
\hspace{10cm} t \geq T, \\[2mm]
V\left(S, L, T\right) = \max\left(\alpha S - L, 0\right), \hspace{2cm} 0 \leq S, \quad 0 \leq L \leq S, \\[2mm]
\dfrac{\partial V}{\partial L}\left(S, S, t\right) = 0, \hspace{4.5cm} 0 \leq S.
\end{cases}
$$

For a European lookback price put option, the only change is that the final condition should be replaced by

$$V\left(S, L, T\right) = \max\left(E - L, 0\right).$$

Readers are asked to prove these results as exercises. Clearly, the constraints for such American options are

$$V\left(S, L, t\right) \geq \max\left(\alpha S - L, 0\right) \quad \text{or} \quad V\left(S, L, t\right) \geq \max\left(E - L, 0\right).$$

3.4.2 Reducing to One-Dimensional Problems

As with Asian options, some lookback option problems can be reduced to one-dimensional problems.[2] For example, a lookback strike put option can be reduced to a one-dimensional problem. We show this here.

Let

$$\eta = \frac{H}{S} \quad \text{and} \quad W = \frac{V}{S}.$$

Suppose W depends only on η and t. Using the first three relations in (3.24) and

$$\frac{\partial V}{\partial H} = \frac{\partial W}{\partial \eta},$$

the first equation in (3.34) can be rewritten as

$$S\frac{\partial W}{\partial t} + \frac{1}{2}\sigma^2 S^2 \frac{\eta^2}{S}\frac{\partial^2 W}{\partial \eta^2} + (r - D_0)\, S\left(W - \eta\frac{\partial W}{\partial \eta}\right) - rSW = 0$$

or

$$\frac{\partial W}{\partial t} + \mathbf{L}_\eta W = 0,$$

where

---

[2]The way to reduce some lookback option problems to one-dimensional problems is not unique (see [71], [49], [2], and [69]).

$$\mathbf{L}_\eta = \frac{1}{2}\sigma^2\eta^2\frac{\partial^2}{\partial\eta^2} + (D_0 - r)\,\eta\frac{\partial}{\partial\eta} - D_0,$$

and from the second and third relations in (3.34), we have

$$
\begin{aligned}
W\left(\eta, t_i^-\right) &= \frac{V\left(S, H, t_i^-\right)}{S} \\
&= \frac{V\left(S, \max\left(S, H\right), t_i^+\right)}{S} \\
&= W\left(\max\left(1, \eta\right), t_i^+\right)
\end{aligned}
$$

and

$$
\begin{aligned}
W\left(\eta, T\right) &= \frac{V\left(S, H, T\right)}{S} \\
&= \frac{\max\left(H - \beta S, 0\right)}{S} \\
&= \max\left(\eta - \beta, 0\right).
\end{aligned}
$$

Therefore, it is true that (3.34) can be reduced to the following one-dimensional problem

$$
\begin{cases}
\dfrac{\partial W}{\partial t} + \mathbf{L}_\eta W = 0, & \text{if } T \ge t \ne t_i,\ i = 1, 2, \cdots, K, \\
W\left(\eta, t_i^-\right) = W\left(\max\left(1, \eta\right), t_i^+\right), & \text{if } t = t_i,\ i = 1, 2, \cdots,\ \text{or } K, \\
W\left(\eta, T\right) = \max\left(\eta - \beta, 0\right), &
\end{cases}
\tag{3.39}
$$

where $0 \le \eta$. This is a formulation of a European lookback strike put option as a one-dimensional problem if sampling is done discretely. Similarly, (3.38) can be written as

$$
\begin{cases}
\dfrac{\partial W}{\partial t} + \mathbf{L}_\eta W = 0, & 1 \le \eta,\ t \le T, \\
W\left(\eta, T\right) = \max\left(\eta - \beta, 0\right), & 1 \le \eta, \\
\dfrac{\partial W}{\partial \eta}\left(1, t\right) = 0, & t \le T.
\end{cases}
\tag{3.40}
$$

This is a formulation of a European lookback strike put option as a one-dimensional problem if sampling is done continuously. The constraint for such an American option, (3.35), can also be rewritten in a form involving η only:

$$W\left(\eta, t\right) \ge \max\left(\eta - \beta, 0\right). \tag{3.41}$$

For a European lookback strike call option with continuous sampling, the corresponding one-dimensional problem is

$$\begin{cases} \dfrac{\partial W}{\partial t} + \mathbf{L}_\eta W = 0, & 0 \le \eta \le 1,\ t \le T, \\[2mm] W(\eta, T) = \max(\alpha - \eta, 0), & 0 \le \eta \le 1, \\[2mm] \dfrac{\partial W}{\partial \eta}(1, t) = 0, & t \le T, \end{cases}$$

where

$$\eta = L/S \quad \text{and} \quad W = V/S$$

and for a corresponding American option, the constraint is

$$W(\eta, t) \ge \max(\alpha - \eta, 0).$$

These are left for the reader to derive as exercises.

It is useful to write an American option problem as a linear complementarity problem. Let us write the American lookback strike put problem as a linear complementarity problem for the case of continuous sampling. Just like other American options we have met, the price of the American lookback strike put option is the solution of the following linear complementarity problem

$$\begin{cases} \left(\dfrac{\partial W}{\partial t} + \mathbf{L}_\eta W \right) [W - \max(\eta - \beta, 0)] = 0, & 1 \le \eta, \quad t \le T, \\[3mm] \dfrac{\partial W}{\partial t} + \mathbf{L}_\eta W \le 0, & 1 \le \eta, \quad t \le T, \\[3mm] W - \max(\eta - \beta, 0) \ge 0, & 1 \le \eta, \quad t \le T, \\[3mm] W(\eta, T) = \max(\eta - \beta, 0), & 1 \le \eta, \\[3mm] \dfrac{\partial W}{\partial \eta}(1, t) = 0, & t \le T. \end{cases}$$

For the lookback strike call option, similar results can be obtained (Problem 14).

3.4.3 Closed-Form Solutions for European Lookback Options

For some European lookback options with continuous sampling, the closed-form solutions have been found. For example, such solutions for lookback strike options with $\alpha = 1$ or $\beta = 1$ were found by Goldman, Sosin, and Gatto (see [31]). Later, Conze and Viswanathan (see [20]) derived explicit solutions for the lookback strike options with $\alpha < 1$ or $\beta > 1$ and the lookback price options. In Kwok's book [49], some details of those derivations are given.

In the following, we will derive the closed-form solution for a lookback strike call option with $\alpha \le 1$. In Section 3.1, we used the method of images to find the explicit solution for a barrier option. Here, we use another way to get an explicit solution for a lookback option.

As before, S stands for the price at time t, and L represents the minimum price during the time period $[0,t]$. Let S_T and L_T denote the price at time T and the minimum price during the time period $[0,T]$, and let L_t^T be the lowest price during the time period $[t,T]$.

Among L, L_T, and L_t^T, there is the relation $L_T = \min(L, L_t^T)$. Thus, the payoff can be written as

$$\max(\alpha S_T - L_T, 0) = \max(\alpha S_T - \min(L, L_t^T), 0).$$

In this case, S_T and L_t^T are two random variables, and the value of the European lookback strike call option, $c_{ls}(S, L, t)$, can be expressed as the expectation of the payoff times a discounting factor:

$$c_{ls}(S, L, t) = e^{-r(T-t)} E\left[\max(\alpha S_T - \min(L, L_t^T), 0)\right]$$

$$= e^{-r(T-t)} \int_0^S \int_{L_t^T}^\infty \max(\alpha S_T - \min(L, L_t^T), 0) \, g(S_T, L_t^T) \, dS_T dL_t^T,$$

where $g(S_T, L_t^T)$ is a two-dimensional probability density function in the "risk-neutral" world, and we have used the fact that the two random variables S_T and L_t^T appear only in the domain (the shaded area in Fig. 3.4): $0 \le L_t^T \le S$ and $L_t^T \le S_T$ because of $L_t^T \le S$ and $L_t^T \le S_T$. It is clear that this domain can also be written as $0 \le S_T$ and $0 \le L_t^T \le \min(S_T, S)$.

From Subsection 3.2.2, the probability of the price at time T being in $[S_T, S_T + dS_T)$ and the price during the time period $[t, T]$ never being lower than L_t^T is

$$f(S_T, L_t^T) \, dS_T,$$

where

$$f(S_T, L_t^T)$$
$$= \frac{1}{S_T \sigma \sqrt{2\pi\tau}} \left[e^{-[\ln(S_T/S) - \mu\tau]^2/2\sigma^2\tau} - (L_t^T/S)^{2\mu/\sigma^2} e^{-\left[\ln\left(S_T S/L_t^{T2}\right) - \mu\tau\right]^2/2\sigma^2\tau} \right].$$

In this expression,

$$\mu = r - D_0 - \sigma^2/2$$

and

$$\tau = T - t.$$

Thus, the probability of the event $S_T \in [S_T, S_T + dS_T)$ and $L_t^T \in [L_t^T, L_t^T + dL_t^T)$ is

$$g(S_T, L_t^T) \, dS_T dL_t^T = f(S_T, L_t^T) \, dS_T - f(S_T, L_t^T + dL_t^T) \, dS_T$$
$$= -\frac{\partial f}{\partial L_t^T} \, dS_T dL_t^T,$$

that is,

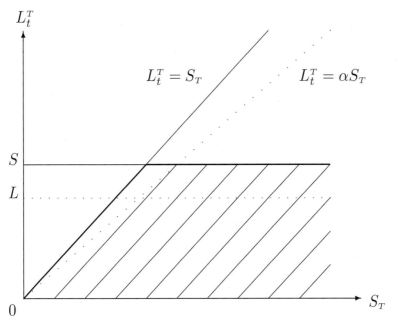

Fig. 3.4. The area of integration for a lookback strike call option

$$g\left(S_T, L_t^T\right) = -\frac{\partial f}{\partial L_t^T}.$$

Because the first term in $f\left(S_T, L_t^T\right)$ does not depend on L_t^T, $g\left(S_T, L_t^T\right)$ actually is equal to the partial derivative of the second term with respect to L_t^T, i.e.,

$$g\left(S_T, L_t^T\right) = \frac{\partial f_1\left(S_T, L_t^T\right)}{\partial L_t^T},$$

where

$$f_1\left(S_T, L_t^T\right) = \frac{1}{S_T \sigma \sqrt{2\pi\tau}} \left(\frac{L_t^T}{S}\right)^{2\mu/\sigma^2} e^{-\left[\ln\left(S_T S/L_t^{T2}\right)-\mu\tau\right]^2/2\sigma^2\tau}.$$

Now let us express $\mathrm{E}\left[\max\left(S_T - \min\left(L, L_t^T\right), 0\right)\right]$ in terms of the cumulative distribution function for the standardized normal variable. Because $L \leq S$, we have

$$\int_0^S \int_{L_t^T}^\infty \max\left(\alpha S_T - \min\left(L, L_t^T\right), 0\right) \frac{\partial f_1}{\partial L_t^T} dS_T dL_t^T$$

$$= \int_0^L \int_{L_t^T}^\infty \max\left(\alpha S_T - L_t^T, 0\right) \frac{\partial f_1}{\partial L_t^T} dS_T dL_t^T$$

$$+ \int\limits_{L}^{S} \int\limits_{L_t^T}^{\infty} \max\left(\alpha S_T - L, 0\right) \frac{\partial f_1}{\partial L_t^T} dS_T dL_t^T$$

$$= \int\limits_{0}^{L} \int\limits_{L_t^T/\alpha}^{\infty} \left(\alpha S_T - L_t^T\right) \frac{\partial f_1}{\partial L_t^T} dS_T dL_t^T + \int\limits_{L}^{S} \int\limits_{L}^{S_T} \max\left(\alpha S_T - L, 0\right) \frac{\partial f_1}{\partial L_t^T} dL_t^T dS_T$$

$$+ \int\limits_{S}^{\infty} \int\limits_{L}^{S} \max\left(\alpha S_T - L, 0\right) \frac{\partial f_1}{\partial L_t^T} dL_t^T dS_T$$

$$= \int\limits_{0}^{L/\alpha} \int\limits_{0}^{\alpha S_T} \left(\alpha S_T - L_t^T\right) \frac{\partial f_1}{\partial L_t^T} dL_t^T dS_T + \int\limits_{L/\alpha}^{\infty} \int\limits_{0}^{L} \left(\alpha S_T - L_t^T\right) \frac{\partial f_1}{\partial L_t^T} dL_t^T dS_T$$

$$+ \int\limits_{L}^{S} \max\left(\alpha S_T - L, 0\right) \left[f_1\left(S_T, S_T\right) - f_1\left(S_T, L\right)\right] dS_T$$

$$+ \int\limits_{S}^{\infty} \max\left(\alpha S_T - L, 0\right) \left[f_1\left(S_T, S\right) - f_1\left(S_T, L\right)\right] dS_T$$

$$= \int\limits_{0}^{L/\alpha} \alpha S_T f_1\left(S_T, \alpha S_T\right) dS_T - \int\limits_{0}^{L/\alpha} \int\limits_{0}^{\alpha S_T} L_t^T \frac{\partial f_1}{\partial L_t^T} dL_t^T dS_T$$

$$+ \int\limits_{L/\alpha}^{\infty} \alpha S_T f_1\left(S_T, L\right) dS_T - \int\limits_{L/\alpha}^{\infty} \int\limits_{0}^{L} L_t^T \frac{\partial f_1}{\partial L_t^T} dL_t^T dS_T$$

$$+ \int\limits_{L}^{S} \max\left(\alpha S_T - L, 0\right) f_1\left(S_T, S_T\right) dS_T$$

$$+ \int\limits_{S}^{\infty} \max\left(\alpha S_T - L, 0\right) f_1\left(S_T, S\right) dS_T - \int\limits_{L/\alpha}^{\infty} \left(\alpha S_T - L\right) f_1\left(S_T, L\right) dS_T$$

$$= \int\limits_{0}^{L/\alpha} \alpha S_T f_1\left(S_T, \alpha S_T\right) dS_T - \int\limits_{0}^{L/\alpha} \left[L_t^T f_1|_0^{\alpha S_T} - \int\limits_{0}^{\alpha S_T} f_1\left(S_T, L_t^T\right) dL_t^T \right] dS_T$$

$$- \int\limits_{L/\alpha}^{\infty} \left[L_t^T f_1|_0^{L} - \int\limits_{0}^{L} f_1\left(S_T, L_t^T\right) dL_t^T \right] dS_T$$

$$+ \int\limits_{L}^{S} \max\left(\alpha S_T - L, 0\right) f_1\left(S_T, S_T\right) dS_T$$

$$+ \int\limits_{S}^{\infty} \max\left(\alpha S_T - L, 0\right) f_1\left(S_T, S\right) dS_T + \int\limits_{L/\alpha}^{\infty} L f_1\left(S_T, L\right) dS_T$$

$$= \int\limits_{0}^{L/\alpha} \int\limits_{0}^{\alpha S_T} f_1\left(S_T, L_t^T\right) dL_t^T dS_T + \int\limits_{L/\alpha}^{\infty} \int\limits_{0}^{L} f_1\left(S_T, L_t^T\right) dL_t^T dS_T$$

$$+ \int\limits_{L}^{S} \max\left(\alpha S_T - L, 0\right) f_1\left(S_T, S_T\right) dS_T$$

$$+ \int\limits_{S}^{\infty} \max\left(\alpha S_T - L, 0\right) f_1\left(S_T, S\right) dS_T$$

$$= \int\limits_{0}^{L} \int\limits_{L_t^T/\alpha}^{\infty} f_1\left(S_T, L_t^T\right) dS_T dL_t^T + \int\limits_{L}^{S} \max\left(\alpha S_T - L, 0\right) f_1\left(S_T, S_T\right) dS_T$$

$$+ \int\limits_{S}^{\infty} \max\left(\alpha S_T - L, 0\right) f_1\left(S_T, S\right) dS_T.$$

Now let us find the result for each integral. The first integral is equal to

$$\int\limits_{0}^{L} \int\limits_{L_t^T/\alpha}^{\infty} f_1\left(S_T, L_t^T\right) dS_T dL_t^T$$

$$= \int\limits_{0}^{L} \int\limits_{L_t^T/\alpha}^{\infty} \frac{1}{S_T \sigma \sqrt{2\pi\tau}} \left(\frac{L_t^T}{S}\right)^{2\mu/\sigma^2} e^{-\left[\ln\left(S_T S/L_t^{T^2}\right) - \mu\tau\right]^2 / 2\sigma^2 \tau} dS_T dL_t^T$$

$$= \int\limits_{0}^{L} \left(\frac{L_t^T}{S}\right)^{2\mu/\sigma^2} N \left(\frac{-\ln\frac{S}{\alpha L_t^T} + \mu\tau}{\sigma\sqrt{\tau}}\right) dL_t^T$$

$$= \frac{S^{-2\mu/\sigma^2}}{2\mu/\sigma^2 + 1} \int\limits_{0}^{L} N \left(\frac{\ln\frac{\alpha L_t^T}{S} + \mu\tau}{\sigma\sqrt{\tau}}\right) d\left(L_t^T\right)^{2\mu/\sigma^2+1}$$

$$= \frac{S^{-2\mu/\sigma^2}}{2\mu/\sigma^2 + 1} \left[\left(L_t^T\right)^{2\mu/\sigma^2+1} N \left(\frac{\ln\frac{\alpha L_t^T}{S} + \mu\tau}{\sigma\sqrt{\tau}}\right)\right]\Bigg|_{0}^{L}$$

$$
-\int_0^L (L_t^T)^{2\mu/\sigma^2+1}\, dN\left(\frac{\ln\frac{\alpha L_t^T}{S}+\mu\tau}{\sigma\sqrt{\tau}}\right)\Bigg]
$$

$$
= \frac{\sigma^2 S^{-2\mu/\sigma^2}}{2\,(r-D_0)} L^{2\mu/\sigma^2+1} N\left(\frac{\ln\frac{\alpha L}{S}+\mu\tau}{\sigma\sqrt{\tau}}\right)
$$

$$
-\frac{\sigma^2 S^{-2\mu/\sigma^2}}{2\,(r-D_0)}\int_0^L (L_t^T)^{2\mu/\sigma^2+1}\frac{1}{\sqrt{2\pi}}e^{-[\ln(\alpha L_t^T/S)+\mu\tau]^2/2\sigma^2\tau}\frac{dL_t^T}{\sigma\sqrt{\tau}L_t^T}
$$

$$
= \frac{\sigma^2 S}{2\,(r-D_0)}\left(\frac{L}{S}\right)^{2(r-D_0)/\sigma^2} N\left(\frac{\ln\frac{\alpha L}{S}+\mu\tau}{\sigma\sqrt{\tau}}\right) - \frac{\sigma^2 S}{2\,(r-D_0)}\alpha^{-2\mu/\sigma^2-1}
$$

$$
\times\, e^{\left[-\mu^2+(\mu+\sigma^2)^2\right]\tau/2\sigma^2}\int_0^L \frac{1}{\sqrt{2\pi}}e^{-[\ln(\alpha L_t^T/S)-(\mu+\sigma^2)\tau]^2/2\sigma^2\tau}\frac{dL_t^T}{\sigma\sqrt{\tau}L_t^T}
$$

$$
= \frac{\sigma^2 S}{2\,(r-D_0)}\left(\frac{L}{S}\right)^{2(r-D_0)/\sigma^2} N\left(\frac{\ln\frac{\alpha L}{S}+\mu\tau}{\sigma\sqrt{\tau}}\right)
$$

$$
-\frac{\sigma^2 S}{2\,(r-D_0)}\alpha^{-2(r-D_0)/\sigma^2}e^{(r-D_0)\tau} N\left(\frac{\ln\frac{\alpha L}{S}-(\mu+\sigma^2)\tau}{\sigma\sqrt{\tau}}\right).
$$

If $S > L/\alpha$, then the second and the third integrals can be written as

$$
\int_L^S \max(\alpha S_T - L, 0)\, f_1(S_T, S_T)\, dS_T + \int_S^\infty \max(\alpha S_T - L, 0)\, f_1(S_T, S)\, dS_T
$$

$$
= \int_{L/\alpha}^S (\alpha S_T - L)\, f_1(S_T, S_T)\, dS_T + \int_S^\infty (\alpha S_T - L)\, f_1(S_T, S)\, dS_T
$$

$$
= \alpha\int_{L/\alpha}^S S_T\frac{1}{S_T\sigma\sqrt{2\pi\tau}}\left(\frac{S_T}{S}\right)^{2\mu/\sigma^2} e^{-[\ln(S/S_T)-\mu\tau]^2/2\sigma^2\tau}\, dS_T
$$

$$
-L\int_{L/\alpha}^S \frac{1}{S_T\sigma\sqrt{2\pi\tau}}\left(\frac{S_T}{S}\right)^{2\mu/\sigma^2} e^{-[\ln(S/S_T)-\mu\tau]^2/2\sigma^2\tau}\, dS_T
$$

$$
+\alpha\int_S^\infty S_T\frac{1}{S_T\sigma\sqrt{2\pi\tau}}e^{-[\ln(S_T/S)-\mu\tau]^2/2\sigma^2\tau}\, dS_T
$$

$$
-L\int_S^\infty \frac{1}{S_T\sigma\sqrt{2\pi\tau}}e^{-[\ln(S_T/S)-\mu\tau]^2/2\sigma^2\tau}\, dS_T
$$

$$= \alpha S e^{\left[-\mu^2 + (\mu + \sigma^2)^2\right]\tau/2\sigma^2} \int_{L/\alpha}^{S} \frac{1}{S_T \sigma \sqrt{2\pi\tau}} e^{-\left[\ln(S_T/S) - (\mu + \sigma^2)\tau\right]^2/2\sigma^2\tau} dS_T$$

$$- L e^{\left[-\mu^2 + \mu^2\right]\tau/2\sigma^2} \int_{L/\alpha}^{S} \frac{1}{S_T \sigma \sqrt{2\pi\tau}} e^{-\left[\ln(S_T/S) - \mu\tau\right]^2/2\sigma^2\tau} dS_T$$

$$+ \alpha S e^{\left[-\mu^2 + (\mu + \sigma^2)^2\right]\tau/2\sigma^2} \int_{S}^{\infty} \frac{1}{S_T \sigma \sqrt{2\pi t}} e^{-\left[\ln(S_T/S) - (\mu + \sigma^2)\tau\right]^2/2\sigma^2\tau} dS_T$$

$$- L \int_{S}^{\infty} \frac{1}{S_T \sigma \sqrt{2\pi\tau}} e^{-\left[\ln(S_T/S) - \mu\tau\right]^2/2\sigma^2\tau} dS_T$$

$$= \alpha S e^{(r - D_0)\tau} \int_{L/\alpha}^{\infty} \frac{1}{S_T \sigma \sqrt{2\pi\tau}} e^{-\left[\ln(S_T/S) - (\mu + \sigma^2)\tau\right]^2/2\sigma^2\tau} dS_T$$

$$- L \int_{L/\alpha}^{\infty} \frac{1}{S_T \sigma \sqrt{2\pi\tau}} e^{-\left[\ln(S_T/S) - \mu\tau\right]^2/2\sigma^2\tau} dS_T$$

$$= \alpha S e^{(r - D_0)\tau} N \left(\frac{-\ln \frac{L}{\alpha S} + (\mu + \sigma^2)\tau}{\sigma \sqrt{\tau}} \right) - L N \left(\frac{-\ln \frac{L}{\alpha S} + \mu\tau}{\sigma \sqrt{\tau}} \right).$$

If $S \leq L/\alpha$, then for the second and the third integrals, we have

$$\int_{L}^{S} \max\left(\alpha S_T - L, 0\right) f_1\left(S_T, S_T\right) dS_T + \int_{S}^{\infty} \max\left(\alpha S_T - L, 0\right) f_1\left(S_T, S\right) dS_T$$

$$= \int_{L/\alpha}^{\infty} \left(\alpha S_T - L\right) f_1\left(S_T, S\right) dS_T$$

$$= \alpha \int_{L/\alpha}^{\infty} S_T \frac{1}{S_T \sigma \sqrt{2\pi\tau}} e^{-\left[\ln(S_T/S) - \mu\tau\right]^2/2\sigma^2\tau} dS_T$$

$$- L \int_{L/\alpha}^{\infty} \frac{1}{S_T \sigma \sqrt{2\pi\tau}} e^{-\left[\ln(S_T/S) - \mu\tau\right]^2/2\sigma^2\tau} dS_T$$

$$= \alpha S e^{(r - D_0)\tau} \int_{L/\alpha}^{\infty} \frac{1}{S_T \sigma \sqrt{2\pi\tau}} e^{-\left[\ln(S_T/S) - (\mu + \sigma^2)\tau\right]^2/2\sigma^2\tau} dS_T$$

$$-L \int_{L/\alpha}^{\infty} \frac{1}{S_T \sigma \sqrt{2\pi\tau}} e^{-[\ln(S_T/S) - \mu\tau]^2/2\sigma^2\tau} dS_T$$

$$= \alpha S e^{(r-D_0)\tau} N \left(\frac{-\ln \frac{L}{\alpha S} + (\mu + \sigma^2)\tau}{\sigma\sqrt{\tau}} \right) - LN \left(\frac{-\ln \frac{L}{\alpha S} + \mu\tau}{\sigma\sqrt{\tau}} \right).$$

Thus, the results are the same for $S \leq L/S$ and $S > L/S$. Consequently, the price of the European lookback strike call option is

$$c_{ls}(S, L, t)$$

$$= e^{-r\tau} S \left[\frac{\sigma^2}{2(r - D_0)} \left(\frac{L}{S} \right)^{2(r-D_0)/\sigma^2} N \left(\frac{\ln \frac{\alpha L}{S} + (r - D_0 - \sigma^2/2)\tau}{\sigma\sqrt{\tau}} \right) \right.$$

$$- \frac{\sigma^2}{2(r - D_0)} \alpha^{-2(r-D_0)\sigma^2} e^{(r-D_0)\tau} N \left(\frac{\ln \frac{\alpha L}{S} - (r - D_0 + \sigma^2/2)\tau}{\sigma\sqrt{\tau}} \right)$$

$$+ \alpha e^{(r-D_0)\tau} N \left(\frac{\ln \frac{\alpha S}{L} + (r - D_0 + \sigma^2/2)\tau}{\sigma\sqrt{\tau}} \right)$$

$$\left. - \frac{L}{S} N \left(\frac{\ln \frac{\alpha S}{L} + (r - D_0 - \sigma^2/2)\tau}{\sigma\sqrt{\tau}} \right) \right].$$

This is a closed-form solution for a lookback strike call option with $\alpha \leq 1$. From this formula, we know that c_{ls}/S is a function of L/S and t, so c_{ls}/S is a solution of a one-dimensional problem.

Using a similar procedure, we can have explicit formulae for lookback strike put options and lookback price options. We leave these for the reader as Problems 10–12.

3.4.4 American Options Formulated as Free-Boundary Problems

First, consider the American lookback strike put option with continuous sampling. In Subsection 3.4.2, this problem was formulated as a LC problem. Here let us formulate this problem as a free-boundary problem. According to Theorem 2.1 in Section 2.5, we need to check

$$\left(\frac{\partial}{\partial t} + \mathbf{L}_\eta \right) \max(\eta - \beta, 0).$$

When $\eta < \beta$, we have

$$\left(\frac{\partial}{\partial t} + \mathbf{L}_\eta \right) \max(\eta - \beta, 0) = 0;$$

and when $\beta < \eta$, we obtain

$$\left(\frac{\partial}{\partial t} + \mathbf{L}_\eta\right) \max\left(\eta - \beta, 0\right) = (D_0 - r)\eta - D_0(\eta - \beta) = -r\eta + \beta D_0.$$

The inequality $-r\eta + \beta D_0 > 0$ is equivalent to $\eta < \beta D_0/r$. Therefore, when $\beta D_0/r > \beta$, there exists an interval $[\beta, \beta D_0/r]$ where

$$\left(\frac{\partial}{\partial t} + \mathbf{L}_\eta\right) \max\left(\eta - \beta, 0\right) > 0;$$

when $\beta D_0/r < \beta$, no such an interval exists. Thus when

$$\eta < \beta \max\left(1, D_0/r\right),$$

we have

$$\left(\frac{\partial}{\partial t} + \mathbf{L}_\eta\right) \max\left(\eta - \beta, 0\right) \geq 0.$$

Otherwise, i.e., when

$$\beta \max\left(1, D_0/r\right) < \eta,$$

we have

$$\left(\frac{\partial}{\partial t} + \mathbf{L}_\eta\right) \max\left(\eta - \beta, 0\right) < 0.$$

Consequently, at time T there is only one free boundary, and its location is $\beta \max\left(1, D_0/r\right)$. Let $V(S, H, t)$ be the price of an American lookback strike put option. In Subsection 2.5.2, for American vanilla options we proved that $V(S, t^*) \geq V(S, t^{**})$ if $t^* < t^{**}$. For American lookback options, the situation is similar. By the same method it can be proven[3] that

$$V(S, H, t^*) \geq V(S, H, t^{**})$$

if

$$t^* < t^{**}.$$

From this inequality, we can further have

$$W(\eta, t^*) \geq W(\eta, t^{**})$$

if

$$t^* < t^{**}.$$

Therefore, no new free boundary can appear at $t < T$. Consequently, the rectangular domain $[1, \infty) \times [0, T]$ can be divided into two parts:

$$[1, \eta_f(t)] \times [0, T] \quad \text{and} \quad (\eta_f(t), \infty) \times [0, T],$$

where the curve $\eta = \eta_f(t)$ is the free boundary with $\eta_f(T) = \beta \max\left(1, D_0/r\right)$. In the left region, the price of the American option, W, satisfies the partial

---

[3] A similar problem is given as Problem 13 for the reader to prove.

differential equation in the linear complementarity problem, and in the right region, W is equal to $\max(\eta - \beta, 0)$. That is, in the left region, the solution of the following free-boundary problem

$$\begin{cases} \dfrac{\partial W}{\partial t} + \dfrac{1}{2}\sigma^2\eta^2\dfrac{\partial^2 W}{\partial \eta^2} + (D_0 - r)\,\eta\dfrac{\partial W}{\partial \eta} - D_0 W = 0, \\[2mm] \qquad\qquad\qquad\qquad\qquad 1 \le \eta \le \eta_f(t),\; t \le T, \\[2mm] W(\eta, T) = \max(\eta - \beta, 0), \quad 1 \le \eta \le \eta_f(T), \\[2mm] \dfrac{\partial W}{\partial \eta}(1, t) = 0, \qquad\qquad t \le T, \\[2mm] W(\eta_f, t) = \eta_f - \beta, \qquad\quad t \le T, \\[2mm] \dfrac{\partial W}{\partial \eta}(\eta_f, t) = 1, \qquad\qquad t \le T, \\[2mm] \eta_f(T) = \beta\max(1, D_0/r) \end{cases} \tag{3.42}$$

provides the price of the American option, and in the domain $(\eta_f(t), \infty) \times [0, T]$, the value of W is $\max(\eta - \beta, 0)$. We will call the problem (3.42) the free-boundary problem for American lookback strike put options. This problem has boundary conditions on the both sides, which is similar to the American barrier options but different from American vanilla options and American Asian options.

Using the same procedure, we can find that in the domain $[\eta_f(t), 1] \times [0, T]$ the price of an American lookback strike call option with continuous sampling is the solution of the free-boundary problem

$$\begin{cases} \dfrac{\partial W}{\partial t} + \dfrac{1}{2}\sigma^2\eta^2\dfrac{\partial^2 W}{\partial \eta^2} + (D_0 - r)\,\eta\dfrac{\partial W}{\partial \eta} - D_0 W = 0, \\[2mm] \qquad\qquad\qquad\qquad\qquad \eta_f(t) \le \eta \le 1,\; t \le T, \\[2mm] W(\eta, T) = \max(\alpha - \eta, 0), \quad \eta_f(T) \le \eta \le 1, \\[2mm] W(\eta_f, t) = \alpha - \eta_f, \qquad\quad t \le T, \\[2mm] \dfrac{\partial W}{\partial \eta}(\eta_f, t) = -1, \qquad\quad t \le T, \\[2mm] \dfrac{\partial W}{\partial \eta}(1, t) = 0, \qquad\qquad t \le T, \\[2mm] \eta_f(T) = \alpha\min(1, D_0/r); \end{cases} \tag{3.43}$$

whereas in the domain $[0, \eta_f(t)) \times [0, T]$, the value of W is $\max(\alpha - \eta, 0)$. This is left for the reader as a part of Problem 14.

3.4.5 A Closed-Form Solution for a Perpetual American Lookback Option

In Section 2.8, closed-form solutions for perpetual American vanilla options were derived. Here, a closed-form solution for a perpetual American lookback option will be derived. As explained in Section 2.8, a perpetual option means an option with $T = \infty$. Because $T = \infty$, the solution does not depend on t and the final condition. Let us look at a perpetual American lookback option that depends on S and H and satisfies the constraint $V \geq H$. This option is called the Russian option (see [25]). We again let $\eta = H/S$ and $W = V/S$. The constraint now should be $W \geq \eta$. Therefore, from the procedure to derive (3.42), we can see that the free-boundary problem for this case is

$$
\begin{cases}
\dfrac{1}{2}\sigma^2\eta^2\dfrac{d^2W}{d\eta^2} + (D_0 - r)\,\eta\dfrac{dW}{d\eta} - D_0 W = 0, \quad 1 \leq \eta \leq \eta_f, \\[2mm]
\dfrac{dW(1)}{d\eta} = 0, \\[2mm]
W(\eta_f) = \eta_f, \\[2mm]
\dfrac{dW(\eta_f)}{d\eta} = 1,
\end{cases}
$$

where η_f is a number representing the location of the free boundary. Let us look for a solution of the ordinary differential equation in the form η^α. After substituting this function into the equation, we know that α is a root of the quadratic equation

$$
\frac{1}{2}\sigma^2\alpha^2 + \left(D_0 - r - \frac{1}{2}\sigma^2\right)\alpha - D_0 = 0.
$$

Therefore, the solution of the problem is in the form

$$
W = C_+\left(\frac{\eta}{\eta_f}\right)^{\alpha_+} + C_-\left(\frac{\eta}{\eta_f}\right)^{\alpha_-},
$$

where

$$
\alpha_+ = \frac{-D_0 + r + \dfrac{1}{2}\sigma^2 + \sqrt{\left(D_0 - r - \dfrac{1}{2}\sigma^2\right)^2 + 2\sigma^2 D_0}}{\sigma^2}
$$

and

$$
\alpha_- = \frac{-D_0 + r + \dfrac{1}{2}\sigma^2 - \sqrt{\left(D_0 - r - \dfrac{1}{2}\sigma^2\right)^2 + 2\sigma^2 D_0}}{\sigma^2}.
$$

Consequently, these boundary conditions become

$$\begin{cases} \dfrac{dW(1)}{d\eta} = C_+\alpha_+\eta_f^{-\alpha_+} + C_-\alpha_-\eta_f^{-\alpha_-} = 0, \\[2mm] W(\eta_f) = C_+ + C_- = \eta_f, \\[2mm] \dfrac{dW(\eta_f)}{d\eta} = C_+\alpha_+\eta_f^{-1} + C_-\alpha_-\eta_f^{-1} = 1. \end{cases}$$

From the last two equations, we have

$$C_+(1-\alpha_+) + C_-(1-\alpha_-) = 0.$$

Comparing this equation with $C_+\alpha_+\eta_f^{-\alpha_+} + C_-\alpha_-\eta_f^{-\alpha_-} = 0$, we know

$$\frac{1-\alpha_+}{1-\alpha_-} = \frac{\alpha_+\eta_f^{-\alpha_+}}{\alpha_-\eta_f^{-\alpha_-}},$$

which gives

$$\eta_f = \left[\frac{\alpha_+(1-\alpha_-)}{\alpha_-(1-\alpha_+)}\right]^{\frac{1}{\alpha_+-\alpha_-}}.$$

From $C_+ + C_- = \eta_f$ and $C_+(1-\alpha_+) + C_-(1-\alpha_-) = 0$, we have

$$C_+ = \frac{\eta_f(1-\alpha_-)}{\alpha_+-\alpha_-},$$

$$C_- = \frac{-\eta_f(1-\alpha_+)}{\alpha_+-\alpha_-}.$$

Therefore, the solution of the free-boundary problem is

$$W(\eta) = \frac{\eta_f}{\alpha_+-\alpha_-}\left[(1-\alpha_-)\left(\frac{\eta}{\eta_f}\right)^{\alpha_+} - (1-\alpha_+)\left(\frac{\eta}{\eta_f}\right)^{\alpha_-}\right]$$

and the price of a Russian option is

$$V(S,H) = \begin{cases} \dfrac{S\eta_f}{\alpha_+-\alpha_-}\left[(1-\alpha_-)\left(\dfrac{H}{S\eta_f}\right)^{\alpha_+} - (1-\alpha_+)\left(\dfrac{H}{S\eta_f}\right)^{\alpha_-}\right], \\[4mm] \qquad\qquad\qquad\qquad\qquad \text{if } 1 \le \dfrac{H}{S} < \eta_f, \\[4mm] H, \qquad\qquad\qquad\qquad\quad \text{if } \eta_f \le \dfrac{H}{S}. \end{cases}$$

If $D_0 = 0$, then $\alpha_- = 0$ and $\eta_f = \infty$, implying that there is no free boundary in this case.

3.4.6 Lookback-Asian Options

Suppose the sampling is done discretely. An option could depend on the average of N largest or smallest sampled prices. For example, an European option could have such a payoff function

$$\max\left(\frac{1}{N}\sum_{n=1}^{N} S_n - S, 0\right), \tag{3.44}$$

where S_n is the n-th largest sampled price during a time period $[0, T]$, T being the expiry. Let us call it a European lookback-Asian option. Consider the following scenario. During the peak season of July–September, the price of electricity might be very high for only a few days, and it is uncertain about how high the price will be. Therefore, there is a risk for consumers of electricity. In October, the price of electricity is normal, so if the expiry of the option is in October, a holder of such an option will receive certain amount of money that is proportional to the difference between the average of uncertain high prices and the normal price. Therefore, a consumer of electricity can be protected from the risk if the consumer holds such an option. Consequently, in May or so, electricity companies sell options with a payoff (3.44) and an expiry in October, which are interesting for many companies who consume a large amount of electricity.

Let us assume that S satisfies

$$dS = \mu S dt + \sigma(t) S dX,$$

where dX is a Wiener process. In this model, we require that $\sigma(t)$ is a function of t so that it is possible to have a big volatility in one period, for example, in July–September, and to have a normal volatility in other periods, for example, in May and in October.

Now let us determine $\dfrac{dS_n}{dt}$. Suppose that sampling is performed at $t = t_i$, $i = 1, 2, \cdots, K$ with $K \geq N$, and $0 < t_1 < t_2 < \cdots < t_K < T$. Let $S_n(t_i^-)$ and $S_n(t_i^+)$ denote the n-th largest price before and after the sampling procedure at $t = t_i$, respectively, and let S be the price at $t = t_i$. In this case, $S_1(t_i^+)$ is determined by S and $S_1(t_i^-)$ according to the following relation:

$$S_1(t_i^+) = \begin{cases} S_1(t_i^-), & \text{if} \quad S \leq S_1(t_i^-), \\ S, & \text{if} \quad S_1(t_i^-) < S \end{cases}$$

or

$$S_1(t_i^+) = \max(S, S_1(t_i^-)).$$

For $1 < n \leq N$, $S_n(t_i^+)$ can be expressed by S, $S_{n-1}(t_i^-)$, and $S_n(t_i^-)$ as follows (see Fig. 3.5):

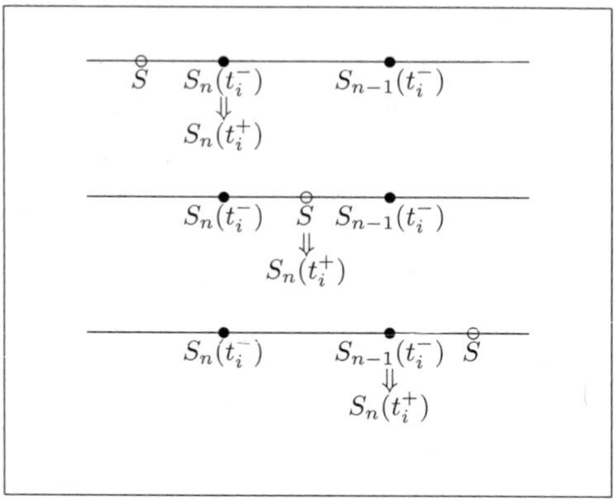

Fig. 3.5. Different $S_n(t_i^+)$ for three cases

$$S_n(t_i^+) = \begin{cases} S_n(t_i^-), & \text{if} \quad S \le S_n(t_i^-), \\ S, & \text{if} \quad S_n(t_i^-) < S \le S_{n-1}(t_i^-), \\ S_{n-1}(t_i^-), & \text{if} \quad S_{n-1}(t_i^-) < S \end{cases}$$

or[4]

$$S_n(t_i^+) = \max(\min(S, S_{n-1}(t_i^-)), \; S_n(t_i^-)).$$

Therefore at $t = t_i$

$$\frac{dS_n}{dt} = \begin{cases} \left(\max(S, S_1(t_i^-)) - S_1(t_i^-) \right) \delta\left(t - t_i\right), & \text{if } n = 1, \\ \left(\max(\min(S, S_{n-1}(t_i^-)), \; S_n(t_i^-)) - S_n(t_i^-) \right) \delta\left(t - t_i\right), \\ & \text{if } n = 2, 3, \cdots, N. \end{cases}$$

If $t \ne t_i$, $i = 1, 2, \cdots, K$, then $\dfrac{dS_n}{dt} = 0$. Let $V(S, S_1, S_2, \cdots, S_N, t)$ denote the price of the option. From Section 2.9, we know that V satisfies

$$\frac{\partial V}{\partial t} + \frac{1}{2}\sigma^2 S^2 \frac{\partial^2 V}{\partial S^2} + (r - D_0)S\frac{\partial V}{\partial S} - rV + \sum_{n=1}^{N} \frac{\partial V}{\partial S_n}\frac{dS_n}{dt} = 0. \qquad (3.45)$$

At $t \ne t_i$, $i = 1, 2, \cdots, K$, because $\dfrac{dS_n}{dt} = 0$, this equation reduces to

$$\frac{\partial V}{\partial t} + \frac{1}{2}\sigma^2 S^2 \frac{\partial^2 V}{\partial S^2} + (r - D_0)S\frac{\partial V}{\partial S} - rV = 0; \qquad (3.46)$$

---

[4] $S_n(t_i^+)$ can also be expressed as $S_n(t_i^+) = \min(S_{n-1}(t_i^-), \max(S, S_n(t_i^-)))$.

whereas at $t = t_i$, $i = 1, 2, \cdots$, or K, according to Section 2.10, V satisfies the following jump condition:

$$V(S, S_1^-, S_2^-, \cdots, S_N^-, t_i^-) = V(S, \max(S, S_1^-), \max(\min(S, S_1^-), S_2^-),$$
$$\cdots, \max(\min(S, S_{N-1}^-), S_N^-), t_i^+), \qquad (3.47)$$

where S_n^- stands for $S_n(t_i^-)$. This is the jump condition for lookback-Asian options.

In order to get the solution of this problem, it might be necessary to use numerical methods. This problem actually is an $(N+1)$-dimensional problem, so solving such a problem is very time-consuming. In order to reduce the time needed to get a numerical solution, we can reduce it to an N-dimensional problem by letting

$$\xi = \frac{S}{S_N}, \ \xi_1 = \frac{S_1}{S_N}, \ \cdots, \ \xi_{N-1} = \frac{S_{N-1}}{S_N},$$

and

$$U(\xi, \xi_1, \cdots, \xi_{N-1}, t) = \frac{V(S, S_1, \cdots, S_{N-1}, S_N, t)}{S_N}.$$

Moreover, in this case we need to solve a problem defined on an infinite domain. As pointed out in Subsection 2.3.2, it is not convenient to solve a problem on an infinite domain numerically. In order to avoid such a problem, we can transform the problem on an infinite domain to a problem on a finite domain by using the transformation given in Subsection 2.3.2.

3.4.7 Some Examples

In this subsection, we give some results for some of those problems formulated in the previous subsections. These results are obtained by using the numerical methods in Part II.

In Fig. 3.6, the function $W(\eta, t)$ of an American lookback strike call option with continuous sampling for $t = 0, 0.2, 0.4, 0.6, 0.8$ is shown. From the figure, we know that $W(\eta, t) = V(S, L, t)/S$ is a decreasing function in $\eta = L/S$, i.e., if S is fixed, then $V(S, L, t)$ is a decreasing function in L. This is because the value of the payoff $\max(S - L, 0)$ decreases or does not change as L increases. The lowest price up to time t is of course less than or equal to the price at time t. Thus, $\eta = L/S$ must be less than or equal to 1. Therefore, for a fixed t, the price has a minimum at $\eta = L/S = 1$. The minimum price at $t = 0$ is 16.37% of S, which is much higher than 9.94% – the value for the vanilla case with $S = E$. In Fig. 3.7, the location of the free boundary on (η, t)-plane is also given, which is a monotone function in t.

In Table 3.5, today's prices of some lookback strike put options with discrete sampling are given. The parameters are given there. The sampling times are $t_k = t_0 + (2k - 1)T/2K$, $k = 1, 2, \cdots, K$, where t_0 is the time today. Both European and American option prices are given. From the table, we see that

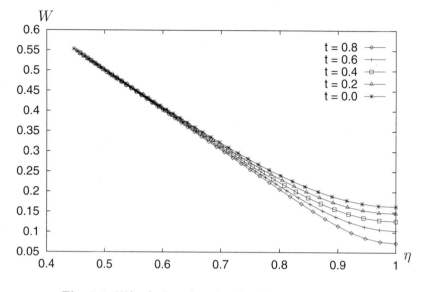

Fig. 3.6. $W(\eta, t)$ of an American lookback strike call option
($r = 0.1$, $D_0 = 0.05$, $\sigma = 0.2$, and $\alpha = 1$)

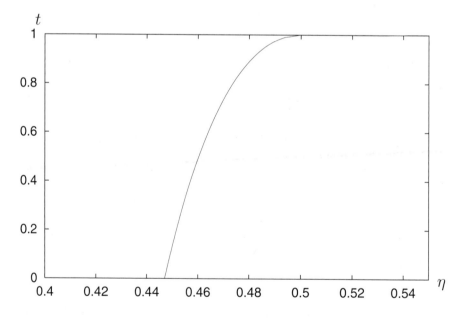

Fig. 3.7. The free boundary of an American lookback strike call option
($r = 0.1$, $D_0 = 0.05$, $\sigma = 0.2$, and $\alpha = 1$)

the American option prices are about 9% higher than the European option prices for these cases. The table also shows that the larger the number K, the higher the price. This is because H and the payoff $\max(H - S, 0)$ increase or do not change as K increases. For comparison, the values of the vanilla options are also given. From those values, we can see that the prices of lookback options are much higher than those of the corresponding vanilla options.

Table 3.5. Lookback strike put option prices

$(r = 0.1, D_0 = 0.05, \sigma = 0.2, S = H = 100, T = 1, \text{ and } \beta = 1)$

	$K = 48$	$K = 24$	$K = 12$	$K = 6$	$K = 3$	Vanilla
European option	11.865	11.283	10.499	9.535	8.456	5.302
American option	12.893	12.280	11.452	10.427	9.277	5.928

3.5 Multi-Asset Options

Sometimes an option involves several assets. Such an option is called an multi-asset option. For example, a U.S. company buys raw materials from foreign country A and sells its product in foreign country B. Its income depends on the exchange rates of the currencies of the foreign countries A and B. The higher the exchange rate of the currency of country A, the higher the cost, and the higher the exchange rate of the currency of country B, the higher the revenue. Here, an exchange rate of a foreign currency is referred to as the price of one unit of the foreign currency in U.S. dollars. Therefore, the company is interested in an option of exchanging the currency of country B into the currency of country A with a fixed rate because such an option protects the company from the exchange rate risk. Such an option involves two assets and is called an exchange option or a cross-currency option. Besides this, there are many options involving more than one asset. For example, options on the maximum or the minimum of several assets, maximum/minimum multi-strike options, and basket options are such options. This section is devoted to such options.

In this section, first we give the equation and Green's formula of solution of European multi-asset options. Then, we study the exchange options and options on the extremum of several assets. If these options are European, it is possible to express their solution in terms of multivariate cumulative normal distribution functions. Derivation of such expressions are given when we study them. Finally, the formulation of multi-asset option problems on a finite domain is given, which is useful when such a option problem has to be solved numerically, and such American option problems are briefly mentioned. Kwok in his book [49] gave an excellent summary on multi-asset options. For more details on this subject, readers are referred to that book.

3.5.1 Equations for Multi-Asset Options and Green's Formula

Consider an option dependent on n assets. Let S_i be the price of the i-th underlying asset and $V(S_1, S_2, \cdots, S_n, t)$ represent the price of the option. For simplicity, $V(S_1, S_2, \cdots, S_n, t)$ is sometimes written as $V(\mathbf{S}, t)$, $\mathbf{S}$ being $(S_1, S_2, \cdots, S_n)^T$. Suppose that S_i satisfies

$$dS_i = \mu_i S_i dt + \sigma_i S_i dX_i, \quad 0 \le S_i, \tag{3.48}$$

and the i-th asset pays a dividend $D_{0i} S_i dt$ in a time step dt. $dX_i, i = 1, 2, \cdots, n$ are the Wiener processes. They are correlated and

$$\mathrm{E}\left[dX_i dX_j\right] = \rho_{ij} dt, \ i, j = 1, 2, \cdots, n. \tag{3.49}$$

It is clear that $\rho_{ij} = \rho_{ji}$ and $\rho_{ii} = 1, i = 1, 2, \cdots, n$. Let us call ρ_{ij} the correlation coefficient between the two standardized normal random variables associated with S_i and S_j, or simply, the correlation coefficient between S_i and S_j. According to Section 2.9, $V(\mathbf{S}, t)$ satisfies

$$\left\{ \begin{array}{l} \dfrac{\partial V}{\partial t} + \dfrac{1}{2} \displaystyle\sum_{i=1}^{n} \sum_{j=1}^{n} \rho_{ij} \sigma_i \sigma_j S_i S_j \dfrac{\partial^2 V}{\partial S_i \partial S_j} + \sum_{i=1}^{n} (r - D_{0i}) S_i \dfrac{\partial V}{\partial S_i} - rV = 0, \\[4mm] \qquad\qquad 0 \le \mathbf{S}, \quad 0 \le t \le T, \end{array} \right.$$

$$\tag{3.50}$$

where the inequality $0 \le \mathbf{S}$ means $0 \le S_i$, $i = 1, 2, \cdots, n$, and similar notation will be used later on. Suppose the payoff function of an option is

$$V(\mathbf{S}, T) = V_T(\mathbf{S}), \ 0 \le \mathbf{S}, \tag{3.51}$$

we need to evaluate the price of the option dependent on n assets and with the payoff above.

In what follows, we assume that μ_i, σ_i and D_{0i} are constants and that the option is European, and we want to find the solution in an integral form. Let $V(\mathbf{S}, t) = \mathrm{e}^{-r(T-t)} \overline{V}(\mathbf{S}, t)$. Because

$$\frac{\partial V}{\partial t} = rV + \mathrm{e}^{-r(T-t)} \frac{\partial \overline{V}}{\partial t},$$

$$\frac{\partial V}{\partial S_i} = \mathrm{e}^{-r(T-t)} \frac{\partial \overline{V}}{\partial S_i},$$

$$\frac{\partial^2 V}{\partial S_i \partial S_j} = \mathrm{e}^{-r(T-t)} \frac{\partial \overline{V}}{\partial S_i \partial S_j},$$

we know that $\overline{V}(\mathbf{S}, t)$ is the solution of the problem

$$\left\{ \begin{array}{l} \dfrac{\partial \overline{V}}{\partial t} + \dfrac{1}{2} \displaystyle\sum_{i=1}^{n} \sum_{j=1}^{n} \rho_{ij} \sigma_i \sigma_j S_i S_j \dfrac{\partial^2 \overline{V}}{\partial S_i \partial S_j} + \sum_{i=1}^{n} (r - D_{0i}) S_i \dfrac{\partial \overline{V}}{\partial S_i} = 0, \\[4mm] \qquad\qquad 0 \le \mathbf{S}, \quad 0 \le t \le T, \\[3mm] \overline{V}(\mathbf{S}, T) = V_T(\mathbf{S}), \qquad\qquad 0 \le \mathbf{S}. \end{array} \right.$$

$$\tag{3.52}$$

Furthermore, we introduce the following transformation

$$\begin{cases} y_i = \dfrac{1}{\sigma_i}(\ln S_i + (r - D_{0i} - \sigma_i^2/2)(T - t)), & i = 1, 2, \cdots, n, \\[2mm] \tau = T - t. \end{cases}$$

Noticing

$$\frac{\partial}{\partial t} = -\frac{\partial}{\partial \tau} - \sum_{i=1}^{n} \frac{1}{\sigma_i}\left(r - D_{0i} - \frac{\sigma_i^2}{2}\right)\frac{\partial}{\partial y_i},$$

$$\frac{\partial}{\partial S_i} = \frac{1}{\sigma_i S_i}\frac{\partial}{\partial y_i},$$

$$\frac{\partial^2}{\partial S_i^2} = -\frac{1}{\sigma_i S_i^2}\frac{\partial}{\partial y_i} + \frac{1}{\sigma_i^2 S_i^2}\frac{\partial^2}{\partial y_i^2},$$

$$\frac{\partial^2}{\partial S_i \partial S_j} = \frac{1}{\sigma_i \sigma_j S_i S_j}\frac{\partial^2}{\partial y_i \partial y_j}, \qquad i \neq j,$$

we can rewrite the problem above as follows:

$$\begin{cases} \dfrac{\partial \overline{V}}{\partial \tau} = \dfrac{1}{2}\sum_{i=1}^{n}\sum_{j=1}^{n}\rho_{ij}\dfrac{\partial \overline{V}}{\partial y_i \partial y_j}, & -\infty < \mathbf{y} < \infty, \quad 0 \leq \tau \leq T, \\[3mm] \overline{V}(\mathbf{y}, 0) = V_{1T}(\mathbf{y}), & -\infty < \mathbf{y} < \infty, \end{cases} \tag{3.53}$$

where $\overline{V}$ means $\overline{V}(\mathbf{y}, \tau)$, $\mathbf{y}$ stands for $(y_1, y_2, \cdots, y_n)^T$ and

$$V_{1T}(\mathbf{y}) = V_T(e^{\sigma_1 y_1}, e^{\sigma_2 y_2}, \cdots, e^{\sigma_n y_n}).$$

Define

$$\mathbf{P} = \begin{bmatrix} 1 & \rho_{12} & \cdots & \rho_{1n} \\ \rho_{21} & 1 & \cdots & \rho_{2n} \\ \vdots & \vdots & \ddots & \vdots \\ \rho_{n1} & \rho_{n2} & \cdots & 1 \end{bmatrix}.$$

Because $\mathbf{P}$ should be a symmetric positive definite matrix, there exists an orthogonal matrix[5] $\mathbf{Q}$ and a diagonal positive definite matrix $\mathbf{\Lambda}$ such that

$$\mathbf{P} = \mathbf{Q}\mathbf{\Lambda}\mathbf{Q}^T = \mathbf{Q}\mathbf{\Lambda}^{1/2}\mathbf{\Lambda}^{1/2}\mathbf{Q}^T, \quad \mathbf{P}^{-1} = \mathbf{Q}\mathbf{\Lambda}^{-1/2}\mathbf{\Lambda}^{-1/2}\mathbf{Q}^T$$

and

$$\mathbf{\Lambda}^{-1/2}\mathbf{Q}^T\mathbf{P}\mathbf{Q}\mathbf{\Lambda}^{-1/2} = \mathbf{I}.$$

Let

---

[5] If $\mathbf{Q}\mathbf{Q}^T = \mathbf{I}$, then $\mathbf{Q}$ is called an orthogonal matrix.

$$\mathbf{R} = \mathbf{\Lambda}^{-1/2}\mathbf{Q}^T \equiv \begin{bmatrix} r_{11} & r_{12} & \cdots & r_{1n} \\ r_{21} & r_{22} & \cdots & r_{2n} \\ \vdots & \vdots & \ddots & \vdots \\ r_{n1} & r_{n2} & \cdots & r_{nn} \end{bmatrix},$$

then the relations above can further be rewritten as

$$\mathbf{P}^{-1} = \mathbf{R}^T\mathbf{R}$$

and

$$\mathbf{R}\mathbf{P}\mathbf{R}^T = \mathbf{I}$$

or in component form

$$\sum_{i=1}^{n}\sum_{j=1}^{n} r_{li}\rho_{ij}r_{kj} = \delta_{lk}, \quad l,k = 1,2,\cdots,n.$$

Now we define new variables as follows:

$$\mathbf{x} \equiv \begin{bmatrix} x_1 \\ x_2 \\ \vdots \\ x_n \end{bmatrix} = \begin{bmatrix} r_{11} & r_{12} & \cdots & r_{1n} \\ r_{21} & r_{22} & \cdots & r_{2n} \\ \vdots & \vdots & \ddots & \vdots \\ r_{n1} & r_{n2} & \cdots & r_{nn} \end{bmatrix} \begin{bmatrix} y_1 \\ y_2 \\ \vdots \\ y_n \end{bmatrix}$$

and further rewrite the partial differential equation in (3.53) as

$$\frac{\partial \overline{V}}{\partial \tau} = \frac{1}{2}\sum_{i=1}^{n}\sum_{j=1}^{n}\rho_{ij}\frac{\partial^2\overline{V}}{\partial y_i \partial y_j}$$

$$= \frac{1}{2}\sum_{i=1}^{n}\sum_{j=1}^{n}\left[\rho_{ij}\sum_{l=1}^{n}\sum_{k=1}^{n} r_{li}r_{kj}\frac{\partial^2\overline{V}}{\partial x_l \partial x_k}\right]$$

$$= \frac{1}{2}\sum_{l=1}^{n}\sum_{k=1}^{n}\delta_{lk}\frac{\partial^2\overline{V}}{\partial x_l \partial x_k}$$

$$= \frac{1}{2}\sum_{l=1}^{n}\frac{\partial^2\overline{V}}{\partial x_l^2}.$$

Consequently, the original problem is now reduced to

$$\begin{cases} \dfrac{\partial \overline{V}(\mathbf{x},\tau)}{\partial \tau} = \dfrac{1}{2}\displaystyle\sum_{l=1}^{n}\dfrac{\partial^2\overline{V}(\mathbf{x},\tau)}{\partial x_l^2}, & -\infty < \mathbf{x} < \infty, \quad 0 \le \tau \le T, \\ \overline{V}(\mathbf{x},0) = V_{2T}(\mathbf{x}), & -\infty < \mathbf{x} < \infty, \end{cases} \tag{3.54}$$

where $V_{2T}(\mathbf{x}) = V_{1T}(\mathbf{R}^{-1}\mathbf{x})$. Similar to the one-dimensional case, the function[6]

---

[6]This function is referred to as Green's function for the n-dimensional heat equation.

$$\phi(\mathbf{x}_0; \mathbf{x}, \tau) = \frac{1}{(2\pi\tau)^{n/2}} e^{-\sum_{i=1}^{n}(x_i - x_{i0})^2/(2\tau)}$$

is a solution to the equation above, where x_{i0}, $i = 1, 2, \cdots, n$ are constants, and $\mathbf{x}_0$ stands for $(x_{10}, x_{20}, \cdots, x_{n0})^T$. This can be verified by finding the derivatives $\dfrac{\partial \phi}{\partial \tau}, \dfrac{\partial^2 \phi}{\partial x_l^2}, l = 1, 2, \cdots, n$ and substituting them in the equation above. It can also be shown straightforwardly that

$$\int_{-\infty}^{\infty} \int_{-\infty}^{\infty} \cdots \int_{-\infty}^{\infty} \phi(\mathbf{x}_0; \mathbf{x}, \tau) dx_{10} dx_{20} \cdots dx_{n0} = 1$$

and

$$\lim_{\tau \to 0} \phi(\mathbf{x}_0; \mathbf{x}, \tau) = \begin{cases} \infty, & \text{at } \mathbf{x} = \mathbf{x}_0, \\ 0, & \text{otherwise.} \end{cases}$$

Therefore, the expression

$$\overline{V}(\mathbf{x}, \tau) = \int_{-\infty}^{\infty} \int_{-\infty}^{\infty} \cdots \int_{-\infty}^{\infty} V_{2T}(\mathbf{x}_0)\phi(\mathbf{x}_0; \mathbf{x}, \tau) dx_{10} dx_{20} \cdots dx_{n0}$$

is a solution to (3.54), which shows how $\overline{V}(\mathbf{x}, \tau)$ depends on the solution at $\tau = 0$, $V_{2T}(\mathbf{x}_0)$. This is left as an exercise to the reader. Consequently, the solution of (3.50) with condition (3.51) is

$$
\begin{aligned}
V&(\mathbf{S}, t) \\
&= e^{-r(T-t)} \int_{-\infty}^{\infty} \int_{-\infty}^{\infty} \cdots \int_{-\infty}^{\infty} V_{2T}(\mathbf{x}_0)\phi(\mathbf{x}_0; \mathbf{x}, \tau) dx_{10} dx_{20} \cdots dx_{n0} \\
&= e^{-r(T-t)} \int_{0}^{\infty} \int_{0}^{\infty} \cdots \int_{0}^{\infty} V_T(\mathbf{S}_T)\phi J_{\mathbf{x}_0\mathbf{y}_0} J_{\mathbf{y}_0\mathbf{S}_T} dS_{1T} dS_{2T} \cdots dS_{nT} \\
&= e^{-r(T-t)} \int_{0}^{\infty} \int_{0}^{\infty} \cdots \int_{0}^{\infty} V_T(\mathbf{S}_T)\psi(\mathbf{S}_T; \mathbf{S}, t) dS_{1T} dS_{2T} \cdots dS_{nT}. \quad (3.55)
\end{aligned}
$$

Here, $\mathbf{y}_0 = (y_{10}, y_{20}, \cdots, y_{n0})^T = \mathbf{R}^{-1}\mathbf{x}_0$. $J_{\mathbf{x}_0\mathbf{y}_0}$ and $J_{\mathbf{y}_0\mathbf{S}_T}$ are the Jacobians of the transformation from $\mathbf{x}_0$ to $\mathbf{y}_0$ and from $\mathbf{y}_0$ to $\mathbf{S}_T$:

$$J_{\mathbf{x}_0\mathbf{y}_0} = \frac{\partial(x_{10}, x_{20}, \cdots, x_{n0})}{\partial(y_{10}, y_{20}, \cdots, y_{n0})} = \det \mathbf{R} = \frac{1}{\sqrt{\det \mathbf{P}}},$$

$$J_{\mathbf{y}_0\mathbf{S}_T} = \frac{\partial(y_{10}, y_{20}, \cdots, y_{n0})}{\partial(S_{1T}, S_{2T}, \cdots, S_{nT})} = \prod_{i=1}^{n} \frac{1}{\sigma_i S_{iT}}$$

and

$$
\begin{aligned}
\psi(\mathbf{S}_T; \mathbf{S}, t) &= \phi J_{\mathbf{x}_0\mathbf{y}_0} J_{\mathbf{y}_0\mathbf{S}_T} \\
&= \frac{1}{(2\pi\tau)^{n/2}\sqrt{\det \mathbf{P}} \prod_{i=1}^{n}(\sigma_i S_{iT})} e^{-(\mathbf{x}_0 - \mathbf{x})^T(\mathbf{x}_0 - \mathbf{x})/(2\tau)}
\end{aligned}
$$

$$= \frac{1}{(2\pi\tau)^{n/2}\sqrt{\det \mathbf{P}}\, \prod_{i=1}^{n}(\sigma_i S_{iT})} e^{-\eta^T \mathbf{R}^T \mathbf{R}\eta/2}$$

$$= \frac{1}{(2\pi\tau)^{n/2}\sqrt{\det \mathbf{P}}\, \prod_{i=1}^{n}(\sigma_i S_{iT})} e^{-\eta^T \mathbf{P}^{-1}\eta/2}, \tag{3.56}$$

where η is a vector whose i-th component is

$$\eta_i = \frac{\ln S_{iT} - \left[\ln S_i + (r - D_{0i} - \sigma_i^2/2)\tau\right]}{\sigma_i\sqrt{\tau}} = \frac{\ln S_{iT} - \sigma_i y_i}{\sigma_i\sqrt{\tau}}.$$

The expression (3.55) is called Green's formula. For some payoff function $V_T(\mathbf{S})$, it can be written in terms of multivariate cumulative distribution functions for standardized normal variables, which will be called multivariate cumulative distribution functions for brevity in what follows, and we can have closed-form solutions. In Subsection 3.5.3, some examples will be given. If σ_i depends on S_i, it might be necessary to use numerical methods to price those options.

3.5.2 Exchange Options

Sometimes, a two-asset option problem can be reduced to a one-asset option, and its closed-form solution can be found for such a European option. Exchange options on two assets is such an example.

An exchange option is a contract that gives its holder a right to exchange certain assets for some other assets. In this subsection, we consider an exchange option to exchange n_B shares of asset B for n_A shares of asset A. Suppose S_A and S_B are the prices of assets A and B, respectively. Let

$$\begin{cases} S_1 = n_A S_A, \\ S_2 = n_B S_B. \end{cases}$$

Then, the payoff function of the exchange option is

$$\max(n_A S_A - n_B S_B, 0) = \max(S_1 - S_2, 0).$$

Therefore, if we consider n_A shares of asset A as an asset $\bar{A}$, and n_B shares of asset B as an asset $\bar{B}$, then the exchange option is an option to exchange an asset $\bar{B}$ for another asset $\bar{A}$. Suppose that S_A and S_B satisfy

$$\begin{cases} dS_A = \mu_1 S_A dt + \sigma_1 S_A dX_1, \\ dS_B = \mu_2 S_B dt + \sigma_2 S_B dX_2, \end{cases} \tag{3.57}$$

where dX_1 and dX_2 satisfy

$$E\left[dX_1 dX_2\right] = \rho_{12} dt.$$

Then, using Itô's lemma, we have

$$\begin{cases} dS_1 = \mu_1 S_1 dt + \sigma_1 S_1 dX_1, \\ dS_2 = \mu_2 S_2 dt + \sigma_2 S_2 dX_2. \end{cases}$$

Thus, S_1 and S_2 have the same volatilities as S_A and S_B, respectively. The dividend yields related to the assets A and $\bar{A}$ are the same, and for the assets B and $\bar{B}$, this is also true. In what follows, D_{01} and D_{02} denote the dividend yields related to the assets A and B, respectively. Let $V(S_1, S_2, t)$ be the value of a European exchange option. According to Subsection 3.5.1, V should be the solution of the problem

$$\begin{cases} \dfrac{\partial V}{\partial t} + \dfrac{1}{2}\sigma_1^2 S_1^2 \dfrac{\partial^2 V}{\partial S_1^2} + \rho_{12}\sigma_1\sigma_2 S_1 S_2 \dfrac{\partial^2 V}{\partial S_1 \partial S_2} + \dfrac{1}{2}\sigma_2^2 S_2^2 \dfrac{\partial^2 V}{\partial S_2^2} \\[2mm] +(r - D_{01})S_1 \dfrac{\partial V}{\partial S_1} + (r - D_{02})S_2 \dfrac{\partial V}{\partial S_2} - rV = 0, \\[2mm] \qquad\qquad\qquad\qquad 0 \le S_1,\ 0 \le S_2,\ 0 \le t \le T, \\[2mm] V(S_1, S_2, T) = \max(S_1 - S_2, 0), \qquad 0 \le S_1,\ 0 \le S_2. \end{cases} \qquad (3.58)$$

This is a two-dimensional problem, but it can be reduced to a one-dimensional problem. In fact, let

$$W = \frac{V}{S_2} \quad \text{and} \quad \xi = \frac{S_1}{S_2},$$

then we have

$$\frac{\partial V}{\partial t} = S_2 \frac{\partial W}{\partial t}, \qquad \frac{\partial V}{\partial S_1} = \frac{\partial W}{\partial \xi}, \qquad \frac{\partial^2 V}{\partial S_1^2} = \frac{\partial^2 W}{\partial \xi^2}\frac{1}{S_2},$$

$$\frac{\partial^2 V}{\partial S_1 \partial S_2} = \frac{-\xi}{S_2}\frac{\partial^2 W}{\partial \xi^2}, \qquad \frac{\partial V}{\partial S_2} = W - \xi\frac{\partial W}{\partial \xi}, \qquad \frac{\partial^2 V}{\partial S_2^2} = \frac{\xi^2}{S_2}\frac{\partial^2 W}{\partial \xi^2},$$

and the problem (3.58) can be rewritten as

$$\begin{cases} \dfrac{\partial W}{\partial t} + \dfrac{1}{2}(\sigma_1^2 - 2\rho_{12}\sigma_1\sigma_2 + \sigma_2^2)\xi^2 \dfrac{\partial^2 W}{\partial \xi^2} + (D_{02} - D_{01})\dfrac{\partial W}{\partial \xi} \\[2mm] -D_{02}\, W = 0, \qquad\qquad 0 \le \xi,\quad 0 \le t \le T, \\[2mm] W(\xi, T) = \max(\xi - 1, 0), \qquad 0 \le \xi. \end{cases} \qquad (3.59)$$

This is a European call option problem with $r = D_{02}$, $D_0 = D_{01}$, $E = 1$, and

$$\sigma = \sigma_{12} \equiv \sqrt{\sigma_1^2 - 2\rho_{12}\sigma_1\sigma_2 + \sigma_2^2}.$$

Hence

$$W(\xi, t) = \xi e^{-D_{01}(T-t)} N(d_{10}) - e^{-D_{02}(T-t)} N(d_{20}),$$

where

$$d_{10} = \frac{\ln \frac{\xi e^{-D_{01}(T-t)}}{e^{-D_{02}(T-t)}} + \frac{\sigma_{12}^2}{2}(T-t)}{\sigma_{12}\sqrt{T-t}}$$

and

$$d_{20} = d_{10} - \sigma_{12}\sqrt{T-t}.$$

Therefore

$$V(S_1, S_2, t) = S_1 e^{-D_{01}(T-t)} N(d_{10}) - S_2 e^{-D_{02}(T-t)} N(d_{20})$$
$$= n_A S_A e^{-D_{01}(T-t)} N(d_{10}) - n_B S_B e^{-D_{02}(T-t)} N(d_{20}), \quad (3.60)$$

where

$$d_{10} = \frac{\ln \frac{n_A S_A e^{-D_{01}(T-t)}}{n_B S_B e^{-D_{02}(T-t)}} + \frac{\sigma_{12}^2}{2}(T-t)}{\sigma_{12}\sqrt{T-t}}$$

and

$$d_{20} = d_{10} - \sigma_{12}\sqrt{T-t}.$$

Maryrabe[55] derived this closed-form solution with $D_{01} = D_{02} = 0$, and Rumsey[63] and Brooks[14] gave this closed-form solution and called this exchange option a cross-currency option because the assets there were foreign currencies.

An exchange option could be an American option. We can also introduce the transformation $W = \frac{V}{S_2}$ and $\xi = \frac{S_1}{S_2}$ and reduce the two-dimensional problem to a one-dimensional problem. W is the solution of an American call option with

$$\sigma = \sqrt{\sigma_1^2 - 2\rho_{12}\sigma_1\sigma_2 + \sigma_2^2}, \ r = D_{02}, \ D_0 = D_{01} \text{ and } E = 1.$$

After finding $W(\xi, t)$, the function V is given by

$$V = S_2 W\left(\frac{S_1}{S_2}, t\right) = n_B S_B W\left(\frac{n_A S_A}{n_B S_B}, t\right).$$

3.5.3 Options on the Extremum of Several Assets

The price of certain European multi-asset options can be expressed in terms of multivariate cumulative distribution functions. Options on the extremum of several assets are such options. In this subsection, we first explain how these options appear in practice. Then, the price expression of such European options in terms of multivariate cumulative distribution functions are derived out if the volatilities and the dividend yields are constants.

Let us consider a zero-coupon bond that at maturity date T pays, at the choice of holder, either Z_0 units of domestic currency, Z_A units of currency of country A, or Z_B units of currency of country B. Let S_A denote the domestic price of currency of country A and S_B that of country B. Then, the payment at maturity (payoff) is

$$\max(Z_0, S_A Z_A, S_B Z_B).$$

Let $S_0 = Z_0$, $S_1 = S_A Z_A$, and $S_2 = S_B Z_B$, then the payment becomes

$$\max(S_0, S_1, S_2).$$

The question is what is the present value of the option. The value depends on S_1, S_2, and t. Let r be the interest rate of domestic currency. Suppose that S_A and S_B are governed by (3.57) and that the interest rates of currencies of the countries A and B are D_{01} and D_{02}, respectively. As shown previously, S_1 and S_2 satisfy the same stochastic differential equations, and the dividend yields for the assets S_1 and S_2 also are D_{01} and D_{02}. Therefore, the value of the bond, $V(S_1, S_2, t)$, is the solution of the problem

$$
\begin{cases}
\dfrac{\partial V}{\partial t} + \dfrac{1}{2}\sigma_1^2 S_1^2 \dfrac{\partial^2 V}{\partial S_1^2} + \rho_{12}\sigma_1\sigma_2 S_1 S_2 \dfrac{\partial^2 V}{\partial S_1 \partial S_2} + \dfrac{1}{2}\sigma_2^2 S_2^2 \dfrac{\partial^2 V}{\partial S_2^2} \\[2mm]
+(r - D_{01})S_1 \dfrac{\partial V}{\partial S_1} + (r - D_{02})S_2 \dfrac{\partial V}{\partial S_2} - rV = 0, \\[2mm]
\hspace{4cm} 0 \le S_1,\ 0 \le S_2,\ 0 \le t \le T, \\[2mm]
V(S_1, S_2, T) = \max(S_0, S_1, S_2), \quad 0 \le S_1,\ 0 \le S_2.
\end{cases}
\tag{3.61}
$$

This problem is usually called a European option on the maximum of assets S_0, S_1, and S_2. From the formula (3.55), the solution of the problem (3.61) is

$$
\begin{aligned}
V(S_1, S_2, t) &= e^{-r(T-t)} \int_0^\infty \int_0^\infty \max(S_0, S_{1T}, S_{2T})\psi(\mathbf{S}_T; \mathbf{S}, t)dS_{1T}dS_{2T} \\
&= e^{-r(T-t)} \Bigg[S_0 \int_0^{S_0} \int_0^{S_0} \psi(\mathbf{S}_T; \mathbf{S}, t)dS_{1T}dS_{2T} \\
&\quad + \int_{S_0}^\infty \int_0^{S_{1T}} S_{1T}\psi(\mathbf{S}_T; \mathbf{S}, t)dS_{2T}dS_{1T} \\
&\quad + \int_{S_0}^\infty \int_0^{S_{2T}} S_{2T}\psi(\mathbf{S}_T; \mathbf{S}, t)dS_{1T}dS_{2T} \Bigg].
\end{aligned}
\tag{3.62}
$$

In the expression, there are three terms that represent the contributions to the solution from the three domains (see Fig. 3.8):

$$0 \le S_{1T} \le S_0,\ 0 \le S_{2T} \le S_0; \quad S_0 \le S_{1T},\ 0 \le S_{2T} \le S_{1T}$$

and

$$S_0 \le S_{2T},\ 0 \le S_{1T} \le S_{2T}.$$

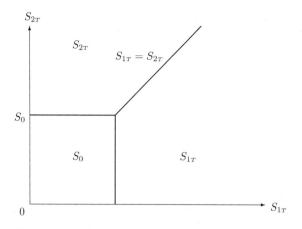

Fig. 3.8. The maximum values on three domains for options
with a payoff $\max(S_0, S_1, S_2)$

Indeed, every term in (3.62) can be expressed in terms of the bivariate cumulative distribution function

$$N_2(x_1, x_2; \rho) = \frac{1}{2\pi\sqrt{1-\rho^2}} \int_{-\infty}^{x_2} \int_{-\infty}^{x_1} e^{-\frac{1}{2}(\eta_1^2 - 2\rho\eta_1\eta_2 + \eta_2^2)/(1-\rho^2)} d\eta_1 d\eta_2,$$

(3.63)

where ρ is a parameter. Stulz gave such a result in [66]. Later, Johnson[45] improved the method for deriving this result. Here, we show how every term in (3.62) can be expressed in terms of (3.63). Let us begin with looking at the first term. Noticing the concrete expression (3.56) for ψ, we have

$$e^{-r(T-t)}S_0 \int_0^{S_0} \int_0^{S_0} \psi(\mathbf{S}_T; \mathbf{S}, t) dS_{1T} dS_{2T}$$

$$= e^{-r\tau} S_0 \int_0^{S_0} \int_0^{S_0} \frac{1}{2\pi\tau\sqrt{\det \mathbf{P}}\sigma_1\sigma_2 S_{1T} S_{2T}} e^{-\eta^T \mathbf{P}^{-1}\eta/2} dS_{1T} dS_{2T}$$

$$= e^{-r\tau} S_0 \int_{-\infty}^{(\ln S_0 - \sigma_2 y_2)/\sigma_2\sqrt{\tau}} \int_{-\infty}^{(\ln S_0 - \sigma_1 y_1)/\sigma_1\sqrt{\tau}} \frac{1}{2\pi\sqrt{\det \mathbf{P}}} e^{-\eta^T \mathbf{P}^{-1}\eta/2} d\eta_1 d\eta_2$$

$$= e^{-r\tau} S_0 N_2 \left(\frac{\ln \dfrac{S_0}{S_1} - \left(r - D_{01} - \dfrac{\sigma_1^2}{2}\right)\tau}{\sigma_1\sqrt{\tau}}, \frac{\ln \dfrac{S_0}{S_2} - \left(r - D_{02} - \dfrac{\sigma_2^2}{2}\right)\tau}{\sigma_2\sqrt{\tau}}; \rho_{12} \right)$$

$$= S_0^* N_2 \left(\frac{\ln \dfrac{S_0^*}{S_1^*} + \dfrac{\sigma_1^2}{2}\tau}{\sigma_1\sqrt{\tau}}, \frac{\ln \dfrac{S_0^*}{S_2^*} + \dfrac{\sigma_2^2}{2}\tau}{\sigma_2\sqrt{\tau}}; \rho_{12} \right),$$

(3.64)

where

$$S_0^* = S_0 e^{-r\tau}, \quad S_1^* = S_1 e^{-D_{01}\tau}, \quad S_2^* = S_2 e^{-D_{02}\tau}.$$

Here, we have used the following formulae

$$\det \mathbf{P} = \begin{vmatrix} 1 & \rho_{12} \\ \rho_{12} & 1 \end{vmatrix} = 1 - \rho_{12}^2,$$

$$\mathbf{P}^{-1} = \begin{bmatrix} 1 & \rho_{12} \\ \rho_{12} & 1 \end{bmatrix}^{-1} = \frac{1}{1 - \rho_{12}^2} \begin{bmatrix} 1 & -\rho_{12} \\ -\rho_{12} & 1 \end{bmatrix}$$

and

$$\eta^T \mathbf{P}^{-1} \eta = \frac{\eta_1^2 - 2\rho_{12}\eta_1\eta_2 + \eta_2^2}{1 - \rho_{12}^2}.$$

The value of the bivariate cumulative distribution function in (3.64) can be interpreted as the probability of the event $\max(S_{1T}, S_{2T}) \leq S_0$ in the "risk-neutral" world.

Now let us work on the second and third terms in (3.62). Actually, we can find their expressions from the first term.

Let S_i be a random variable, σ_i be its volatility, and ν_i be its dividend yield. For S_1 and S_2 given in this subsection, the notation for volatilities is unchanged, and $\nu_1 = D_{01}$ and $\nu_2 = D_{02}$. We can understand S_0 as a random variable with $\sigma_0 = 0$ and $\nu_0 = r$. Because S_0 is not a random variable, we have $\rho_{i0} = \rho_{0i} = 0$ for any $i \neq 0$. Define

$$\xi_{ij} = S_i/S_j \quad \text{for any } i \neq j.$$

Let σ_{ij}^2 be the variance of the random variable ξ_{ij} and ρ_{ijk} be the correlation coefficient between ξ_{ik} and ξ_{jk}. Using Itô's lemma, we have

$$\sigma_{ij} = \sqrt{\sigma_i^2 - 2\rho_{ij}\sigma_i\sigma_j + \sigma_j^2} \tag{3.65}$$

and

$$\rho_{ijk} = \frac{\sigma_k^2 - \rho_{ik}\sigma_i\sigma_k - \rho_{jk}\sigma_j\sigma_k + \rho_{ij}\sigma_i\sigma_j}{\sigma_{ik}\sigma_{jk}}. \tag{3.66}$$

The dividend yield of ξ_{ij} is denoted by ν_{ij}. It is clear that

$$\nu_{ij} = \nu_i - \nu_j.$$

Let us take $\xi_{10} = S_1/S_0$ and $\xi_{20} = S_2/S_0$ as independent variables and $V_0 = V/S_0$ as the unknown function. In this case, problem (3.61) becomes

$$\begin{cases} \dfrac{\partial V_0}{\partial t} + \dfrac{1}{2}\sigma_1^2\xi_{10}^2 \dfrac{\partial^2 V_0}{\partial \xi_{10}^2} + \rho_{12}\sigma_1\sigma_2\xi_{10}\xi_{20} \dfrac{\partial^2 V_0}{\partial \xi_{10}\partial \xi_{20}} + \dfrac{1}{2}\sigma_2^2\xi_{20}^2 \dfrac{\partial^2 V_0}{\partial \xi_{20}^2} \\ +(r - D_{01})\xi_{10}\dfrac{\partial V_0}{\partial \xi_{10}} + (r - D_{02})\xi_{20}\dfrac{\partial V_0}{\partial \xi_{20}} - rV_0 = 0, \\ \qquad\qquad\qquad\qquad\qquad 0 \leq \xi_{10}, \quad 0 \leq \xi_{20}, \quad 0 \leq t \leq T, \\ V_0(\xi_{10}, \xi_{20}, T) = \max(1, \xi_{10}, \xi_{20}), \quad 0 \leq \xi_{10}, \quad 0 \leq \xi_{20}. \end{cases}$$

Because $\sigma_{10} = \sigma_1$, $\sigma_{20} = \sigma_1$, $\rho_{120} = \rho_{12}$, $\nu_0 = r$, $\nu_{10} = \nu_1 - \nu_0 = D_{01} - r$, and $\nu_{20} = \nu_2 - \nu_0 = D_{02} - r$, the problem above can be rewritten as

$$
\begin{cases}
\dfrac{\partial V_0}{\partial t} + \dfrac{1}{2}\sigma_{10}^2\xi_{10}^2\dfrac{\partial^2 V_0}{\partial \xi_{10}^2} + \rho_{120}\sigma_{10}\sigma_{20}\xi_{10}\xi_{20}\dfrac{\partial^2 V_0}{\partial \xi_{10}\partial \xi_{20}} + \dfrac{1}{2}\sigma_{20}^2\xi_{20}^2\dfrac{\partial^2 V_0}{\partial \xi_{20}^2} \\[2mm]
-\nu_{10}\xi_{10}\dfrac{\partial V_0}{\partial \xi_{10}} - \nu_{20}\xi_{20}\dfrac{\partial V_0}{\partial \xi_{20}} - \nu_0 V_0 = 0, \\[2mm]
\hspace{5cm} 0 \le \xi_{10}, \ \ 0 \le \xi_{20}, \ \ 0 \le t \le T, \\[2mm]
V_0(\xi_{10}, \xi_{20}, T) = \max(1, \xi_{10}, \xi_{20}), \ \ 0 \le \xi_{10}, \ \ 0 \le \xi_{20}.
\end{cases}
\tag{3.67}
$$

By using the integral expression of the solution of (3.67), $V(S_1, S_2, t)$ can be expressed as

$$
V(S_1, S_2, t) = S_0 V(\xi_{10}, \xi_{20}, t)
$$

$$
= S_0 e^{-r\tau} \int_0^\infty \int_0^\infty \max(1, \xi_{10T}, \xi_{20T}) \psi d\xi_{10T} d\xi_{20T}
$$

$$
= S_0 e^{-r\tau} \left\{ \int_0^1 \int_0^1 \psi d\xi_{10} d\xi_{20T} + \int_1^\infty \int_0^{\xi_{10T}} \xi_{10T}\psi d\xi_{20T} d\xi_{10T} \right.
$$

$$
\left. + \int_1^\infty \int_0^{\xi_{20T}} \xi_{20T}\psi d\xi_{10T} d\xi_{20T} \right\}.
\tag{3.68}
$$

Clearly, the first term in (3.68) is the same as the first term in (3.62). Thus, the first term in (3.68) is equal to

$$
S_0 e^{-r\tau} N_2 \left(\dfrac{\ln \frac{S_0}{S_1} - \left(r - D_{01} - \frac{\sigma_1^2}{2}\right)\tau}{\sigma_1\sqrt{\tau}}, \dfrac{\ln \frac{S_0}{S_2} - \left(r - D_{02} - \frac{\sigma_2^2}{2}\right)\tau}{\sigma_2\sqrt{\tau}}; \rho_{12} \right)
$$

$$
= S_0 e^{-\nu_0\tau} N_2 \left(\dfrac{-\ln \xi_{10} + \left(\nu_{10} + \frac{\sigma_{10}^2}{2}\right)\tau}{\sigma_{10}\sqrt{\tau}}, \dfrac{-\ln \xi_{20} + \left(\nu_{20} + \frac{\sigma_{20}^2}{2}\right)\tau}{\sigma_{20}\sqrt{\tau}}; \rho_{120} \right).
\tag{3.69}
$$

Now we take $\xi_{21} = S_2/S_1$ and $\xi_{01} = S_0/S_1$ as independent variables and $V_1 = V/S_1$ as the unknown function. Because

$$
\dfrac{\partial V}{\partial t} = S_1 \dfrac{\partial V_1}{\partial t},
$$

$$
\dfrac{\partial V}{\partial S_1} = V_1 - \xi_{21}\dfrac{\partial V_1}{\partial \xi_{21}} - \xi_{01}\dfrac{\partial V_1}{\partial \xi_{01}},
$$

$$\frac{\partial^2 V}{\partial S_1^2} = \frac{1}{S_1}\left(\xi_{21}^2 \frac{\partial^2 V_1}{\partial \xi_{21}^2} + 2\xi_{21}\xi_{01}\frac{\partial^2 V_1}{\partial \xi_{21}\partial \xi_{01}} + \xi_{01}^2\frac{\partial^2 V_1}{\partial \xi_{01}^2}\right),$$

$$\frac{\partial V}{\partial S_2} = \frac{\partial V_1}{\partial \xi_{21}},$$

$$\frac{\partial^2 V}{\partial S_2^2} = \frac{1}{S_1}\frac{\partial^2 V_1}{\partial \xi_{21}^2},$$

$$\frac{\partial^2 V}{\partial S_1 \partial S_2} = -\frac{1}{S_1}\left(\xi_{21}\frac{\partial^2 V_1}{\partial \xi_{21}^2} + \xi_{01}\frac{\partial^2 V_1}{\partial \xi_{21}\partial \xi_{01}}\right),$$

we have

$$
\begin{cases}
\dfrac{\partial V_1}{\partial t} + \dfrac{1}{2}\sigma_{21}^2\xi_{21}^2\dfrac{\partial^2 V_1}{\partial \xi_{21}^2} + \rho_{201}\sigma_{21}\sigma_1\xi_{21}\xi_{01}\dfrac{\partial^2 V_1}{\partial \xi_{21}\partial \xi_{01}} + \dfrac{1}{2}\sigma_1^2\xi_{01}^2\dfrac{\partial^2 V_1}{\partial \xi_{01}^2} \\[2mm]
\quad +(D_{01}-D_{02})\xi_{21}\dfrac{\partial V_1}{\partial \xi_{21}} + (D_{01}-r)\xi_{01}\dfrac{\partial V_1}{\partial \xi_{01}} - D_{01}V_1 = 0, \\[2mm]
\qquad\qquad\qquad\qquad\qquad\qquad 0 \le \xi_{21}, \quad 0 \le \xi_{01}, \quad 0 \le t \le T, \\[2mm]
V_1(\xi_{21},\xi_{01},T) = \max(1,\xi_{21},\xi_{01}), \qquad 0 \le \xi_{21}, \quad 0 \le \xi_{01},
\end{cases}
$$

where σ_{21} and ρ_{201} are defined by (3.65) and (3.66), namely,

$$\sigma_{21} = \sqrt{\sigma_1^2 - 2\rho_{12}\sigma_1\sigma_2 + \sigma_2^2}$$

and

$$\rho_{201} = \frac{\sigma_1^2 - \rho_{12}\sigma_1\sigma_2 - \rho_{01}\sigma_0\sigma_1 + \rho_{20}\sigma_2\sigma_0}{\sigma_{21}\sigma_{01}} = \frac{\sigma_1 - \rho_{12}\sigma_2}{\sigma_{21}}.$$

Because $\sigma_{01} = \sigma_1$, $\nu_1 = D_{01}$, $\nu_{21} = \nu_2 - \nu_1 = D_{02} - D_{01}$, and $\nu_{01} = \nu_0 - \nu_1 = r - D_{01}$, the problem above can be rewritten as

$$
\begin{cases}
\dfrac{\partial V_1}{\partial t} + \dfrac{1}{2}\sigma_{21}^2\xi_{21}^2\dfrac{\partial^2 V_1}{\partial \xi_{21}^2} + \rho_{201}\sigma_{21}\sigma_{01}\xi_{21}\xi_{01}\dfrac{\partial^2 V_1}{\partial \xi_{21}\partial \xi_{01}} + \dfrac{1}{2}\sigma_{01}^2\xi_{01}^2\dfrac{\partial^2 V_1}{\partial \xi_{01}^2} \\[2mm]
\quad -\nu_{21}\xi_{21}\dfrac{\partial V_1}{\partial \xi_{21}} - \nu_{01}\xi_{01}\dfrac{\partial V_1}{\partial \xi_{01}} - \nu_1 V_1 = 0, \\[2mm]
\qquad\qquad\qquad\qquad\qquad\qquad 0 \le \xi_{21}, \ 0 \le \xi_{01}, \ 0 \le t \le T, \\[2mm]
V_1(\xi_{21},\xi_{01},T) = \max(1,\xi_{21},\xi_{01}), \quad 0 \le \xi_{21}, \ 0 \le \xi_{01}.
\end{cases}
\tag{3.70}
$$

Therefore, we have

$$V(S_1,S_2,t) = S_1 V_1\left(\xi_{21},\xi_{01},t\right)$$

$$= S_1 e^{-D_{01}\tau}\int_0^\infty\int_0^\infty \max\left(1,\xi_{21T},\xi_{01T}\right)\psi d\xi_{21T}d\xi_{01T}$$

$$= S_1 e^{-D_{01}\tau} \left\{ \int_0^1 \int_0^1 \psi d\xi_{21T} d\xi_{01T} + \int_1^\infty \int_0^{\xi_{21T}} \xi_{21T} \psi d\xi_{01T} d\xi_{21T} \right.$$

$$\left. + \int_1^\infty \int_0^{\xi_{01T}} \xi_{01T} \psi d\xi_{21T} \xi_{01T} \right\}. \tag{3.71}$$

Because we can have (3.70) from (3.67) by the rule of substitution of subscripts:

$$0 \to 1, \quad 1 \to 2, \quad 2 \to 0,$$

we can obtain the result of the first term in (3.71) from the result of the first term in (3.68) by the same rule. The result of the first term in (3.68) is (3.69), so for the first term in (3.71) we have

$$S_1 e^{-D_{01}\tau} \int_0^1 \int_0^1 \psi d\xi_{21T} d\xi_{01T}$$

$$= S_1 e^{-\nu_1 \tau} N_2 \left(\frac{-\ln \xi_{21} + \left(\nu_{21} + \frac{\sigma_{21}^2}{2}\right)\tau}{\sigma_{21}\sqrt{\tau}}, \frac{-\ln \xi_{01} + \left(\nu_{01} + \frac{\sigma_{01}^2}{2}\right)\tau}{\sigma_{01}\sqrt{\tau}}; \rho_{201} \right)$$

$$= S_1^* N_2 \left(\frac{\ln \frac{S_1^*}{S_2^*} + \frac{\sigma_{12}^2}{2}\tau}{\sigma_{12}\sqrt{\tau}}, \frac{\ln \frac{S_1^*}{S_0^*} + \frac{\sigma_1^2}{2}\tau}{\sigma_1\sqrt{\tau}}; \frac{\sigma_1 - \rho_{12}\sigma_2}{\sigma_{12}} \right). \tag{3.72}$$

Because

$$\max(\xi_{21T}, \xi_{01T}) = \max\left(\frac{S_{2T}}{S_{1T}}, \frac{S_0}{S_{1T}} \right) \le 1,$$

which is equivalent to

$$S_{1T} \ge \max(S_0, S_{2T}),$$

both the first term in (3.71) and the second term in (3.62) represent the contribution to the solution from the same domain where $S_{1T} \ge \max(S_0, S_{2T})$. This domain is $S_0 \le S_{1T}$, $0 \le S_{2T} \le S_{1T}$. Thus, the second term in (3.62) should be equal to (3.72).

Similarly, we can prove that the result of the third term in (3.62) can be obtained from the second term in (3.62) by the same rule of substitution of subscripts, namely, it is equal to

$$S_2 e^{-\nu_2 \tau} N_2 \left(\frac{-\ln \xi_{02} + \left(\nu_{02} + \frac{\sigma_{02}^2}{2}\right)\tau}{\sigma_{02}\sqrt{\tau}}, \frac{-\ln \xi_{12} + \left(\nu_{12} + \frac{\sigma_{12}^2}{2}\right)\tau}{\sigma_{12}\sqrt{\tau}}; \rho_{012} \right)$$

$$
= S_2^* N_2 \left(\frac{\ln \frac{S_2^*}{S_0^*} + \frac{\sigma_2^2}{2}\tau}{\sigma_2\sqrt{\tau}}, \frac{\ln \frac{S_2^*}{S_1^*} + \frac{\sigma_{12}^2}{2}\tau}{\sigma_{12}\sqrt{\tau}}; \frac{\sigma_2 - \rho_{12}\sigma_1}{\sigma_{12}} \right).
$$

Therefore, we finally arrive at

$$
V(S_1, S_2, t)
$$

$$
= S_0^* N_2 \left(\frac{\ln \frac{S_0^*}{S_1^*} + \frac{\sigma_1^2}{2}\tau}{\sigma_1\sqrt{\tau}}, \frac{\ln \frac{S_0^*}{S_2^*} + \frac{\sigma_2^2}{2}\tau}{\sigma_2\sqrt{\tau}}; \rho_{12} \right)
$$

$$
+ S_1^* N_2 \left(\frac{\ln \frac{S_1^*}{S_2^*} + \frac{\sigma_{12}^2}{2}\tau}{\sigma_{12}\sqrt{\tau}}, \frac{\ln \frac{S_1^*}{S_0^*} + \frac{\sigma_1^2}{2}\tau}{\sigma_1\sqrt{\tau}}; \frac{\sigma_1 - \rho_{12}\sigma_2}{\sigma_{12}} \right)
$$

$$
+ S_2^* N_2 \left(\frac{\ln \frac{S_2^*}{S_0^*} + \frac{\sigma_2^2}{2}\tau}{\sigma_2\sqrt{\tau}}, \frac{\ln \frac{S_2^*}{S_1^*} + \frac{\sigma_{12}^2}{2}\tau}{\sigma_{12}\sqrt{\tau}}; \frac{\sigma_2 - \rho_{12}\sigma_1}{\sigma_{12}} \right). \qquad (3.73)
$$

This expression will also be referred to as $V_{max}(S_1, S_2, t)$ later because the payoff of the option is $\max(S_0, S_1, S_2)$. In the expression (3.73), it seems that the rule of substitution of subscripts does not work. Actually, the rule should be used in the following way. First, the rule is applied to S_i^*. Then, you should determine the volatilities and correlation coefficient in a function as follows. If S_i^*/S_j^* appears in an argument expression, then the volatility in the expression is the volatility of S_i/S_j. If S_i^*/S_k^* and S_j^*/S_k^* (or equivalently, S_k^*/S_i^* and S_k^*/S_j^*) appear in the function, then the third argument is the correlation coefficient between S_i/S_k and S_j/S_k.

Here, we would like to make the following four remarks:

1. Through a procedure similar to what we used to derive (3.73), it can be shown that for a European option on the minimum of two assets S_1, S_2 and cash S_0, whose payoff is $\min(S_0, S_1, S_2)$, the price is

$$
V_{\min}(S_1, S_2, t)
$$

$$
= S_0^* N_2 \left(\frac{\ln \frac{S_1^*}{S_0^*} - \frac{\sigma_1^2}{2}\tau}{\sigma_1\sqrt{\tau}}, \frac{\ln \frac{S_2^*}{S_0^*} - \frac{\sigma_2^2}{2}\tau}{\sigma_2\sqrt{\tau}}; \rho_{12} \right)
$$

$$
+ S_1^* N_2 \left(\frac{\ln \frac{S_2^*}{S_1^*} - \frac{\sigma_{12}^2}{2}\tau}{\sigma_{12}\sqrt{\tau}}, \frac{\ln \frac{S_0^*}{S_1^*} - \frac{\sigma_1^2}{2}\tau}{\sigma_1\sqrt{\tau}}; \frac{\sigma_1 - \rho_{12}\sigma_2}{\sigma_{12}} \right)
$$

$$+ S_2^* N_2 \left(\frac{\ln \frac{S_0^*}{S_2^*} - \frac{\sigma_2^2}{2}\tau}{\sigma_2 \sqrt{\tau}}, \frac{\ln \frac{S_1^*}{S_2^*} - \frac{\sigma_{12}^2}{2}\tau}{\sigma_{12}\sqrt{\tau}}; \frac{\sigma_2 - \rho_{12}\sigma_1}{\sigma_{12}} \right).$$

This is left for the reader as Problem 17.

2. Because

$$\max(E, S) = \max(S - E, 0) + E,$$

for a European option with a payoff $\max(S, E)$, the solution is

$$
\begin{aligned}
c(S, t) + Ee^{-r(T-t)} &= Se^{-D_0(T-t)}N(d_1) - Ee^{-r(T-t)}N(d_2) + Ee^{-r(T-t)} \\
&= Ee^{-r(T-t)}N(-d_2) + Se^{-D_0(T-t)}N(d_1) \\
&= S_0^* N \left(\frac{\ln \frac{S_0^*}{S_1^*} + \frac{\sigma^2}{2}\tau}{\sigma\sqrt{\tau}} \right) + S_1^* N \left(\frac{\ln \frac{S_1^*}{S_0^*} + \frac{\sigma^2}{2}\tau}{\sigma\sqrt{\tau}} \right),
\end{aligned}
$$

where

$$S_0^* = Ee^{-r(T-t)}, \qquad S_1^* = Se^{-D_0(T-t)}.$$

For this case, obtaining the second term from the first term follows the rule:

$$0 \to 1, \quad 1 \to 0.$$

Therefore, (3.73) can be understood as a generalization of this expression. For an option with a payoff $\max(S_0, S_1, S_2, \cdots, S_n)$, where S_0 is a constant and $S_1, S_2, \cdots, S_n$ are random variables, a term can be obtained from another term by the following rule

$$0 \to 1, \quad 1 \to 2, \quad \cdots, \quad n-1 \to n, \quad n \to 0$$

if the solution is written in a similar form to (3.73). The solution of the option with a payoff $\max(S_0, S_1, S_2, S_3)$ can be found in Problem 18, and the reader is asked to show this result. In Problem 19, the reader is asked to guess the solution for an option with a payoff $\max(S_0, S_1, S_2, \cdots, S_n)$. For options with a payoff $\min(S_0, S_1, S_2, \cdots, S_n)$, the situation is similar, and Problems 18–19 also involve these options.

3. We have pointed out that the value of the bivariate cumulative distribution function appearing in the first term in (3.73) denotes the probability of the event $\max(S_{1T}, S_{2T}) \leq S_0$ in the so-called "risk-neutral" world. Now we look at

$$N_2 \left(\frac{\ln \frac{S_1^*}{S_2^*} + \frac{\sigma_{12}^2}{2}\tau}{\sigma_{12}\sqrt{\tau}}, \frac{\ln \frac{S_1^*}{S_0^*} + \frac{\sigma_1^2}{2}\tau}{\sigma_1\sqrt{\tau}}; \frac{\sigma_1 - \rho_{12}\sigma_{12}}{\sigma_{12}} \right).$$

As we saw, it is the value of the integral

$$\int_0^1 \int_0^1 \psi\left(\xi_{21T}, \xi_{01T}; \xi_{21}, \xi_{01}, t\right) d\xi_{21T} d\xi_{01T}.$$

Therefore, the value of the bivariate cumulative distribution function appearing in the second term of $V\left(S_1, S_2, t\right)$ can be interpreted as the probability of the event $\max(\xi_{21T}, \xi_{01T}) = \max\left(\dfrac{S_{2T}}{S_{1T}}, \dfrac{S_0}{S_{1T}}\right) \leq 1$, which is equivalent to $S_{1T} \geq \max(S_0, S_{2T})$, in another "risk-neutral" world with the probability density function

$$\psi(\xi_{21T}, \xi_{01T}; \xi_{21}, \xi_{01}, t) = \frac{1}{2\pi\tau\sqrt{\det \overline{\mathbf{P}}}\sigma_1\sigma_{12}\xi_{21T}\xi_{01T}} e^{-\zeta^T \overline{\mathbf{P}}^{-1}\zeta/2},$$

where

$$\overline{\mathbf{P}} = \begin{bmatrix} 1 & \rho_{201} \\ \rho_{201} & 1 \end{bmatrix}$$

and

$$\zeta = \begin{bmatrix} \zeta_1 \\ \zeta_2 \end{bmatrix} = \begin{bmatrix} \dfrac{\ln\xi_{21T} - \left(\ln\xi_{21} + (D_{01} - D_{02} - \sigma_{12}^2/2)\tau\right)}{\sigma_{12}\sqrt{\tau}} \\ \dfrac{\ln\xi_{01T} - \left(\ln\xi_{01} + (D_{01} - r - \sigma_1^2/2)\tau\right)}{\sigma_1\sqrt{\tau}} \end{bmatrix}.$$

Similarly, the value of the bivariate cumulative distribution function appearing in the third term of $V\left(S_1, S_2, t\right)$ represents the probability of the event $\max\left(\dfrac{S_0}{S_{2T}}, \dfrac{S_{1T}}{S_{2T}}\right) \leq 1$ or $S_{2T} \geq \max(S_0, S_{1T})$ in a third "risk-neutral" world.

4. As the option with a payoff $\max(S, E)$ is related to the European vanilla call option, the option with a payoff $\max(E, S_1, S_2)$ is related to the European call option on the maximum of two assets, whose payoff is $\max(\max(S_1, S_2) - E, 0)$. Because

$$\max(\max(S_1, S_2) - E, 0) = \max(E, S_1, S_2) - E,$$

the price of this option is

$$V_{max}(S_1, S_2, t) - Ee^{-r(T-t)},$$

where $V_{max}(S_1, S_2, t)$ is the expression given by (3.73). Similarly, we can show that the price of a European put option on the minimum of two assets is

$$Ee^{-r(T-t)} - V_{min}(S_1, S_2, t)$$

because its payoff is

$$\max(E - \min(S_1, S_2), 0) = E - \min(E, S_1, S_2).$$

Using a similar procedure, we can prove some other relations, which are left for the reader to show as Problems 20–21.

3.5.4 Formulation of Multi-Asset Option Problems on a Finite Domain

In the last two subsections, we studied some options on multi-assets whose solutions can be expressed in terms of multivariate cumulative distribution functions if every σ_i is constant and pricing of which is reduced to find the value of these functions. If σ_i depends on S_i, it may not be possible to express their solutions in terms of such functions. Moreover, for some other options, it might be hard to express their solutions in such a form even though σ_i are constants. Here we give some examples:

A. Maximum multi-strike options. For such an option, the payoff function is

$$\max(S_1 - E_1, S_2 - E_2, \cdots, S_n - E_n, 0).$$

B. Minimum multi-strike options. The payoff function is in the form

$$\min(S_1 - E_1, S_2 - E_2, \cdots, S_n - E_n, 0).$$

C. Basket options and index options. The payoff function of a basket option (see [41]) is

$$V(S_1, S_2, \cdots, S_n, T) = \max\left(\sum_{i=1}^{n} \lambda_i S_i - E, 0\right),$$

where $\lambda_i \geq 0$ and $\sum_{i=1}^{n} \lambda_i = 1$. The payoff function of an index option (see [15]) is

$$V(S_1, S_2, \cdots, S_n, T) = \max\left(\sum_{i=1}^{n} n_i S_i - E, 0\right),$$

where n_i is the number of shares of asset i held in the index. An index option is equivalent to a basket option because

$$\max\left(\sum_{i=1}^{n} n_i S_i - E, 0\right) = \left(\sum_{i=1}^{n} n_i\right) \max\left(\sum_{i=1}^{n} \lambda_i S_i - \overline{E}, 0\right),$$

where

$$\lambda_i = \frac{n_i}{\sum\limits_{i=1}^{n} n_i}, \quad \overline{E} = \frac{E}{\sum\limits_{i=1}^{n} n_i}$$

and λ_i satisfies the relation $\sum_{i=1}^{n} \lambda_i = 1$.

From Smithson's paper [65], the reader can find these and some other options on multi-assets whose solutions might not be expressed in terms of multivariate cumulative distribution functions. For these cases, it might be necessary to use numerical methods. It will become much easier to get numerical solutions if the problem can be reformulated on a finite domain. In this section, we reformulate the two-asset option problems on a finite domain, even though the method can be generalized to multi-asset option problems.

Let us first introduce a new coordinate system:

$$\begin{cases} S_1 = P_1 S \cos\theta, \\ S_2 = P_2 S \sin\theta, \end{cases}$$

that is,

$$\begin{cases} S = \sqrt{\left(\dfrac{S_1}{P_1}\right)^2 + \left(\dfrac{S_2}{P_2}\right)^2}, \\ \theta = \tan^{-1}\dfrac{P_1 S_2}{P_2 S_1}. \end{cases}$$

Under the transformation, the domain $[0,\infty) \times [0,\infty)$ on the (S_1, S_2)-plane becomes the domain $[0,\infty) \times \left[0, \dfrac{\pi}{2}\right]$ on the (S, θ)-plane. Noticing

$$\frac{\partial S}{\partial S_1} = \frac{\cos\theta}{P_1}, \qquad \frac{\partial S}{\partial S_2} = \frac{\sin\theta}{P_2},$$

$$\frac{\partial \theta}{\partial S_1} = \frac{-\sin\theta}{P_1 S}, \qquad \frac{\partial \theta}{\partial S_2} = \frac{\cos\theta}{P_2 S},$$

we have the following relations

$$\frac{\partial V}{\partial S_1} = \frac{1}{P_1}\left(\frac{\partial V}{\partial S}\cos\theta - \frac{1}{S}\frac{\partial V}{\partial \theta}\sin\theta\right),$$

$$\begin{aligned} \frac{\partial^2 V}{\partial S_1^2} = \frac{1}{P_1^2}\bigg(&\frac{\partial^2 V}{\partial S^2}\cos^2\theta - \frac{2}{S}\frac{\partial^2 V}{\partial S\partial\theta}\sin\theta\cos\theta + \frac{1}{S^2}\frac{\partial^2 V}{\partial \theta^2}\sin^2\theta \\ &+ \frac{1}{S}\frac{\partial V}{\partial S}\sin^2\theta + \frac{2}{S^2}\frac{\partial V}{\partial \theta}\sin\theta\cos\theta\bigg), \end{aligned}$$

$$\frac{\partial V}{\partial S_2} = \frac{1}{P_2}\left(\frac{\partial V}{\partial S}\sin\theta + \frac{1}{S}\frac{\partial V}{\partial \theta}\cos\theta\right),$$

$$\begin{aligned} \frac{\partial^2 V}{\partial S_2^2} = \frac{1}{P_2^2}\bigg(&\frac{\partial^2 V}{\partial S^2}\sin^2\theta + \frac{2}{S}\frac{\partial^2 V}{\partial S\partial\theta}\sin\theta\cos\theta + \frac{1}{S^2}\frac{\partial^2 V}{\partial \theta^2}\cos^2\theta \\ &+ \frac{1}{S}\frac{\partial V}{\partial S}\cos^2\theta - \frac{2}{S^2}\frac{\partial V}{\partial \theta}\sin\theta\cos\theta\bigg), \end{aligned}$$

$$\begin{aligned} \frac{\partial^2 V}{\partial S_1 \partial S_2} = \frac{1}{P_1 P_2}\bigg[&\frac{\partial^2 V}{\partial S^2}\sin\theta\cos\theta + \frac{1}{S}\frac{\partial^2 V}{\partial S\partial\theta}(\cos^2\theta - \sin^2\theta) \\ &- \frac{1}{S^2}\frac{\partial^2 V}{\partial \theta^2}\sin\theta\cos\theta - \frac{1}{S}\frac{\partial V}{\partial S}\sin\theta\cos\theta - \frac{1}{S^2}\frac{\partial V}{\partial \theta}(\cos^2\theta - \sin^2\theta)\bigg]. \end{aligned}$$

From these, we can rewrite (3.50) with $n = 2$ as

$$\begin{cases} \dfrac{\partial V}{\partial t} + a_1 S^2 \dfrac{\partial^2 V}{\partial S^2} + a_{12} S \dfrac{\partial^2 V}{\partial S \partial \theta} + a_2 \dfrac{\partial^2 V}{\partial \theta^2} + b_1 S \dfrac{\partial V}{\partial S} + b_2 \dfrac{\partial V}{\partial \theta} - rV = 0, \\[2mm] 0 \le S, \quad 0 \le \theta \le \dfrac{\pi}{2}, \quad 0 \le t \le T, \end{cases}$$

$$(3.74)$$

where

$$a_1 = \frac{1}{2}\left(\sigma_1^2 \cos^4 \theta + 2\rho_{12}\sigma_1\sigma_2 \sin^2 \theta \cos^2 \theta + \sigma_2^2 \sin^4 \theta\right),$$

$$\begin{aligned} a_{12} &= -\sigma_1^2 \sin\theta \cos^3 \theta + \rho_{12}\sigma_1\sigma_2 \cos\theta \sin\theta (\cos^2 \theta - \sin^2 \theta) + \sigma_2^2 \sin^3 \theta \cos\theta \\ &= \left[-\sigma_1(\sigma_1 - \rho_{12}\sigma_2) \cos^2 \theta + \sigma_2 (\sigma_2 - \rho_{12}\sigma_1) \sin^2 \theta\right] \sin\theta \cos\theta, \end{aligned}$$

$$a_2 = \frac{1}{2}(\sigma_1^2 - 2\rho_{12}\sigma_1\sigma_2 + \sigma_2^2) \cos^2 \theta \sin^2 \theta,$$

$$\begin{aligned} b_1 &= (r - D_{01}) \cos^2 \theta + (r - D_{02}) \sin^2 \theta + \frac{1}{2}(\sigma_1^2 - 2\rho_{12}\sigma_1\sigma_2 + \sigma_2^2) \sin^2 \theta \cos^2 \theta \\ &= r + \left[-D_{01} \cos^2 \theta - D_{02} \sin^2 \theta + \frac{1}{2}(\sigma_1^2 - 2\rho_{12}\sigma_1\sigma_2 + \sigma_2^2) \sin^2 \theta \cos^2 \theta\right], \end{aligned}$$

$$\begin{aligned} b_2 &= [-(r - D_{01}) + (r - D_{02})] \sin\theta \cos\theta + \sigma_1^2 \sin\theta \cos^3 \theta \\ &\quad - \rho_{12}\sigma_1\sigma_2 \sin\theta \cos\theta (\cos^2 \theta - \sin^2 \theta) - \sigma_2^2 \sin^3 \theta \cos\theta \\ &= \left[D_{01} - D_{02} + \sigma_1^2 \cos^2 \theta - \rho_{12}\sigma_1\sigma_2(\cos^2 \theta - \sin^2 \theta) - \sigma_2^2 \sin^2 \theta\right] \sin\theta \cos\theta. \end{aligned}$$

Now let us introduce another transformation

$$\begin{cases} \xi = \dfrac{S}{S + P_m}, \\[2mm] \theta = \theta, \\[2mm] w = \dfrac{V}{S + P_m}. \end{cases}$$

Under this transformation, the domain $[0, \infty)$ on the S-axis is transformed into the domain $[0, 1)$ on the ξ-axis. Because

$$S = \frac{\xi P_m}{1 - \xi},$$

$$S + P_m = \frac{P_m}{1 - \xi},$$

$$\frac{d\xi}{dS} = \frac{P_m}{(S + P_m)^2} = \frac{(1 - \xi)^2}{P_m},$$

we get

$$\frac{\partial V}{\partial t} = (S + P_m) \frac{\partial w}{\partial t} = \frac{P_m}{1 - \xi} \frac{\partial w}{\partial t},$$

$$\frac{\partial V}{\partial S} = w + (S + P_m)\frac{\partial w}{\partial \xi}\frac{d\xi}{dS} = w + (1 - \xi)\frac{\partial w}{\partial \xi},$$

$$\frac{\partial V}{\partial \theta} = \frac{P_m}{1 - \xi}\frac{\partial w}{\partial \theta},$$

$$\frac{\partial^2 V}{\partial S^2} = \frac{(1 - \xi)^3}{P_m}\frac{\partial^2 w}{\partial \xi^2},$$

$$\frac{\partial^2 V}{\partial \theta \partial S} = \frac{\partial w}{\partial \theta} + (1 - \xi)\frac{\partial^2 w}{\partial \theta \partial \xi},$$

$$\frac{\partial^2 V}{\partial \theta^2} = \frac{P_m}{1 - \xi}\frac{\partial^2 w}{\partial \theta^2}.$$

Consequently, we arrive at the final equation

$$
\begin{cases}
\dfrac{\partial w}{\partial t} + a_1\xi^2(1 - \xi)^2\dfrac{\partial^2 w}{\partial \xi^2} + a_{12}\xi(1 - \xi)\dfrac{\partial^2 w}{\partial \theta \partial \xi} + a_2\dfrac{\partial^2 w}{\partial \theta^2} \\[3mm]
+ b_1\xi(1 - \xi)\dfrac{\partial w}{\partial \xi} + (b_2 + a_{12}\xi)\dfrac{\partial w}{\partial \theta} - (r - b_1\xi)w = 0, \\[3mm]
\qquad 0 \le \xi \le 1, \quad 0 \le \theta \le \dfrac{\pi}{2}, \quad 0 \le t \le T.
\end{cases}
\tag{3.75}
$$

The combination of the above two transformations is

$$
\begin{cases}
\xi = \dfrac{\sqrt{\left(\dfrac{S_1}{P_1}\right)^2 + \left(\dfrac{S^2}{P_2}\right)^2}}{\sqrt{\left(\dfrac{S_1}{P_1}\right)^2 + \left(\dfrac{S_2}{P_2}\right)^2} + P_m}, \\[8mm]
\theta = \tan^{-1}\dfrac{P_1 S_2}{P_2 S_1}, \\[5mm]
w = \dfrac{V}{\sqrt{\left(\dfrac{S_1}{P_1}\right)^2 + \left(\dfrac{S_2}{P_2}\right)^2} + P_m}
\end{cases}
$$

and we can derive (3.75) directly from (3.50) with $n = 2$ by using this transformation. Here, we do it through two steps in order to make the idea clear.

It can be proven that in order to determine a unique solution, only the final condition

$$w(\xi, \theta, T) = f(\xi, \theta), \qquad 0 \le \xi \le 1, \quad 0 \le \theta \le \frac{\pi}{2} \tag{3.76}$$

is needed. The proof is similar to that given in the paper by Zhu and Li [85]. Here, we do not give the proof but an explanation on this issue. At $\theta = 0$ and $\theta = \dfrac{\pi}{2}$, the equation (3.75) becomes

$$\frac{\partial w}{\partial t} + \frac{1}{2}\sigma_1^2 \xi^2 (1-\xi)^2 \frac{\partial^2 w}{\partial \xi^2} + (r - D_{01})\xi(1-\xi)\frac{\partial w}{\partial \xi} - [r(1-\xi) + D_{01}\xi]\, w = 0,$$

$$0 \le \xi \le 1, \qquad 0 \le t \le T$$

and

$$\frac{\partial w}{\partial t} + \frac{1}{2}\sigma_2^2 \xi^2 (1-\xi)^2 \frac{\partial^2 w}{\partial \xi^2} + (r - D_{02})\xi(1-\xi)\frac{\partial w}{\partial \xi} - [r(1-\xi) + D_{02}\xi]\, w = 0,$$

$$0 \le \xi \le 1, \qquad 0 \le t \le T$$

respectively. These are one-dimensional parabolic equations that degenerate into ordinary differential equations at $\xi = 0$ and $\xi = 1$. Therefore, if a final condition is given for each equation, the solution for each equation is unique. These two solutions give the value of the solution to (3.75) at the boundaries $\theta = 0$ and $\theta = \pi/2$. At $\xi = 0$ and $\xi = 1$, (3.75) becomes

$$\frac{\partial w}{\partial t} + a_2 \frac{\partial^2 w}{\partial \theta^2} + b_2 \frac{\partial w}{\partial \theta} - rw = 0, \qquad 0 \le \theta \le \frac{\pi}{2}, \qquad 0 \le t \le T$$

and

$$\frac{\partial w}{\partial t} + a_2 \frac{\partial^2 w}{\partial \theta^2} + (b_2 + a_{12}) \frac{\partial w}{\partial \theta} - (r - b_1)w = 0, \qquad 0 \le \theta \le \frac{\pi}{2}, \qquad 0 \le t \le T$$

respectively. These two parabolic partial differential equations also degenerate to ordinary differential equations at $\theta = 0$ and $\theta = \pi/2$, so in order for them to have unique solutions, only final conditions are enough. Just like the situation on the boundaries $\theta = 0$ and $\theta = \pi/2$, the solutions of these two equations provide the value of the solution to the equation (3.75) at $\xi = 0$ and $\xi = 1$. Consequently, the final condition (3.76) determines a unique solution to the equation (3.75).

For an American multi-asset option, when the problem is formulated as a linear complementarity problem and this transformation given in this subsection is adopted, the problem is defined on a finite rectangular domain. Such a linear complementarity problem is not difficult to solve numerically.

3.6 Some Other Exotic Options

In this section, we introduce some other exotic options, namely, binary options, forward start options, compound options, and chooser options. In what follows, we discuss each case in each subsection.

3.6.1 Binary Options

Binary options are options with discontinuous payoffs. A simple example is a cash-or-nothing call. For this case, the payoff is

$$V(S,T) = B \times H(S - E),$$

where B is a constant and H is the Heaviside function:

$$H(S - E) = \begin{cases} 0, & S < E, \\ 1, & S > E. \end{cases}$$

From the formula (2.35) in Subsection 2.4.2, we know that $V(S,t)$ is given by

$$\begin{aligned} V(S,t) &= e^{-r(T-t)} \int_0^{\infty} B \times H(S' - E) G(S',T;S,t)\, dS' \\ &= Be^{-r(T-t)} \int_E^{\infty} G(S',T;S,t)\, dS' \\ &= Be^{-r(T-t)} N(d_2), \end{aligned}$$

where $N(z)$ is defined by (2.40) in Subsection 2.4.2 and

$$d_2 = \frac{\ln(S/E) + (r - D_0 - \sigma^2/2)(T - t)}{\sigma\sqrt{T - t}}.$$

Besides cash-or-nothing calls, we can also have cash-or-nothing puts, asset-or-nothing calls, and asset-or-nothing puts. Their payoffs are

$$B \times H(E - S), \quad S \times H(S - E), \quad \text{and} \quad S \times H(E - S)$$

respectively. Their closed-form solutions are left for the reader to derive as an exercise.

3.6.2 Forward Start Options (Delayed Strike Options)

Forward start (delayed strike) options are options that are paid for now but will start at some time T_1 in the future, and the exercise price of which depends on the stock price at time T_1. Here, we assume that the exercise price $E = \alpha S_{T_1}$, where S_{T_1} is the stock price at time T_1 and α is a positive number.

Suppose that the current time is t_0 and consider a forward start American call option that will start at time $T_1 > t_0$ and mature at time $T_2 > T_1$. In order to have the value of the forward start option, we first need to find the solution of the American call problem with a payoff

$$C(S,T_2) = \max(S - \alpha S_{T_1}, 0)$$

at time T_2. Actually, the solution of this problem $C(S,t)$ is equal to

$$\alpha S_{T_1} C^* (S/\alpha S_{T_1}, t),$$

where $C^* (\bar{S},t)$ is the solution of the American option with the following standard payoff

$$C^* (\bar{S}, T_2) = \max (\bar{S} - 1, 0)$$

at time T_2. Therefore,

$$C(S_{T_1}, T_1) = \alpha S_{T_1} C^* \left(\frac{1}{\alpha}, T_1 \right).$$

At time t_0, we do not know the stock price at time T_1, implying that S_{T_1} is a random variable. Therefore, the value of the forward start option at time t is

$$e^{-r(T_1-t_0)} \mathrm{E} \left[C(S_{T_1}, T_1) \right] = \alpha e^{-r(T_1-t_0)} C^* \left(\frac{1}{\alpha}, T_1 \right) \mathrm{E} \left[S_{T_1} \right]$$

$$= \alpha e^{-r(T_1-t_0)} C^* \left(\frac{1}{\alpha}, T_1 \right) S e^{(r-D_0)(T_1-t_0)} = \alpha S e^{-D_0(T_1-t_0)} C^* \left(\frac{1}{\alpha}, T_1 \right).$$

Here, we assume that we are in a "risk-neutral" world. If the option is European, we can have a similar formula, and $C^* \left(\frac{1}{\alpha}, T_1 \right)$ should be replaced by $c^* \left(\frac{1}{\alpha}, T_1 \right)$, which is the solution of the standard European call option and has an analytic expression.

3.6.3 Compound Options

Compound options are options on options. There are four main types of compound options: a call on a call, a put on a call, a call on a put, and a put on a put. Compound options have two strike prices and two exercise dates. For example, the holder of a call on call option is entitled to pay the first strike price E_1 and receive a call option on the first date T_1, and is given the right to buy the underlying asset for the second strike price E_2 on the second exercise date $T_2 > T_1$ if the first option is exercised. Here we assume that both options are European. Let c_1 denote the value of the option when the first option is a European call, and let p_1 denote the value of the option when the first one is a put. Furthermore, let $c_1(S, t; C_2)$ and $c_1(S, t; c_2)$ denote the price of a European call option on an American call option and a European call option, respectively. For the other cases, we adopt similar notation. When the first option is a European call, the compound option will only be exercised on the first exercise date if the value of the option on that date is greater than the first strike price. Assume that the second call option is American. Then, the price of the compound option is

$$c_1\left(S,t;C_2\right) = e^{-r(T_1-t)} \int_0^\infty c_1\left(S',T_1;C_2\right) G\left(S',T_1;S,t\right) dS',$$

where $c_1\left(S',T_1;C_2\right) = \max\left(C_2\left(S',T_1\right) - E_1, 0\right)$ and $G\left(S',T_1;S,t\right)$ is given by (2.36) in Subsection 2.4.2.

Clearly, in this case, the price can be written as

$$c_1\left(S,t;C_2\right) = e^{-r(T_1-t)} \int_{S^*}^\infty \left[C_2\left(S',T_1\right) - E_1\right] G\left(S',T_1;S,t\right) dS',$$

where S^* satisfies the condition $C_2\left(S^*,T_1\right) - E_1 = 0$. In order to get $c_1\left(S,t;C_2\right)$, we can find the value of the integral numerically or we can solve the following final value problem by a numerical method:

$$\begin{cases} \dfrac{\partial c_1}{\partial t} + \dfrac{1}{2}\sigma^2 S^2 \dfrac{\partial^2 c_1}{\partial S^2} + (r - D_0)\, S \dfrac{\partial c_1}{\partial S} - rc_1 = 0, & 0 \le S, \quad 0 \le t \le T_1, \\ c_1\left(S,T_1;C_2\right) = \max\left(C_2(S,T_1) - E_1, 0\right), & 0 \le S. \end{cases}$$

Before that, we need to solve an American call option with a payoff function

$$C_2\left(S,T_2\right) = \max\left(S - E_2, 0\right)$$

and satisfying the constraint

$$C_2\left(S,t\right) \ge \max\left(S - E_2, 0\right)$$

from T_2 to T_1, so that we can have the value of the function $C_2\left(S,T_1\right)$ for any S. Because there is no analytic expression for $C_2\left(S,T_1\right)$ if $D_0 \ne 0$, we also have to get $C_2\left(S,T_1\right)$ numerically. In Fig. 3.9, the price of a European call on an American call option at time $t = 0$ is given. The result is obtained by solving the partial differential equation problem, and the parameters of this problem are given in the figure. For a European call on an American put, the price can be determined similarly.

The price of a European put on an American put is the solution of the problem

$$\begin{cases} \dfrac{\partial p_1}{\partial t} + \dfrac{1}{2}\sigma^2 S^2 \dfrac{\partial^2 p_1}{\partial S^2} + (r - D_0)\, S \dfrac{\partial p_1}{\partial S} - rp_1 = 0, & 0 \le S, \quad 0 \le t \le T_1, \\ p_1\left(S,T_1;P_2\right) = \max\left(E_1 - P_2(S,T_1), 0\right), & 0 \le S, \end{cases}$$

where P_2 is the price of an American put option with a payoff $P_2(S,T_2) = \max(E_2 - S, 0)$ and satisfying the constraint $P_2(S,t) \ge \max(E_2 - S, 0)$. The price can be obtained numerically. In Fig. 3.10, the price of a European put on an American put at time $t = 0$ is shown. The price of a European put on an American call option can be obtained in a similar way.

For a European option on a European option, the compound option can be valued analytically in terms of integrals of the bivariate standardized normal distribution (see [29], [39], or [49]). These closed-form solutions are

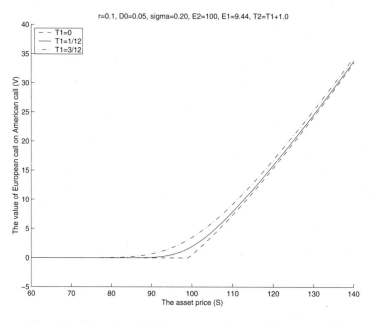

Fig. 3.9. The price of a European call on an American call

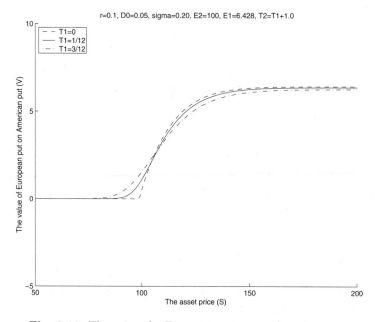

Fig. 3.10. The price of a European put on an American put

$$c_1\left(S, t; c_2\right) = Se^{-D_0\left(T_2-t\right)}N_2\left(d_{11}, d_{12}; \rho\right) - E_2 e^{-r\left(T_2-t\right)}N_2\left(d_{21}, d_{22}; \rho\right)$$
$$-E_1 e^{-r\left(T_1-t\right)}N_2\left(d_{21}\right),$$
$$c_1\left(S, t; p_2\right) = E_2 e^{-r\left(T_2-t\right)}N_2\left(-d_{23}, -d_{22}; \rho\right) - Se^{-D_0\left(T_2-t\right)}N_2\left(-d_{13}, -d_{12}; \rho\right)$$
$$-E_1 e^{-r\left(T_1-t\right)}N\left(-d_{23}\right),$$
$$p_1\left(S, t; c_2\right) = E_1 e^{-r\left(T_1-t\right)}N\left(-d_{21}\right) - Se^{-D_0\left(T_2-t\right)}N_2\left(-d_{11}, d_{12}; -\rho\right)$$
$$+E_2 e^{-r\left(T_2-t\right)}N_2\left(-d_{21}, d_{22}; -\rho\right),$$
$$p_1\left(S, t; p_2\right) = E_1 e^{-r\left(T_1-t\right)}N\left(d_{23}\right) - E_2 e^{-r\left(T_2-t\right)}N_2\left(d_{23}, -d_{22}; -\rho\right)$$
$$+Se^{-D_0\left(T_2-t\right)}N_2\left(d_{13}, -d_{12}; -\rho\right),$$

where

$$d_{11} = \frac{\ln(S/S^*) + \left(r - D_0 + \sigma^2/2\right)\left(T_1 - t\right)}{\sigma\sqrt{T_1 - t}},$$

$$d_{21} = \frac{\ln(S/S^*) + \left(r - D_0 - \sigma^2/2\right)\left(T_1 - t\right)}{\sigma\sqrt{T_1 - t}},$$

$$d_{12} = \frac{\ln(S/E_2) + \left(r - D_0 + \sigma^2/2\right)\left(T_2 - t\right)}{\sigma\sqrt{T_2 - t}},$$

$$d_{22} = \frac{\ln(S/E_2) + \left(r - D_0 - \sigma^2/2\right)\left(T_2 - t\right)}{\sigma\sqrt{T_2 - t}},$$

$$d_{13} = \frac{\ln(S/S^{**}) + \left(r - D_0 + \sigma^2/2\right)\left(T_1 - t\right)}{\sigma\sqrt{T_1 - t}},$$

$$d_{23} = \frac{\ln(S/S^{**}) + \left(r - D_0 - \sigma^2/2\right)\left(T_1 - t\right)}{\sigma\sqrt{T_1 - t}},$$

$$\rho = \sqrt{\frac{T_1 - t}{T_2 - t}}.$$

Here, S^* and S^{**} are the solutions of the following equations:

$$c_2\left(S^*, T_1\right) = E_1$$

and

$$p_2\left(S^{**}, T_1\right) = E_1.$$

Let us derive the formula for the price of a put on a call, and other formulae are left for the reader to prove. For a put on a call, we have

$$p_1\left(S, t; c_2\right) = e^{-r\left(T_1-t\right)}\int_0^\infty \max\left(E_1 - c_2(S', T_1), 0\right)G\left(S', T_1; S, t\right)dS'$$

$$= e^{-r\left(T_1-t\right)}\int_0^{S^*}\left[E_1 - c_2\left(S', T_1\right)\right]G\left(S', T_1; S, t\right)dS'$$

$$= e^{-r(T_1-t)}E_1 \int_0^{S^*} G\left(S',T_1;S,t\right)dS'$$

$$-e^{-r(T_1-t)}\int_0^{S^*} e^{-r(T_2-T_1)}\int_{E_2}^{\infty}(S''-E_2)G\left(S'',T_2;S',T_1\right)$$

$$\times G\left(S',T_1;S,t\right)dS''dS'$$

$$= e^{-r(T_1-t)}E_1 \int_0^{S^*} G\left(S',T_1;S,t\right)dS'$$

$$-e^{-r(T_2-t)}\int_0^{S^*}\int_{E_2}^{\infty} S''G\left(S'',T_2;S',T_1\right)G\left(S',T_1;S,t\right)dS''dS'$$

$$+E_2 e^{-r(T_2-t)}\int_0^{S^*}\int_{E_2}^{\infty} G\left(S'',T_2;S',T_1\right)G\left(S',T_1;S,t\right)dS''dS',$$

where

$$G\left(S',T_1;S,t\right) = \frac{1}{\sigma\sqrt{2\pi\left(T_1-t\right)}S'}e^{-\left[\ln S'-\ln S-\left(r-D_0-\sigma^2/2\right)(T_1-t)\right]^2/2\sigma^2(T_1-t)}$$

and $G\left(S'',T_2;S',T_1\right)$ is defined in the same way.

Now let us express the three integrals above by cumulative distribution functions. Noticing

$$\int_0^{S^*} G\left(S',T_1;S,t\right)dS' = N\left(\frac{\ln(S^*/a)+b^2/2}{b}\right),$$

where

$$a = Se^{(r-D_0)(T_1-t)}$$

and

$$b = \sigma\sqrt{T_1-t},$$

we know that the first term is equal to $E_1 e^{-r(T_1-t)}N\left(-d_{21}\right)$.

Now we find out the result of the integral in the third term. Let

$$x = \frac{\ln S' - \ln S - \left(r-D_0-\sigma^2/2\right)\left(T_1-t\right)}{\sigma\sqrt{T_1-t}},$$

$$y = \frac{\ln S'' - \ln S - \left(r-D_0-\sigma^2/2\right)\left(T_2-t\right)}{\sigma\sqrt{T_2-t}},$$

$$\rho = \sqrt{\frac{T_1-t}{T_2-t}}.$$

Then

$$\sqrt{1-\rho^2} = \sqrt{\frac{T_2 - T_1}{T_2 - t}},$$

$$\frac{y - \rho x}{\sqrt{1-\rho^2}} = \frac{\ln S'' - \ln S' - \left(r - D_0 - \sigma^2/2\right)\left(T_2 - T_1\right)}{\sigma\sqrt{T_2 - T_1}}.$$

Therefore

$$\int_0^{S^*}\int_{E_2}^{\infty} G\left(S'', T_2; S', T_1\right) G\left(S', T_1; S, t\right) dS'' dS'$$

$$= \int_0^{S^*}\int_{E_2}^{\infty} \frac{1}{\sigma\sqrt{2\pi\left(T_2 - T_1\right)}S''} e^{-\left[\ln S'' - \ln S' - \left(r - D_0 - \sigma^2/2\right)\left(T_2 - T_1\right)\right]^2/2\sigma^2\left(T_2 - T_1\right)}$$

$$\times \frac{1}{\sigma\sqrt{2\pi\left(T_1 - t\right)}S'} e^{-\left[\ln S' - \ln S - \left(r - D_0 - \sigma^2/2\right)\left(T_1 - t\right)\right]^2/2\sigma^2\left(T_1 - t\right)} dS'' dS'$$

$$= \int_{-\infty}^{-d_{21}}\int_{-d_{22}}^{\infty} \frac{1}{2\pi\sqrt{1-\rho^2}} e^{-\frac{1}{2}(y-\rho x)^2/(1-\rho^2) - \frac{1}{2}x^2} dy dx$$

$$= \frac{1}{2\pi\sqrt{1-\rho^2}} \int_{-\infty}^{-d_{21}}\int_{-d_{22}}^{\infty} e^{-(y^2 - 2\rho xy + x^2)/2(1-\rho^2)} dy dx$$

$$= \frac{1}{2\pi\sqrt{1-\rho^2}} \int_{-\infty}^{-d_{21}}\int_{-\infty}^{d_{22}} e^{-(y^2 + 2\rho xy + x^2)/2(1-\rho^2)} dy dx$$

$$= N_2\left(-d_{21}, d_{22}; -\rho\right).$$

Now let us calculate the integral in the second term. Because

$$S'' e^{-\left[\ln S'' - \ln S' - \left(r - D_0 - \sigma^2/2\right)\left(T_2 - T_1\right)\right]^2/2\sigma^2\left(T_2 - T_1\right)}$$
$$= S' e^{(r - D_0)(T_2 - T_1)} e^{-\left[\ln S'' - \ln S' - \left(r - D_0 + \sigma^2/2\right)\left(T_2 - T_1\right)\right]^2/2\sigma^2\left(T_2 - T_1\right)}$$

and

$$S' e^{-\left[\ln S' - \ln S - \left(r - D_0 - \sigma^2/2\right)\left(T_1 - t\right)\right]^2/2\sigma^2\left(T_1 - t\right)}$$
$$= S e^{(r - D_0)(T_1 - t)} e^{-\left[\ln S' - \ln S - \left(r - D_0 + \sigma^2/2\right)\left(T_1 - t\right)\right]^2/2\sigma^2\left(T_1 - t\right)},$$

we have

$$\int_0^{S^*}\int_{E_2}^{\infty} S'' G\left(S'', T_2; S', T_1\right) G\left(S', T_1; S, t\right) dS'' dS'$$

$$= \int_0^{S^*} \int_{E_2}^\infty \frac{S''}{\sigma\sqrt{2\pi(T_2-T_1)}S''} e^{-\left[\ln S'' - \ln S' - (r-D_0-\sigma^2/2)(T_2-T_1)\right]^2/2\sigma^2(T_2-T_1)}$$

$$\times \frac{1}{\sigma\sqrt{2\pi(T_1-t)}S'} e^{-\left[\ln S' - \ln S - (r-D_0-\sigma^2/2)(T_1-t)\right]^2/2\sigma^2(T_1-t)} dS'' dS'$$

$$= Se^{(r-D_0)(T_1-t)} e^{(r-D_0)(T_2-T_1)}$$

$$\times \int_0^{S^*} \int_{E_2}^\infty \frac{1}{\sigma\sqrt{2\pi(T_2-T_1)}S''} e^{-\left[\ln S'' - \ln S' - (r-D_0+\sigma^2/2)(T_2-T_1)\right]^2/2\sigma^2(T_2-T_1)}$$

$$\times \frac{1}{\sigma\sqrt{2\pi(T_1-t)}S'} e^{-\left[\ln S' - \ln S - (r-D_0+\sigma^2/2)(T_1-t)\right]^2/2\sigma^2(T_1-t)} dS'' dS'$$

$$= Se^{(r-D_0)(T_2-t)} N_2\left(-d_{11}, d_{12}; -\rho\right).$$

Here, we use the fact that the only difference between the last integral above and the integral related to the third term is that $(r - D_0 - \sigma^2/2)$ is replaced by $(r - D_0 + \sigma^2/2)$, so replacing d_{21} and d_{22} by d_{11} and d_{12} yields the result here. Consequently, we arrive at

$$p_1(S,t;c_2) = E_1 e^{-r(T_1-t)} N(-d_{21}) - Se^{-D_0(T_2-t)} N_2\left(-d_{11}, d_{12}; -\rho\right)$$
$$+ E_2 e^{-r(T_2-t)} N_2\left(-d_{21}, d_{22}; -\rho\right).$$

3.6.4 Chooser Options

Chooser (as-you-like-it) options are only slightly more complicated than compound options (see [72]). A chooser option gives its owner the right to purchase either a call for an amount E_{1c} or a put for an amount E_{1p} at time T_1. We suppose that both the call and put options are expired at time T_2 and with an exercise price E_2. We still assume that the first option is European and the second options are American. Therefore, in order to find the price of such a chooser option, we need to do the following. First, find the price functions at time T_1 of the American call option and the American put option with exercise price E_2 and expiry T_2. Then, calculate the price of the chooser option by

$$c_1(S,t) = e^{-r(T_1-t)} \int_0^\infty \max\left(C_2(S',T_1) - E_{1c}, P_2(S',T_1) - E_{1p}, 0\right)$$
$$\times G(S', T_1; S, t) \, dS',$$

where $C_2(S',T_1)$ and $P_2(S',T_1)$ are the values of the second American call and put options, respectively. Just like the case of compound options, $c_1(S,t)$ can be obtained by numerical integration or by solving a final value problem.

A chooser option could be more complicated (see [39] and [73]). It could also be less complicated. If $E_{1c} = E_{1p} = 0$, then the chooser option is called a standard chooser option (see [62], [49], or [39]). In this case, if both the underlying options are European, then the chooser option price is given by

$$V(S,t) = Se^{-D_0(T_2-t)}N(d_1) - E_2e^{-r(T_2-t)}N(d_2)$$
$$+E_2e^{-r(T_2-t)}N(-d_4) - Se^{-D_0(T_2-t)}N(-d_3)$$

where

$$d_1 = \left[\ln(S/E_2) + \left(r - D_0 + \frac{1}{2}\sigma^2\right)(T_2 - t)\right]/\sigma\sqrt{T_2 - t},$$

$$d_2 = d_1 - \sigma\sqrt{T_2 - t},$$

$$d_3 = \left[\ln(S/E_2) + (r - D_0)(T_2 - t) + \frac{1}{2}\sigma^2(T_1 - t)\right]/\sigma\sqrt{T_1 - t},$$

$$d_4 = d_3 - \sigma\sqrt{T_1 - t}.$$

This is left for the reader to prove as an exercise.

Problems

1. *Suppose that $V(S,t)$ is the solution of the problem

$$\begin{cases} \dfrac{\partial V}{\partial t} + \dfrac{1}{2}\sigma^2 S^2\dfrac{\partial^2 V}{\partial S^2} + (r - D_0)S\dfrac{\partial V}{\partial S} - rV = 0, \\[2mm] \qquad\qquad\qquad\qquad 0 \le S, \quad t \le T, \\[2mm] V(S,T) = V_T(S), \qquad 0 \le S. \end{cases}$$

Let

$$x = \ln S, \quad \bar{\tau} = \sigma^2(T - t)/2$$

and

$$V(S,t) = e^{\alpha x + \beta\bar{\tau}}u(x, \bar{\tau}),$$

where α and β are constants. Find α and β such that $u(x, \bar{\tau})$ is the solution of the problem

$$\begin{cases} \dfrac{\partial u}{\partial\bar{\tau}} = \dfrac{\partial^2 u}{\partial x^2}, \qquad -\infty < x < \infty, \quad 0 \le \bar{\tau}, \\[2mm] u(x,0) = e^{-\alpha x}V_T(e^x), \quad -\infty < x < \infty. \end{cases}$$

2. As we know, the solution of the problem

$$\begin{cases} \dfrac{\partial u}{\partial\bar{\tau}} = \dfrac{\partial^2 u}{\partial x^2}, \qquad -\infty < x < \infty, \quad 0 \le \bar{\tau}, \\[2mm] u(x,0) = u_0(x), \quad -\infty < x < \infty \end{cases}$$

is

$$u\left(x, \bar{\tau}\right) = \frac{1}{2\sqrt{\pi\bar{\tau}}} \int_{-\infty}^{\infty} u_0\left(\xi\right) e^{-(x-\xi)^2/4\bar{\tau}} d\xi.$$

By using the result above, find the solution of the following problem

$$\begin{cases} \dfrac{\partial u}{\partial \bar{\tau}} = \dfrac{\partial^2 u}{\partial x^2}, & -\infty < x \leq a, \quad 0 \leq \bar{\tau}, \\[2mm] u\left(x, 0\right) = u_0(x), & -\infty < x \leq a, \\[2mm] u\left(a, \bar{\tau}\right) = 0, & 0 \leq \bar{\tau}. \end{cases}$$

3. Based on the result given in Problem 2, show that for the European up-and-out put option problem

$$\begin{cases} \dfrac{\partial p_o}{\partial t} + \dfrac{1}{2}\sigma^2 S^2 \dfrac{\partial^2 p_o}{\partial S^2} + (r - D_0)\, S \dfrac{\partial p_o}{\partial S} - r p_o = 0, & 0 \leq S \leq B_u, \ t \leq T, \\[2mm] p_o\left(S, T\right) = \max(E - S, 0), & 0 \leq S \leq B_u, \\[2mm] p_o\left(B_u, t\right) = 0, & t \leq T, \end{cases}$$

the solution is

$$p_o = e^{-r(T-t)} \int_0^{B_u} \max(E - S', 0) G_1(S', T; S, t, B_u) dS'$$

and when $E < B_u$, the expression can be further rewritten as

$$p_o = p(S, t) - \left(\frac{B_u}{S}\right)^{2(r-D_0-\sigma^2/2)/\sigma^2} p\left(\frac{B_u^2}{S}, t\right),$$

where $p(S, t)$ is the price of a European put option and

$$G_1\left(S', T; S, t, B_u\right)$$
$$= \frac{1}{S'\sigma\sqrt{2\pi\left(T - t\right)}} \left[e^{-\left[\ln(S'/S) - \left(r - D_0 - \sigma^2/2\right)(T-t)\right]^2/2\sigma^2(T-t)} \right.$$
$$\left. - (B_u/S)^{2(r-D_0-\sigma^2/2)/\sigma^2} e^{-\left[\ln(S'S/B_u^2) - \left(r - D_0 - \sigma^2/2\right)(T-t)\right]^2/2\sigma^2(T-t)} \right].$$

4. Show that a European up-and-out put option with $B_u > E$ plus a European up-and-in put option with the same parameters is equal to a vanilla European put option.

5. Find the solution of the European down-and-out call option

$$\begin{cases} \dfrac{\partial c_o}{\partial t} + \dfrac{1}{2}\sigma^2 S^2 \dfrac{\partial^2 c_o}{\partial S^2} + (r - D_0)\, S \dfrac{\partial c_o}{\partial S} - r c_o = 0, & B_l(t) \leq S, \ t \leq T, \\[2mm] c_o\left(S, T\right) = \max(S - E, 0), & B_l(t) \leq S, \\[2mm] c_o\left(B_l(t), t\right) = 0, & t \leq T, \end{cases}$$

where $B_l(t) = bEe^{-\alpha(T-t)}$ with $b \in [0,1]$ and $\alpha \geq 0$. (Hint: Let $\eta = Se^{\alpha(T-t)}$, the moving barrier becomes a fixed barrier in the (η, t)-plane. Then, solve a barrier option problem with a fixed barrier.)

6. Show that the closed-form solutions for the European down-and-out call option with $B_l \geq E$ and the European up-and-out put option with $B_u \leq E$ are

$$c_o = Se^{-D_0(T-t)}N\left(\tilde{d}_1(B_l)\right) - Ee^{-r(T-t)}N\left(\tilde{d}_1(B_l) - \sigma\sqrt{T-t}\right)$$

$$- (B_l/S)^{2(r-D_0-\sigma^2/2)/\sigma^2}$$

$$\times \left[\frac{B_l^2}{S}e^{-D_0(T-t)}N\left(\bar{d}_1(B_l)\right) - Ee^{-r(T-t)}N\left(\bar{d}_1(B_l) - \sigma\sqrt{T-t}\right)\right]$$

and

$$p_o = Ee^{-r(T-t)}N\left(-\tilde{d}_1(B_u) + \sigma\sqrt{T-t}\right) - Se^{-D_0(T-t)}N\left(-\tilde{d}_1(B_u)\right)$$

$$- (B_u/S)^{2(r-D_0-\sigma^2/2)/\sigma^2}$$

$$\times \left[Ee^{-r(T-t)}N\left(-\bar{d}_1(B_u) + \sigma\sqrt{T-t}\right) - \frac{B_u^2}{S}e^{-D_0(T-t)}N\left(-\bar{d}_1(B_u)\right)\right]$$

respectively. Here, the functions $\tilde{d}_1(B)$ and $\bar{d}_1(B)$ are defined by

$$\tilde{d}_1(B) = \left[\ln\frac{Se^{(r-D_0)(T-t)}}{B} + \frac{1}{2}\sigma^2(T-t)\right] \bigg/ \left(\sigma\sqrt{T-t}\right),$$

$$\bar{d}_1(B) = \left[\ln\frac{Be^{(r-D_0)(T-t)}}{S} + \frac{1}{2}\sigma^2(T-t)\right] \bigg/ \left(\sigma\sqrt{T-t}\right).$$

7. Consider an Asian option with discrete arithmetic averaging. Assume that the stock pays dividends and that during the time step $[t, t+dt]$, the dividend payment is $D(S,t)dt$. Take S and

$$I = \frac{1}{K}\int_0^t S(\tau)f(\tau)d\tau$$

as state variables, where

$$f(t) = \sum_{i=1}^{K}\delta(t-t_i).$$

a) Derive the equation for such an option directly by using a portfolio $\Pi = V - \Delta S$.

b) Then, through comparing the equation you obtain and the general PDE for derivatives, determine the market price of risk for S.

c) Finally, under the assumption $D(S,t) = D_0 S$, reduce the equation to an equation with only two independent variables.

8. Let $V(S, A, t)$ be the price of a European Asian option with continuous arithmetic averaging, where A is the average of the price during the time period $[0, t]$. As we know, the equation for European Asian option with continuous arithmetic averaging is

$$\frac{\partial W(\eta, t)}{\partial t} + \mathbf{L}_{a,t} W(\eta, t) = 0,$$

where $W = V(S, A, t)/S$, $\eta = A/S$ and $\mathbf{L}_{a,t}$ is the time-dependent operator related to Asian options and given by

$$\mathbf{L}_{a,t} = \frac{1}{2}\sigma^2\eta^2 \frac{\partial^2}{\partial\eta^2} + \left[(D_0 - r)\eta + \frac{1-\eta}{t}\right] \frac{\partial}{\partial\eta} - D_0.$$

a) Write down the LC problem for an American Asian put option with a continuous arithmetic average strike price.
b) Determine where the PDE can always be used and a free boundary cannot appear and where a free boundary may appear.
c) Derive the free-boundary problem for this case. (Assume that there exists at most one free boundary.)

9. *Consider a lookback option with continuous sampling. Derive the partial differential equation and the boundary condition for such an option involving the maximum realized price H and reduce the partial differential equation and the boundary condition to a partial differential equation and a boundary condition involving only two independent variables.

10. Suppose that the payoff of a lookback strike put option is

$$\max(H - \beta S, 0),$$

where $\beta \geq 1$. Show that its solution is

$$
\begin{aligned}
& p_{ls}\,(S, H, t) \\
&= \mathrm{e}^{-r(T-t)} S \Bigg[\frac{H}{S} N \left(\frac{\ln\frac{H}{\beta S} - \mu\,(T-t)}{\sigma\sqrt{T-t}} \right) \\
&\quad - \beta \mathrm{e}^{(r-D_0)(T-t)} N \left(\frac{\ln\frac{H}{\beta S} - (\mu + \sigma^2)\,(T-t)}{\sigma\sqrt{T-t}} \right) \\
&\quad - \frac{\sigma^2}{2\,(r - D_0)} \left(\frac{H}{S}\right)^{2(r-D_0)/\sigma^2} N \left(\frac{\ln\frac{S}{\beta H} - \mu\,(T-t)}{\sigma\sqrt{T-t}} \right) \\
&\quad + \frac{\sigma^2 \beta^{-2(r-D_0)/\sigma^2} \mathrm{e}^{(r-D_0)(T-t)}}{2\,(r - D_0)} N \left(\frac{\ln\frac{S}{\beta H} + (\mu + \sigma^2)\,(T-t)}{\sigma\sqrt{T-t}} \right) \Bigg],
\end{aligned}
$$

where

$$\mu = r - D_0 - \sigma^2/2.$$

(Hint: As the first step, show

$$E\left[\max\left(H_T-\beta S_T\right),0\right]$$

$$= \int_H^\infty \int_0^{H_t^T/\beta} f_1\left(S_T,H_t^T\right)dS_T dH_t^T + \int_0^S \max\left(H-\beta S_T,0\right)f_1\left(S_T,S\right)dS_T$$

$$+ \int_S^H \max\left(H-\beta S_T,0\right)f_1\left(S_T,S_T\right)dS_T,$$

where

$$f_1\left(S_T,H_t^T\right) = \frac{1}{S_T\sigma\sqrt{2\pi\tau}}\left(\frac{H_t^T}{S}\right)^{2\mu/\sigma^2}e^{-\left[\ln\left(S_T S/H_t^{T2}\right)-\mu\tau\right]^2/2\sigma^2\tau}.$$

)

11. Suppose the payoff of a lookback price call option is

$$\max(H-E,0).$$

Show that if $H>E$, the price is

$$c_{lp}\left(S,H,t\right)$$
$$= e^{-r(T-t)}\left\{HN\left(\frac{\ln\frac{H}{S}-\mu(T-t)}{\sigma\sqrt{T-t}}\right)-E\right.$$
$$+\left[1+\frac{\sigma^2}{2\left(r-D_0\right)}\right]Se^{(r-D_0)(T-t)}N\left(\frac{\ln\frac{S}{H}+\left(\mu+\sigma^2\right)(T-t)}{\sigma\sqrt{T-t}}\right)$$
$$\left.-\frac{\sigma^2 S}{2\left(r-D_0\right)}\left(\frac{H}{S}\right)^{2(r-D_0)/\sigma^2}N\left(\frac{\ln\frac{S}{H}-\mu(T-t)}{\sigma\sqrt{T-t}}\right)\right\};$$

and that if $H\leq E$, the price is

$$c_{lp}(S,H,t)$$
$$= e^{-r(T-t)}\left\{-EN\left(\frac{\ln\frac{S}{E}+\mu(T-t)}{\sigma\sqrt{T-t}}\right)\right.$$
$$+\left[1+\frac{\sigma^2}{2\left(r-D_0\right)}\right]Se^{(r-D_0)(T-t)}N\left(\frac{\ln\frac{S}{E}+\left(\mu+\sigma^2\right)(T-t)}{\sigma\sqrt{T-t}}\right)$$
$$\left.-\frac{\sigma^2 S}{2\left(r-D_0\right)}\left(\frac{E}{S}\right)^{2(r-D_0)/\sigma^2}N\left(\frac{\ln\frac{S}{E}-\mu(T-t)}{\sigma\sqrt{T-t}}\right)\right\},$$

where

$$\mu = r - D_0 - \sigma^2/2.$$

12. Suppose the payoff of a lookback price put option is

$$\max(E - L, 0).$$

Show that for the case $E > L$, the price is

$$
\begin{aligned}
&p_{lp}\,(S, L, t) \\
&= e^{-r(T-t)} \Bigg\{ E - LN\left(\frac{\ln\frac{S}{L} + \mu\,(T-t)}{\sigma\sqrt{T-t}}\right) \\
&\quad - \left[1 + \frac{\sigma^2}{2\,(r - D_0)}\right] Se^{(r-D_0)(T-t)}\,N\left(\frac{\ln\frac{L}{S} - (\mu + \sigma^2)\,(T-t)}{\sigma\sqrt{T-t}}\right) \\
&\quad + \frac{\sigma^2 S}{2\,(r - D_0)}\left(\frac{L}{S}\right)^{2(r-D_0)/\sigma^2} N\left(\frac{\ln\frac{L}{S} + \mu\,(T-t)}{\sigma\sqrt{T-t}}\right) \Bigg\};
\end{aligned}
$$

and that for $E \leq L$, the price is

$$
\begin{aligned}
&p_{lp}\,(S, L, t) \\
&= e^{-r(T-t)} \Bigg\{ EN\left(\frac{\ln\frac{E}{S} - \mu\,(T-t)}{\sigma\sqrt{T-t}}\right) \\
&\quad - \left[1 + \frac{\sigma^2}{2\,(r - D_0)}\right] Se^{(r-D_0)(T-t)}\,N\left(\frac{\ln\frac{E}{S} - (\mu + \sigma^2)\,(T-t)}{\sigma\sqrt{T-t}}\right) \\
&\quad + \frac{\sigma^2 S}{2\,(r - D_0)}\left(\frac{E}{S}\right)^{2(r-D_0)/\sigma^2} N\left(\frac{\ln\frac{E}{S} + \mu\,(T-t)}{\sigma\sqrt{T-t}}\right) \Bigg\},
\end{aligned}
$$

where

$$\mu = r - D_0 - \sigma^2/2.$$

13. Let $V(S, L, t)$ be the price of the American lookback strike call option with continuous sampling. Show that $V(S, L, t^*) \geq V(S, L, t^{**})$ if $t^* < t^{**}$.

14. As we know, for a European lookback strike call option with continuous sampling, the corresponding one-dimensional problem is

$$
\begin{cases}
\dfrac{\partial W}{\partial t} + \mathbf{L}_\eta W = 0, & 0 \leq \eta \leq 1, \quad t \leq T, \\[2mm]
W(\eta, T) = \max(\alpha - \eta, 0), & 0 \leq \eta \leq 1, \\[2mm]
\dfrac{\partial W}{\partial \eta}(1, t) = 0, & t \leq T,
\end{cases}
$$

where

$$\eta = L/S, \quad W(\eta, t) = V(S, L, t)/S,$$

$$\mathbf{L}_\eta = \frac{1}{2}\sigma^2\eta^2\frac{\partial^2}{\partial\eta^2} + (D_0 - r)\eta\frac{\partial}{\partial\eta} - D_0,$$

and $0 < \alpha \leq 1$.

a) Let $V(S, L, t)$ also denote the price of the American lookback strike call option with continuous sampling and define $W = V/S$. Derive the linear complementarity problem for W.

b) Assume that we have proven $V(S, L, t^*) \geq V(S, L, t^{**})$ for any $t^* \leq t^{**}$. Derive the free-boundary problem for W.

15. a) Show that

$$\phi(\mathbf{x}_0; \mathbf{x}, \tau) = \frac{1}{(2\pi\tau)^{n/2}} e^{-\sum\limits_{i=1}^{n}(x_i - x_{i0})^2/(2\tau)}$$

is a solution to

$$\frac{\partial \phi}{\partial \tau} = \frac{1}{2} \sum_{i=1}^{n} \frac{\partial^2 \phi}{\partial x_i^2}, \qquad -\infty < \mathbf{x} < \infty, \qquad 0 \leq \tau,$$

where

$$\mathbf{x} = \begin{bmatrix} x_1 \\ x_2 \\ \vdots \\ x_n \end{bmatrix}, \quad \mathbf{x}_0 = \begin{bmatrix} x_{10} \\ x_{20} \\ \vdots \\ x_{n0} \end{bmatrix}$$

and $-\infty < \mathbf{x} < \infty$ means

$$-\infty < x_i < \infty, \qquad i = 1, 2, \cdots, n.$$

b) Show that the function $\phi(\mathbf{x}_0; \mathbf{x}, \tau)$ satisfies the conditions

$$\int_{-\infty}^{\infty} \int_{-\infty}^{\infty} \cdots \int_{-\infty}^{\infty} \phi(\mathbf{x}_0; \mathbf{x}, \tau) dx_{10} dx_{20} \cdots dx_{n0} = 1$$

and

$$\lim_{\tau \to 0} \phi(\mathbf{x}_0; \mathbf{x}, \tau) = \begin{cases} \infty, & \text{at } \mathbf{x} = \mathbf{x}_0, \\ 0, & \text{otherwise.} \end{cases}$$

c) Show that

$$V(\mathbf{x}, \tau) = \int_{-\infty}^{\infty} \int_{-\infty}^{\infty} \cdots \int_{-\infty}^{\infty} V_0(\mathbf{x}_0)\phi(\mathbf{x}_0; \mathbf{x}, \tau) dx_{10} dx_{20} \cdots dx_{n0}$$

is the solution of the problem

$$\begin{cases} \dfrac{\partial V}{\partial \tau} = \dfrac{1}{2} \sum\limits_{i=1}^{n} \dfrac{\partial^2 V}{\partial x_i^2}, & -\infty < \mathbf{x} < \infty, \quad 0 \leq \tau, \\ V(\mathbf{x}, 0) = V_0(\mathbf{x}), & -\infty < \mathbf{x} < \infty. \end{cases}$$

16. Suppose that S_1, S_2, and S_3 are three random variables satisfying the following stochastic differential equations

$$dS_i = \mu_i S_i dt + \sigma_i S_i dX_i, \quad i = 1, 2, 3,$$

where $\mu_i, \sigma_i, i = 1, 2, 3$, are constants, $dX_i, i = 1, 2, 3$, are three Wiener processes, i.e., $dX_i = \phi_i \sqrt{dt}$, ϕ_i being distinct standardized normal random variables, $i = 1, 2, 3$. ϕ_1, ϕ_2, ϕ_3 could be correlated and

$$E[\phi_i \phi_j] = \rho_{ij}, \quad i, j = 1, 2, 3,$$

where $-1 \le \rho_{ij} \le 1$. Consider three new random variables ξ_1, ξ_2, and ξ_3 defined by

$$\begin{cases} \xi_1 = \dfrac{1}{S_1}, \\[2mm] \xi_2 = \dfrac{S_2}{S_1}, \\[2mm] \xi_3 = \dfrac{S_3}{S_1}. \end{cases}$$

a) Show that ξ_1, ξ_2, ξ_3 satisfy the following stochastic differential equations

$$d\xi_1 = (-\mu_1 + \sigma_1^2)\xi_1 dt + \sigma_1 \xi_1 dY_1,$$
$$d\xi_2 = (-\mu_1 + \mu_2 + \sigma_1^2 - \rho_{12}\sigma_1\sigma_2)\xi_2 dt + \sigma_{12}\xi_2 dY_2,$$
$$d\xi_3 = (-\mu_1 + \mu_3 + \sigma_1^2 - \rho_{13}\sigma_1\sigma_3)\xi_3 dt + \sigma_{13}\xi_3 dY_3,$$

where dY_1, dY_2, dY_3 are the Wiener processes and

$$dY_1 = -dX_1,$$
$$dY_2 = \frac{-\sigma_1 dX_1 + \sigma_2 dX_2}{\sigma_{12}},$$
$$dY_3 = \frac{-\sigma_1 dX_1 + \sigma_3 dX_3}{\sigma_{13}},$$
$$\sigma_{ij} = \sqrt{\sigma_i^2 - 2\rho_{ij}\sigma_i\sigma_j + \sigma_j^2}.$$

b) Let S_0 be equal to one, σ_0 be the volatility of S_0, and σ_{01} denote the volatility of $\xi_1 = S_0/S_1$. Show

$$\sigma_0 = 0$$

and

$$\sigma_{01} = \sigma_1.$$

c) Define

$$\rho_{ijk} = \frac{\sigma_k^2 - \rho_{ik}\sigma_i\sigma_k - \rho_{jk}\sigma_j\sigma_k + \rho_{ij}\sigma_i\sigma_j}{\sigma_{ik}\sigma_{jk}}.$$

Show

$$E\left[dY_1 dY_2\right] = \frac{\sigma_1 - \rho_{12}\sigma_2}{\sigma_{12}}dt = \rho_{021}dt = \rho_{201}dt,$$

$$E\left[dY_1 dY_3\right] = \frac{\sigma_1 - \rho_{13}\sigma_3}{\sigma_{13}}dt = \rho_{031}dt = \rho_{301}dt,$$

$$E\left[dY_2 dY_3\right] = \frac{\sigma_1^2 - \rho_{12}\sigma_1\sigma_2 - \rho_{13}\sigma_1\sigma_3 + \rho_{23}\sigma_2\sigma_3}{\sigma_{12}\sigma_{13}}dt$$

$$= \rho_{231}dt = \rho_{321}dt.$$

17. Suppose S_1 and S_2 are the prices of two assets A and B, respectively. The random variables S_1 and S_2 satisfy

$$dS_1 = \mu_1 S_1 dt + \sigma_1 S_1 dX_1,$$
$$dS_2 = \mu_2 S_2 dt + \sigma_2 S_2 dX_2,$$

where μ_1, μ_2, σ_1, and σ_2 are constants, and dX_1 and dX_2 are two Wiener processes with

$$E\left[dX_1 dX_2\right] = \rho_{12}dt.$$

Also, suppose that the two assets pay dividends continuously and that the dividend yields of the assets A and B are D_{01} and D_{02}, respectively. Consider a European option on the minimum of S_1, S_2, and S_0, i.e., its payoff function is

$$\min(S_0, S_1, S_2),$$

where S_0 is a constant. Let $V_{\min}(S_1, S_2, t)$ be the price of the option. Show that

$$V_{\min}(S_1, S_2, t)$$

$$= S_0^* N_2\left(\frac{\ln\frac{S_1^*}{S_0^*} - \frac{\sigma_1^2}{2}\tau}{\sigma_1\sqrt{\tau}}, \frac{\ln\frac{S_2^*}{S_0^*} - \frac{\sigma_2^2}{2}\tau}{\sigma_2\sqrt{\tau}}; \rho_{12}\right)$$

$$+ S_1^* N_2\left(\frac{\ln\frac{S_2^*}{S_1^*} - \frac{\sigma_{12}^2}{2}\tau}{\sigma_{12}\sqrt{\tau}}, \frac{\ln\frac{S_0^*}{S_1^*} - \frac{\sigma_1^2}{2}\tau}{\sigma_1\sqrt{\tau}}; \frac{\sigma_1 - \rho_{12}\sigma_2}{\sigma_{12}}\right)$$

$$+ S_2^* N_2\left(\frac{\ln\frac{S_0^*}{S_2^*} - \frac{\sigma_2^2}{2}\tau}{\sigma_2\sqrt{\tau}}, \frac{\ln\frac{S_1^*}{S_2^*} - \frac{\sigma_{12}^2}{2}\tau}{\sigma_{12}\sqrt{\tau}}; \frac{\sigma_2 - \rho_{12}\sigma_1}{\sigma_{12}}\right),$$

where $S_0^* = S_0 e^{-r\tau}$, $S_1^* = S_1 e^{-D_{01}\tau}$ and $S_2^* = S_2 e^{-D_{02}\tau}$.

18. Suppose that S_1, S_2, and S_3 are the prices of three assets satisfying

$$dS_i = \mu_i S_i dt + \sigma_i S_i dX_i, \quad i = 1, 2, 3,$$

where μ_i, σ_i, $i = 1, 2, 3$ are constants and dX_i, $i = 1, 2, 3$ are the Wiener processes with

$$\mathrm{E}\,[dX_i dX_j] = \rho_{ij} dt, \quad i, j = 1, 2, 3.$$

Also, suppose that the three assets pay dividends continuously and that the dividend yields of the three assets are D_{0i}, $i = 1, 2, 3$.

a) Consider a European option on the maximum of S_1, S_2, S_3, and S_0, i.e., its payoff is

$$\max(S_0, S_1, S_2, S_3),$$

where S_0 is a certain amount of cash. If we understand S_0 as a random variable, then its volatility, σ_0, is equal to zero. Let $V_{\max}(S_1, S_2, S_3, t)$ be the price of such a T-year option. Show that

$$\begin{aligned}
V_{\max}(S_1, S_2, S_3, t) = &\ S_0^* N_3 \left(A_{10}, A_{20}, A_{30}; \rho_{120}, \rho_{130}, \rho_{230}\right) \\
&+ S_1^* N_3 \left(A_{21}, A_{31}, A_{01}; \rho_{231}, \rho_{201}, \rho_{301}\right) \\
&+ S_2^* N_3 \left(A_{32}, A_{02}, A_{12}; \rho_{302}, \rho_{312}, \rho_{012}\right) \\
&+ S_3^* N_3 \left(A_{03}, A_{13}, A_{23}; \rho_{013}, \rho_{023}, \rho_{123}\right),
\end{aligned}$$

where

$$S_0^* = S_0 e^{-r(T-t)},$$
$$S_i^* = S_i e^{-D_{0i}(T-t)}, \quad i = 1, 2, 3,$$
$$A_{ij} = \frac{\ln \dfrac{S_j^*}{S_i^*} + \dfrac{\sigma_{ij}^2(T-t)}{2}}{\sigma_{ij}\sqrt{T-t}}, \quad i, j = 0, 1, 2, 3 \text{ but } i \neq j,$$

σ_{ij} being $\sqrt{\sigma_i^2 - 2\rho_{ij}\sigma_i\sigma_j + \sigma_j^2}$,

$$\rho_{ijk} = \frac{\sigma_k^2 - \rho_{ik}\sigma_i\sigma_k - \rho_{jk}\sigma_j\sigma_k + \rho_{ij}\sigma_i\sigma_j}{\sigma_{ik}\sigma_{jk}},$$

$i, j, k = 0, 1, 2, 3$ but i, j, k being distinct,

and $N_3(x_1, x_2, x_3; \rho_{12}, \rho_{13}, \rho_{23})$ is the trivariate cumulative distribution function:

$$\begin{aligned}
&N_3(x_1, x_2, x_3; \rho_{12}, \rho_{13}, \rho_{23}) \\
&= \frac{1}{(2\pi)^{3/2}\sqrt{\det \mathbf{P}}} \int_{-\infty}^{x_3} \int_{-\infty}^{x_2} \int_{-\infty}^{x_1} e^{-\frac{1}{2}\eta^T \mathbf{P}^{-1}\eta} d\eta_1 d\eta_2 d\eta_3,
\end{aligned}$$

$$\mathbf{P} \text{ being } \begin{bmatrix} 1 & \rho_{12} & \rho_{13} \\ \rho_{12} & 1 & \rho_{23} \\ \rho_{13} & \rho_{23} & 1 \end{bmatrix} \text{ and } \eta \text{ being } \begin{bmatrix} \eta_1 \\ \eta_2 \\ \eta_3 \end{bmatrix}.$$

b) Consider a European option on the minimum of S_1, S_2, S_3, and S_0, i.e., its payoff is

$$\min(S_0, S_1, S_2, S_3),$$

where S_0 is a certain amount of cash. Find the expression of the price for this option in terms of the trivariate cumulative distribution function.

19. Suppose that $S_1, S_2, \cdots, S_n$ are the prices of n assets and that each asset pays a dividend continuously, the dividend yield for S_i being D_{0i}, $i = 1, 2, \cdots, n$. Each price S_i satisfies the stochastic equation

$$dS_i = \mu_i S_i dt + \sigma_i S_i dX_i,$$

where μ_i, σ_i are constants and dX_i is a Wiener process, and

$$E\left[dX_i dX_j\right] = \rho_{ij} dt, \quad i, j = 1, 2, \cdots, n.$$

a) Guess the expression of price of the European option on the maximum of $S_1, S_2, \cdots, S_n$ and S_0 according to the result given in part (a) of Problem 18, where S_0 is a certain amount of cash.
b) Guess the expression of price of the European option on the minimum of $S_1, S_2, \cdots, S_n$ and S_0 according to the result given in part (b) of Problem 18.

20. Suppose that $c_{\max}(S_1, S_2, t)$, $c_{\min}(S_1, S_2, t)$, $c(S_1, t)$, and $c(S_2, t)$ are the prices of four call options with payoff functions

$$\max(\max(S_1, S_2) - E, 0), \quad \max(\min(S_1, S_2) - E, 0), \quad \max(S_1 - E, 0)$$

and

$$\max(S_2 - E, 0)$$

respectively. Show

$$c_{\max}(S_1, S_2, t) + c_{\min}(S_1, S_2, t) = c(S_1, t) + c(S_2, t).$$

21. Let $p_{\max}(S_1, S_2, t)$ and $p_{\min}(S_1, S_2, t)$ be the prices of two European put options with payoff functions

$$\max(E - \max(S_1, S_2), 0) \quad \text{and} \quad \max(E - \min(S_1, S_2), 0)$$

respectively. Suppose that $c_{\max}(S_1, S_2, t)$ and $c_{\min}(S_1, S_2, t)$ are the prices of two European call options with payoff functions

$$\max(\max(S_1, S_2) - E, 0) \quad \text{and} \quad \max(\min(S_1, S_2) - E, 0)$$

respectively. $\bar{c}_{\max}(S_1, S_2, t)$ and $\bar{c}_{\min}(S_1, S_2, t)$ denote the prices of European options with payoff functions

$$\max(S_1, S_2) \quad \text{and} \quad \min(S_1, S_2).$$

Show

a)

$$p_{\max}(S_1, S_2, t) = Ee^{-r(T-t)} - \bar{c}_{\max}(S_1, S_2, t) + c_{\max}(S_1, S_2, t);$$

b)

$$p_{\min}(S_1, S_2, t) = Ee^{-r(T-t)} - \bar{c}_{\min}(S_1, S_2, t) + c_{\min}(S_1, S_2, t).$$

22. Show that the closed-form solutions of cash-or-nothing puts, asset-or-nothing calls, and asset-or-nothing puts are

$$Be^{-r(T-t)}N(-d_2), \quad Se^{-D_0(T-t)}N(d_1) \quad \text{and} \quad Se^{-D_0(T-t)}N(-d_1)$$

respectively. Here

$$d_1 = \frac{\ln\left(Se^{(r-D_0)(T-t)}/E\right) + \sigma^2(T-t)/2}{\sigma\sqrt{T-t}},$$

$$d_2 = \frac{\ln\left(Se^{(r-D_0)(T-t)}/E\right) - \sigma^2(T-t)/2}{\sigma\sqrt{T-t}}.$$

23. Show that the value of a forward start American put option with exercise price $E = \alpha S_{T_1}$ at time $t_0 < T_1$ is

$$\alpha Se^{-D_0(T_1-t_0)}P^*\left(\frac{1}{\alpha}, T_1\right),$$

where $P^*\left(\frac{1}{\alpha}, T_1\right)$ is the value of a standard American put option.

24. Consider compound options and assume that both options are European. Let $c_1(S, t; c_2)$, $c_1(S, t; p_2)$, and $p_1(S, t; p_2)$ denote the prices of a call on a call, a call on a put, and a put on a put, respectively. Show that their closed-form solutions are

$$c_1(S, t; c_2) = Se^{-D_0(T_2-t)}N_2(d_{11}, d_{12}; \rho) - E_2e^{-r(T_2-t)}N_2(d_{21}, d_{22}; \rho)$$
$$-E_1e^{-r(T_1-t)}N_2(d_{21}),$$
$$c_1(S, t; p_2) = E_2e^{-r(T_2-t)}N_2(-d_{23}, -d_{22}; \rho)$$
$$-Se^{-D_0(T_2-t)}N_2(-d_{13}, -d_{12}; \rho)$$
$$-E_1e^{-r(T_1-t)}N(-d_{23}),$$
$$p_1(S, t; p_2) = E_1e^{-r(T_1-t)}N(d_{23}) - E_2e^{-r(T_2-t)}N_2(d_{23}, -d_{22}; -\rho)$$
$$+Se^{-D_0(T_2-t)}N_2(d_{13}, -d_{12}; -\rho),$$

where

$$d_{11} = \frac{\ln(S/S^*) + \left(r - D_0 + \sigma^2/2\right)(T_1 - t)}{\sigma\sqrt{T_1 - t}},$$

$$d_{21} = \frac{\ln(S/S^*) + \left(r - D_0 - \sigma^2/2\right)(T_1 - t)}{\sigma\sqrt{T_1 - t}},$$

$$d_{12} = \frac{\ln(S/E_2) + \left(r - D_0 + \sigma^2/2\right)(T_2 - t)}{\sigma\sqrt{T_2 - t}},$$

$$d_{22} = \frac{\ln(S/E_2) + \left(r - D_0 - \sigma^2/2\right)(T_2 - t)}{\sigma\sqrt{T_2 - t}},$$

$$d_{13} = \frac{\ln(S/S^{**}) + \left(r - D_0 + \sigma^2/2\right)(T_1 - t)}{\sigma\sqrt{T_1 - t}},$$

$$d_{23} = \frac{\ln(S/S^{**}) + \left(r - D_0 - \sigma^2/2\right)(T_1 - t)}{\sigma\sqrt{T_1 - t}},$$

$$\rho = \sqrt{\frac{T_1 - t}{T_2 - t}}.$$

Here, S^* and S^{**} are the solutions of the following equations:

$$c_2\left(S^*, T_1\right) = E_1$$

and

$$p_2\left(S^{**}, T_1\right) = E_1.$$

25. How do we determine the price of a European put option on an American put option?

26. Show

$$\int_0^\infty G\left(S'', T_2; S', T_1\right) G\left(S', T_1; S, t\right) dS' = G\left(S'', T_2; S, t\right),$$

where

$$G\left(S', T_1; S, t\right)$$
$$= \frac{1}{\sigma\sqrt{2\pi\left(T_1 - t\right)}S'} e^{-\left[\ln S' - \ln S - \left(r - D_0 - \sigma^2/2\right)(T_1 - t)\right]^2 / 2\sigma^2(T_1 - t)}$$

and $G\left(S'', T_2; S', T_1\right)$ and $G\left(S'', T_2; S, t\right)$ are defined in the same way.

27. The payoff of a standard chooser option is

$$V\left(S, T_1\right) = \max\left(c\left(S, T_1\right),\ p\left(S, T_1\right)\right),$$

where $c\left(S, T_1\right)$ and $p\left(S, T_1\right)$ are the prices of European call and put options with the same exercise price E_2 and the same expiration date T_2. Find its closed-form solution. (Hint: Use the put–call parity relation.)

4

Interest Rate Derivative Securities

4.1 Introduction

This chapter is devoted to interest rate derivatives. Interest rate derivatives are financial products derived from interest rates. There are various interest rates that will be mentioned in this chapter. Here we first give the meaning of each interest rate and derive some relations among them.

An N-year zero-coupon yield or an N-year spot interest rate is the interest rate on an investment starting at time t and lasting for N years. The investment is a "pure" N-year investment with no intermediate payments. Assume that the interest is compounded continuously. In this case, suppose that at time t the N-year zero-coupon yield is $Y(t, t+N)$, then the investor will get

$$e^{Y(t,t+N)N}$$

at the end of year N for each dollar invested. A zero-coupon yield curve is a curve showing the relation between $Y(t, t+N)$ and N.

A zero-coupon bond with a face value or a par value of one dollar is a contract whose holder will get one dollar at the maturity of the contract from its issuer. Let $Z(t;T)$ denote the money a person needs to pay in order to have the contract with maturity date T at time t. Then, between $Y(t, T)$ and $Z(t;T)$, there is the following relation

$$Z(t;T) = e^{-Y(t,T)(T-t)}, \tag{4.1}$$

or

$$Y(t, T) = \frac{-\ln Z(t;T)}{T-t}.$$

Suppose $t \leq T_1 \leq T_2$. An interest rate determined at time t for a period $[T_1, T_2]$ and paid at time T_2 is called a forward interest rate. Let us denote this rate by $f(t, T_1, T_2)$ and again assume that the interest is compounded continuously. Among $f(t, T_1, T_2)$, $Z(t;T_1)$, and $Z(t;T_2)$, there is the following relation:

$$Z(t; T_1) = Z(t; T_2) e^{f(t, T_1, T_2)(T_2 - T_1)},$$

or

$$f(t, T_1, T_2) = \frac{1}{T_2 - T_1} \ln \frac{Z(t; T_1)}{Z(t; T_2)}. \tag{4.2}$$

The reason is the following. If we borrow one dollar at time T_1, then we need to return $e^{f(t, T_1, T_2)(T_2 - T_1)}$ dollars at time T_2 according to the forward interest rate at time t. At time t, the values of one dollar at time T_1 and $e^{f(t, T_1, T_2)(T_2 - T_1)}$ dollars at time T_2 should be the same, otherwise there is an arbitrage opportunity.

An instantaneous forward interest rate $F(t, T_1)$ is the limit of $f(t, T_1, T_2)$ as $T_2 \to T_1$, written as

$$
\begin{aligned}
F(t, T_1) &= \lim_{T_2 \to T_1} f(t, T_1, T_2) = \lim_{T_2 \to T_1} \frac{-[\ln Z(t; T_2) - \ln Z(t; T_1)]}{T_2 - T_1} \\
&= \frac{-1}{Z(t; T_1)} \frac{\partial Z(t; T_1)}{\partial T_1}. \tag{4.3}
\end{aligned}
$$

This gives

$$Z(t; T) = Z(t; t) e^{-\int_t^T F(t, u) du} = e^{-\int_t^T F(t, u) du}.$$

Furthermore, combining this expression for $Z(t, T)$ with (4.1) yields

$$Y(t, T) = \frac{1}{T - t} \int_t^T F(t, u) du. \tag{4.4}$$

The limit of $Y(t, T)$ as $T \to t$ is called the instantaneous spot interest rate, the spot interest rate, or the short-term interest rate, denoted by $r(t)$, so

$$r(t) = \lim_{T \to t} Y(t, T) = Y(t, t).$$

Because from (4.4) we also have

$$\lim_{T \to t} Y(t, T) = \lim_{T \to t} \frac{1}{T - t} \int_t^T F(t, u) du = F(t, t),$$

we get

$$r(t) = Y(t, t) = F(t, t). \tag{4.5}$$

Clearly, if $Y(t, T)$ is equal to a constant r, then

$$Z(t; T) = e^{-r(T - t)},$$

and

$$f(t, T_1, T_2) = F(t, T_1) = F(t, t) = Y(t, t) = r(t) = r.$$

In practice, the interest is often compounded discretely. If a loan of one dollar is required to pay at an interest rate $\bar{r}$ compounded m times per year, then the amount of each payment is

$$\frac{\bar{r}}{m}.$$

For an investment with an interest rate r compounded continuously, the interest payment for a period $\frac{1}{m}$ years is

$$e^{r/m} - 1.$$

If

$$e^{r/m} - 1 = \frac{\bar{r}}{m},$$

that is,

$$r = m\ln(1 + \bar{r}/m),$$

then the two investments are equivalent. Suppose that a forward interest rate at time t for the period $[T_1, T_1 + 1/m]$ is an interest rate compounded m times per year and we use $\bar{f}(t, T_1, T_1 + 1/m)$ to denote this forward interest rate. Let $f(t, T_1, T_1 + 1/m)$ be equivalent to the interest rate $\bar{f}(t, T_1, T_1 + 1/m)$. Then we have

$$f(t, T_1, T_1 + 1/m) = m\ln\left(1 + \frac{\bar{f}(t, T_1, T_1 + 1/m)}{m}\right)$$

and (4.2) can be rewritten as

$$m\ln\left(1 + \frac{\bar{f}(t, T_1, T_1 + 1/m)}{m}\right) = m\ln\left(\frac{Z(t; T_1)}{Z(t; T_1 + 1/m)}\right)$$

or

$$\bar{f}(t, T_1, T_1 + 1/m) = m\left[\frac{Z(t; T_1)}{Z(t; T_1 + 1/m)} - 1\right]. \tag{4.6}$$

This is the counterpart of (4.2) for an interest rate compounded m times per year. Actually, this relation can also be derived directly. For the formulae (4.1) and (4.3)–(4.5), we can also have their counterparts for interest rates compounded discretely.

As we know, the value of a bond is related to interest rates. There are many other financial contracts related to interest rates, which are signed between two parties, for example, a bank and a company. These are called interest rate derivatives. For an equity option, a typical life span is nine months or less. In this case, the assumption of a spot interest rate being a deterministic function of t, even a constant, is acceptable. If this is not the case, it may be necessary to consider a spot interest rate as a random variable. For example, a life span of a bond may be 5 years, 10 years, even 30 years. Therefore, it is more realistic to deal with a spot interest rate as a random variable. An interest rate cannot be traded on the market. In Chapter 2, we pointed out that there is a unknown function called the market price of risk for a spot interest rate in the governing partial differential equation (PDE) for interest rate derivatives. Before using such an equation to price a derivative security, one has to find this function. From the mathematical point of view, to find a unknown function in the partial differential equation is to solve an inverse problem. This function in the PDE is determined by some data associated with solutions of the equation. The values of zero-coupon bonds with various maturity dates on the market or some other data can be taken as the data needed. Moreover, reducing the randomness of a zero-coupon bond curve to the randomness of the spot interest rate is not a good approximation in many cases. Thus, describing the randomness of a zero-coupon bond curve by the randomness of several interest rates, namely, considering multi-factor models, is necessary.

Therefore, the rest of this chapter is organized as follows. In Sections 4.2 and 4.3, the problem for a bond is formulated and for four special models, explicit solutions are derived. In Section 4.4, we discuss the inverse problem of determining the market price of risk and give a formulation of the inverse problem so that the determination of the unknown function can be reduced to solving such a problem. Then, we discuss bond options, swaps, swaptions, and so forth in Section 4.5. Section 4.6 is devoted to multi-factor interest rate models, especially, a three-factor model that can used in practice easily. Finally, two-factor convertible bonds are discussed in Section 4.7.

4.2 Bonds

A bond is a long-term contract under which the issuer promises to pay the holder a specified amount of money on a specified date. The specified amount is called the face value of the bond, which is denoted by Z in this chapter, and the specified date is named the maturity date T. Usually, the holder is also paid a specified amount at fixed times during the life of the contract. Such a specified amount is called a coupon. If there is no coupon payment, the bond is known as a zero-coupon bond. Clearly, the bondholder must pay a certain amount of money to the issuer when the bond is purchased. This amount is called the upfront premium. In this section, we will mainly derive the equations by which one can determine a fair value of the bond for any time t, including the upfront premium.

4.2.1 Bond Values for Deterministic Spot Rates

Let r be the interest rate for the shortest possible deposit, which is commonly called the spot interest rate or the spot rate. For a short period, r may be assumed to be a constant. For a long period, for example, a few years, it is unreasonable to consider r as a constant. As a starting point, we assume that the spot interest rate is a known function of t, i.e., $r = r(t)$. Let $V(t)$ stand for the value of a bond with coupon rate $k(t)$ at time t. Assume that the return rate of a bond during the time interval $[t, t + dt]$ be the risk-free spot rate, so we have

$$dV + Zk(t)dt = r(t)V dt,$$

where $Zk(t)dt$ is the coupon payment the bondholder receives during the time interval. If the coupon is paid continuously, $k(t)$ is a continuous function of t. If it is paid at fixed times, $k(t)$ is a linear combination of Dirac delta functions, i.e., $k(t) = \sum_i k_i \delta(t - t_i)$, $t_i \leq T$. The relation above can also be written as

$$dV - r(t)V dt = -Zk(t)dt.$$

Multiplying both sides of the equation by $e^{\int_t^T r(\tau)d\tau}$, which is usually referred to as the integrating factor, yields

$$e^{\int_t^T r(\tau)d\tau} [dV - r(t)V dt] = -Zk(t)e^{\int_t^T r(\tau)d\tau} dt.$$

The left-hand side actually is $d\left(e^{\int_t^T r(\tau)d\tau}V\right)$. Therefore, we have

$$\int_t^T d\left(e^{\int_t^T r(\tau)d\tau}V\right) = V(T) - e^{\int_t^T r(\tau)d\tau}V(t) = -Z\int_t^T k(\bar{\tau})e^{\int_{\bar{\tau}}^T r(\tau)d\tau}d\bar{\tau}$$

and

$$V(t) = e^{-\int_t^T r(\tau)d\tau}\left[V(T) + Z\int_t^T k(\bar{\tau})e^{\int_{\bar{\tau}}^T r(\tau)d\tau}d\bar{\tau}\right]$$

$$= V(T)e^{-\int_t^T r(\tau)d\tau}\left[1 + \int_t^T k(\bar{\tau})e^{\int_{\bar{\tau}}^T r(\tau)d\tau}d\bar{\tau}\right], \tag{4.7}$$

where we have used the condition $Z = V(T)$. For a zero-coupon bond, $k(t) = 0$ and

$$V(t) = V(T)e^{-\int_t^T r(\tau)d\tau} = Ze^{-\int_t^T r(\tau)d\tau}.$$

From the right-hand side, we see that the value of $V(t)$ depends on T. However, this dependence is suppressed in this expression. In order to express this dependence explicitly, the relation above can be rewritten as

$$V(t; T) = V(T; T)e^{-\int_t^T r(\tau)d\tau}, \tag{4.8}$$

where $V(T; T) = Z$.

At time t, the values of zero-coupon bonds with various maturities can be obtained from the market, i.e., $V(t; T)$ with a fixed t and various T is observable. Suppose we have such a function. Differentiating (4.8) with respect to T yields

$$\frac{\partial V(t; T)}{\partial T} = -V(T; T)e^{-\int_t^T r(\tau)d\tau}r(T) = -V(t; T)r(T)$$

and

$$r(T) = \frac{-1}{V(t;T)} \frac{\partial V(t;T)}{\partial T}.$$

This means that the spot rate at time T can be determined by the value and the slope of the function $V(t;T)$. It is clear that $r(T)$ does not depend on Z. Let $Z = 1$, then comparing the expression for $r(T)$ and (4.3) yields

$$F(t,T) = r(T)$$

and

$$Z(t;T) = e^{-\int_t^T r(u)du} \tag{4.9}$$

if $Z = 1$. Also for a zero-coupon bond,

$$\frac{V(t;T)}{V(T;T)} = Z(t;T).$$

Thus, from (4.1) we have

$$Y(t,T) = \frac{-\ln Z(t;T)}{T-t} = \frac{-\ln\left(V(t;T)/V(T;T)\right)}{T-t}, \tag{4.10}$$

which is usually called the yield of a bond during the time interval $[t,T]$. A plot of Y against the time to maturity, $T-t$, is called the yield curve. The dependence of the yield on $T-t$ is called the term structure of interest rates. The historical data on bonds are usually given in the form of yields for various $T-t$.

4.2.2 Bond Equations for Random Spot Rates

It will be more realistic to consider the spot interest rate r as a random variable. Suppose

$$dr = u(r,t)dt + w(r,t)dX. \tag{4.11}$$

From Section 2.9, we know that the value of a bond as a spot interest rate derivative, $V(r,t)$, satisfies the equation (2.81) with only one random variable r:

$$\frac{\partial V}{\partial t} + \frac{1}{2}w^2\frac{\partial^2 V}{\partial r^2} + (u - \lambda w)\frac{\partial V}{\partial r} - rV + kZ = 0, \tag{4.12}$$

where kZ is the coupon payment. For a bond the value at maturity date T is a constant Z, i.e.,

$$V(r,T) = Z. \tag{4.13}$$

If the spot rate model satisfies the conditions (2.78) and (2.79), then no boundary condition is needed, i.e., (4.12) with the final condition (4.13) has a unique solution.

4.3 Some Explicit Solutions of Bond Equations

There exist many spot rate models. Here, we discuss the following model (see [71]):

$$dr = [\bar{\mu}(t) - \bar{\gamma}(t)r]\,dt + \sqrt{\alpha(t)r - \beta(t)}\,dX, \qquad (4.14)$$

where $\alpha(t)$, $\beta(t)$, $\bar{\gamma}(t)$, and $\bar{\mu}(t)$ are given functions of t. Several important models, for example, the Vasicek model (see [70]), the Cox–Ingersoll–Ross model (see [22]), the Ho–Lee model (see [38]), and the Hull–White model (see [40]) possess this form. For the models in the form (4.14), the determination of the value of a zero-coupon bond can be reduced to solving two ordinary differential equations. Sometimes we can find analytic solutions or the solution can be expressed in terms of integrals with known integrands. Such a solution is referred to as an explicit solution here.

If a spot interest rate model is in the form (4.14) and we take

$$\lambda(r,t) = \bar{\lambda}(t)\sqrt{\alpha(t)r - \beta(t)}, \qquad (4.15)$$

then the equation (4.12) can be written as

$$\frac{\partial V}{\partial t} + \frac{1}{2}[\alpha(t)r - \beta(t)]\frac{\partial^2 V}{\partial r^2} + [\mu(t) - \gamma(t)r]\frac{\partial V}{\partial r} - rV = 0, \qquad (4.16)$$

where

$$\mu(t) = \bar{\mu}(t) + \bar{\lambda}(t)\beta(t) \qquad (4.17)$$

and

$$\gamma(t) = \bar{\gamma}(t) + \bar{\lambda}(t)\alpha(t). \qquad (4.18)$$

Here, we let $k = 0$ because we are going to determine the value of a zero-coupon bond. Because the coefficients of $\dfrac{\partial^2 V}{\partial r^2}$ and $\dfrac{\partial V}{\partial r}$ are linear functions in r, the solution of (4.16) with the condition (4.13) has the following form

$$V(r,t) = Ze^{A(t,T) - rB(t,T)} \qquad (4.19)$$

with

$$A(T,T) = 0 \qquad (4.20)$$

and

$$B(T,T) = 0. \qquad (4.21)$$

In fact, because (4.20) and (4.21) hold, we have

$$V(r,T) = Z.$$

Substituting (4.19) into the equation (4.16) yields

$$\frac{dA}{dt} - r\frac{dB}{dt} + \frac{1}{2}[\alpha(t)r - \beta(t)]B^2 - [\mu(t) - \gamma(t)r]B - r = 0.$$

If the sum of the terms independent of r is equal to zero, i.e.,

$$\frac{dA}{dt} - \frac{1}{2}\beta(t)B^2 - \mu(t)B = 0$$

and the sum of all coefficients of r is equal to zero, i.e.,

$$-\frac{dB}{dt} + \frac{1}{2}\alpha B^2 + \gamma(t)B - 1 = 0,$$

then (4.19) is a solution to a zero-coupon bond. These two equations above, which can be rewritten as

$$\frac{dA}{dt} = \frac{1}{2}\beta(t)B^2 + \mu(t)B \tag{4.22}$$

and

$$\frac{dB}{dt} = \frac{1}{2}\alpha(t)B^2 + \gamma(t)B - 1, \tag{4.23}$$

have unique solutions satisfying (4.20) and (4.21). Thus, it is true that (4.16) with the condition (4.13) has a solution in the form (4.19) satisfying the conditions (4.20) and (4.21), and the solution of the problem can be reduced to solving the two ordinary differential equations (4.22) and (4.23) with the conditions (4.20) and (4.21).

4.3.1 Analytic Solutions for the Vasicek and Cox–Ingersoll–Ross Models

If $\alpha, \beta, \gamma, \mu$ in equations (4.22) and (4.23) are constant, then we can find analytic expressions for A and B. When A and B have such expressions, (4.19) gives an analytic solution for a zero-coupon bond. In this case, the equation (4.23) can be rewritten as

$$\frac{dB}{B^2 + \dfrac{2\gamma}{\alpha}B - \dfrac{2}{\alpha}} = \frac{\alpha}{2}dt. \tag{4.24}$$

Since

$$B^2 + \frac{2\gamma}{\alpha}B - \frac{2}{\alpha} = \left(B + \frac{\gamma - \psi}{\alpha}\right)\left(B + \frac{\gamma + \psi}{\alpha}\right),$$

where

$$\psi = \sqrt{\gamma^2 + 2\alpha}, \tag{4.25}$$

using the method of partial fraction decomposition, we can have

$$\frac{1}{B^2 + \dfrac{2\gamma}{\alpha}B - \dfrac{2}{\alpha}} = \frac{\dfrac{\alpha}{2\psi}}{B + \dfrac{\gamma - \psi}{\alpha}} - \frac{\dfrac{\alpha}{2\psi}}{B + \dfrac{\gamma + \psi}{\alpha}}.$$

Noticing this relation, we can easily find the solution to (4.24) by integrating both sides of the equation:

$$\int_{B(t,T)}^{B(T,T)} \frac{dB}{B^2 + \dfrac{2\gamma}{\alpha} B - \dfrac{2}{\alpha}}$$

$$= \frac{\alpha}{2\psi} \left[\int_{B(t,T)}^{0} \frac{dB}{B + (\gamma - \psi)/\alpha} - \int_{B(t,T)}^{0} \frac{dB}{B + (\gamma + \psi)/\alpha} \right]$$

$$= \frac{\alpha}{2\psi} \left[\ln \frac{(\gamma - \psi)/\alpha}{B + (\gamma - \psi)/\alpha} - \ln \frac{(\gamma + \psi)/\alpha}{B + (\gamma + \psi)/\alpha)} \right]$$

$$= \frac{\alpha}{2} \int_{t}^{T} dt = \frac{\alpha}{2} (T - t).$$

From this we have

$$\frac{B + (\gamma + \psi)/\alpha}{B + (\gamma - \psi)/\alpha} = \frac{\gamma + \psi}{\gamma - \psi} e^{\psi(T-t)}$$

or

$$B = \frac{\dfrac{\gamma + \psi}{\alpha} e^{\psi(T-t)} - \dfrac{\gamma + \psi}{\alpha}}{1 - \dfrac{\gamma + \psi}{\gamma - \psi} e^{\psi(T-t)}}$$

$$= \frac{2 \left[e^{\psi(T-t)} - 1 \right]}{(\gamma + \psi) e^{\psi(T-t)} - (\gamma - \psi)}, \tag{4.26}$$

where we have used the relation $\psi^2 - \gamma^2 = 2\alpha$. After we find B, from (4.22) we have

$$\int_{A(t,T)}^{A(T,T)} dA = A(T,T) - A(t,T)$$

$$= \int_{t}^{T} \left(\frac{1}{2} \beta B^2 + \mu B \right) dt$$

or

$$A(t,T) = -\frac{1}{2} \beta \int_{t}^{T} B^2 dt - \mu \int_{t}^{T} B dt.$$

Using the relation (4.24), we can obtain the results of $\int_{t}^{T} B dt$ and $\int_{t}^{T} B^2 dt$ easily as follows:

$$\int_{t}^{T} B dt = \int_{B(t,T)}^{0} \frac{2B/\alpha}{B^2 + 2\gamma B/\alpha - 2/\alpha} dB$$

$$= \frac{2}{\alpha} \int_{B(t,T)}^{0} \left[\frac{-(\gamma - \psi)/(2\psi)}{B + (\gamma - \psi)/\alpha} + \frac{(\gamma + \psi)/(2\psi)}{B + (\gamma + \psi)/\alpha} \right] dB$$

$$= -\frac{\gamma - \psi}{\alpha\psi} \ln \frac{(\gamma - \psi)/\alpha}{B + (\gamma - \psi)/\alpha} + \frac{\gamma + \psi}{\alpha\psi} \ln \frac{(\gamma + \psi)/\alpha}{B + (\gamma + \psi)/\alpha}$$

and

$$\int_t^T B^2 dt = \frac{2}{\alpha} \int_{B(t,T)}^0 \frac{B^2}{B^2 + 2\gamma B/\alpha - 2/\alpha} dB$$

$$= \frac{2}{\alpha} \int_{B(t,T)}^0 \left(1 - \frac{2\gamma B/\alpha}{B^2 + 2\gamma B/\alpha - 2/\alpha} + \frac{2/\alpha}{B^2 + 2\gamma B/\alpha - 2/\alpha}\right) dB$$

$$= \frac{2}{\alpha} \left\{ -B + \left[\frac{\gamma(\gamma - \psi)}{\alpha\psi} + \frac{1}{\psi}\right] \ln \frac{(\gamma - \psi)/\alpha}{B + (\gamma - \psi)/\alpha} \right.$$

$$\left. - \left[\frac{\gamma(\gamma + \psi)}{\alpha\psi} + \frac{1}{\psi}\right] \ln \frac{(\gamma + \psi)/\alpha}{B + (\gamma + \psi)/\alpha} \right\}$$

$$= \frac{2}{\alpha} \left\{ -B - \frac{\gamma - \psi}{(\gamma + \psi)\psi} \ln \frac{(\gamma - \psi)/\alpha}{B + (\gamma - \psi)/\alpha} \right.$$

$$\left. + \frac{(\gamma + \psi)}{(\gamma - \psi)\psi} \ln \frac{(\gamma + \psi)/\alpha}{B + (\gamma + \psi)/\alpha} \right\}.$$

Therefore

$$A = \frac{\beta}{\alpha} B + \left[\frac{\beta(\gamma - \psi)}{\alpha(\gamma + \psi)\psi} + \mu\frac{\gamma - \psi}{\alpha\psi}\right] \ln \frac{(\gamma - \psi)/\alpha}{B + (\gamma - \psi)/\alpha}$$

$$- \left[\frac{\beta(\gamma + \psi)}{\alpha(\gamma - \psi)\psi} + \mu\frac{\gamma + \psi}{\alpha\psi}\right] \ln \frac{(\gamma + \psi)/\alpha}{B + (\gamma + \psi)/\alpha} \qquad (4.27)$$

and

$$V(r,t) = Z \left[\frac{B + (\gamma - \psi)/\alpha}{(\gamma - \psi)\alpha}\right]^{\beta(\psi - \gamma)/\alpha(\gamma + \psi)\psi + \mu(\psi - \gamma)/\alpha\psi}$$

$$\times \left[\frac{B + (\gamma + \psi)/\alpha}{(\gamma + \psi)/\alpha}\right]^{\beta(\gamma + \psi)/\alpha(\gamma - \psi)\psi + \mu(\gamma + \psi)/\alpha\psi} e^{(\beta/\alpha - r)B}. \quad (4.28)$$

This is a solution of a zero-coupon bond suitable for all the models (4.14) with constant $\alpha, \beta, \bar{\gamma}$, and $\bar{\mu}$ as long as we choose the market price of risk in the form $\lambda(r,t) = \bar{\lambda}\sqrt{\alpha r - \beta}$. The parameters $\alpha, \beta, \bar{\gamma}$, and $\bar{\mu}$ can be obtained from the data on the spot interest rate on the market. However $\bar{\lambda}$, a parameter in the expression of the market price of risk, cannot be determined from the data on the spot rate and should be obtained from the other data on the market. For example, $\bar{\lambda}$ can be determined from the yield function on the market by the least squares method, i.e., by choosing $\bar{\lambda}$ so that

$$\int_t^T \left[Y(t,T;\bar{\lambda}) - \tilde{Y}(t - T)\right]^2 dT \text{ is minimized, or}$$

$$\int_t^T \left[Y(t,T;\bar{\lambda}) - \tilde{Y}(t,T)\right] \frac{\partial Y(t,T;\bar{\lambda})}{\partial\bar{\lambda}} dT = 0. \qquad (4.29)$$

Here, $\tilde{Y}(t,T)$ is the yield function observed on the market, whereas according to (4.10) and (4.19), the function $Y(t,T;\bar{\lambda})$ is given by

$$Y\left(t,T;\bar{\lambda}\right) = \frac{rB(t,T;\bar{\lambda}) - A(t,T;\bar{\lambda})}{T-t}, \qquad (4.30)$$

where $A(t,T;\bar{\lambda})$ and $B(t,T;\bar{\lambda})$ are given by (4.27) and (4.26), respectively, but the dependence of A and B on $\bar{\lambda}$ is expressed explicitly here. If the value of yield is only available discretely on the market, then we can find a $\bar{\lambda}$ such that $\sum_i \left[Y(t,T_i;\bar{\lambda}) - \tilde{Y}(t,T_i)\right]^2$ is minimized, or

$$\sum_i \left[Y(t,T_i;\bar{\lambda}) - \tilde{Y}(t,T_i)\right] \frac{\partial Y(t,T_i;\bar{\lambda})}{\partial \bar{\lambda}} = 0. \qquad (4.31)$$

As soon as we have $\bar{\lambda}$, the equation (4.16) with constant α, β, γ, and μ can be used to determine the value of any other spot interest rate derivative. Generally speaking, it is impossible to fit the entire yield curve by choosing only one parameter. This is a drawback of such a model.

For some special models, for example, the Vasicek model (see [70]) and the Cox–Ingersoll–Ross model (see [22]), the expression can be simplified. Let us do this for these two models.

The Vasicek model is in the form

$$dr = (\bar{\mu} - \bar{\gamma}r)\,dt + \sqrt{-\beta}dX, \quad \beta < 0, \quad \bar{\gamma} > 0.$$

Therefore, (4.26) and (4.27) with

$$\alpha = 0, \quad \mu = \bar{\mu} + \bar{\lambda}\beta$$

and

$$\gamma = \bar{\gamma}$$

give B and A for this model. In this case, (4.26) becomes[1]

$$B = \frac{e^{\gamma(T-t)} - 1}{\gamma\, e^{\gamma(T-t)}} = \frac{1}{\gamma}(1 - e^{-\gamma(T-t)}). \qquad (4.32)$$

However, the expression (4.27) cannot be used for calculation because of $\alpha = 0$. In order to have an expression for A that can be used for calculation, we need to find the limit of (4.27) as $\alpha \to 0$ or solve (4.22) with B given by (4.32) directly. Let us solve (4.22) directly. Putting (4.32) into (4.22), we have:

$$A(T,T) - A(t,T) = \int_{A(t,T)}^{A(T,T)} dA$$

$$= \int_t^T \left[\frac{\beta}{2\gamma^2}\left(1 - e^{-\gamma(T-t)}\right)^2 + \frac{\mu}{\gamma}\left(1 - e^{-\gamma(T-t)}\right)\right] dt$$

---

[1]This expression can also be obtained by integrating (4.23) with $\alpha = 0$ directly, and for the case $\alpha = 0$, this direct way of finding the solution is easier.

$$= \int_t^T \left[\frac{\beta}{2\gamma^2} \left(1 - e^{-\gamma(T-t)} \right) \left(-e^{-\gamma(T-t)} \right) \right.$$
$$\left. + \left(\frac{\beta}{2\gamma^2} + \frac{\mu}{\gamma} \right) \left(1 - e^{-\gamma(T-t)} \right) \right] dt$$

$$= \left[\frac{\beta}{4\gamma^3} \left(1 - e^{-\gamma(T-t)} \right)^2 + \left(\frac{\beta}{2\gamma^2} + \frac{\mu}{\gamma} \right) \left(t - \frac{1}{\gamma} e^{-\gamma(T-t)} \right) \right] \Big|_t^T$$

$$= -\frac{\beta}{4\gamma^3} \left(1 - e^{-\gamma(T-t)} \right)^2 + \left(\frac{\beta}{2\gamma^2} + \frac{\mu}{\gamma} \right) (T - t)$$
$$- \left(\frac{\beta}{2\gamma^2} + \frac{\mu}{\gamma} \right) \frac{1}{\gamma} \left(1 - e^{-\gamma(T-t)} \right).$$

Because of $A(T,T) = 0$, we obtain

$$A = - \left(\frac{\beta}{2\gamma^2} + \frac{\mu}{\gamma} \right) (T - t) + \left(\frac{\beta}{2\gamma^2} + \frac{\mu}{\gamma} \right) B + \frac{\beta B^2}{4\gamma}. \qquad (4.33)$$

Consequently

$$V(r,t) = Z e^{-(\beta/2\gamma^2 + \mu/\gamma)(T-t) + (\beta/2\gamma^2 + \mu/\gamma - r)B + \beta B^2/4\gamma}$$
$$= Z e^{-(\beta/2\gamma^2 + \mu/\gamma)(T-t) + (\beta/2\gamma^2 + \mu/\gamma - r)(1 - e^{-\gamma(T-t)})/\gamma + \beta(1 - e^{-\gamma(T-t)})^2/4\gamma^3}.$$
$$(4.34)$$

This is the value of a zero-coupon bond if the Vasicek model is adopted. As pointed out above, (4.34) can also be obtained by finding the limit of (4.28). This is left to the reader as Problem 5.

Noticing that B does not depend on $\bar{\lambda}$ in this case, we have

$$Y(t,T;\bar{\lambda}) = \frac{rB(t,T) - A(t,T;\bar{\lambda})}{T-t}$$

$$= \frac{\left(\dfrac{\beta}{2\gamma^2} + \dfrac{\mu}{\gamma} \right)(T-t) - \left(\dfrac{\beta}{2\gamma^2} + \dfrac{\mu}{\gamma} - r \right) B - \dfrac{\beta B^2}{4\gamma}}{T-t}$$

$$= \frac{\left(\dfrac{\beta}{2\gamma^2} + \dfrac{\bar{\mu}}{\gamma} \right)(T-t) - \left(\dfrac{B}{2\gamma^2} + \dfrac{\bar{\mu}}{\gamma} - r \right) B - \dfrac{\beta B^2}{4\gamma}}{T-t}$$

$$+ \frac{\bar{\lambda}\beta(T-t-B)}{\gamma(T-t)}$$

and

$$\frac{\partial Y}{\partial \bar{\lambda}}(t,T;\bar{\lambda}) = \frac{\beta(T-t-B)}{\gamma(T-t)}.$$

Hence, (4.29) becomes a linear equation for $\bar{\lambda}$. From the linear equation, we see that $\bar{\lambda}$ is given by

$$\frac{\int_t^T \left[\dfrac{\left(\dfrac{\beta}{2\gamma^2} + \dfrac{\bar{\mu}}{\gamma}\right)(T-t) - \left(\dfrac{B}{2\gamma^2} + \dfrac{\bar{\mu}}{\gamma} - r\right)B - \dfrac{\beta B^2}{4\gamma}}{T-t} - \tilde{Y}\dfrac{T-t-B}{(T-t)} dT \right]}{-\dfrac{\beta}{\gamma}\int_t^T \dfrac{(T-t-B)^2}{(T-t)^2}dT}.$$

$$(4.35)$$

Because only $\bar{\lambda}$ is chosen, the yield curve cannot be fitted entirely. Another problem of this model is that r may be negative.

In order to rectify this problem, Cox, Ingersoll, and Ross (see [22]) proposed another model:

$$dr = (\bar{\mu} - \bar{\gamma}r)dt + \sqrt{\alpha r}dX. \qquad (4.36)$$

This is also in the form (4.14) and $\beta = 0$ here. In this case, the solution for a zero-coupon bond is

$$V(r,t) = Z\left[\frac{B + (\gamma-\psi)/\alpha}{(\gamma-\psi)/\alpha}\right]^{\mu(\psi-\gamma)/\alpha\psi}\left[\frac{B + (\gamma+\psi)/\alpha}{(\gamma+\psi)/\alpha}\right]^{\mu(\gamma+\psi)/\alpha\psi}e^{-rB}.$$

Here, B is given by (4.26), i.e.,

$$B = \frac{2\left(e^{\psi(T-t)} - 1\right)}{(\gamma+\psi)\,e^{\psi(T-t)} - (\gamma-\psi)},$$

$$\mu = \bar{\mu}$$

and

$$\gamma = \bar{\gamma} + \bar{\lambda}\alpha,$$

where $\bar{\lambda}$ is a parameter in the expression (4.15) for the market price of risk. However, the solution can have another form. Because

$$A(T,T) - A(t,T) = \int_{A(t,T)}^{A(T,T)} dA = \int_t^T \mu B dt$$

$$= \mu \int_t^T \frac{2\left(e^{\psi(T-t)} - 1\right)}{(\gamma+\psi)\,e^{\psi(T-t)} - (\gamma-\psi)}dt,$$

noticing $A(T,T) = 0$ and setting $\xi = e^{\psi(T-t)}$, we have

$$A(t,T)$$

$$= -\mu\int_t^T \frac{2\left(e^{\psi(T-t)} - 1\right)}{(\gamma+\psi)\,e^{\psi(T-t)} - (\gamma-\psi)}dt = \mu\int_\xi^1 \frac{2(\xi-1)d\xi}{[(\gamma+\psi)\xi - (\gamma-\psi)]\psi\xi}$$

$$= \frac{2\mu}{\psi(\gamma+\psi)}\int_\xi^1 \left[\frac{-2\psi/(\gamma-\psi)}{\xi - (\gamma-\psi)/(\gamma+\psi)} + \frac{(\gamma+\psi)/(\gamma-\psi)}{\xi}\right]d\xi$$

$$= \frac{-4\mu}{\gamma^2 - \psi^2} \left[\ln(\xi - (\gamma - \psi)/(\gamma + \psi)) - (\gamma + \psi) \ln \xi/2\psi \right]\big|_\xi^1$$

$$= \frac{2\mu}{\alpha} \left[\ln(1 - (\gamma - \psi)/(\gamma + \psi)) - \ln(\xi - (\gamma - \psi)/(\gamma + \psi)) \right.$$
$$\left. + (\gamma + \psi) \ln \xi/2\psi \right]$$

$$= \ln \left(\frac{2\psi e^{(\gamma + \psi)(T-t)/2}}{(\gamma + \psi)e^{\psi(T-t)} - (\gamma - \psi)} \right)^{2\mu/\alpha}.$$

Therefore, we have a solution

$$V(r,t) = Z \left[\frac{2\psi e^{(\gamma + \psi)(T-t)/2}}{(\gamma + \psi)e^{\psi(T-t)} - (\gamma - \psi)} \right]^{2\mu/\alpha} e^{-rB}, \qquad (4.37)$$

which is the form given in the paper by Cox, Ingersoll, and Ross. It can be proven that the two expressions are identical. This is left to the reader to prove as Problem 7.

In this case

$$Y(t,T;\lambda) = \frac{\dfrac{2\left(e^{\psi(T-t)} - 1\right)r}{(\gamma + \psi)e^{\psi(T-t)} - (\gamma - \psi)} - \dfrac{2\mu}{\alpha} \ln \left(\dfrac{2\psi e^{(\gamma + \psi)(T-t)/2}}{(\gamma + \psi)e^{\psi(T-t)} - (\gamma - \psi)} \right)}{T - t},$$

where $\gamma = \bar\gamma + \bar\lambda\alpha$ and $\psi = \sqrt{\gamma^2 + 2\alpha}$, so the dependence of $Y(t,T;\bar\lambda)$ on $\bar\lambda$ is quite complicated.

As we have stated, in order to use the partial differential equation (4.16) to price the value of other derivatives, we need to determine $\bar\lambda$ so that we can have $\gamma = \bar\gamma + \bar\lambda\alpha$. For example, we can obtain $\bar\lambda$ by solving (4.29). Because the dependence of $Y(t,T;\bar\lambda)$ on $\bar\lambda$ in this case is quite complicated, (4.29) has to be solved numerically. Just like the Vasicek model, generally speaking, it is impossible to "build" the entire term structure of the spot interest rate into a parameter $\bar\lambda$.

4.3.2 Explicit Solutions for the Ho–Lee and Hull–White Models

In order to fit the entire term structure of interest rates, it seems to be necessary to require $\bar\lambda$ to be dependent on t or r. If $\bar\lambda$ depends on t, then for some models in the form (4.14), the solution of a zero-coupon bond can explicitly be expressed by elemental functions and integrals with known integrands. We refer to such a solution as an explicit solution or a closed-form solution. The Ho–Lee model (see [38])

$$dr = \bar\mu(t)dt + \sqrt{-\beta}dX \qquad (4.38)$$

and the Hull–White model (see [40])

$$dr = (\bar\mu(t) - \bar\gamma r)dt + \sqrt{-\beta}dX \qquad (4.39)$$

are such models. We note that the Hull–White model is an extension of the Ho–Lee model and the Vasicek model. For the Hull–White model, $B(t,T)$ is the same as for the Vasicek model, given by (4.32):

$$B(t,T) = \frac{1}{\gamma}(1 - e^{-\gamma(T-t)}),$$

where

$$\gamma = \bar{\gamma}.$$

Let $\gamma \to 0$, we have

$$B(t,T) = T - t, \tag{4.40}$$

which is the expression of $B(t,T)$ for the Ho–Lee model. For both of them, the equation (4.22) is in the form

$$\frac{dA}{dt} = \frac{1}{2}\beta B^2 + \mu(t)B,$$

where $\mu(t)$ is given by (4.17):

$$\mu(t) = \bar{\mu}(t) + \bar{\lambda}(t)\beta.$$

Here, we assume that the market price of risk is $\lambda(r,t) = \bar{\lambda}(t)\sqrt{-\beta}$. From the ordinary differential equation above, we can find

$$A(t,T) = -\frac{1}{2}\beta \int_t^T B^2(\tau,T)d\tau - \int_t^T \mu(\tau)B(\tau,T)d\tau$$

and

$$V(r,t) = Ze^{-\frac{1}{2}\beta \int_t^T B^2(\tau,T)d\tau - \int_t^T \mu(\tau)B(\tau,T)d\tau - rB(t,T)}. \tag{4.41}$$

Here, B is given by (4.32) or (4.40), depending on which model is used. Therefore, if $\bar{\lambda}$ is given, we can find $V(r,t)$ without any difficulties.

In practice, we need to find $\bar{\lambda}(t)$ from some data on the market, for example, a given yield function $Y(t,T)$. In order to do this, we rewrite (4.41) as

$$\ln V(r,t) = \ln Z - \frac{1}{2}\beta \int_t^T B^2(\tau,T)d\tau - \int_t^T \mu(\tau)B(\tau,T)d\tau - rB(t,T)$$

or if we require that the solution (4.41) fits the yield function on the market, we furthermore have

$$\int_t^T \mu(\tau)B(\tau,T)d\tau = Y(t,T)(T-t) - \frac{1}{2}\beta \int_t^T B^2(\tau,T)d\tau - rB(t,T), \tag{4.42}$$

where we have used the definition of the yield (4.10). If we define

$$F_1(t,T) \equiv Y(t,T)(T-t) - \frac{1}{2}\beta \int_t^T B^2(\tau,T)d\tau - rB(t,T) \tag{4.43}$$

and substitute $(1 - e^{-\gamma(T-\tau)})/\gamma$ for B on the left-hand side of the equation (4.42), it becomes

$$\frac{1}{\gamma} \int_t^T \mu(\tau)(1 - e^{-\gamma(T-\tau)})d\tau = F_1(t, T).$$

Differentiating both sides of this relation with respect to T twice yields

$$\mu(T) = \frac{\partial^2 F_1(t, T)}{\partial T^2} + \gamma \frac{\partial F_1(t, T)}{\partial T}, \tag{4.44}$$

which is the function $\mu(t)$ for the Hull–White model. After having the function $\mu(t)$, we can obtain $\bar{\lambda}(t)$ immediately by

$$\bar{\lambda}(t) = \frac{1}{\beta} [\mu(t) - \bar{\mu}(t)]$$

if we want. Therefore, for the Hull–White model, we can find a function for the market price of risk for r such that the entire term structure of interest rate can be fitted. For the Ho–Lee model, in order to do this, we can use the same formula with $\gamma = 0$, so in (4.43) $B = T - t$. Because in these models the entire term structure of interest rate is built into the function $\bar{\lambda}(t)$, these two models are often referred to as no-arbitrage interest rate models. The difference between them is that the Hull–White model has the mean reversion property that an interest rate model should have, whereas the Ho–Lee model does not.

4.4 Inverse Problem on the Market Price of Risk

As we saw in Section 4.3, for some special interest rate models and some special function of the market price of risk, we can find an explicit solution for a zero-coupon bond and furthermore explicit expressions for the market price of risk for which the entire term structure of interest rate or the entire zero-coupon bond price curve is fitted. However, even though the model is in the form (4.14) and solving the partial differential equation (4.16) can be reduced to solving ordinary differential equations (4.22) and (4.23), we still may not be able to find an explicit expression for the market price of risk if $\alpha(t)$ really depends on t or even if α is a nonzero constant. In this case, the unknown function $\bar{\lambda}(t)$ appears in both (4.22) and (4.23) and it may be necessary to use numerical methods.

Also, there are other models, for example, the Black–Derman–Toy model (see [8]):

$$d\ln r = \left[\bar{\mu}(t) - \frac{\sigma_r'(t)}{\sigma_r(t)} \ln r\right] dt + \sigma_r(t)dX$$

and the Black–Karasinski model (see [9]):

Fig. 4.1. One month LIBOR on U.S. dollar during 1988–2000

$$d \ln r = [\bar{\mu}(t) - \bar{\gamma}(t) \ln r] \, dt + \sigma_r(t) dX.$$

For these models, it might even be impossible to reduce solving a partial differential equation into solving two ordinary differential equations. In addition, it may be necessary to consider interest rate models (4.11):

$$dr = u(r,t)dt + w(r,t)dX$$

with more general functions $u(r,t)$ and $w(r,t)$. For example, a model might be more useful if the forms of $u(r,t)$ and $w(r,t)$ are determined from the data of the spot rate on the market. In Fig. 4.1, the one-month LIBOR (London Interbank Offer Rate) on U.S. dollar for years 1988–2000 is given. According to the figure, it is reasonable to assume that r has a lower bound r_l and an upper bound r_u. According to Section 2.9, if u and w satisfy

$$\begin{cases} u(r_l, t) - w(r_l, t)\dfrac{\partial}{\partial r}w(r_l, t) \geq 0, \\ w(r_l, t) = 0 \end{cases} \tag{4.45}$$

and

$$\begin{cases} u(r_u, t) - w(r_u, t)\dfrac{\partial}{\partial r}w(r_u, t) \leq 0, \\ w(r_u, t) = 0, \end{cases} \tag{4.46}$$

then the random variable r is always in $[r_l, r_u]$. In this section, we will require (4.45) and (4.46), so r will be in $[r_l, r_u]$. The dependence of u and w on r in the interval (r_l, r_u) can be determined by using statistics, and we can choose more complicated functions so that the important features of these functions are incorporated into the models.

As pointed out in Subsection 4.2.2, if we use the model (4.11), then any interest rate derivative, $V(r, t)$, satisfies the equation (4.12):

$$\frac{\partial V}{\partial t} + \frac{1}{2} w^2 \frac{\partial^2 V}{\partial r^2} + (u - \lambda w) \frac{\partial V}{\partial r} - rV = 0, \qquad 0 \le t \le T,$$

where we assume that there is no coupon related to the derivative, so $kZ = 0$. This parabolic partial differential equation degenerates to a hyperbolic partial differential equation or an ordinary differential equation at $r = r_l$ and r_u when $w(r_l, t) = 0$ and $w(r_u, t) = 0$. Moreover, if (4.45) holds, from Subsection 2.9.5, we see that no extra boundary condition at $r = r_l$ is needed in order to find a unique solution. Similarly, if (4.46) holds, then no extra boundary condition at $r = r_u$ is needed. Consequently, the final value problem without any boundary conditions

$$\begin{cases} \dfrac{\partial V}{\partial t} + \dfrac{1}{2} w^2 \dfrac{\partial^2 V}{\partial r^2} + (u - \lambda w) \dfrac{\partial V}{\partial r} - rV = 0, & r_l \le r \le r_u, \quad t \le T, \\ V(r, T) = f(r), & r_l \le r \le r_u \end{cases}$$

has a unique solution if the conditions (4.45) and (4.46) hold. As we have discussed, u and w can be determined from the historical data of the spot rate on the market. However, in order to use this equation to price any derivatives, we need to know λ. As soon as such a λ is determined, an interest rate model (4.11) becomes a no-arbitrage interest rate model. Thus, it is important in practice. Suppose λ is a function of t, i.e., $\lambda = \lambda(t)$. Then this function, as the solution of the following inverse problem, can be determined by the term structure of interest rates or, equivalently, by the zero-coupon bond price curve. Suppose that $t = 0$ corresponds to today and today's spot interest rate is r^*. Let $V(r, t; T^*)$ be the solution of the problem

$$\begin{cases} \dfrac{\partial V}{\partial t} + \dfrac{1}{2} w^2 \dfrac{\partial^2 V}{\partial r^2} + [u - \lambda(t)w] \dfrac{\partial V}{\partial r} - rV = 0, \\ \qquad\qquad\qquad\qquad r_l \le r \le r_u, \quad 0 \le t \le T^*, \\ V(r, T^*; T^*) = 1, \qquad r_l \le r \le r_u. \end{cases} \qquad (4.47)$$

Here T^* is a parameter. We need to find a function $\lambda(t)$ defined on $[0, T^*_{max}]$ such that $V(r^*, 0; T^*)$ is equal to the today's value of the zero-coupon bond maturing at time T^* and with a face value of one dollar for any $T^* \in [0, T^*_{max}]$, where T^*_{max} in the longest maturity of zero-coupon bonds on the market.

In this problem, the value of $\lambda(t)$ for $t \in [0, T^*_1]$ is determined by the value of zero-coupon bonds maturing at time $T^* \in [0, T^*_1]$. If $\lambda(t)$ for $t \in [0, T^*_1]$

has been obtained and $T_2^* > T_1^*$, then the value of $\lambda(t)$ for $t \in [T_1^*, T_2^*]$ will be found by letting $V(r^*, 0; T^*)$ be equal to the value of a zero-coupon bond maturing at time T^* for any $T^* \in [T_1^*, T_2^*]$. Therefore, we can roughly say that the value of $\lambda(t)$ at $t = T^*$ is determined by the value of a zero-coupon bond maturing at time T^*. Suppose the value of $\lambda(t)$ for $t \in [0, T^*)$ has been obtained. In order to find the value of $\lambda(T^*)$, we need to make a guess about it and solve (4.47) from $t = T^*$ to $t = 0$ and then check if $V(r^*, 0; T^*)$ is equal to the value of the zero-coupon bond maturing at time T^*. If T^* is 20 or 30 years, then the procedure of solving (4.47) is quite long.

4.5 Application of Bond Equations

The bond equation (4.12) can be applied to evaluating not only bonds but also bond options, options on bond futures contracts, swaps, caps, floors, collars, and even options on them. In what follows, we describe these applications.

4.5.1 Bond Options and Options on Bond Futures Contracts

A bond option is similar to an equity option except that the underlying asset is a bond. A bond depends on the interest r, and consequently, a bond option will also depend on r. Suppose that a bond will mature at time T_b, and for simplicity, let the face value of the bond be equal to one. Thus, the bond price is the solution of the problem

$$
\begin{cases}
\dfrac{\partial V_b}{\partial t} + \dfrac{1}{2} w^2 \dfrac{\partial^2 V_b}{\partial r^2} + (u - \lambda w) \dfrac{\partial V_b}{\partial r} - r V_b + k = 0, \\[2mm]
\qquad\qquad r_l \le r \le r_u, \quad t \le T_b, \\[2mm]
V_b(r, T_b; T_b) = 1, \quad r_l \le r \le r_u,
\end{cases}
\tag{4.48}
$$

where we consider a coupon-bearing bond with a coupon payment $k(t)dt$ during a time period $[t, t + dt]$ and use V_b to represent the cash price (dirty price) of a bond, not the quoted price (clean price). Here, we assume that (4.45) and (4.46) hold, so at $r = r_l$ and $r = r_u$ the equation degenerates to a hyperbolic equation and does not require any boundary conditions. Every model can be modified locally, so (4.45) and (4.46) hold. Therefore, this assumption is realistic. We also assume that $\lambda(t)$ is known. A European call bond option is a contract whose holder has a right to purchase a bond at time $T < T_b$ at a price E. Let $V(r, t)$ be the price of the option. Clearly, $V(r, t)$ should be the solution of the problem

$$
\begin{cases}
\dfrac{\partial V}{\partial t} + \dfrac{1}{2} w^2 \dfrac{\partial^2 V}{\partial r^2} + (u - \lambda w) \dfrac{\partial V}{\partial r} - r V = 0, \quad r_l \le r \le r_u, \\[2mm]
\qquad\qquad t \le T, \\[2mm]
V(r, T) = \max(V_b(r, T; T_b) - E, 0), \quad r_l \le r \le r_u.
\end{cases}
\tag{4.49}
$$

For a European put bond option, the final condition is

$$V(r, T) = \max \left(E - V_b(r, T; T_b), 0 \right).$$

For American call and put bond options, we need to require

$$V(r, t) \geq \max \left(V_b \left(r, t; t + N \right) - E, 0 \right)$$

and

$$V(r, t) \geq \max(E - V_b(r, t; t + N), 0)$$

respectively, where N is the maturity of the bond. For example, if the option is on a three-year bond, then $N = 3$. In this case, in order to determine the solution, we need to solve a problem involving free boundaries, and the constraint is a function of t. Therefore, this free-boundary problem is more complicated than that in equity option cases.

We can also determine the value of an option on a bond futures contract, which is denoted by $V(r, t)$ in what follows. Again, let T_b be the maturity date of the bond and T be the expiry of the option and the date the futures contract is initiated. Also, suppose that the futures contract is matured at time $T_f \in (T, T_b)$ and that the delivery price given in the option – the exercise price of the option is K. When $V(r, T)$ is given, we can obtain the value of the option today by solving a problem similar to the problem (4.49). How do we find $V(r, T)$?

Let $V_{b0}(r, t; T_f)$ be the value of the zero-coupon bond with maturity date T_f, which is the solution of the following problem

$$
\begin{cases}
\dfrac{\partial V_{b0}}{\partial t} + \dfrac{1}{2} w^2 \dfrac{\partial^2 V_{b0}}{\partial r^2} + (u - \lambda w) \dfrac{\partial V_{b0}}{\partial r} - r V_{b0} = 0, \\
\qquad\qquad\qquad\qquad\qquad\qquad r_l \leq r \leq r_u, \quad t \leq T_f, \qquad (4.50) \\
V_{b0}(r, T_f; T_f) = 1, \qquad\qquad r_l \leq r \leq r_u.
\end{cases}
$$

Then the value of the bond futures contract with a delivery price K given in the option can be expressed as

$$V_f(r, t; T_f) = V_b(r, t; T_b) - K V_{b0}(r, t; T_f) \qquad (4.51)$$

for any $t \leq T_f$. Let K^* be the futures price for the futures contract with maturity date T_f at time T. K^* should be determined by the condition that the value of the futures contract is equal to zero when it is initiated at time T, i.e.,

$$V_f(r, T; T_f) = V_b(r, T; T_b) - K^* V_{b0}(r, T; T_f) = 0.$$

From this condition, we immediately know that the futures price K^* is equal to $V_b(r, T; T_b)/V_{b0}(r, T; T_f)$. If

$$K < K^* = V_b(r, T; T_b)/V_{b0}(r, T; T_f),$$

the holder of the option will exercise the option because the value of the bond futures contract

$$V_f(r,T;T_f) = V_b(r,T;T_b) - KV_{b0}(r,T;T_f) > 0.$$

Actually this is the the value of the option for this case. If

$$K \geq K^* = V_b(r,T;T_b)/V_{b0}(r,T;T_f),$$

the value of the bond futures contract

$$V_f(r,T;T_f) = V_b(r,T;T_b) - KV_b(r,T;T_f) \leq 0,$$

and the holder will not exercise the option, which means $V(r,T) = 0$. Putting the two cases together, for $V(r,T)$ we have the following expression

$$V(r,T) = \max\left(V_b(r,T;T_b) - KV_{b0}(r,T;T_f), 0\right). \tag{4.52}$$

Therefore, we can first solve the problem (4.48) from T_b to T to get $V_b(r,T;T_b)$ and solve the problem (4.50) from T_f to T to get $V_{b0}(r,T;T_f)$, and then use the formula (4.52) in order to get $V(r,T)$. As soon as we have $V(r,T)$, we can solve (4.49) with $V(r,T)$ as the final condition in order to find the price of the option on a bond futures contract today.

It is possible to consider V_b as a state variable and let the bond option price depend on V_b and t. For example, suppose

$$dV_b = \mu V_b dt + \sigma V_b dX,$$

where μ and σ is constant. In this case, we get the Black–Scholes equation with independent variables t and V_b, and use the Black–Scholes formulae to find the prices of European bond options. However, because the bond price must be equal to the face value at time T_b, which is often referred to as the pull-to-par phenomenon, a bond has different features from an equity, especially when $t \approx T_b$ (see Fig. 1.3). Therefore, even though a model in the form $dV_b = \mu V_b dt + \sigma V_b dX$ can describe the dynamics of an equity well, it could not state that of a bond. Consequently, the bond price obtained in this way is expected to have a large error, especially when $T \approx T_b$. If the model is in the form

$$dV_b = \alpha(t)(1 - V_b)dt + \sigma(t)V_b dX,$$

where $\alpha(t) \to \infty$ and $\sigma(t) \to 0$ as $t \to T_b$, then the result might be much better because such a model guarantees that V_b has a unique value one at time T_b. Of course, in this case it might be necessary to get solutions by numerical methods. Another problem of pricing a bond option in this way is to assume that the spot rate is constant throughout the whole life of the option. If T is not small, it is not a good assumption.

Promising to pay an amount E at time T is equivalent to issuing a bond maturing at time T with a face value E. Thus, a right to pay E for a bond with a maturity date T_b at time T is the same right to exchange a bond of a face value E with a maturity date T for another bond with a maturity date T_b at time T. Therefore, a bond option can be understood as an exchange option that allows the holder to exchange a bond maturing at time T for another bond maturing at time T_b. If a bond option is dealt with in this way, it may be necessary to choose a model so that at least the random variable for the bond maturing at time T has the property of "pull-to-par."

4.5.2 Interest Rate Swaps and Swaptions

As an example, let us look at the following N-year swap on a notional principal Q between a bank and a company. In the swap, the bank and the company agree that during the next N years, the company will pay the bank the interest payment on the notional principal Q at a fixed rate $r_s(N)$ semiannually and in return, the bank will pay the company the interest payment on the same principal at a floating rate at the same times. Here, the floating rate is the six-month London Interbank Offer Rate (LIBOR) prevailing six months before the payment date. When the swap is initiated, both parties do not need to pay any money. Thus, the contract has no value at initiation. The fixed rate $r_s(N)$ is called the swap rate for an N year swap and determined through negotiation by the two parties. Clearly, the company wants $r_s(N)$ to be as small as possible, and the bank prefers a higher $r_s(N)$. What is the value of $r_s(N)$ both parties can accept? $r_s(N)$ should be a rate such that the value of the swap at initiation is zero. In order to know what equation $r_s(N)$ should satisfy, we need to find out how the value of the swap is related to r_s, where r_s denotes a swap rate that might not equal $r_s(N)$.

Suppose the swap is initiated at time T and today's time is $t^* \geq T$. The interest payments are exchanged semiannually at time $t_k = T + k/2$, $k = k^* + 1, k^* + 2, \cdots, 2N$, where k^* is the integer part of $(t^* - T)/2$. Suppose today the price of the zero-coupon bond with a face value of one dollar and with maturity date t_k is $Z(t^*; t_k)$. In the swap given above, the company will pay cash $Qr_s/2$ at time t_k, $k = k^* + 1, k^* + 2, \cdots, 2N$. The present value of this cash flow is

$$\sum_{k=k^*+1}^{2N} \frac{Qr_s}{2} Z(t^*; t_k).$$

At the same times, the bank will pay the company an amount of cash $\frac{Q}{2} \bar{f}(t_{k-1}, t_{k-1}, t_k)$ at time t_k, $k = k^*+1, k^*+2, \cdots, 2N$, where $\bar{f}(t_{k-1}, t_{k-1}, t_k)$ is the forward rate for the period $[t_{k-1}, t_k]$ determined at time t_{k-1} and we define $t_{k^*} = T + k^*/2$. Because $t_{k^*} \leq t$, $\bar{f}(t_{k^*}, t_{k^*}, t_{k^*+1})$ is known today and the present value of the first payment is

$$\frac{Q}{2} Z(t^*; t_{k^*+1}) \bar{f}(t_{k^*}, t_{k^*}, t_{k^*+1}).$$

What is the present value of the other payments? Suppose we deposit Q in the bank at time t_{k^*+1} for a period $[t_{k^*+1}, t_{k^*+2}]$ at a floating rate $f(t_{k^*+1}, t_{k^*+1}, t_{k^*+2})$. At time t_k, $k = k^* + 2, k^* + 3, \cdots, 2N - 1$, we take the interest payment away and still leave Q in the bank for the next half year. In this way, we can generate a cash flow $\dfrac{Q}{2}\bar{f}(t_{k-1}, t_{k-1}, t_k)$ at time t_k, $k = k^* + 2, k^* + 3, \cdots, 2N - 1$ and cash $\dfrac{Q}{2}\bar{f}(t_{2N-1}, t_{2N-1}, t_{2N}) + Q$ at time t_{2N}. Therefore, the value of the other payments is the difference between Q at time t_{k^*+1} and Q at time $t_{2N} = T + N$. Written mathematically, the present value of the other payments is

$$QZ(t^*; t_{k^*+1}) - QZ(t^*; T + N).$$

This result also can be obtained analytically. In fact, from (4.6) we know that the forward interest rate compounded semiannually at time t^* during a period $[t_k, t_{k+1}]$ is

$$\bar{f}(t^*, t_k, t_{k+1}) = 2\left[\frac{Z(t^*; t_k)}{Z(t^*; t_{k+1})} - 1\right],$$

where $t_{k+1} = t_k + 1/2$. Therefore at time t^*, the value of the cash flow $\dfrac{Q}{2}\bar{f}(t^*, t_k, t_{k+1})$ at time t_{k+1}, $k = k^* + 1, k^* + 2, \cdots, 2N - 1$, is

$$\sum_{k=k^*+1}^{2N-1} \frac{Q}{2}\bar{f}(t^*, t_k, t_{k+1})Z(t^*; t_{k+1})$$

$$= Q\sum_{k=k^*+1}^{2N-1} \left[\frac{Z(t^*; t_k)}{Z(t^*; t_{k+1})} - 1\right] Z(t^*; t_{k+1})$$

$$= Q(Z(t^*; t_{k^*+1}) - Z(t^*; t_{2N}))$$

$$= QZ(t^*; t_{k^*+1}) - QZ(t^*; T + N).$$

Let $V_s(t^*, r_s)$ be the present value of the swap to the company, which is the present value of the cash flow the company will receive minus the present value of the cash flow it will pay. From previous results, we arrive at

$$V_s(t^*; r_s) = \frac{Q}{2}Z(t^*; t_{k^*+1})\bar{f}(t_{k^*}, t_{k^*}, t_{k^*+1}) + QZ(t^*; t_{k^*+1}) - QZ(t^*; T + N)$$

$$- \sum_{k=k^*+1}^{2N} \frac{Qr_s}{2}Z(t^*; t_k)$$

$$= QZ(t^*; t_{k^*+1})\left[1 + \frac{1}{2}\bar{f}(t_{k^*}, t_{k^*}, t_{k^*+1})\right]$$

$$- Q\left[\sum_{k=k^*+1}^{2N} \frac{r_s}{2}Z(t^*; t_k) + Z(t^*; T + N)\right]. \tag{4.53}$$

The expression $Q \left[\sum\limits_{k=k^*+1}^{2N} \dfrac{r_s}{2} Z(t^*; t_k) + Z(t^*; T+N) \right]$ can be understood as the present value of a coupon-bearing bond, and the expression $QZ(t^*; t_{k^*+1})$ $\times \left[1 + \dfrac{1}{2} \bar{f}(t_{k^*}, t_{k^*}, t_{k^*+1}) \right]$ is the present value of another coupon-bearing bond. Therefore, a swap can be seen as a combination of a long position in one coupon-bearing bond with a short position in another coupon-bearing bond.

Here, we also need to point out that the values of a swap to two parties have the same magnitude but opposite signs. Thus, the value of the swap mentioned above to the bank is

$$Q \left[\sum_{k=k^*+1}^{2N} \frac{r_s}{2} Z(t^*; t_k) + Z(t^*; T_s + N) \right]$$
$$- QZ(t^*; t_{k^*+1}) \left[1 + \frac{1}{2} \bar{f}(t_{k^*}, t_{k^*}, t_{k^*+1}) \right].$$

In the case $t^* = T$, we have $k^* = 0, t_{k^*+1} = T + 1/2$ and

$$\bar{f}(T, T, T+1/2) = 2 \left[\frac{1}{Z(T; T+1/2)} - 1 \right],$$

that is,

$$Z(T; T+1/2) \left[1 + \frac{1}{2} \bar{f}(T, T, T+1/2) \right] = 1,$$

so we have

$$V_s(T; r_s) = Q \left[1 - \frac{r_s}{2} \sum_{k=1}^{2N} Z(T; T+k/2) - Z(T; T+N) \right], \tag{4.54}$$

where we have used the relation $t_k = T + k/2$. As we have stated, when the swap is initiated, the value of the swap should be zero. Therefore, for the fixed rate in the contract we obtain

$$r_s(N) = 2 \frac{1 - Z(T; T+N)}{\sum\limits_{k=1}^{2N} Z(T; T+k/2)}. \tag{4.55}$$

Therefore, between the swap rate for an N-year swap and $Z(T; T+k/2), k = 1, 2, \cdots, 2N$, there is a simple relation: $r_s(N)$ can be determined by $Z(T; T+k/2), k = 1, 2, \cdots 2N$. This relation is true for $N = 1/2, 1, 3/2, \cdots$. Actually, $Z(T; T+k/2), k = 1, 2, \cdots, 2N$, can also be obtained recursively by

$$Z(T; T+k/2) = \frac{1 - \dfrac{r_s(k/2)}{2} \sum\limits_{i=1}^{k-1} Z(T; T+i/2)}{1 + \dfrac{r_s(k/2)}{2}} \tag{4.56}$$

if $r_s(k/2), k = 1, 2, \cdots, 2N$ are given. Therefore, knowing $r_s(k/2)$ for different k is the same as knowing the yield curve.

As we have mentioned, a swap can be understood as the difference between two different coupon-bearing bonds. From (4.53), we know that the face values of both bonds are Q. The expiration date of one bond is t_{k^*+1} and it pays a coupon $\dfrac{Q}{2}\bar{f}(t_{k^*}, t_{k^*}, t_{k^*+1})$ at $t = t_{k^*+1}$. Let V_i denote the value of this bond. The expiration date of the other bond is $T + N$, and it pays coupons $\dfrac{Qr_s}{2}$ semiannually starting at $t = t_{k^*+1}$. Let V_o represent the value of the other bond. The value of swap $V_s(t)$ is equal to $V_i - V_o$. Any bond can be priced by the bond equation. In fact, $V_i(r, t)$ is the solution of the problem

$$\begin{cases} \dfrac{\partial V_i}{\partial t} + \dfrac{1}{2}w^2\dfrac{\partial^2 V_i}{\partial r^2} + (u - \lambda w)\dfrac{\partial V_i}{\partial r} - rV_i = 0, & r_l \leq r \leq r_u, \quad t \leq t_{k^*+1}, \\ V_i(r, t_{k^*+1}) = Q[1 + \bar{f}(t_{k^*}, t_{k^*}, t_{k^*+1})/2], & r_l \leq r \leq r_u \end{cases}$$

and $V_o(r, t)$ is the solution of the problem

$$\begin{cases} \dfrac{\partial V_o}{\partial t} + \dfrac{1}{2}w^2\dfrac{\partial^2 V_o}{\partial r^2} + (u - \lambda w)\dfrac{\partial V_o}{\partial r} - rV_o + \displaystyle\sum_{k=k^*+1}^{2N-1}\dfrac{Qr_s}{2}\delta(t - t_k) = 0, \\ \qquad\qquad\qquad\qquad\qquad\qquad r_l \leq r \leq r_u, \quad t \leq T + N, \\ V_o(r, T + N) = Q(1 + r_s/2), \qquad r_l \leq r \leq r_u. \end{cases}$$

Let $r = r^*$ today and let $\lambda(t)$ be chosen so that $V(r^*, t^*; t_k) = Z(t^*; t_k)$, $k = k^* + 1, k^* + 2, \cdots, 2N$, where $V(r, t; t_k)$ is the solution of the following problem

$$\begin{cases} \dfrac{\partial V}{\partial t} + \dfrac{1}{2}w^2\dfrac{\partial^2 V}{\partial r^2} + (u - \lambda w)\dfrac{\partial V}{\partial r} - rV = 0, \\ \qquad\qquad\qquad\qquad r_l \leq r \leq r_u, \quad t \leq t_k, \\ V(r, t_k; t_k) = 1, \qquad r_l \leq r \leq r_u, \end{cases} \tag{4.57}$$

then

$$V_o(r^*, t^*; r_s) = \dfrac{Qr_s}{2}\sum_{k=k^*+1}^{2N-1} V(r^*, t^*; t_k) + Q(1 + r_s/2)V(r^*, t^*; t_{2N})$$

$$= Q\left[\dfrac{r_s}{2}\sum_{k=k^*+1}^{2N-1} Z(t^*; t_k) + (1 + r_s/2)Z(t^*; t_{2N})\right]$$

and

$$V_i(r^*, t^*) = Q\left[1 + \bar{f}(t_{k^*}, t_{k^*}, t_{k^*+1})/2\right] V(r^*, t^*; t_{k^*+1})$$
$$= Q\left[1 + \bar{f}(t_{k^*}, t_{k^*}, t_{k^*+1})/2\right] Z(t^*; t_{k^*+1}).$$

From these two expressions, we can show that

$$V_s(r^*, t^*; r_s) = V_i(r^*, t^*) - V_o(r^*, t^*; r_s)$$

will have the same value as that given by (4.53). In this case, the value of the swap is not only given at $r = r^*$, and V_s should be considered as a function of r and t.

Indeed, in order to find $V_s(r, t)$, it is not necessary to find $V_i(r, t)$ and $V_o(r, t)$ separately; instead, we only need to solve

$$
\begin{cases}
\dfrac{\partial V_s}{\partial t} + \dfrac{1}{2} w^2 \dfrac{\partial^2 V_s}{\partial r^2} + (u - \lambda w) \dfrac{\partial V_s}{\partial r} - r V_s - \displaystyle\sum_{k=k^*+1}^{2N-1} \dfrac{Q r_s}{2} \delta(t - t_k) \\[3mm]
\qquad + Q \left[1 + \dfrac{\bar{f}(t_{k^*}, t_{k^*}, t_{k^*+1})}{2} \right] \delta(t - t_{k^*+1}) = 0, \qquad (4.58) \\[3mm]
\qquad\qquad r_l \le r \le r_u, \quad t^* \le t \le T + N, \\[2mm]
V_s(r, T + N) = -Q(1 + r_s/2), \quad r_l \le r \le r_u.
\end{cases}
$$

It is not difficult to show this conclusion, and we leave this proof to the reader as Problem 13. Now we can find the value of a swap either using the formula (4.53) or solving the problem (4.58) and get the same answer. Many people will choose to calculate the value of the swap by (4.53) because it is simple. Why do we need to consider problem (4.58)? It can provide some information on $\dfrac{\partial V_s(r^*, t)}{\partial r}$ and the bond equation will be useful when pricing a swaption.

A swap option, or a swaption, is a contract to give the holder the right to enter into a certain interest rate swap by a certain time in the future. Consider a European swaption. Its holder has the right to choose if he should have an N-year swap at time T under which he will pay interest at a fixed rate r_{se} (the so-called exercise swap rate) and receive interest payment at a floating rate. Let r'_s be the N-year swap rate at time T, which is unknown today. If $r_{se} < r'_s$ at time T, then the holder will choose to exercise the swaption because the value of a swap with a swap rate r'_s at time T is zero and the value of a swap with a swap rate $r_{se} < r'_s$ should be positive, but the holder can enter into such a swap without paying any money. If $r_{se} > r'_s$, then the holder will choose not to exercise the option because the swap rate is lower on the market.

Such an option interests companies who plan to enter into a swap as a fixed rate payer because the swaption provides the companies with a guarantee that the fixed rate of interest they will pay on a loan will not exceed r_{se}.

According to (4.54), at time T, the values of the swaps with swap rates r'_s and r_{se} to the company are

$$V_s(T; r'_s) = Q \left[1 - \sum_{k=1}^{2N} \frac{r'_s}{2} Z(T; t_k) - Z(T; T + N) \right]$$

and

$$V_s(T; r_{se}) = Q\left[1 - \sum_{k=1}^{2N} \frac{r_{se}}{2} Z(T; t_k) - Z(T; T+N)\right]$$

respectively. If $r_{se} \le r'_s$, then the value of the swaption V at time T is

$$V(r'_s, T) = V_s(T; r_{se}) - V_s(T; r'_s) = Q\frac{r'_s - r_{se}}{2} \sum_{k=1}^{2N} Z(T; t_k);$$

while if $r_{se} > r'_s$, then $V(r'_s, T) = 0$. Consequently, the payoff of the swaption is

$$V(r'_s, T) = \frac{Q}{2} \sum_{k=1}^{2N} Z(T; t_k) \max(r'_s - r_{se}, 0).$$

Suppose that at time T, r'_s has a lognormal distribution with the following probability density function

$$G(r'_s) = \frac{1}{r'_s \sigma \sqrt{2\pi(T-t)}} e^{-\left[\ln(r'_s/r_s) + \sigma^2(T-t)/2\right]^2 / 2\sigma^2(T-t)},$$

where r_s is the swap rate at time t. This model is often referred to as Black's model (see [7]). This probability density function is the probability density function (2.36) with $r - D_0 = 0$. Thus, the expectation of $\max(r'_s - r_{se}, 0)$ is $e^{r(T-t)}$ times the price of a call option with $r - D_0 = 0$. That is

$$E\left[\max(r'_s - r_{se}, 0)\right] = r_s N\left(\frac{\ln(r_s/r_{se}) + \sigma^2(T-t)/2}{\sigma\sqrt{T-t}}\right)$$
$$-r_{se} N\left(\frac{\ln(r_s/r_{se}) - \sigma^2(T-t)/2}{\sigma\sqrt{T-t}}\right).$$

Therefore at time t, the value of the payoff is

$$Z(t; T)\frac{Q}{2} \sum_{k=1}^{2N} Z(T; t_k)$$
$$\times \left[r_s N\left(\frac{\ln\frac{r_s}{r_{se}} + \frac{\sigma^2(T-t)}{2}}{\sigma\sqrt{T-t}}\right) - r_{se} N\left(\frac{\ln\frac{r_s}{r_{se}} - \frac{\sigma^2(T-t)}{2}}{\sigma\sqrt{T-t}}\right)\right]$$
$$= \frac{Q}{2} \sum_{k=1}^{2N} Z(t; t_k)$$
$$\times \left[r_s N\left(\frac{\ln\frac{r_s}{r_{se}} + \frac{\sigma^2(T-t)}{2}}{\sigma\sqrt{T-t}}\right) - r_{se} N\left(\frac{\ln\frac{r_s}{r_{se}} - \frac{\sigma^2(T-t)}{2}}{\sigma\sqrt{T-t}}\right)\right],$$

where $Z(t;T)$ is the discounting factor between t and T and we have used the relation $Z(t;T)Z(T;t_k) = Z(t;t_k)$. European swaptions are frequently valued in this way. Obviously, it is an approximate method.

We may also evaluate the European swaption in the following way. Because the value of a swap can be determined by (4.54) and bonds can be considered as interest rate derivatives, the value of a swap can also be understood as an interest rate derivative. Because it also depends on r_s, we can denote the value of a swap by $V_s(r,t;r_s)$, where r_s is a parameter. According to the expression (4.54), we can furthermore represent it by

$$V_s(r,t;r_s) = Q[1 - r_s V_{s1}(r,t)/2 - V_{s2}(r,t)], \tag{4.59}$$

where V_{s1} and V_{s2} are solutions of the problem

$$\begin{cases} \dfrac{\partial V_{s1}}{\partial t} + \dfrac{1}{2}w^2\dfrac{\partial^2 V_{s1}}{\partial r^2} + (u - \lambda w)\dfrac{\partial V_{s1}}{\partial r} - rV_{s1} \\ \qquad\qquad\qquad + \displaystyle\sum_{k=1}^{2N}\delta(t - T - k/2) = 0, \\ \qquad\qquad\qquad r_l \le r \le r_u, \quad t \le T + N, \\ V_{s1}(r,T+N) = 0, \qquad r_l \le r \le r_u \end{cases} \tag{4.60}$$

and the problem

$$\begin{cases} \dfrac{\partial V_{s2}}{\partial t} + \dfrac{1}{2}w^2\dfrac{\partial^2 V_{s2}}{\partial r^2} + (u - \lambda w)\dfrac{\partial V_{s2}}{\partial r} - rV_{s2} = 0, \\ \qquad\qquad\qquad r_l \le r \le r_u, \quad t \le T + N, \\ V_{s2}(r,T+N) = 1, \qquad r_l \le r \le r_u \end{cases} \tag{4.61}$$

respectively. As soon as we get the expression $V_s(r,T;r_s)$, we can have the value of $V_s(r,T;r_{se})$. If for a given r, the value $V_s(r,T;r_{se})$ is greater than zero, then the holder of the swaption will get an amount $V_s(r,T;r_{se})$ through exercising the swaption, otherwise he will give up the swaption. Therefore, the payoff of the swaption is

$$\max\left(V_s(r,T;r_{se}),0\right) = Q\max\left(1 - r_{se}V_{s1}(r,T)/2 - V_{s2}(r,T),0\right).$$

Consequently, the value of the swaption today is $V_{so}(r^*,t^{**};r_{se})$, where r^* is the interest rate today, t^{**} is the time today, and $V_{so}(r,t;r_{se})$ is the solution of the following problem:

$$\begin{cases} \dfrac{\partial V_{so}}{\partial t} + \dfrac{1}{2}w^2\dfrac{\partial^2 V_{so}}{\partial r^2} + (u - \lambda w)\dfrac{\partial V_{so}}{\partial r} - rV_{so} = 0, \\ \qquad\qquad\qquad r_l \le r \le r_u, \quad t \le T, \\ V_{so}(r,T) = Q\max\left(1 - r_{se}V_{s1}(r,T)/2 - V_{s2}(r,T),0\right), \\ \qquad\qquad\qquad r_l \le r \le r_u. \end{cases} \tag{4.62}$$

Through first solving the problems (4.60) and (4.61) and then calculating the problem (4.62), we can find the value of the European swaption. For American swaptions we can use a similar treatment to find their prices.

4.5.3 Interest Rate Caps, Floors, and Collars

An interest rate cap is a contract whose holder will receive some money from the issuer if the floating rate exceeds a certain level r_c, where r_c is called the cap rate. Therefore, interest rate caps can be used to provide insurance against the rate of interest on a floating-rate loan rising above a certain level.

For example, someone is going to get an N-year floating-rate loan with a principal amount Q at time t^*. The borrower will pay interest quarterly at time

$$t_k = t^* + k/4, \quad k = 1, 2, \cdots, 4N.$$

The rate is a three-month LIBOR determined at time t_{k-1} for the period $[t_{k-1}, t_k]$, where we define $t_0 = t^*$. The LIBOR is a forward interest rate. According to the notation given in Section 4.1, $\bar{f}(t_{k-1}, t_{k-1}, t_k)$ stands for this rate. In what follows, we use the notation $\bar{f}_{k-1}$ instead of $\bar{f}(t_{k-1}, t_{k-1}, t_k)$ for brevity. The borrower is worrying that he will pay too much interest if the three-month LIBOR becomes very high during the period $[t^*, t^* + N]$. Therefore, he is interested in such a cap: it starts from t^* and lasts N years, and at time t_k, the issuer of the cap will pay the holder an amount of cash $Q \max(\bar{f}_{k-1} - r_c, 0)/4$. Suppose he purchases this cap. Then when $\bar{f}_{k-1} < r_c$, he will pay interest payment on the loan $Q\bar{f}_{k-1}/4$ and receive zero from the issuer of the cap; whereas $\bar{f}_{k-1} > r_c$, his actual payment is $Qr_c/4$ because he receives $Q \max(\bar{f}_{k-1} - r_c, 0)/4$ from the cap. Hence the cap provides insurance against the interest rate on the floating-rate loan rising above an upper bound r_c.

How much should be paid in order to obtain such an insurance? The present value of the payment $Q \max(\bar{f}_{k-1} - r_c, 0)/4$ at time t_k is actually the value of a call option with expiry t_k. This call option is usually called the k-th caplet. The LIBOR $\bar{f}_{k-1}$ is a forward rate determined at time t_{k-1} for the period $[t_{k-1}, t_k]$, so an amount $Q \max(\bar{f}_{k-1} - r_c, 0)/4$ at time t_k is equivalent to the amount

$$\frac{Q}{4(1 + \bar{f}_{k-1}/4)} \max(\bar{f}_{k-1} - r_c, 0) = \max\left(Q - Q\frac{1 + r_c/4}{1 + \bar{f}_{k-1}/4}, 0\right)$$

at time t_{k-1}. A loan with a face value $Q(1 + r_c/4)$ and maturity t_k is worth $Q(1 + r_c/4)/(1 + \bar{f}_{k-1}/4)$ at time t_{k-1} for any $\bar{f}_{k-1}$. Therefore, a caplet with a payoff $Q \max(\bar{f}_{k-1} - r_c, 0)/4$ at time t_k is equivalent to a put option with maturity t_{k-1} and a strike price Q on a zero-coupon bond with maturity t_k and a face value $Q(1 + r_c/4)$. At time $t^*(= t_0)$, the value of the first caplet is equal to a known value $\dfrac{Q}{4(1 + \bar{f}_0/4)} \max(\bar{f}_0 - r_c, 0)$. Usually, this value is excluded from the premium and there is no payment at time t_1 even if the LIBOR is greater than r_c. Thus, a cap comprises $4N - 1$ put options on zero-coupon bonds. Because a bond or an option on a bond can be seen as a derivative on the spot rate r, their values can be calculated by the bond equation. Let the value of the bond with maturity t_k be $V_{bk}(r, t)$. Then, $V_{bk}(r, t)$ is the solution of the problem

$$
\begin{cases}
\dfrac{\partial V_{bk}}{\partial t} + \dfrac{1}{2}w^2\dfrac{\partial^2 V_{bk}}{\partial r^2} + (u - \lambda w)\dfrac{\partial V_{bk}}{\partial r} - rV_{bk} = 0, & r_l \le r \le r_u, \\
& t_{k-1} \le t \le t_k, \\
V_{bk}(r, t_k) = (1 + r_c/4)\, Q, & r_l \le r \le r_u,
\end{cases} \tag{4.63}
$$

where $k = 2, 3, \cdots, 4N$. Let $V_c(r, t)$ be the solution of the problem

$$
\begin{cases}
\dfrac{\partial V_c}{\partial t} + \dfrac{1}{2}w^2\dfrac{\partial^2 V_c}{\partial r^2} + (u - \lambda w)\dfrac{\partial V_c}{\partial r} - rV_c \\
\qquad + \displaystyle\sum_{k=2}^{4N} \max(Q - V_{bk}(r, t_{k-1}), 0)\delta(t - t_{k-1}) = 0, \\
\qquad\qquad\qquad\qquad r_l \le r \le r_u, \quad t^* \le t \le t_{4N-1}, \\
V_c(r, t_{4N-1}) = 0, \qquad\qquad r_l \le r \le r_u.
\end{cases} \tag{4.64}
$$

Then, $V_c(r, t^*)$ gives the value of the cap – the total value of the $4N - 1$ put options at time t^* and the premium of the cap is given by

$$
V_c(r^*, t^*),
$$

where r^* is the spot interest rate at time t^*.

There are some other derivatives analogous to interest rate caps, such as interest rate floors and collars. A holder of a floor will receive some money from the issuer if the floating rate is below a certain level r_f, which is called the floor rate. If a borrower of a floating-rate loan believes that the floating rate will never be less than the lower bound r_f, then he may want to write such a floor. This is because he will get some money from writing a floor but, according to his opinion, he will not actually pay any money to the holder of the floor. Therefore, he hopes that he can reduce his expenses on the loan through writing a floor. If we assume that the floor starts at t^* and lasts N years, that the floating rate is three-month LIBOR, and that the money will be paid quarterly at time t_k, $k = 2, 3, \cdots, 4N$, then the value of the floor is the sum of $4N - 1$ floorlets that are call options on zero-coupon bonds. In order to determine the premium, we can first solve the problem

$$
\begin{cases}
\dfrac{\partial V_{bk}}{\partial t} + \dfrac{1}{2}w^2\dfrac{\partial^2 V_{bk}}{\partial r^2} + (u - \lambda w)\dfrac{\partial V_{bk}}{\partial r} - rV_{bk} = 0, & r_l \le r \le r_u, \\
& t_{k-1} \le t \le t_k, \\
V_{bk}(r, t_k) = (1 + r_f/4)\, Q, & r_l \le r \le r_u
\end{cases}
$$

and get $V_{bk}(r, t_{k-1})$. Based on $V_{bk}(r, t_{k-1})$, we then can determine the solution of the problem

$$
\begin{cases}
\dfrac{\partial V_f}{\partial t} + \dfrac{1}{2} w^2 \dfrac{\partial^2 V_f}{\partial r^2} + (u - \lambda w) \dfrac{\partial V_f}{\partial r} - r V_f \\[2mm]
\qquad + \displaystyle\sum_{k=2}^{4N} \max(V_{bk}(r, t_{k-1}) - Q, 0) \delta(t - t_{k-1}) = 0, \\[2mm]
\qquad\qquad\qquad\qquad\qquad r_l \le r \le r_u, \quad t^* \le t \le t_{4N-1}, \\[2mm]
V_f(r, t_{4N-1}) = 0, \qquad\qquad r_l \le r \le r_u
\end{cases}
$$

and the value $V_f(r^*, t^*)$ gives the premium of the floor. The derivation of this conclusion is left for the reader as Problem 14.

A collar specifies both the upper bound r_c and the lower bound r_f. It may be understood as a combination of a long position in a cap with a short position in a floor. The value of a collar V_{co} is

$$ V_{co} = V_c - V_f. $$

Usually, we choose r_c and r_f such that

$$ V_c = V_f $$

or

$$ V_{co} = 0. $$

It is clear that a portfolio of a collar and the original floating-rate loan is equivalent to a new loan with a floating rate in $[r_c, r_f]$. If

$$ r_c = r_f, $$

then the collar actually becomes a swap based on three-month LIBOR and with $4N - 1$ exchanges of payments. There exist other interest rate derivatives such as captions and floortions. Their evaluations are similar to what we have discussed.

4.6 Multi-Factor Interest Rate Models

4.6.1 Brief Description of Several Multi-Factor Interest Rate Models

Sometimes, it is necessary to assume that interest rate derivatives depend on not only the spot rate r, but also some other random state variables. Because volatility is always a dominant factor in determining the prices of bonds and options, we need to have a more accurate model for volatility. It may be necessary to consider the interest rate volatility as a random variable.

Fong and Vasicek[27] proposed such a two-factor model. In their model, they postulated that both the spot rate r and the variance v of the spot rate are stochastic state variables and assumed

$$dr = (\bar{\mu} - \gamma r)dt + \sqrt{v}dX,$$

$$dv = (\nu - \eta v)dt + \xi\sqrt{v}dX_v,$$

$$E\left[dX dX_v\right] = \rho dt,$$

where $\bar{\mu}, \gamma, \nu, \eta, \xi$ are constants and dX and dX_v are two standard Wiener processes. As we can see in this model, the stochastic equation for r is the same as that in the Vasicek model, and r could become negative. Here, not only the spot rate but also the variance possess the mean reversion property. In this case, (2.81) can be written as

$$\frac{\partial V}{\partial t} + \frac{1}{2}v\frac{\partial^2 V}{\partial r^2} + \rho\xi v\frac{\partial^2 V}{\partial r\partial v} + \frac{1}{2}\xi^2 v\frac{\partial^2 V}{\partial v^2} + (\bar{\mu} - \gamma r - \bar{\lambda}v)\frac{\partial V}{\partial r}$$

$$+ \left[\nu - (\eta + \bar{\lambda}_v\xi)v\right]\frac{\partial V}{\partial v} - rV = 0,$$

where the market prices of risk for r and v are $\bar{\lambda}\sqrt{v}$ and $\bar{\lambda}_v\sqrt{v}$, respectively, $\bar{\lambda}$ and $\bar{\lambda}_v$ being constants.

Brenman and Schwartz[12] considered another two-factor model. In their model, the two random state variables are the short-term interest rate r and the long-term interest rate l. They assumed

$$dr = u(r, l, t)dt + w(r, l, t)dX,$$

$$dl = u_l(r, l, t)dt + w_l(r, l, t)dX_l,$$

$$E\left[dX dX_l\right] = \rho(r, l, t)dt,$$

where dX and dX_l are the standard Wiener processes. According to the equation (2.81), any derivative dependent on r and l should satisfy

$$\frac{\partial V}{\partial t} + \frac{1}{2}w^2\frac{\partial^2 V}{\partial r^2} + \rho w w_l\frac{\partial^2 V}{\partial r\partial l} + \frac{1}{2}w_l^2\frac{\partial^2 V}{\partial l^2} + (u - \lambda w)\frac{\partial V}{\partial r} + (u_l - \lambda_l w_l)\frac{\partial V}{\partial l} - rV = 0.$$

For other models. for example, see [1], [18], and [52]. From these models, we can have the corresponding partial differential equations. Any reader who is interested in knowing more about these models and other models is suggested to consult these papers and the book [43] by James and Webber.

In order to use these models to price derivatives, we need to determine these market prices of risk, which is similar to what we have done for one-factor models. Also, if we make some modifications on these models so that some conditions similar to (4.45) and (4.46) hold, then unique solutions of these equations can be obtained only by requiring final conditions.

Not only can the interest rates and their variances be taken as state variables. Heath, Jarrow, and Morton[35],[36],[37] suggested a model where the driving state variable of the model is $F(t,T)$, the forward rate at time t for instantaneous borrowing at a later time T. They assume

$$dF(t,T) = \alpha_F(t,T)dt + \sum_{i=1}^{n} \sigma_F^i(t,T)dX_i,$$

where dX_i is the i-th Wiener process, and the n Wiener processes are independent. In this sense, it can be called a multi-factor model. Jarrow wrote a monograph on this method in 1996 (see [44]). Any reader who wants to know its details is referred to that book.

4.6.2 Reducing the Randomness of a Zero-Coupon Bond Curve to That of a Few Zero-Coupon Bonds

As we know, if we have an effective way to describe the randomness of a zero-coupon bond curve, then we can have an effective model for interest rate derivatives such as bond options or swaptions. In this and the next subsections, we discuss a three-factor model, which can be easily used in practice and generalized to the cases with more factors without any difficulty.

As we have done in Section 4.1, let $Z(t; t+T)$ denote the price of a T-year zero-coupon bond with a face value of one dollar at time t, and we use the notation $Z_i(t) = Z(t; t+T_i)$ for any T_i, $i = 0, 1, \cdots, N$. Here, we also assume $T_i < T_{i+1}$, for $i = 0, 1, \cdots, N-1$, and $T_0 = 0$. According to $Z_i(t), i = 0, 1, \cdots, N$, we can have an interpolation function $\bar{Z}(T; t)$ for $T \in [0, T_N]$ by requiring $\bar{Z}(T; t)$ to be a continuous function with continuous first and second derivatives in the form:

$$\bar{Z}(T; t) = \begin{cases} a_{0,1} + a_{1,1}T + a_{2,1}T^2, & 0 \leq T \leq T_1, \\ a_{0,i} + a_{1,i}T + a_{2,i}T^2 + a_{3,i}T^3, & T_{i-1} \leq T \leq T_i, \\ & i = 2, \cdots, N-1, \\ a_{0,N} + a_{1,N}T + a_{2,N}T^2, & T_{N-1} \leq T \leq T_N. \end{cases} \tag{4.65}$$

In this function, there are $4(N-2)+6 = 4N-2$ coefficients. Because we have $N+1$ conditions on the value of the function

$$\bar{Z}(T_i; t) = Z_i(t), \quad i = 0, 1, \cdots, N$$

and $3(N-1)$ continuity conditions on the function, first and second derivatives at $T_1, T_2, \cdots, T_{N-1}$, the total number of conditions is also $4N-2$. Therefore, it is possible that those coefficients in (4.65) can be determined by these conditions uniquely. This interpolation method is called a cubic spline interpolation, and the way of determining the coefficients in (4.65) will be given in Subsection 5.1.1. A zero-coupon bond curve is a monotone function with respect to T. If for a set of $Z_i(t)$, $i = 0, 1, \cdots, N$, (4.65) does not possess this property, the approximation needs to be modified so that the monotonicity is guaranteed. This is important in practice.

We assume that $\bar{Z}(T; t)$ is a very good approximation to the zero-coupon bond curve $Z(t; t + T)$. In this way, a random curve is reduced to N random variables with a small error.

Now let us reduce the number of random variables from N to K by the principal component analysis. Suppose that we have N random variables

$$S_i, \quad i = 1, 2, \cdots, N$$

and the covariance between S_i and S_j is

$$\mathrm{Cov}[S_i S_j] = b_i b_j \rho_{i,j}, \quad i, j = 1, 2, \cdots, N,$$

where $-1 \le \rho_{i,j} = \rho_{j,i} \le 1$ and $\rho_{i,i} = 1$. Let

$$c_i^2 \quad \text{and} \quad \mathbf{a}_i = \begin{bmatrix} a_{i,1} \\ a_{i,2} \\ \vdots \\ a_{i,N} \end{bmatrix}, \quad i = 1, 2, \cdots, N,$$

be the eigenvalues and unit eigenvectors of the covariance matrix

$$\mathbf{B} = \begin{bmatrix} b_1^2 & b_1 b_2 \rho_{1,2} & \cdots & b_1 b_N \rho_{1,N} \\ b_2 b_1 \rho_{2,1} & b_2^2 & \cdots & b_2 b_N \rho_{2,N} \\ \vdots & \vdots & \ddots & \vdots \\ b_N b_1 \rho_{N,1} & b_N b_2 \rho_{N,2} & \cdots & b_N^2 \end{bmatrix}.$$

That is, there is the following relation:

$$\mathbf{B} \mathbf{A}^T = \mathbf{A}^T \mathbf{C},$$

where $\mathbf{A}^T$ is the transpose of $\mathbf{A}$ and

$$\mathbf{A} = \begin{bmatrix} a_{1,1} & a_{1,2} & \cdots & a_{1,N} \\ a_{2,1} & a_{2,2} & \cdots & a_{2,N} \\ \vdots & \vdots & \ddots & \vdots \\ a_{N,1} & a_{N,2} & \cdots & a_{N,N} \end{bmatrix}, \quad \mathbf{C} = \begin{bmatrix} c_1^2 & 0 & \cdots & 0 \\ 0 & c_2^2 & \cdots & 0 \\ \vdots & \vdots & \ddots & \vdots \\ 0 & 0 & \cdots & c_N^2 \end{bmatrix}.$$

Here $\mathbf{A}$ is an orthogonal matrix, i.e., $\mathbf{A}\mathbf{A}^T = \mathbf{I}$ because $\mathbf{B}$ is a symmetric matrix.

Let $\bar{S}_1, \bar{S}_2, \cdots, \bar{S}_N$ be N other random variables defined by

$$\begin{bmatrix} \bar{S}_1 \\ \bar{S}_2 \\ \vdots \\ \bar{S}_N \end{bmatrix} = \mathbf{A} \begin{bmatrix} S_1 \\ S_2 \\ \vdots \\ S_N \end{bmatrix}.$$

For simplicity, this relation can be written as

$$\bar{\mathbf{S}} = \mathbf{AS},$$

where

$$\bar{\mathbf{S}} = \begin{bmatrix} \bar{S}_1 \\ \bar{S}_2 \\ \vdots \\ \bar{S}_N \end{bmatrix}, \quad \mathbf{S} = \begin{bmatrix} S_1 \\ S_2 \\ \vdots \\ S_N \end{bmatrix}.$$

Then

$$\begin{aligned} \mathrm{Cov}\left[\bar{S}_i\bar{S}_j\right] &= \mathrm{E}\left[\left(\bar{S}_i - \mathrm{E}\left[\bar{S}_i\right]\right)\left(\bar{S}_j - \mathrm{E}\left[\bar{S}_j\right]\right)\right] \\ &= \mathrm{E}\left[\left(\sum_{k=1}^{N} a_{ik}\left(S_k - \mathrm{E}\left[S_k\right]\right)\right)\left(\sum_{l=1}^{N} a_{jl}\left(S_l - \mathrm{E}\left[S_l\right]\right)\right)\right] \\ &= \sum_{k=1}^{N}\sum_{l=1}^{N} a_{ik}a_{jl}\,\mathrm{Cov}\left[S_kS_l\right] \\ &= \begin{cases} 0, & i \neq j, \\ c_i^2, & i = j. \end{cases} \end{aligned}$$

That is, $\mathbf{C}$ is the covariance matrix of the random vector $\bar{\mathbf{S}}$. We furthermore suppose that $c_i^2 \geq c_j^2$ for $i < j$ and $c_i^2 \ll c_K^2$, $i = K+1, \cdots, N$. Assume that on some day

$$\mathbf{S} = \begin{bmatrix} S_1^* \\ S_2^* \\ \vdots \\ S_N^* \end{bmatrix} \equiv \mathbf{S}^*$$

and

$$\bar{\mathbf{S}} = \mathbf{A}\begin{bmatrix} S_1^* \\ S_2^* \\ \vdots \\ S_N^* \end{bmatrix} = \begin{bmatrix} \bar{S}_1^* \\ \bar{S}_2^* \\ \vdots \\ \bar{S}_N^* \end{bmatrix} \equiv \bar{\mathbf{S}}^*.$$

Because c_i^2, $i = K+1, \cdots, N$ are very small, for a period starting from that day, we neglect the uncertainty caused by the last $N - K$ components of $\bar{\mathbf{S}}$. That is, we assume that in this period $\bar{\mathbf{S}}$ has the following form:

$$\bar{\mathbf{S}} = \begin{bmatrix} \bar{S}_1 \\ \vdots \\ \bar{S}_K \\ \bar{S}_{K+1}^* \\ \vdots \\ \bar{S}_N^* \end{bmatrix},$$

where $\bar{S}_1, \cdots, \bar{S}_K$ can take all possible values. In this case

$$\mathbf{S} = \mathbf{A}^T \begin{bmatrix} \bar{S}_1 \\ \vdots \\ \bar{S}_K \\ \bar{S}^*_{K+1} \\ \vdots \\ \bar{S}^*_N \end{bmatrix}. \tag{4.66}$$

Under this assumption, among $S_1, S_2, \cdots, S_N$, only K components are independent. Suppose

$$\begin{vmatrix} a_{1,1} & a_{2,1} & \cdots & a_{K,1} \\ a_{1,2} & a_{2,2} & \cdots & a_{K,2} \\ \vdots & \vdots & \ddots & \vdots \\ a_{1,K} & a_{2,K} & \cdots & a_{K,K} \end{vmatrix} \neq 0.$$

Then, we can choose $S_1, S_2, \cdots, S_K$ as independent components. Rewrite (4.66) as

$$\begin{bmatrix} S_1 \\ \vdots \\ S_K \end{bmatrix} = \mathbf{A}_1^T \begin{bmatrix} \bar{S}_1 \\ \vdots \\ \bar{S}_K \end{bmatrix} + \mathbf{A}_2^T \begin{bmatrix} \bar{S}^*_{K+1} \\ \vdots \\ \bar{S}^*_N \end{bmatrix},$$

$$\begin{bmatrix} S_{K+1} \\ \vdots \\ S_N \end{bmatrix} = \mathbf{A}_3^T \begin{bmatrix} \bar{S}_1 \\ \vdots \\ \bar{S}_K \end{bmatrix} + \mathbf{A}_4^T \begin{bmatrix} \bar{S}^*_{K+1} \\ \vdots \\ \bar{S}^*_N \end{bmatrix},$$

where

$$\mathbf{A}_1^T = \begin{bmatrix} a_{1,1} & \cdots & a_{K,1} \\ \vdots & \ddots & \vdots \\ a_{1,K} & \cdots & a_{K,K} \end{bmatrix}, \quad \mathbf{A}_2^T = \begin{bmatrix} a_{K+1,1} & \cdots & a_{N,1} \\ \vdots & \ddots & \vdots \\ a_{K+1,K} & \cdots & a_{N,K} \end{bmatrix},$$

$$\mathbf{A}_3^T = \begin{bmatrix} a_{1,K+1} & \cdots & a_{K,K+1} \\ \vdots & \ddots & \vdots \\ a_{1,N} & \cdots & a_{K,N} \end{bmatrix}, \quad \mathbf{A}_4^T = \begin{bmatrix} a_{K+1,K+1} & \cdots & a_{N,K+1} \\ \vdots & \ddots & \vdots \\ a_{K+1,N} & \cdots & a_{N,N} \end{bmatrix}.$$

Then, for $S_{K+1}, \cdots, S_N$, we have

$$\begin{bmatrix} S_{K+1} \\ \vdots \\ S_N \end{bmatrix} = \mathbf{A}_3^T (\mathbf{A}_1^T)^{-1} \left(\begin{bmatrix} S_1 \\ \vdots \\ S_K \end{bmatrix} - \mathbf{A}_2^T \begin{bmatrix} \bar{S}^*_{K+1} \\ \vdots \\ \bar{S}^*_N \end{bmatrix} \right)$$

$$+ \mathbf{A}_4^T \begin{bmatrix} \bar{S}^*_{K+1} \\ \vdots \\ \bar{S}^*_N \end{bmatrix}. \tag{4.67}$$

Thus, for given $S_1, \cdots, S_K$, using (4.67) we can get all other components of a vector $\mathbf{S}$. Consequently, (4.67) defines a class of vectors with K parameters. That is, by (4.67), we actually determine a class of $\mathbf{S}$, where only $S_1, \cdots, S_K$ are independent. Here, we take $S_1, \cdots, S_K$ as independent components. However, it is also possible to choose other K components as independent components.

Letting $S_i = Z_i/T_i$, $i = 1, 2, \cdots, N$, by the principal component analysis described above, we can find a class of vectors $[Z_1/T_1, \cdots, Z_N/T_N]^T$ with K parameters[2] and using the cubic spline interpolation given at the beginning of this subsection, we can further determine the curve $\bar{Z}(T;t)$ for $T \in [0, T_N]$. From the books by Jarrow[44], Hull[39], James and Webber[43], and Wilmott[73], we know that K usually is equal to three or four for the random curves related to interest rates. Thus, all the curves determined by (4.67) form a class of curves with three or four parameters. The zero-coupon bond curve at that day is one of such curves, and the projections of any vector $\mathbf{S}$ determined by (4.67) on the eigenvectors corresponding to the eigenvalues $c_{K+1}, \cdots, c_N$ are the same as those of $\mathbf{S}^*$. Those projections are different for different $\mathbf{S}^*$, so this is a feature belonging to $\mathbf{S}^*$. It is clear that the class of curves with such a feature needs to be considered most for derivative-pricing problems. Hence, when $K = 3$ or 4, the class contains all possible and need-to-be-considered-most zero-coupon bond curves. As soon as we have a zero-coupon bond curve, we can determine various interest rates at t, including the spot interest rate at time t: $-\dfrac{\partial \bar{Z}(T;t)}{\partial T}\bigg|_{T=0}$. In what follows, we denote

$$-\frac{\partial \bar{Z}(T;t)}{\partial T}\bigg|_{T=0} \quad \text{by } r(Z_1, \cdots, Z_K, t), \text{ or simply, by } r.$$

4.6.3 A Three-Factor Interest Rate Model and the Equation for Interest Rate Derivatives

Suppose Z_1, Z_2 and Z_3 are prices of zero-coupon bonds with maturities T_1, T_2, and T_3, respectively. Assume $T_1 < T_2 < T_3$, which implies the relations $1 \geq Z_1 \geq Z_2 \geq Z_3$. Furthermore, we assume $Z_1 \geq Z_{1,l}$, $Z_2 \geq Z_{2,l}$ and $Z_3 \geq Z_{3,l}$, where $Z_{1,l} \geq Z_{2,l} \geq Z_{3,l} \geq 0$. Z_1, Z_2 and Z_3 are random variables and satisfy the system of stochastic differential equations:

$$dZ_i = \mu_i(Z_1, Z_2, Z_3, t)\, dt + \sigma_i(Z_1, Z_2, Z_3, t)\, dX_i, \quad i = 1, 2, 3$$

on the domain Ω: $\{Z_{1,l} \leq Z_1 \leq 1, \ Z_{2,l} \leq Z_2 \leq Z_1, \ Z_{3,l} \leq Z_3 \leq Z_2\}$. dX_i are the Wiener processes and $\mathrm{E}\,[dX_i dX_j] = \rho_{ij} dt$ with $-1 \leq \rho_{ij} \leq 1$. The coefficients μ_i, σ_i and their first- and second-order derivatives are assumed to be bounded on the domain Ω. On the six boundaries of Ω, the following conditions hold:

---

[2]If the conditions $Z_i \geq Z_{i+1}$, $i = 0, 1, \cdots N - 1$ are not satisfied, then some modification needs to be done in order to guarantee the monotonicity.

i) on surface I: $\{Z_1 = Z_{1,l},\ Z_{2,l} \leq Z_2 \leq Z_1,\ Z_{3,l} \leq Z_3 \leq Z_2\}$,

$$\begin{cases} \mu_1\left(Z_{1,l}, Z_2, Z_3, t\right) \geq 0, \\ \sigma_1\left(Z_{1,l}, Z_2, Z_3, t\right) = 0; \end{cases} \tag{4.68}$$

ii) on surface II: $\{Z_1 = 1,\ Z_{2,l} \leq Z_2 \leq Z_1,\ Z_{3,l} \leq Z_3 \leq Z_2\}$,

$$\begin{cases} \mu_1\left(1, Z_2, Z_3, t\right) \leq 0, \\ \sigma_1\left(1, Z_2, Z_3, t\right) = 0; \end{cases} \tag{4.69}$$

iii) on surface III: $\{Z_{1,l} \leq Z_1 \leq 1,\ Z_2 = Z_{2,l},\ Z_{3,l} \leq Z_3 \leq Z_2\}$,

$$\begin{cases} \mu_2\left(Z_1, Z_{2,l}, Z_3, t\right) \geq 0, \\ \sigma_2\left(Z_1, Z_{2,l}, Z_3, t\right) = 0; \end{cases} \tag{4.70}$$

iv) on surface IV: $\{Z_{1,l} \leq Z_1 \leq 1,\ Z_2 = Z_1,\ Z_{3,l} \leq Z_3 \leq Z_2\}$,

$$\begin{cases} -\mu_1\left(Z_1, Z_1, Z_3, t\right) + \mu_2\left(Z_1, Z_1, Z_3, t\right) \leq 0, \\ \sigma_1\left(Z_1, Z_1, Z_3, t\right) = \sigma_2\left(Z_1, Z_1, Z_3, t\right),\ \rho_{1,2}\left(Z_1, Z_1, Z_3, t\right) = 1; \end{cases} \tag{4.71}$$

v) on surface V: $\{Z_{1,l} \leq Z_1 \leq 1,\ Z_{2,l} \leq Z_2 \leq Z_1,\ Z_3 = Z_{3,l}\}$,

$$\begin{cases} \mu_3\left(Z_1, Z_2, Z_{3,l}, t\right) \geq 0, \\ \sigma_3\left(Z_1, Z_2, Z_{3,l}, t\right) = 0; \end{cases} \tag{4.72}$$

vi) on surface VI: $\{Z_{1,l} \leq Z_1 \leq 1,\ Z_{2,l} \leq Z_2 \leq Z_1,\ Z_3 = Z_2\}$,

$$\begin{cases} -\mu_2\left(Z_1, Z_2, Z_2, t\right) + \mu_3\left(Z_1, Z_2, Z_2, t\right) \leq 0, \\ \sigma_2\left(Z_1, Z_2, Z_2, t\right) = \sigma_3\left(Z_1, Z_2, Z_2, t\right),\ \rho_{2,3}\left(Z_1, Z_2, Z_2, t\right) = 1. \end{cases} \tag{4.73}$$

This model will be called the three-factor interest rate model in this book.
As you can see, conditions (4.68)–(4.70) and (4.72) have the same form as (4.45) or (4.46), and (4.71) and (4.73) are in a similar form. They are the reversion conditions on the non-rectangular domain Ω. Intuitively, when

$$n_1 dZ_1 + n_2 dZ_2 + n_3 dZ_3 \leq 0 \tag{4.74}$$

holds at any point on the boundary, where n_1, n_2 and n_3 are the three components of the outer normal vector of the boundary at the point, it is guaranteed that if a point is in Ω at time t^*, then the point is still in Ω at $t = t^* + dt$, where $dt > 0$. In fact, if these conditions (4.68)–(4.73) are satisfied, then (4.74) holds at every point on the boundary of the domain Ω. For example, on surface I (see Fig. 4.2), $n_1 = -1$, $n_2 = 0$ and $n_3 = 0$, so the conditions $\sigma_1 = 0$ and $\mu_1 \geq 0$ cause

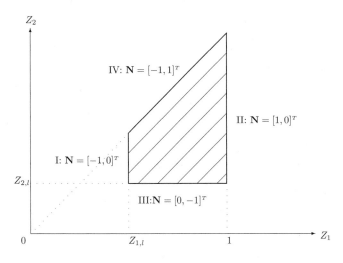

Fig. 4.2. Projection of the domain Ω on the (Z_1, Z_2)-plane

$$n_1 dZ_1 + n_2 dZ_2 + n_3 dZ_3 = -dZ_1 = -\mu_1 dt - \sigma_1 dX_i = -\mu_1 dt \leq 0.$$

On surface IV, $-\mu_1 + \mu_2 \leq 0$, $\sigma_1 = \sigma_2$, $\rho_{1,2} = 1$ and $n_1 = -1$, $n_2 = 1$, $n_3 = 0$ (see Fig. 4.2). From $\rho_{1,2} = 1$, we have $dX_1 = dX_2$. Thus, in this case we have

$$
\begin{aligned}
n_1 dZ_1 + n_2 dZ_2 + n_3 dZ_3 &= -dZ_1 + dZ_2 \\
&= -\mu_1 dt + \mu_2 dt - \sigma_1 dX_1 + \sigma_2 dX_2 \\
&= (-\mu_1 + \mu_2)dt - \sigma_1 dX_1 + \sigma_1 dX_1 \\
&= (-\mu_1 + \mu_2)dt \leq 0.
\end{aligned}
$$

Therefore, conditions (4.68)–(4.73) guarantee that a point (Z_1, Z_2, Z_3) will never move from inside of the domain Ω to its outside. This is a natural property of a stochastic model for interest rates when $Z_{1,l}$, $Z_{2,l}$ and $Z_{3,l}$ are given properly.

Let $V(Z_1, Z_2, Z_3, t)$ be the value of a derivative security depending on Z_1, Z_2, Z_3, t. According to Subsection 2.9.3, $V(Z_1, Z_2, Z_3, t)$ should satisfy

$$\frac{\partial V}{\partial t} + \frac{1}{2} \sum_{i=1}^{3} \sum_{j=1}^{3} \sigma_i \sigma_j \rho_{i,j} \frac{\partial^2 V}{\partial Z_i \partial Z_j} + r \sum_{i=1}^{3} Z_i \frac{\partial V}{\partial Z_i} - rV = 0.$$

Here, we have used the fact that because zero-coupon bonds can be traded on the market, the market prices of risk for these bonds are known and (2.82) with $D_0 = 0$ holds:

$$\mu_i(Z_1, Z_2, Z_3, t) - \lambda_i(Z_1, Z_2, Z_3, t)\sigma_i(Z_1, Z_2, Z_3, t) = r(Z_1, Z_2, Z_3, t)Z_i,$$
$$i = 1, 2, 3.$$

Let

$$\mathbf{L}_{3z} = \frac{1}{2} \sum_{i=1}^{3} \sum_{j=1}^{3} \sigma_i \sigma_j \rho_{i,j} \frac{\partial^2}{\partial Z_i \partial Z_j} + r \sum_{i=1}^{3} Z_i \frac{\partial}{\partial Z_i} - r. \qquad (4.75)$$

The equation above can be written as

$$\frac{\partial V}{\partial t} + \mathbf{L}_{3z} V = 0.$$

For a derivative security, at the maturity date T, its price should be equal to its payoff $V_T(Z_1, Z_2, Z_3)$. Therefore, any European interest rate derivatives under this model should be solutions of the problem

$$\begin{cases} \dfrac{\partial V}{\partial t} + \mathbf{L}_{3z} V = 0 & \text{on } \Omega \times [0, T], \\[2mm] V(Z_1, Z_2, Z_3, T) = V_T(Z_1, Z_2, Z_3) & \text{on } \Omega. \end{cases} \qquad (4.76)$$

Introduce the following transformation:

$$\begin{cases} \xi_1 = \dfrac{Z_1 - Z_{1,l}}{1 - Z_{1,l}}, \\[3mm] \xi_2 = \dfrac{Z_2 - Z_{2,l}}{Z_1 - Z_{2,l}}, \\[3mm] \xi_3 = \dfrac{Z_3 - Z_{3,l}}{Z_2 - Z_{3,l}}. \end{cases} \qquad (4.77)$$

Through this transformation, the domain Ω in the (Z_1, Z_2, Z_3)-space is transformed into the domain $\tilde{\Omega}$: $[0, 1] \times [0, 1] \times [0, 1]$ in the (ξ_1, ξ_2, ξ_3)-space. Because

$$\frac{\partial \xi_1}{\partial Z_1} = \frac{1}{1 - Z_{1,l}},$$

$$\frac{\partial \xi_2}{\partial Z_1} = \frac{-\xi_2}{Z_1 - Z_{2,l}}, \qquad \frac{\partial \xi_2}{\partial Z_2} = \frac{1}{Z_1 - Z_{2,l}},$$

$$\frac{\partial \xi_3}{\partial Z_2} = \frac{-\xi_3}{Z_2 - Z_{3,l}}, \qquad \frac{\partial \xi_3}{\partial Z_3} = \frac{1}{Z_2 - Z_{3,l}},$$

we have

$$\frac{\partial V}{\partial Z_1} = \frac{1}{1 - Z_{1,l}} \frac{\partial V}{\partial \xi_1} - \frac{\xi_2}{Z_1 - Z_{2,l}} \frac{\partial V}{\partial \xi_2},$$

$$\frac{\partial V}{\partial Z_2} = \frac{1}{Z_1 - Z_{2,l}} \frac{\partial V}{\partial \xi_2} - \frac{\xi_3}{Z_2 - Z_{3,l}} \frac{\partial V}{\partial \xi_3},$$

$$\frac{\partial V}{\partial Z_3} = \frac{1}{Z_2 - Z_{3,l}} \frac{\partial V}{\partial \xi_3},$$

$$\frac{\partial^2 V}{\partial Z_1^2} = \frac{1}{(1 - Z_{1,l})^2} \frac{\partial^2 V}{\partial \xi_1^2} - \frac{2\xi_2}{(1 - Z_{1,l})(Z_1 - Z_{2,l})} \frac{\partial^2 V}{\partial \xi_1 \partial \xi_2}$$

$$+\frac{\xi_2^2}{(Z_1-Z_{2,l})^2}\frac{\partial^2 V}{\partial\xi_2^2}+\frac{2\xi_2}{(Z_1-Z_{2,l})^2}\frac{\partial V}{\partial\xi_2},$$

$$\frac{\partial^2 V}{\partial Z_2^2}=\frac{1}{(Z_1-Z_{2,l})^2}\frac{\partial^2 V}{\partial\xi_2^2}-\frac{2\xi_3}{(Z_1-Z_{2,l})(Z_2-Z_{3,l})}\frac{\partial^2 V}{\partial\xi_2\partial\xi_3}$$

$$+\frac{\xi_3^2}{(Z_2-Z_{3,l})^2}\frac{\partial^2 V}{\partial\xi_3^2}+\frac{2\xi_3}{(Z_2-Z_{3,l})^2}\frac{\partial V}{\partial\xi_3},$$

$$\frac{\partial^2 V}{\partial Z_3^2}=\frac{1}{(Z_2-Z_{3,l})^2}\frac{\partial^2 V}{\partial\xi_3^2},$$

$$\frac{\partial^2 V}{\partial Z_1\partial Z_2}=\frac{-1}{(Z_1-Z_{2,l})^2}\frac{\partial V}{\partial\xi_2}+\frac{1}{Z_1-Z_{2,l}}\left(\frac{1}{1-Z_{1,l}}\frac{\partial^2 V}{\partial\xi_1\partial\xi_2}-\frac{\xi_2}{Z_1-Z_{2,l}}\frac{\partial^2 V}{\partial\xi_2^2}\right)$$

$$-\frac{\xi_3}{Z_2-Z_{3,l}}\left(\frac{1}{1-Z_{1,l}}\frac{\partial^2 V}{\partial\xi_1\partial\xi_3}-\frac{\xi_2}{Z_1-Z_{2,l}}\frac{\partial^2 V}{\partial\xi_2\partial\xi_3}\right),$$

$$\frac{\partial^2 V}{\partial Z_1\partial Z_3}=\frac{1}{Z_2-Z_{3,l}}\left(\frac{1}{1-Z_{1,l}}\frac{\partial^2 V}{\partial\xi_1\partial\xi_3}-\frac{\xi_2}{Z_1-Z_{2,l}}\frac{\partial^2 V}{\partial\xi_2\partial\xi_3}\right),$$

$$\frac{\partial^2 V}{\partial Z_2\partial Z_3}=\frac{-1}{(Z_2-Z_{3,l})^2}\frac{\partial V}{\partial\xi_3}+\frac{1}{Z_2-Z_{3,l}}\left(\frac{1}{Z_1-Z_{2,l}}\frac{\partial^2 V}{\partial\xi_2\partial\xi_3}-\frac{\xi_3}{Z_2-Z_{3,l}}\frac{\partial^2 V}{\partial\xi_3^2}\right).$$

Therefore, the operator $\mathbf{L}_{3z}$ defined by (4.75) can be rewritten as

$$\mathbf{L}_{3\xi}=\frac{1}{2}\tilde\sigma_1^2\frac{\partial^2}{\partial\xi_1^2}+\frac{1}{2}\tilde\sigma_2^2\frac{\partial^2}{\partial\xi_2^2}+\frac{1}{2}\tilde\sigma_3^2\frac{\partial^2}{\partial\xi_3^2}$$

$$+\tilde\sigma_1\tilde\sigma_2\tilde\rho_{1,2}\frac{\partial^2}{\partial\xi_1\partial\xi_2}+\tilde\sigma_1\tilde\sigma_3\tilde\rho_{1,3}\frac{\partial^2}{\partial\xi_1\partial\xi_3}+\tilde\sigma_2\tilde\sigma_3\tilde\rho_{2,3}\frac{\partial^2}{\partial\xi_2\partial\xi_3}$$

$$+b_1\frac{\partial}{\partial\xi_1}+b_2\frac{\partial}{\partial\xi_2}+b_3\frac{\partial}{\partial\xi_3}-r,\qquad(4.78)$$

where

$$\begin{cases}\dfrac{1}{2}\tilde\sigma_1^2=\dfrac{\frac{1}{2}\sigma_1^2}{(1-Z_{1,l})^2},\\[2ex]\dfrac{1}{2}\tilde\sigma_2^2=\dfrac{\frac{1}{2}\left(\sigma_1^2\xi_2^2-2\sigma_1\sigma_2\xi_2\rho_{1,2}+\sigma_2^2\right)}{(Z_1-Z_{2,l})^2},\\[2ex]\dfrac{1}{2}\tilde\sigma_3^2=\dfrac{\frac{1}{2}\left(\sigma_2^2\xi_3^2-2\sigma_2\sigma_3\xi_3\rho_{2,3}+\sigma_3^2\right)}{(Z_2-Z_{3,l})^2},\end{cases}\qquad(4.79)$$

$$\begin{cases}\tilde\sigma_1\tilde\sigma_2\tilde\rho_{1,2}=\dfrac{\sigma_1\left(\sigma_2\rho_{1,2}-\sigma_1\xi_2\right)}{(1-Z_{1,l})(Z_1-Z_{2,l})},\\[2ex]\tilde\sigma_1\tilde\sigma_3\tilde\rho_{1,3}=\dfrac{\sigma_1\left(\sigma_3\rho_{1,3}-\sigma_2\rho_{1,2}\xi_3\right)}{(1-Z_{1,l})(Z_2-Z_{3,l})},\\[2ex]\tilde\sigma_2\tilde\sigma_3\tilde\rho_{2,3}=\dfrac{\sigma_1\xi_2\left(\sigma_2\rho_{1,2}\xi_3-\sigma_3\rho_{1,3}\right)+\sigma_2\left(\sigma_3\rho_{2,3}-\sigma_2\xi_3\right)}{(Z_1-Z_{2,l})(Z_2-Z_{3,l})},\end{cases}\qquad(4.80)$$

and

$$
\begin{cases}
b_1 = \dfrac{rZ_1}{1 - Z_{1,l}}, \\[4mm]
b_2 = \dfrac{r\left(Z_2 - Z_1\xi_2\right)}{Z_1 - Z_{2,l}} + \dfrac{\sigma_1\left(\sigma_1\xi_2 - \sigma_2\rho_{1,2}\right)}{\left(Z_1 - Z_{2,l}\right)^2}, \\[4mm]
b_3 = \dfrac{r\left(Z_3 - Z_2\xi_3\right)}{Z_2 - Z_{3,l}} + \dfrac{\sigma_2\left(\sigma_2\xi_3 - \sigma_3\rho_{2,3}\right)}{\left(Z_2 - Z_{3,l}\right)^2}.
\end{cases}
\tag{4.81}
$$

Consequently, (4.76) can be rewritten as

$$
\begin{cases}
\dfrac{\partial V}{\partial t} + \mathbf{L}_{3\xi}V = 0 \quad \text{on } \tilde{\Omega} \times [0,T], \\[3mm]
V(\xi_1, \xi_2, \xi_3, T) = V_T\left(Z_1(\xi_1), Z_2(\xi_1, \xi_2), Z_3(\xi_1, \xi_2, \xi_3)\right) \quad \text{on } \tilde{\Omega},
\end{cases}
\tag{4.82}
$$

where $\mathbf{L}_{3\xi}$ is defined by (4.78) and

$$
\begin{cases}
Z_1\left(\xi_1\right) = Z_{1,l} + \xi_1\left(1 - Z_{1,l}\right), \\[2mm]
Z_2\left(\xi_1, \xi_2\right) = Z_{2,l} + \xi_2\left[Z_{1,l} + \xi_1\left(1 - Z_{1,l}\right) - Z_{2,l}\right], \\[2mm]
Z_3\left(\xi_1, \xi_2, \xi_3\right) = Z_{3,l} + \xi_3\left\{Z_{2,l} + \xi_2\left[Z_{1,l} + \xi_1\left(1 - Z_{1,l}\right) - Z_{2,l}\right] - Z_{3,l}\right\}.
\end{cases}
\tag{4.83}
$$

This is a final-value problem on a rectangular domain. Thus, when the three-factor interest rate model is used, evaluating an interest rate derivative is reduced to solving a final-value problem on a rectangular domain. It has been proven that this problem has a unique solution (see [81]). In such a case, this problem can be solved by numerical methods without any difficulty.

We would like to point out the relations among $\tilde{\sigma}_1, \tilde{\sigma}_2, \tilde{\sigma}_3, \tilde{\rho}_{1,2}, \tilde{\rho}_{1,3}, \tilde{\rho}_{2,3}$ and $d\xi_1, d\xi_2, d\xi_3$. Using Itô's lemma, from the definitions of ξ_1, ξ_2, ξ_3, we can have

$$
\begin{aligned}
d\xi_1 &= \tilde{\mu}_1 dt + \tilde{\sigma}_1 dX_1, \\
d\xi_2 &= \tilde{\mu}_2 dt + \tilde{\sigma}_2 d\tilde{X}_2, \\
d\xi_3 &= \tilde{\mu}_3 dt + \tilde{\sigma}_3 d\tilde{X}_3,
\end{aligned}
$$

where $d\tilde{X}_2$ and $d\tilde{X}_3$ are two new Wiener processes. Therefore

$$
\tilde{\sigma}_i^2 = \text{Var}[d\xi_i]/dt, \quad j = 1, 2, 3.
$$

It can also be shown that

$$
\text{Cov}[dX_1 d\tilde{X}_2]/dt = \tilde{\rho}_{1,2}, \quad \text{Cov}[dX_1 d\tilde{X}_3]/dt = \tilde{\rho}_{1,3}
$$

and

$$
\text{Cov}[d\tilde{X}_2 d\tilde{X}_3]/dt = \tilde{\rho}_{2,3}.
$$

These are left for the reader to prove as Problem 17.

From (4.79), it is easy to see that the equality conditions in (4.68)–(4.73) can be rewritten as

$$
\begin{cases}
\tilde{\sigma}_1(0,\xi_2,\xi_3,t) = \tilde{\sigma}_1(1,\xi_2,\xi_3,t) = 0, & 0 \le \xi_2 \le 1,\ 0 \le \xi_3 \le 1, \\
\tilde{\sigma}_2(\xi_1,0,\xi_3,t) = \tilde{\sigma}_2(\xi_1,1,\xi_3,t) = 0, & 0 \le \xi_1 \le 1,\ 0 \le \xi_3 \le 1, \\
\tilde{\sigma}_3(\xi_1,\xi_2,0,t) = \tilde{\sigma}_3(\xi_1,\xi_2,1,t) = 0, & 0 \le \xi_1 \le 1,\ 0 \le \xi_2 \le 1.
\end{cases}
\tag{4.84}
$$

Therefore, in order for the equality conditions in (4.68)–(4.73) to hold, we just require that the volatilities of $d\xi_1$, $d\xi_2$, and $d\xi_3$ satisfy (4.84), which is easier to be implemented than the equality conditions in (4.68)–(4.73). Suppose that $\tilde{\sigma}_1$, $\tilde{\sigma}_2$, $\tilde{\sigma}_3$, $\tilde{\rho}_{1,2}$, $\tilde{\rho}_{1,3}$, and $\tilde{\rho}_{2,3}$ are given functions. In this case, in order to be able to use the expressions of b_1, b_2, and b_3 conveniently, we express $\sigma_1 (\sigma_1 \xi_2 - \sigma_2 \rho_{1,2})$ and $\sigma_2 (\sigma_2 \xi_3 - \sigma_3 \rho_{2,3})$ in the expressions of b_2 and b_3 in terms of $\tilde{\sigma}_1$, $\tilde{\sigma}_2$, $\tilde{\sigma}_3$, $\tilde{\rho}_{1,2}$, $\tilde{\rho}_{1,3}$, and $\tilde{\rho}_{2,3}$. From (4.80) we have

$$
\tilde{\sigma}_1\tilde{\sigma}_2\tilde{\rho}_{1,2} (1 - Z_{1,l})(Z_1 - Z_{2,l}) = \sigma_1 (\sigma_2 \rho_{1,2} - \sigma_1 \xi_2),
$$
$$
\tilde{\sigma}_1\tilde{\sigma}_3\tilde{\rho}_{1,3} (1 - Z_{1,l})(Z_2 - Z_{3,l}) = \sigma_1 (\sigma_3 \rho_{1,3} - \sigma_2 \rho_{1,2}\xi_3),
$$
$$
\tilde{\sigma}_2\tilde{\sigma}_3\tilde{\rho}_{2,3} (Z_1 - Z_{2,l})(Z_2 - Z_{3,l}) = \sigma_1 \xi_2 (\sigma_2 \rho_{1,2}\xi_3 - \sigma_3 \rho_{1,3})
$$
$$
+\sigma_2 (\sigma_3 \rho_{2,3} - \sigma_2 \xi_3)
$$

and from the second and third relations we further obtain

$$
\sigma_2 (\sigma_3 \rho_{2,3} - \sigma_2 \xi_3) = \tilde{\sigma}_1\tilde{\sigma}_3\tilde{\rho}_{1,3}\xi_2 (1 - Z_{1,l})(Z_2 - Z_{3,l})
$$
$$
+\tilde{\sigma}_2\tilde{\sigma}_3\tilde{\rho}_{2,3} (Z_1 - Z_{2,l})(Z_2 - Z_{3,l}).
$$

Therefore, the expressions of b_1, b_2 and b_3 can be rewritten as

$$
\begin{cases}
b_1 = \dfrac{rZ_1}{1 - Z_{1,l}}, \\[2ex]
b_2 = \dfrac{r(Z_2 - Z_1\xi_2)}{Z_1 - Z_{2,l}} - \dfrac{\tilde{\sigma}_1\tilde{\sigma}_2\tilde{\rho}_{1,2}(1 - Z_{1,l})}{Z_1 - Z_{2,l}}, \\[2ex]
b_3 = \dfrac{r(Z_3 - Z_2\xi_3)}{Z_2 - Z_{3,l}} - \dfrac{\tilde{\sigma}_1\tilde{\sigma}_3\tilde{\rho}_{1,3}\xi_2(1 - Z_{1,l})}{Z_2 - Z_{3,l}} \\[2ex]
\quad - \dfrac{\tilde{\sigma}_2\tilde{\sigma}_3\tilde{\rho}_{2,3}(Z_1 - Z_{2,l})}{Z_2 - Z_{3,l}}.
\end{cases}
\tag{4.85}
$$

By this relation, we can easily calculate b_1, b_2, and b_3 when the values of $\tilde{\sigma}_1$, $\tilde{\sigma}_2$, $\tilde{\sigma}_3$, $\tilde{\rho}_{1,2}$, $\tilde{\rho}_{1,3}$, and $\tilde{\rho}_{2,3}$ are given.

Finally, we say a few words about how to use this model to evaluate interest rate derivatives. First, we need to choose Z_1, Z_2, and Z_3 and find $\sigma_1, \sigma_2, \sigma_3$, $\rho_{1,2}$, $\rho_{1,3}$, and $\rho_{2,3}$ satisfying conditions (4.68)–(4.73), or find $\tilde{\sigma}_1$, $\tilde{\sigma}_2$, $\tilde{\sigma}_3$, $\tilde{\rho}_{1,2}$, $\tilde{\rho}_{1,3}$, and $\tilde{\rho}_{2,3}$ satisfying conditions (4.84). Finding these functions can be done from the data on markets by statistics. After that, (4.82) needs to be solved.

Let $t = 0$ denote today, and suppose the derivative security is European style. On the maturity date T, for each point (ξ_1, ξ_2, ξ_3) in $\tilde{\Omega}$, we can have Z_1, Z_2, and Z_3 by (4.83). Then, we determine a zero-coupon bond curve by using the method given in Subsection 4.6.2. When we obtain such a curve, the value of the payoff and r for the point can be determined. This can be done for all points (ξ_1, ξ_2, ξ_3) in the domain $\tilde{\Omega}$ for $t = T$. When we have the final value and all the coefficients of the partial differential equation in (4.82), we can solve the final-value problem (4.82) from $t = T$ to $t = 0$ and get the value of the derivative security today for all the points in $\tilde{\Omega}$.

For American-style derivatives, the situation is similar. The only difference is that the value of derivative must be greater than the constraint. Because the value of the constraint can be obtained by the zero-coupon bond curves at all points in $\tilde{\Omega} \times [0, T]$, the value of an American-style derivative can be determined without any difficulty. However, free boundaries will usually appear in this case.

From what we have described, we see that this model has the following features:

- The state variables are prices of three zero-coupon bonds with different maturities that can be traded on markets, so the coefficients of the first derivatives with respect to the bond prices Z_i in the partial differential equation simply are rZ_i.
- The volatilities of these zero-coupon bonds and their correlation coefficients can be found directly from the real markets by statistics, so the model will have the real major feature of the markets.
- All the zero-coupon bond curves having appeared in the real market can be reproduced quite accurately. This is the basis of a model giving correct results. If taking three random variables is not good enough, four-factor models can be adopted. Generalizing three-factor models to four-factor models is straightforward.
- In other models, the partial differential equation is defined on an infinite domain. For this model, the corresponding partial differential equation is defined on a finite domain. It has been proven that no boundary condition is needed in order for its final-value problem to have a unique solution. Thus, it is not difficult to design correct and efficient numerical methods to price interest rate derivatives.

For the details on how to determine models from the market data and how to solve the final-value problem of the partial differential equation, see Section 8.3 and [77]. There, some numerical results are also given.

4.7 Two-Factor Convertible Bonds

Until now, we discussed derivatives depending on either equities or interest rates. This section deals with a derivative dependent on both equity prices and interest rates. This derivative security is a bond that may, at any time chosen by the holder, be converted into n shares of stocks of the company who issues the bond. Such a bond is commonly known as a convertible bond. As a bond, its price depends on the spot rate r. It can be exchanged for n shares of stocks, so its value is also a function of the stock price S. Because its typical life span is about 3 to 10 years, both S and r are considered as random state variables. Therefore, this bond is called a two-factor convertible bond. In this section, we discuss how to price such a bond.

Let $B_c = B_c(S, r, t)$ be the value of a two-factor convertible bond. As usual, we assume that S is governed by

$$dS = \mu(S, t)S dt + \sigma(S, t)S dX_1, \qquad 0 \le S \tag{4.86}$$

and the interest rate by

$$dr = u(r, t)dt + w(r, t)dX_2, \qquad r_l \le r \le r_u, \tag{4.87}$$

where dX_1 and dX_2 are different Wiener processes though they can be correlated. Suppose that

$$\mathrm{E}\left[dX_1 dX_2\right] = \rho dt,$$

where ρ is a constant belonging to $[-1, 1]$ and for S and r, ρ usually is a negative number. According to Section 2.9, such a derivative satisfies

$$\frac{\partial B_c}{\partial t} + \mathbf{L}_{\mathbf{s},\mathbf{r}} B_c + kZ = 0, \qquad 0 \le S, \ r_l \le r \le r_u, \ 0 \le t \le T, \tag{4.88}$$

where

$$\mathbf{L}_{\mathbf{s},\mathbf{r}} = \frac{1}{2}\sigma^2 S^2 \frac{\partial^2}{\partial S^2} + \rho\sigma S w \frac{\partial^2}{\partial S \partial r} + \frac{1}{2} w^2 \frac{\partial^2}{\partial r^2} + (r - D_0)S \frac{\partial}{\partial S} + (u - \lambda w) \frac{\partial}{\partial r} - r.$$

Here, D_0 is the dividend yield a holder of the stock receives per unit time, and kZ is the coupon payment a holder of the bond receives per unit time, Z being the face value of the bond. λ is the market price of risk for the spot rate. T is the maturity date of the bond.

At maturity time T, the holder of the bond should get the face value if $nS \le Z$. If $nS > Z$, then the bond must be exchanged for n shares, and the value is nS. Therefore,

$$B_c(S, r, T) = \max(Z, nS), \qquad 0 \le S, \qquad r_l \le r \le r_u. \tag{4.89}$$

This is the final condition for a convertible bond. We assume that for the interest rate, the conditions (4.45) and (4.46) hold, i.e.,

$$\begin{cases} u(r_l, t) - w(r_l, t)\dfrac{\partial}{\partial r} w(r_l, t) \ge 0, \\ w(r_l, t) = 0, \end{cases}$$

and

$$\begin{cases} u(r_u, t) - w(r_u, t)\dfrac{\partial}{\partial r}w(r_u, t) \le 0, \\ w(r_u, t) = 0. \end{cases}$$

Because $w^2(r, t) \ge 0$ and $w(r_l, t) = 0$, on $[r_l, r_u]$ we conclude $w(r_l, t)\dfrac{\partial}{\partial r}w(r_l, t)$
$= \dfrac{1}{2}\dfrac{\partial}{\partial r}w^2(r_l, t) \ge 0$. Similarly, $w(r_u, t)\dfrac{\partial}{\partial r}w(r_u, t) \le 0$. Therefore, the conditions above can be rewritten as

$$\begin{cases} u(r_l, t) \ge w(r_l, t)\dfrac{\partial}{\partial r}w(r_l, t) \ge 0, \\ w(r_l, t) = 0, \end{cases}$$

$$\begin{cases} u(r_u, t) \le w(r_u, t)\dfrac{\partial}{\partial r}w(r_u, t) \le 0, \\ w(r_u, t) = 0. \end{cases}$$

Because of $w(r_l, t) = 0$, the equation (4.88) at $r = r_l$ degenerates into

$$\frac{\partial B_c}{\partial t} + \frac{1}{2}\sigma^2 S^2 \frac{\partial^2 B_c}{\partial S^2} + (r - D_0)S\frac{\partial B_c}{\partial S} + u\frac{\partial B_c}{\partial r} - rB_c + kZ = 0.$$

This equation has hyperbolic properties in the r-direction. Thus, if $u(r_l, t) \ge 0$, then the value $B_c(S, r_l, t)$ is determined by the value $B_c(S, r, t)$ in the domain $[0, \infty) \times [r_l, r_u] \times [t, T]$ and no extra boundary condition at $r = r_l$ is needed. Similarly, no boundary condition should be required at $r = r_u$ because $u(r_u, t) \le 0$ and $w(r_u, t) = 0$. At $S = 0$, the equation (4.88) becomes

$$\frac{\partial B_c}{\partial t} + \frac{1}{2}w^2\frac{\partial^2 B_c}{\partial r^2} + (u - \lambda w)\frac{\partial B_c}{\partial r} - rB_c + kZ = 0.$$

This is the bond equation, and the value $B_c(0, r, t)$ is determined by this equation and the final condition at $S = 0$. Just like the Black–Scholes equation, there is no need for specifying a condition as $S \to \infty$. Therefore, if (4.45) and (4.46) hold, then we could expect that the problem

$$\begin{cases} \dfrac{\partial B_c}{\partial t} + \mathbf{L}_{s,r}B_c + kZ = 0, & 0 \le S, \quad r_l \le r \le r_u, \quad 0 \le t \le T, \\ B_c(S, r, T) = \max(Z, nS), & 0 \le S, \quad r_l \le r \le r_u \end{cases} \quad (4.90)$$

has a unique solution. If $\dfrac{\partial}{\partial r}w(r_l, t)$ and $\dfrac{\partial}{\partial r}w(r_u, t)$ are bounded, which usually is true, then the uniqueness of solution of (4.90) can be obtained from the results given in the paper by Zhu and Li (see [85]).

A convertible bond can be exchanged for n shares of stocks at any time, so the value $V(S, r, t)$ must satisfy the constraint on convertible bonds

$$B_c(S, r, t) \ge nS, \quad 0 \le S, \quad 0 \le t \le T. \quad (4.91)$$

Sometimes, the solution of the problem (4.90) satisfies the constraint (4.91), so (4.90) determines the solution of a convertible bond. For example, if $D_0 = 0$, then (4.90) gives the price of a convertible bond, which will be explained later. If

$$D_0 > 0,$$

then the price of a convertible bond is the solution of the following linear complementarity problem on the domain $[0, \infty) \times [r_l, r_u] \times [0, T]$:

$$
\begin{cases}
\left(\dfrac{\partial B_c}{\partial t} + \mathbf{L}_{s,r} B_c + kZ \right) [B_c(S, r, t) - nS] = 0, \\[2mm]
\dfrac{\partial B_c}{\partial t} + \mathbf{L}_{s,r} B_c + kZ \leq 0, \\[2mm]
B_c(S, r, t) - nS \geq 0, \\[2mm]
B_c(S, r, T) = \max(Z, nS) \geq nS.
\end{cases}
$$

Let us reformulate this problem as a free-boundary problem if $D_0 > 0$. We cannot directly apply Theorem 2.1 in Section 2.5 to this case because there are three major differences between the problem in the theorem and the problem here. Here, the problem is two-dimensional, the partial differential equation has a nonhomogeneous term, and the payoff is different from the constraint. However, the main idea is still true. For $S < Z/n$, $B_c(S, r, T) = Z > nS$. Therefore, on $[0, Z/n)$, $B_c(S, r, T - \Delta t)$ must be greater than nS if Δt is small enough, and no free boundary can appear in that region at time T. Now let us check the region $(Z/n, \infty)$. In this case, we need to check where

$$\left(\frac{\partial}{\partial t} + \mathbf{L}_{s,r} \right) nS + kZ \geq 0$$

and where

$$\left(\frac{\partial}{\partial t} + \mathbf{L}_{s,r} \right) nS + kZ < 0.$$

Because

$$\left(\frac{\partial}{\partial t} + \mathbf{L}_{s,r} \right) nS + kZ = (r - D_0)nS - rnS + kZ$$

$$= kZ - D_0 nS,$$

when $S > Z/n$ and $S > kZ/D_0 n$, namely, $S > \max(Z/n, kZ/D_0 n)$,

$$\left(\frac{\partial}{\partial t} + \mathbf{L}_{s,r} \right) nS + kZ < 0$$

and the solution is nS. Otherwise, we can use the partial differential equation to determine the solution. Therefore, there is a free boundary starting at

$S = \max(Z/n, kZ/D_0 n)$ and $t = T$. Let $S_f(r,t)$ be the location of the free boundary, then

$$S_f(r,T) = \max\left(\frac{Z}{n}, \frac{kZ}{D_0 n}\right), \qquad r_l \leq r \leq r_u. \tag{4.92}$$

At the free boundary, the solution and its derivatives are continuous. In the subdomain II where $B_c = nS$,

$$\frac{\partial B_c}{\partial S} = n$$

and

$$\frac{\partial B_c}{\partial r} = 0.$$

Thus, it seems that in the subdomain I where the partial differential equation is used, we need to require

$$B_c\left(S_f(r,t), r, t\right) = nS_f(r,t), \qquad r_l \leq r \leq r_u, \qquad 0 \leq t \leq T, \tag{4.93}$$

$$\frac{\partial B_c}{\partial S}\left(S_f(r,t), r, t\right) = n, \qquad r_l \leq r \leq r_u, \qquad 0 \leq t \leq T \tag{4.94}$$

and

$$\frac{\partial B_c}{\partial r}\left(S_f(r,t), r, t\right) = 0, \qquad r_l \leq r \leq r_u, \qquad 0 \leq t \leq T$$

on the free boundary. Differentiating both sides of (4.93) with respect to r in subdomain I yields

$$\frac{\partial B_c}{\partial S}\left(S_f(r,t), r, t\right)\frac{\partial S_f}{\partial r}(r,t) + \frac{\partial B_c}{\partial r}\left(S_f(r,t), r, t\right) = n\frac{\partial S_f}{\partial r}(r,t).$$

Using the condition (4.94), we arrive at

$$\frac{\partial B_c}{\partial r}\left(S_f(r,t), r, t\right) = 0.$$

Consequently, the conditions (4.93) and (4.94) guarantee that all the first derivatives are continuous at the free boundary and we only need to impose (4.93) and (4.94) on the solution in subdomain I.

We assume that there is only one free boundary. From numerical solutions, we know that it is true at least for some cases. Thus, when $D_0 > 0$, the domain $[0, \infty) \times [r_l, r_u] \times [0, T]$ in (S, r, t)-space is divided into subdomains

$$I : [0, S_f(r,t)] \times [r_l, r_u] \times [0, T]$$

and

$$II : (S_f(r,t), \infty) \times [r_l, r_u] \times [0, T].$$

The solution $B_c(S, r, t)$ in subdomain I and the free boundary $S = S_f(r,t)$ are the solution of the problem:

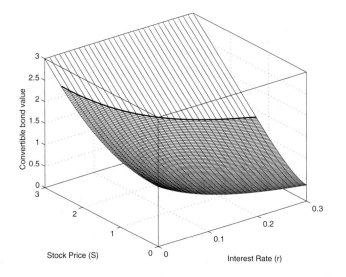

Fig. 4.3. The price of a two-factor convertible bond with $T = 30$ years

$$
\begin{cases}
\dfrac{\partial B_c}{\partial t} + \mathbf{L_{s,r}} B_c + kZ = 0, & 0 \leq S \leq S_f(r,t), \quad r_l \leq r \leq r_u, \\[2mm]
& 0 \leq t \leq T, \\[2mm]
B_c(S,r,T) = \max(Z, nS), & 0 \leq S \leq S_f(r,T), \quad r_l \leq r \leq r_u, \\[2mm]
B_c(S_f(r,t),r,t) = nS_f(r,t), & r_l \leq r \leq r_u, \quad 0 \leq t \leq T, \\[2mm]
\dfrac{\partial B_c}{\partial S}(S_f(r,t),r,t) = n, & r_l \leq r \leq r_u, \quad 0 \leq t \leq T, \\[2mm]
S_f(r,T) = \max\left(\dfrac{Z}{n}, \dfrac{kZ}{D_0 n} \right), & r_l \leq r \leq r_u;
\end{cases}
\tag{4.95}
$$

whereas in subdomain II, $B_c(S,r,t) = nS$. The problem (4.95) is usually
called a free-boundary problem for convertible bonds.

The starting location of the free boundary $S_f(r,T) = \max(Z/n, kZ/D_0 n)$
goes to infinity as D_0 tends to zero, so there is no free boundary at time T.
Because

$$
\left(\frac{\partial}{\partial t} + \mathbf{L_{s,r}} \right) nS + kZ \geq 0
$$

always holds when $D_0 = 0$, no free boundary can appear at any time. This
means that there is no free boundary when $D_0 = 0$. Thus, the value of a
convertible bond in this case is determined by the problem (4.90).

In Fig. 4.3, the price of a two-factor convertible bond with $D_0 = 0.05$ is shown. For this case, there is only one free boundary, which confirms our assumption. The result there is obtained by the singularity-separating finite-difference method, which will be described in Chapter 7.

A convertible bond can also have a call feature that gives the company the right to purchase back the bond at any time (or during specified periods) for a fixed amount M_1. In this case, the price of the bond must not exceed M_1 because no one will spend an amount more than M_1 to buy a bond that can be purchased back for an amount M_1 at any time. When we evaluate the price of such a bond, the constraint

$$B_c(S, r, t) \le M_1, \ 0 \le S, \quad r_l \le r \le r_u, \quad 0 \le t \le T \qquad (4.96)$$

is required. Because of this condition, the price of a convertible bond with a call feature can be less than a convertible bond without this feature. Because the company gets more rights, the buyer of the bond should be asked to pay less money.

A convertible bond can also incorporate a put feature, which means that the owner of the convertible bond can return the bond to the company for an amount M_2 at any time. Now we must impose the constraint

$$B_c(S, r, t) \ge M_2, \quad 0 \le S, \quad r_l \le r \le r_u, \quad 0 \le t \le T. \qquad (4.97)$$

This condition might increase the value of the bond. The owner of the bond has more rights, so he usually needs to pay more money in order to purchase such a bond.

Just like the constraint (4.89), the constraint (4.96) or the constraint (4.97) may induce a free boundary or make the free boundary more complicated. For example, for a convertible bond with a call feature, the location of the free boundary at $t = T$ is

$$S_f(r, T) = \min\left(\frac{M_1}{n}, \ \max\left(\frac{Z}{n}, \frac{kZ}{D_0 n}\right)\right), \quad r_l \le r \le r_u. \qquad (4.98)$$

If we assume that r is a constant, then the bond is a one-factor convertible bond. When $D_0 > 0$, the free-boundary problem is

$$
\begin{cases}
\dfrac{\partial B_c}{\partial t} + \dfrac{1}{2}\sigma^2 S^2 \dfrac{\partial^2 B_c}{\partial S^2} + (r - D_0)S\dfrac{\partial B_c}{\partial S} - rB_c + kZ = 0, \\[2mm]
\qquad\qquad\qquad\qquad\qquad\qquad 0 \le S \le S_f(t), \quad 0 \le t \le T, \\[2mm]
B_c(S, T) = \max(Z, nS), \qquad\qquad 0 \le S \le S_f(T), \\[2mm]
B_c(S_f(t), t) = nS_f(t), \qquad\qquad 0 \le t \le T, \\[2mm]
\dfrac{\partial B_c}{\partial S}(S_f(t), t) = n, \qquad\qquad\quad 0 \le t \le T, \\[2mm]
S_f(T) = \max\left(\dfrac{Z}{n}, \dfrac{kZ}{D_0 n}\right).
\end{cases}
\qquad (4.99)
$$

This problem is only a little different from an American call option problem. Similar to an American option, it can be proved rigorously that there is no free boundary if $D_0 = 0$. This is left as Problem 21 for the reader.

Problems

1. a) *Suppose the spot interest rate is a known function $r(t)$. Consider a bond with a face value Z and assume that it pays a coupon with a coupon rate $k(t)$, that is, during a time interval $[t, t+dt]$, the coupon payment is $Zk(t)dt$. Show that the value of the bond is

$$V(t) = Ze^{-\int_t^T r(\tau)d\tau}\left[1 + \int_t^T k(\bar{\tau})e^{\int_{\bar{\tau}}^T r(\tau)d\tau}d\bar{\tau}\right].$$

   b) Suppose that the bond pays coupon payments at two specified dates T_1 and T_2 before the maturity date T and the payments are Zk_1 and Zk_2, respectively. According to the formula given in (a), find the value of the bond at $t < \min(T_1, T_2)$ and give a financial interpretation of the expression.

2. Suppose that $a(r,t) = \sum_{i=0}^{\infty} a_i(t)r^i$ and $b(r,t) = \sum_{i=0}^{\infty} b_i(t)r^i$ and require that the problem

$$\begin{cases} \dfrac{\partial V}{\partial t} + a(r,t)\dfrac{\partial^2 V}{\partial r^2} + b(r,t)\dfrac{\partial V}{\partial r} - rV = 0, & 0 \le t \le T, \\ V(r,T) = 1 \end{cases}$$

   has a solution in the form

$$V(r,t) = e^{A(t)-rB(t)}.$$

   Show that in order to fulfill this requirement, between a_i and $b_i, i = 2, 3, \cdots$, there must exist the following relations:

$$a_iB - b_i = 0, \quad i = 2, 3, \cdots.$$

   This means that in order to choose $a(r,t)$ and $b(r,t)$ independently and for the solution to be in the form $e^{A(t)-rB(t)}$, we have to assume $a(r,t) = a_0(t) + a_1(t)r$ and $b(r,t) = b_0(t) + b_1(t)r$.

3. Suppose that $a(r,t) = a_0(t) + a_1(t)r$ and $b(r,t) = b_0(t) + b_1(t)r$. Show that the problem

$$\begin{cases} \dfrac{\partial V}{\partial t} + a(r,t)\dfrac{\partial^2 V}{\partial r^2} + b(r,t)\dfrac{\partial V}{\partial r} - rV = 0, & 0 \le t \le T, \\ V(r,T) = 1 \end{cases}$$

has a solution in the form

$$V(r,t) = e^{A(t)-rB(t)}$$

with $A(T) = B(T) = 0$ and determine the system of ordinary differential equations the functions $A(t)$ and $B(t)$ should satisfy.

4. *In the Vasicek model, the spot interest rate is assumed to satisfy

$$dr = (\bar{\mu} - \gamma r)dt + \sqrt{-\beta}dX, \qquad \beta < 0, \qquad \gamma > 0,$$

where $\bar{\mu}$, γ, and β are constants, and dX is a Wiener process. Let the market price of risk $\lambda(r,t) = \bar{\lambda}\sqrt{-\beta}$. Then, the price $V(r,t;T)$ of a zero-coupon bond maturing at time T with a face value Z is the solution of the problem

$$
\begin{cases}
\dfrac{\partial V}{\partial t} + \dfrac{1}{2}(-\beta)\dfrac{\partial^2 V}{\partial r^2} + (\mu - \gamma r)\dfrac{\partial V}{\partial r} - rV = 0, \\[2mm]
\qquad\qquad\qquad\qquad -\infty < r < \infty, \quad 0 \le t \le T, \\[2mm]
V(r,T;T) = Z, \qquad -\infty < r < \infty, \quad 0 \le t \le T,
\end{cases}
$$

where

$$\mu = \bar{\mu} + \bar{\lambda}\beta.$$

a) Show that this problem has a solution in the form

$$V(r,t;T) = Ze^{A(t,T)-rB(t,T)}$$

and A and B are the solution of the system of ordinary differential equations

$$
\begin{cases}
\dfrac{dA}{dt} = \dfrac{1}{2}\beta B^2 + \mu B, \\[3mm]
\dfrac{dB}{dt} = \gamma B - 1
\end{cases}
$$

with the conditions

$$A(T,T) = 0,$$
$$B(T,T) = 0.$$

b) Show that the solution of the above problem of ordinary differential equations is

$$
\begin{cases}
B = \dfrac{1}{\gamma}\left(1 - e^{-\gamma(T-t)}\right), \\[3mm]
A = -\left(\dfrac{\beta}{2\gamma^2} + \dfrac{\mu}{\gamma}\right)(T-t) + \left(\dfrac{\beta}{2\gamma^2} + \dfrac{\mu}{\gamma}\right)B + \dfrac{\beta}{4\gamma}B^2.
\end{cases}
$$

5. Show

$$\lim_{\alpha \to 0} \left\{ \frac{\beta}{\alpha} B + \left[\frac{\beta(\gamma - \psi)}{\alpha(\gamma + \psi)\psi} + \mu \frac{\gamma - \psi}{\alpha\psi} \right] \ln \frac{(\gamma - \psi)/\alpha}{B + (\gamma - \psi)/\alpha} \right.$$
$$\left. - \left[\frac{\beta(\gamma + \psi)}{\alpha(\gamma - \psi)\psi} + \mu \frac{\gamma + \psi}{\alpha\psi} \right] \ln \frac{(\gamma + \psi)/\alpha}{B + (\gamma + \psi)/\alpha} \right\}$$
$$= -\left(\frac{\beta}{2\gamma^2} + \frac{\mu}{\gamma} \right)(T - t) + \left(\frac{\beta}{2\gamma^2} + \frac{\mu}{\gamma} \right) B + \frac{\beta B^2}{4\gamma},$$

where

$$B(t, T) = \frac{1}{\gamma}\left(1 - e^{-\gamma(T-t)}\right) \quad \text{and} \quad \psi = \sqrt{\gamma^2 + 2\alpha}.$$

(The two sides are two expressions for A associated with the Vasicek model obtained by different approaches. This confirms that the two different approaches give the same answer.)

6. *In the Cox–Ingersoll–Ross model, the spot interest rate is assumed to satisfy

$$dr = (\mu - \bar{\gamma}r)dt + \sqrt{\alpha r}dX,$$

where μ, $\bar{\gamma}$, and α are constants, and dX is a Wiener process. Let the market price of risk $\lambda(r, t)$ be $\bar{\lambda}\sqrt{\alpha r}$. Then, the price $V(r, t; T)$ of a zero-coupon bond maturing at time T with a face value Z is the solution of the problem

$$\begin{cases} \dfrac{\partial V}{\partial t} + \dfrac{1}{2}\alpha r \dfrac{\partial^2 V}{\partial r^2} + (\mu - \gamma r)\dfrac{\partial V}{\partial r} - rV = 0, & 0 \le r, \quad 0 \le t \le T, \\ V(r, T; T) = Z, & 0 \le r, \end{cases}$$

where $\gamma = \bar{\gamma} + \bar{\lambda}\alpha$.

a) Show that this problem has a solution in the form

$$V(r, t; T) = Ze^{A(t,T)-rB(t,T)} \quad '$$

and A and B are solutions of the system of ordinary differential equations

$$\begin{cases} \dfrac{dA}{dt} = \mu B, \\ \dfrac{dB}{dt} = \dfrac{1}{2}\alpha B^2 + \gamma B - 1 \end{cases}$$

with the conditions

$$A(T, T) = 0$$

and

$$B(T, T) = 0;$$

b) Show that the solution of the above problem of ordinary differential equations is

$$B = \frac{2(e^{\psi(T-t)} - 1)}{(\gamma + \psi)e^{\psi(T-t)} - (\gamma - \psi)},$$

$$A = \ln \left(\frac{2\psi e^{(\gamma+\psi)(T-t)/2}}{(\gamma + \psi)e^{\psi(T-t)} - (\gamma - \psi)} \right)^{2\mu/\alpha},$$

where

$$\psi = \sqrt{\gamma^2 + 2\alpha}.$$

7. Show

$$Z \left[\frac{B + (\gamma - \psi)/\alpha}{(\gamma - \psi)/\alpha} \right]^{\mu(\psi-\gamma)/\alpha\psi} \left[\frac{B + (\gamma + \psi)/\alpha}{(\gamma + \psi)/\alpha} \right]^{\mu(\gamma+\psi)/\alpha\psi} e^{-rB}$$

$$\equiv Z \left[\frac{2\psi e^{(\gamma+\psi)(T-t)/2}}{(\gamma + \psi)e^{\psi(T-t)} - (\gamma - \psi)} \right]^{2\mu/\alpha} e^{-rB},$$

where

$$B = \frac{2\left(e^{\psi(T-t)} - 1\right)}{(\gamma + \psi)\,e^{\psi(T-t)} - (\gamma - \psi)}.$$

(The two sides are two expressions for the zero-coupon bond price associated with the Cox–Ingersoll–Ross model obtained by different approaches. This confirms that the two different approaches give the same answer.)

8. *Describe a way to determine the market price of risk for the spot interest rate.

9. *Suppose that any European-style interest rate derivative with a continuous coupon satisfies the equation:

$$\frac{\partial V}{\partial t} + \frac{1}{2}w^2\frac{\partial^2 V}{\partial r^2} + (u - \lambda w)\frac{\partial V}{\partial r} - rV + k = 0, \quad r_l \leq r \leq r_u, \quad t \leq T,$$

where k is the coupon rate corresponding to the derivative, the coefficients u and w satisfy the reversion conditions on the boundaries $r = r_l$, $r = r_u$, and λ is a given bounded function. Describe how to evaluate the price of a European call option on a bond with coupon by using this equation.

10. Show that the solution of the problem

$$\begin{cases} \dfrac{\partial V}{\partial t} + \dfrac{1}{2}w^2\dfrac{\partial^2 V}{\partial r^2} + (u - \lambda w)\dfrac{\partial V}{\partial r} - rV = 0, \\[2mm] \qquad\qquad\qquad\qquad r_l \leq r \leq r_u, \quad t \leq T, \\[2mm] V(r, T; T) = 1, \qquad\qquad r_l \leq r \leq r_u \end{cases}$$

is the same as that of the problem

$$\begin{cases} \dfrac{\partial V}{\partial t} + \dfrac{1}{2}w^2\dfrac{\partial^2 V}{\partial r^2} + (u - \lambda w)\dfrac{\partial V}{\partial r} - rV + \delta(t - T) = 0, \\[2mm] \qquad\qquad\qquad\qquad r_l \le r \le r_u, \quad t \le T, \\[2mm] V(r, T^+; T) = 0, \qquad r_l \le r \le r_u \end{cases}$$

for any

$$t < T.$$

11. Consider the problem:

$$\begin{cases} \dfrac{\partial V_{s1}}{\partial t} + \dfrac{1}{2}w^2\dfrac{\partial^2 V_{s1}}{\partial r^2} + (u - \lambda w)\dfrac{\partial V_{s1}}{\partial r} - rV_{s1} \\[2mm] \qquad\qquad\qquad\qquad + \sum\limits_{k=1}^{2N}\delta(t - T - k/2) = 0, \\[2mm] \qquad\qquad\qquad\qquad r_l \le r \le r_u, \quad t \le T + N, \\[2mm] V_{s1}(r, T + N) = 0, \qquad r_l \le r \le r_u. \end{cases}$$

Show that $V_{s1}(r, T)$ gives the sum of values of $2N$ zero-coupon bonds with maturities $1/2, 1, 3/2, \cdots, N$ years.

12. *Suppose that any European-style interest rate derivative satisfies the equation:

$$\dfrac{\partial V}{\partial t} + \dfrac{1}{2}w^2\dfrac{\partial^2 V}{\partial r^2} + (u - \lambda w)\dfrac{\partial V}{\partial r} - rV = 0, \quad r_l \le r \le r_u,$$

where all the coefficients in the equation are known. The value of N-year swap at time T is given by

$$V_s(T; r_s) = Q\left[1 - \dfrac{r_s}{2}\sum\limits_{k=1}^{2N} Z(T; T + k/2) - Z(T; T + N)\right],$$

where Q is the notional principal, r_s is the N-year swap rate, and $Z(T; T + k/2)$ is the value of zero-coupon bond with maturity $k/2$ at time T. Describe how to find the price of a swaption with exercise swap rate r_{se} and maturity T by using this equation.

13. Suppose that the solution of

$$\begin{cases} \dfrac{\partial V}{\partial t} + \dfrac{1}{2}w^2\dfrac{\partial^2 V}{\partial r^2} + (u - \lambda w)\dfrac{\partial V}{\partial r} - rV = 0, \quad r_l \le r \le r_u, \quad t \le t_k, \\[2mm] V(r, t_k; t_k) = 1, \qquad\qquad\qquad\qquad r_l \le r \le r_u \end{cases}$$

is $V(r, t; t_k)$ and that $V(r^*, t^*; t_k) = Z(t^*; t_k)$. Also assume that $V_s(r, t; r_s)$ is the solution of

$$
\left\{
\begin{array}{l}
\dfrac{\partial V_s}{\partial t} + \dfrac{1}{2}w^2\dfrac{\partial^2 V_s}{\partial r^2} + (u - \lambda w)\dfrac{\partial V_s}{\partial r} - rV_s - \displaystyle\sum_{k=k^*+1}^{2N-1}\dfrac{Qr_s}{2}\delta(t - t_k) \\[4mm]
\qquad + Q\left[1 + \dfrac{\bar{f}(t_{k^*}, t_{k^*}, t_{k^*+1})}{2}\right]\delta(t - t_{k^*+1}) = 0, \\[6mm]
\qquad\qquad\qquad\qquad\qquad\qquad r_l \le r \le r_u, \qquad t^* \le t \le T + N, \\[2mm]
V_s(r, T + N) = -Q(1 + r_s/2), \qquad r_l \le r \le r_u.
\end{array}
\right.
$$

Here, $V_s(r, t; r_s)$ actually is the value of a swap. Q and r_s are the notional principal and the swap rate, respectively. t^*, T, and N denote the time today, the time the swap is initiated, and the duration of the swap with the relation $T \le t^* < T + N$. k^* is the integer part of $(t^* - T)/2$, and $t_k = T + k/2$, $k = k^* + 1, k^* + 2, \cdots, 2N$. $\bar{f}(t_{k^*}, t_{k^*}, t_{k^*+1})$ is the six-month LIBOR for the period $[t_{k^*}, t_{k^*+1}]$ determined at time t_{k^*}. Then

$$
V_s(r^*, t^*; r_s) = QZ(t^*; t_{k^*+1})\left[1 + \frac{1}{2}\bar{f}(t_{k^*}, t_{k^*}, t_{k^*+1})\right]
$$

$$
- Q\left[\sum_{k=k^*+1}^{2N}\frac{r_s}{2}Z(t^*; t_k) + Z(t^*; T + N)\right].
$$

14. Consider an N-year floor with a floor rate r_f. Suppose that the money will be paid quarterly at time $t_k = t^* + k/4$, $k = 2, 3, \cdots, 4N$, and the floating rate is the three-month LIBOR. Suppose that $V_{bk}(r, t)$ is the solution of the problem

$$
\left\{
\begin{array}{l}
\dfrac{\partial V_{bk}}{\partial t} + \dfrac{1}{2}w^2\dfrac{\partial^2 V_{bk}}{\partial r^2} + (u - \lambda w)\dfrac{\partial V_{bk}}{\partial r} - rV_{bk} = 0, \\[4mm]
\qquad\qquad\qquad\qquad\qquad r_l \le r \le r_u, \qquad t_{k-1} \le t \le t_k, \\[2mm]
V_{bk}(r, t_k) = Q\left(1 + \dfrac{r_f}{4}\right), \qquad r_l \le r \le r_u,
\end{array}
\right.
$$

where $k = 2, 3, \cdots, 4N$ and $V_f(r, t)$ is the solution of the problem

$$
\left\{
\begin{array}{l}
\dfrac{\partial V_f}{\partial t} + \dfrac{1}{2}w^2\dfrac{\partial^2 V_f}{\partial r^2} + (u - \lambda w)\dfrac{\partial V_f}{\partial r} - rV_f \\[4mm]
\qquad + \displaystyle\sum_{k=2}^{4N}\max(V_{bk}(r, t_{k-1}) - Q, 0)\delta(t - t_{k-1}) = 0, \\[6mm]
\qquad\qquad\qquad\qquad\qquad r_l \le r \le r_u, \qquad t^* \le t \le t_{4N-1}, \\[2mm]
V_f(r, t_{4N-1}) = 0, \qquad r_l \le r \le r_u.
\end{array}
\right.
$$

Show that the premium of the floor should be

$$
V_f(r^*, t^*),
$$

where r^* is the spot interest rate at time t^*.

15. *Assume that Z_1, Z_2, Z_3 are random variables and satisfy the system of stochastic differential equations:

$$dZ_i = \mu_i\,(Z_1, Z_2, Z_3, t)\,dt + \sigma_i\,(Z_1, Z_2, Z_3, t)\,dX_i, \quad i = 1, 2, 3,$$

where dX_i are the Wiener processes and $\mathrm{E}\,[dX_i dX_j] = \rho_{ij} dt$ with $-1 \leq \rho_{ij} \leq 1$. At time t, a point (Z_1, Z_2, Z_3) is in the domain Ω: $[Z_{1,l}, 1] \times [Z_{2,l}, Z_1] \times [Z_{3,l}, Z_2]$, where $Z_{1,l}$, $Z_{2,l}$ and $Z_{3,l}$ are constants and satisfy the relation $Z_{1,l} \geq Z_{2,l} \geq Z_{3,l}$. Give an intuitive explanation on what conditions on μ_i and σ_i are needed in order to guarantee that the point will move only in the domain Ω.

16. *Show that under the transformation

$$\begin{cases} \xi_1 = \dfrac{Z_1 - Z_{1,l}}{1 - Z_{1,l}}, \\[2mm] \xi_2 = \dfrac{Z_2 - Z_{2,l}}{Z_1 - Z_{2,l}}, \\[2mm] \xi_3 = \dfrac{Z_3 - Z_{3,l}}{Z_2 - Z_{3,l}}, \end{cases}$$

the partial differential equation

$$\frac{\partial V}{\partial t} + \frac{1}{2} \sum_{i=1}^{3} \sum_{j=1}^{3} \sigma_i \sigma_j \rho_{i,j} \frac{\partial^2 V}{\partial Z_i \partial Z_j} + r \sum_{i=1}^{3} Z_i \frac{\partial V}{\partial Z_i} - rV = 0$$

becomes

$$\frac{\partial V}{\partial t} + \frac{1}{2} \sum_{i=1}^{3} \sum_{j=1}^{3} \tilde{\sigma}_i \tilde{\sigma}_j \tilde{\rho}_{i,j} \frac{\partial^2 V}{\partial \xi_i \partial \xi_j} + \sum_{i=1}^{3} b_i \frac{\partial V}{\partial \xi_i} - rV = 0,$$

where

$$\begin{cases} b_1 = \dfrac{rZ_1}{1 - Z_{1,l}}, \\[3mm] b_2 = \dfrac{r(Z_2 - Z_1\xi_2)}{Z_1 - Z_{2,l}} + \dfrac{\sigma_1(\sigma_1\xi_2 - \sigma_2\rho_{1,2})}{(Z_1 - Z_{2,l})^2}, \\[3mm] b_3 = \dfrac{r(Z_3 - Z_2\xi_3)}{Z_2 - Z_{3,l}} + \dfrac{\sigma_2(\sigma_2\xi_3 - \sigma_3\rho_{2,3})}{(Z_2 - Z_{3,l})^2}, \end{cases}$$

and $\tilde{\sigma}_1$, $\tilde{\sigma}_2$, $\tilde{\sigma}_3$, $\tilde{\rho}_{1,2}$, $\tilde{\rho}_{1,3}$, $\tilde{\rho}_{2,3}$ are determined by

$$\begin{cases} \dfrac{1}{2}\tilde{\sigma}_1^2 = \dfrac{\frac{1}{2}\sigma_1^2}{(1 - Z_{1,l})^2}, \\[3mm] \dfrac{1}{2}\tilde{\sigma}_2^2 = \dfrac{\frac{1}{2}\left(\sigma_1^2\xi_2^2 - 2\sigma_1\sigma_2\xi_2\rho_{1,2} + \sigma_2^2\right)}{(Z_1 - Z_{2,l})^2}, \\[3mm] \dfrac{1}{2}\tilde{\sigma}_3^2 = \dfrac{\frac{1}{2}\left(\sigma_2^2\xi_3^2 - 2\sigma_2\sigma_3\xi_3\rho_{2,3} + \sigma_3^2\right)}{(Z_2 - Z_{3,l})^2} \end{cases}$$

and

$$
\begin{cases}
\tilde{\sigma}_1 \tilde{\sigma}_2 \tilde{\rho}_{1,2} = \dfrac{\sigma_1 \left(\sigma_2 \rho_{1,2} - \sigma_1 \xi_2 \right)}{\left(1 - Z_{1,l} \right) \left(Z_1 - Z_{2,l} \right)}, \\[3mm]
\tilde{\sigma}_1 \tilde{\sigma}_3 \tilde{\rho}_{1,3} = \dfrac{\sigma_1 \left(\sigma_3 \rho_{1,3} - \sigma_2 \rho_{1,2} \xi_3 \right)}{\left(1 - Z_{1,l} \right) \left(Z_2 - Z_{3,l} \right)}, \\[3mm]
\tilde{\sigma}_2 \tilde{\sigma}_3 \tilde{\rho}_{2,3} = \dfrac{\sigma_1 \xi_2 \left(\sigma_2 \rho_{1,2} \xi_3 - \sigma_3 \rho_{1,3} \right) + \sigma_2 \left(\sigma_3 \rho_{2,3} - \sigma_2 \xi_3 \right)}{\left(Z_1 - Z_{2,l} \right) \left(Z_2 - Z_{3,l} \right)}.
\end{cases}
$$

17. Assume that Z_1, Z_2, Z_3 are random variables and satisfy the system of stochastic differential equations:

$$
dZ_i = \mu_i \left(Z_1, Z_2, Z_3, t \right) dt + \sigma_i \left(Z_1, Z_2, Z_3, t \right) dX_i, \quad i = 1, 2, 3,
$$

where dX_i are the Wiener processes and $\mathrm{E}\left[dX_i dX_j \right] = \rho_{ij} dt$ with $-1 \le \rho_{ij} \le 1$, and that ξ_1, ξ_2 and ξ_3 are governed by

$$
d\xi_i = \tilde{\mu}_i (\xi_1, \xi_2, \xi_3, t) dt + \tilde{\sigma}_i (\xi_1, \xi_2, \xi_3, t) d\tilde{X}_i, \quad i = 1, 2, 3,
$$

where $d\tilde{X}_i$ are the Wiener processes and $\mathrm{E}\left[d\tilde{X}_i d\tilde{X}_j \right] = \tilde{\rho}_{ij} dt$ with $-1 \le \tilde{\rho}_{ij} \le 1$. Furthermore, we suppose that ξ_1, ξ_2 and ξ_3 are defined by

$$
\begin{cases}
\xi_1 = \dfrac{Z_1 - Z_{1,l}}{1 - Z_{1,l}}, \\[3mm]
\xi_2 = \dfrac{Z_2 - Z_{2,l}}{Z_1 - Z_{2,l}}, \\[3mm]
\xi_3 = \dfrac{Z_3 - Z_{3,l}}{Z_2 - Z_{3,l}},
\end{cases}
$$

where $Z_{1,l}$, $Z_{2,l}$, and $Z_{3,l}$ are constants. Find the expressions of $\tilde{\sigma}_1$, $\tilde{\sigma}_2$, $\tilde{\sigma}_3$, $\tilde{\rho}_{12}$, $\tilde{\rho}_{13}$, $\tilde{\rho}_{23}$ as functions of σ_1, σ_2, σ_3, $\rho_{1,2}$, $\rho_{1,3}$, $\rho_{2,3}$, Z_1, Z_2, and Z_3 by using Itô's lemma.

18. Consider a two-factor convertible bond paying coupons with a rate k. For such a convertible bond, derive directly the partial differential equation that contains only the unknown market price of risk for the spot interest rate. "Directly" means "without using the general PDE for derivatives." (Hint: Take a portfolio in the form

$$
\Pi = \Delta_1 V_1 + \Delta_2 V_2 + S,
$$

where V_1 and V_2 are two different convertible bonds.)

19. *Formulate the two-factor convertible coupon-paying bond problem as a linear complementarity problem.

20. *Formulate a two-factor convertible coupon-paying bond problem as a free-boundary problem if $D_0 > 0$. (Assume that there exists at most one free boundary.)

21. Consider the problem

$$
\begin{cases}
\dfrac{\partial B_c}{\partial t} + \dfrac{1}{2}\sigma^2 S^2 \dfrac{\partial^2 B_c}{\partial S^2} + (r - D_0)S\dfrac{\partial B_c}{\partial S} - rB_c = 0, \\[2mm]
\qquad\qquad\qquad\qquad\qquad\qquad 0 \le S, \quad 0 \le t \le T, \\[2mm]
B_c(S,T) = \max(Z, nS), \qquad\quad 0 \le S,
\end{cases}
$$

where σ, r and D_0 are constants. If the solution of this problem fulfills the constraint condition

$$B_c(S,t) \ge nS, \qquad 0 \le S, \qquad 0 \le t \le T,$$

then the solution of the problem above represents the price of a one-factor convertible bond. In this case, the solution of a one-factor convertible bond does not involve any free boundary. Using this fact, show that no free boundary will be encountered when one prices a one-factor convertible bond with $D_0 \le 0$.

22. Consider the problem

$$
\begin{cases}
\dfrac{\partial B_c}{\partial t} + \dfrac{1}{2}\sigma^2 S^2 \dfrac{\partial^2 B_c}{\partial S^2} + (r - D_0)S\dfrac{\partial B_c}{\partial S} - rB_c = 0, \\[2mm]
\qquad\qquad\qquad\qquad\qquad\qquad 0 \le S, \quad 0 \le t \le T, \\[2mm]
B_c(S,T) = \max(Z, nS) = n\max(S - Z/n, 0) + Z, \quad 0 \le S,
\end{cases}
$$

where σ, r and D_0 are constants. Show that its solution is

$$nc(S,t; Z/n) + e^{-r(T-t)}Z,$$

where $c(S, t; Z/n)$ is the price of a European call option with an exercise price $E = Z/n$. This means that the problem can be understood as a problem to determine the value of an investment consisting of n units of European call options with $E = Z/n$ and a cash amount of $e^{-r(T-t)}Z$. By Problem 21, if $D_0 \le 0$, then it is the price of a convertible bond. Therefore, when $D_0 \le 0$, the value of a one-factor convertible bond is equal to the price of n units of European call options with $E = Z/n$ plus a cash amount of $e^{-r(T-t)}Z$.

Part II

Numerical Methods for Derivative Securities

Basic Numerical Methods

This chapter is devoted to the basic numerical methods. We first discuss various approximations, solution of systems, and eigenvalue problems. Then, we deal with finite-difference methods for parabolic partial differential equations, including algorithms, stability and convergence analysis, and extrapolation techniques of numerical solutions. Finally, we discuss how to determine the parameters in stochastic models.

5.1 Approximations

5.1.1 Interpolation

Linear interpolation. Suppose that the values of a function $f(x)$ are given on the grid points x_m, $m = 0, 1, \cdots, M$, where $x_0 < x_1 < \cdots < x_M$. Sometimes, we may need to find the value of the function at other points. A simple way to do this is to interpolate the function by using the known values of the function. Let f_m denote the value of the function $f(x)$ at a point x_m, $m = 0, 1, \cdots, M$. We want to approximate the value $f(x^*)$ for $x^* \in (x_m, x_{m+1})$. The simplest interpolation is to use a linear function to approximate the function $f(x)$ on the subinterval $[x_m, x_{m+1}]$. Let

$$p_1(x) = a_0 + a_1 x.$$

Using the conditions

$$p_1(x_m) = f_m, \quad p_1(x_{m+1}) = f_{m+1},$$

we find

$$a_0 = \frac{x_{m+1} f_m - x_m f_{m+1}}{x_{m+1} - x_m}, \quad a_1 = \frac{f_{m+1} - f_m}{x_{m+1} - x_m}.$$

Then, we have

$$p_1(x) = \frac{x_{m+1} - x}{x_{m+1} - x_m} f_m + \frac{x - x_m}{x_{m+1} - x_m} f_{m+1}.$$

Thus, we have the approximate value:

$$f(x^*) \approx p_1(x^*) = \frac{x_{m+1} - x^*}{x_{m+1} - x_m} f_m + \frac{x^* - x_m}{x_{m+1} - x_m} f_{m+1}.$$

This is called the linear interpolation. If we do the interpolation for all subintervals, then we obtain a piecewise linear function on the interval $[x_0, x_M]$.

Higher order interpolation. If the function data indicates that the function is smooth, then we can use a quadratic or N-th order interpolation to get a better approximation. Assume that we have obtained the values f_{m-1}, f_m, and f_{m+1}. Let

$$p_2(x) = a_0 + a_1 x + a_2 x^2.$$

Using the conditions

$$p_2(x_{m-1}) = f_{m-1}, \quad p_2(x_m) = f_m, \quad p_2(x_{m+1}) = f_{m+1},$$

we find

$$p_2(x) = \frac{(x_m - x)(x_{m+1} - x)}{(x_m - x_{m-1})(x_{m+1} - x_{m-1})} f_{m-1} + \frac{(x - x_{m-1})(x_{m+1} - x)}{(x_m - x_{m-1})(x_{m+1} - x_m)} f_m$$
$$+ \frac{(x - x_{m-1})(x - x_m)}{(x_{m+1} - x_{m-1})(x_{m+1} - x_m)} f_{m+1}.$$

Then, for any $x^* \in (x_{m-1}, x_{m+1})$, $f(x^*)$ can be approximated by $p_2(x^*)$. This is called the quadratic interpolation.

In general, if f_m, $m = 0, 1, \cdots, N$, are known, then an N-th Lagrange interpolating polynomial can be obtained:

$$p_N(x) = \varphi_0(x) f_0 + \varphi_1(x) f_1 + \cdots + \varphi_N(x) f_N,$$

where

$$\varphi_k(x) = \frac{(x - x_0)(x - x_1) \cdots (x - x_{k-1})(x - x_{k+1}) \cdots (x - x_N)}{(x_k - x_0)(x_k - x_1) \cdots (x_k - x_{k-1})(x_k - x_{k+1}) \cdots (x_k - x_N)}$$

for $k = 0, 1, \cdots, N$. This approximation can be used for any $x \in (x_0, x_N)$. It is clear that the linear and quadratic interpolating polynomials are the Lagrange interpolating polynomials with $N = 1$ and 2, respectively. For an N-th Lagrange interpolating polynomial, the error is given by the following theorem:

Theorem 5.1 *If x_m, $m = 0, 1, \cdots, N$, are distinct numbers in the interval $[x_0, x_N]$, and $f(x)$ has $N + 1$ continuous derivatives on $[x_0, x_N]$, then for any $x \in [x_0, x_N]$, there exists a $\xi \in [x_0, x_N]$, such that*

$$f(x) - p_N(x) = \frac{f^{(N+1)}(\xi)}{(N + 1)!} (x - x_0)(x - x_1) \cdots (x - x_N).$$

Therefore, the error of linear interpolation is $O(\Delta x^2)$, and the error of quadratic interpolation is $O(\Delta x^3)$, where $\Delta x = \max\limits_{m}(x_{m+1} - x_m)$.

Cubic spline interpolation. As we can see, linear interpolations result in piecewise linear functions on the interval $[x_0, x_M]$: the function is smooth in each subinterval $[x_m, x_{m+1}]$, continuous in $[x_0, x_M]$, but may not be smooth in $[x_0, x_M]$. For quadratic interpolations, the situation is similar. Cubic spline interpolation is the most commonly used piecewise polynomial approximation, which is a cubic polynomial on each subinterval $[x_m, x_{m+1}]$ and has a continuous second derivative on the whole interval. The cubic spline interpolation $S(x)$ satisfies the following conditions:

A) On the subinterval $[x_m, x_{m+1}]$, $S(x) = S_m(x)$ is a cubic polynomial, $m = 0, 1, \cdots, M - 1$;
B) $S(x_m) = f_m$, $m = 0, 1, \cdots, M$;
C) $S_{m+1}(x_{m+1}) = S_m(x_{m+1})$, $S'_{m+1}(x_{m+1}) = S'_m(x_{m+1})$, $S''_{m+1}(x_{m+1}) = S''_m(x_{m+1})$, $m = 0, 1, \cdots, M - 2$;
D) $S''(x_0) = S''(x_M) = 0$.

Let

$$S_m(x) = a_m + b_m(x - x_m) + c_m(x - x_m)^2 + d_m(x - x_m)^3, \quad m = 0, 1, \cdots, M-1.$$

a_m is obtained directly by

$$a_m = S_m(x_m) = f_m, \quad m = 0, 1, \cdots, M - 1.$$

Using condition C, we get

$$\begin{cases} a_{m+1} = a_m + b_m h_m + c_m h_m^2 + d_m h_m^3, \\ b_{m+1} = b_m + 2c_m h_m + 3d_m h_m^2, \\ c_{m+1} = c_m + 3d_m h_m, \\ \qquad m = 0, 1, \cdots, M - 2, \end{cases} \tag{5.1}$$

where $h_m = x_{m+1} - x_m$. Define

$$a_M = f_M$$

and

$$c_M = S''(x_M)/2.$$

Then, we further have

$$\begin{cases} a_M = a_{M-1} + b_{M-1} h_{M-1} + c_{M-1} h_{M-1}^2 + d_{M-1} h_{M-1}^3, \\ c_M = c_{M-1} + 3d_{M-1} h_{M-1}. \end{cases} \tag{5.2}$$

Eliminating b_m and d_m from these equations we obtain

$$h_{m-1}c_{m-1} + 2(h_{m-1} + h_m)c_m + h_m c_{m+1} = \frac{3(a_{m+1} - a_m)}{h_m} - \frac{3(a_m - a_{m-1})}{h_{m-1}},$$
$$m = 1, 2, \cdots, M - 1.$$

From $S''(x_0) = S''(x_M) = 0$, we have $c_0 = c_M = 0$. Consequently, these equations can be written in the following matrix form:

$$A\mathbf{c} = \mathbf{h}, \tag{5.3}$$

where

$$A = \begin{bmatrix} 1 & 0 & \cdots & \cdots & \cdots & 0 \\ h_0 & 2(h_0 + h_1) & h_1 & 0 & \cdots & 0 \\ 0 & h_1 & 2(h_1 + h_2) & h_2 & 0 & \cdots \\ \vdots & \ddots & \ddots & \ddots & \ddots & \ddots \\ 0 & \cdots & 0 & h_{M-2} & 2(h_{M-2} + h_{M-1}) & h_{M-1} \\ 0 & \cdots & \cdots & \cdots & 0 & 1 \end{bmatrix},$$

$$\mathbf{h} = \begin{bmatrix} 0 \\ \dfrac{3(a_2 - a_1)}{h_1} - \dfrac{3(a_1 - a_0)}{h_0} \\ \vdots \\ \dfrac{3(a_M - a_{M-1})}{h_{M-1}} - \dfrac{3(a_{M-1} - a_{M-2})}{h_{M-2}} \\ 0 \end{bmatrix}, \quad \mathbf{c} = \begin{bmatrix} c_0 \\ c_1 \\ \vdots \\ \vdots \\ c_M \end{bmatrix}.$$

Solving this linear system, we obtain c_m, $m = 0, 1, \cdots, M$. Then d_m, $m = 0, 1, \cdots, M - 1$, can be obtained from the last equations in (5.1) and (5.2) and b_m, $m = 0, 1, \cdots, M - 1$, from the first equations in (5.1) and (5.2).

The condition $S''(x_0) = 0$ could be replaced by $S'(x_0) = f'(x_0)$ or $d_0 = 0$, namely, assuming $S_0(x) = a_0 + b_0(x - x_0) + c_0(x - x_0)^2$. At $x = x_M$, the situation is similar. If such a case appears, then the way to determine these coefficients needs to be changed slightly. Here, assuming $S'(x_0) = f'(x_0)$ and $d_{M-1} = 0$, we explain how to modify the way to determine these coefficients. Because $S'_0(x_0) = b_0$, the coefficient b_0 is known in this case, namely, $b_0 = f'(x_0)$. From

$$\begin{cases} a_1 = a_0 + b_0 h_0 + c_0 h_0^2 + d_0 h_0^3, \\ c_1 = c_0 + 3d_0 h_0, \end{cases}$$

we eliminate d_0 and obtain

$$2c_0 + c_1 = 3\left(\frac{a_1 - a_0}{h_0^2} - \frac{b_0}{h_0}\right).$$

This equation should replace the first equation in (5.3). From $d_{M-1} = 0$ and the second equation in (5.2), we have

$$c_{M-1} - c_M = 0.$$

This equation should replace the last equation in (5.3). Solving the modified (5.3) yields c_m, $m = 0, 1, \cdots, M$, for this case. As soon as all the c_m are obtained, d_m, $m = 0, 1, \cdots, M - 2$, can be obtained from the last equation in (5.1) and b_m, $m = 1, 2, \cdots, M - 1$, from the first equations in (5.1) and (5.2).

5.1.2 Approximation of Partial Derivatives

Finite-difference approximation. Here, we will discuss how derivatives of a function $u(x, t)$ at a point can be approximated by a linear combination of values of the function at adjacent points. Let $x_m = a + m\Delta x$ and $\tau^n = n\Delta\tau$, where m is an integer and n is an integer or an integer plus a half.

Using the Taylor expansion, we have[1]

$$u(x_m, \tau^{n+1}) = u(x_m, \tau^n) + \Delta\tau \frac{\partial u}{\partial \tau}(x_m, \tau^n) + \frac{\Delta\tau^2}{2}\frac{\partial^2 u}{\partial \tau^2}(x_m, \eta),$$

where $\tau^n < \eta < \tau^{n+1}$. Then,

$$\frac{\partial u}{\partial \tau}(x_m, \tau^n) = \frac{u(x_m, \tau^{n+1}) - u(x_m, \tau^n)}{\Delta\tau} - \frac{\Delta\tau}{2}\frac{\partial^2 u}{\partial \tau^2}(x_m, \eta).$$

If $\Delta\tau$ is small, we have

$$\frac{\partial u}{\partial \tau}(x_m, \tau^n) \approx \frac{u(x_m, \tau^{n+1}) - u(x_m, \tau^n)}{\Delta\tau}.$$

This approximation is called the forward finite-difference approximation or the **forward difference** for $\dfrac{\partial u}{\partial \tau}$. Similarly, we can obtain the **backward difference**

$$\frac{\partial u}{\partial \tau}(x_m, \tau^n) \approx \frac{u(x_m, \tau^n) - u(x_m, \tau^{n-1})}{\Delta\tau}.$$

Both forward and backward finite-difference approximations have errors of first order in $\Delta\tau$ (first-order accurate). To obtain a second-order accurate finite-difference approximation, we use the following Taylor expansions:

$$u(x_m, \tau^{n+1}) = u(x_m, \tau^{n+1/2}) + \frac{\Delta\tau}{2}\frac{\partial u}{\partial \tau}(x_m, \tau^{n+1/2})$$
$$+ \frac{\Delta\tau^2}{8}\frac{\partial^2 u}{\partial \tau^2}(x_m, \tau^{n+1/2}) + \frac{\Delta\tau^3}{48}\frac{\partial^3 u}{\partial \tau^3}(x_m, \eta_1),$$

$$u(x_m, \tau^n) = u(x_m, \tau^{n+1/2}) - \frac{\Delta\tau}{2}\frac{\partial u}{\partial \tau}(x_m, \tau^{n+1/2})$$
$$+ \frac{\Delta\tau^2}{8}\frac{\partial^2 u}{\partial \tau^2}(x_m, \tau^{n+1/2}) - \frac{\Delta\tau^3}{48}\frac{\partial^3 u}{\partial \tau^3}(x_m, \eta_2),$$

---

[1] In this book $\Delta\tau^2$ stands for $(\Delta\tau)^2$. For $\Delta\tau^3$, Δx^2, Δx^3 etc., the situation is similar.

where $\tau^{n+1/2} < \eta_1 < \tau^{n+1}$ and $\tau^n < \eta_2 < \tau^{n+1/2}$. Subtracting the second equation from the first one, we get

$$\frac{\partial u}{\partial \tau}(x_m, \tau^{n+1/2}) = \frac{u(x_m, \tau^{n+1}) - u(x_m, \tau^n)}{\Delta \tau} - \frac{\Delta \tau^2}{24} \frac{\partial^3 u}{\partial \tau^3}(x_m, \eta_3),$$

where $\tau^n < \eta_3 < \tau^{n+1}$. Then, we have

$$\frac{\partial u}{\partial \tau}(x_m, \tau^{n+1/2}) \approx \frac{u(x_m, \tau^{n+1}) - u(x_m, \tau^n)}{\Delta \tau}.$$

This approximation is called the central finite-difference approximation or a **central difference** for $\dfrac{\partial u}{\partial \tau}$.

Similarly, for $\dfrac{\partial u}{\partial x}(x_m, \tau^n)$ we can have the following approximations

$$\frac{\partial u}{\partial x}(x_m, \tau^n) \approx \frac{u(x_{m+1}, \tau^n) - u(x_{m-1}, \tau^n)}{2\Delta x},$$

$$\frac{\partial u}{\partial x}(x_m, \tau^n) \approx \frac{u(x_m, \tau^n) - u(x_{m-1}, \tau^n)}{\Delta x}$$

and

$$\frac{\partial u}{\partial x}(x_m, \tau^n) \approx \frac{u(x_{m+1}, \tau^n) - u(x_m, \tau^n)}{\Delta x}.$$

The first one is second order and called the **second-order central difference** for first derivatives. The second and third approximations are first order and called the **first-order one-sided difference**. Sometimes, we also need the following **second-order one-sided differences**:

$$\frac{\partial u}{\partial x}(x_m, \tau^n) \approx \frac{3u(x_m, \tau^n) - 4u(x_{m-1}, \tau^n) + u(x_{m-2}, \tau^n)}{2\Delta x}$$

and

$$\frac{\partial u}{\partial x}(x_m, \tau^n) \approx \frac{-3u(x_m, \tau^n) + 4u(x_{m+1}, \tau^n) - u(x_{m+2}, \tau^n)}{2\Delta x}.$$

For the approximation of the second-order partial derivative with respect to x, we use the following Taylor expansions:

$$u(x_{m+1}, \tau^n) = u(x_m, \tau^n) + \Delta x \frac{\partial u}{\partial x}(x_m, \tau^n) + \frac{\Delta x^2}{2} \frac{\partial^2 u}{\partial x^2}(x_m, \tau^n)$$
$$+ \frac{\Delta x^3}{6} \frac{\partial^3 u}{\partial x^3}(x_m, \tau^n) + \frac{\Delta x^4}{24} \frac{\partial^4 u}{\partial x^4}(\xi_1, \tau^n),$$

$$u(x_{m-1}, \tau^n) = u(x_m, \tau^n) - \Delta x \frac{\partial u}{\partial x}(x_m, \tau^n) + \frac{\Delta x^2}{2} \frac{\partial^2 u}{\partial x^2}(x_m, \tau^n)$$
$$- \frac{\Delta x^3}{6} \frac{\partial^3 u}{\partial x^3}(x_m, \tau^n) + \frac{\Delta x^4}{24} \frac{\partial^4 u}{\partial x^4}(\xi_2, \tau^n),$$

where $x_m < \xi_1 < x_{m+1}$ and $x_{m-1} < \xi_2 < x_m$. Adding these two equations, we obtain

$$\frac{\partial^2 u}{\partial x^2}(x_m, \tau^n) = \frac{u(x_{m+1}, \tau^n) - 2u(x_m, \tau^n) + u(x_{m-1}, \tau^n)}{\Delta x^2} - \frac{\Delta x^2}{12}\frac{\partial^4 u}{\partial x^4}(\xi_3, \tau^n),$$

where $x_{m-1} < \xi_3 < x_{m+1}$. Thus, we have the **second-order central difference** for second derivatives:

$$\frac{\partial^2 u}{\partial x^2}(x_m, \tau^n) \approx \frac{u(x_{m+1}, \tau^n) - 2u(x_m, \tau^n) + u(x_{m-1}, \tau^n)}{\Delta x^2}.$$

Sometimes, we also need to have an approximation to mixed second-order partial derivatives. For $\dfrac{\partial^2 u}{\partial x \partial y}$, we have

$$\frac{\partial^2 u}{\partial x \partial y}(x_m, y_l, \tau^n) \approx \frac{1}{2\Delta x}\left[\frac{u(x_{m+1}, y_{l+1}, \tau^n) - u(x_{m+1}, y_{l-1}, \tau^n)}{2\Delta y}\right.$$
$$\left. - \frac{u(x_{m-1}, y_{l+1}, \tau^n) - u(x_{m-1}, y_{l-1}, \tau^n)}{2\Delta y}\right],$$

where $y_l = b + l\Delta y$, Δy being a small number. It is clear that this is a second-order scheme, and this formula is called the **second-order central difference** for mixed second-order partial derivatives.

Pseudo-spectral approximation. By using more points, we can also construct higher order finite-difference approximations for the partial derivatives. An alternative way to obtain higher order approximations for partial derivatives is to use a pseudo-spectral method. To illustrate the method, we consider the approximation to the partial derivatives with respect to x for a fixed τ. Assume that we want to find the solution u in the interval $0 \le x \le 1$. Suppose we use non-equidistant nodes. For example, we can use the following grid points

$$x_m = \frac{1}{2}\left(1 - \cos\frac{m\pi}{M}\right), \quad m = 0, 1, \cdots, M. \tag{5.4}$$

These points are in $[0, 1]$ and equal to $(1 - x_m^*)/2$, x_m^* being the extrema of the M-th order Chebyshev polynomial $T_M(x)$. Here, the M-th order Chebyshev polynomial is defined by $T_M(x) = \cos(M\cos^{-1}x)$. Assume that the solution for a fixed τ is a polynomial in x with degree M. If we require the polynomial to have a value $u(x_m)$ at $x = x_m$, then we can determine the coefficients of the polynomial, each of which is a linear combination of $u(x_i)$, $i = 0, 1, \cdots, M$. Thus, the derivatives of the polynomial at the point x_m is also a linear combination of $u(x_i)$, $i = 0, 1, \cdots, M$, with coefficients depending on x_m and x_i. Therefore,

$$\frac{\partial u}{\partial x}(x_m) = \sum_{i=0}^{M} D_{x,m,i}u(x_i). \tag{5.5}$$

This is an **M-th order approximation to the derivative** with respect to x used in the pseudo-spectral method. If the grid points are given by (5.4), then $D_{x,m,i}$ has the following expression:

$$
D_{x,m,i} = \begin{cases}
\dfrac{c_m(-1)^{m+i}}{c_i(x_m - x_i)}, & m \neq i, \\[2mm]
-\dfrac{2M^2 + 1}{3}, & m = i = 0, \\[2mm]
\dfrac{1 - 2x_i}{4x_i(1 - x_i)}, & m = i = 1, 2, \cdots, M-1, \\[2mm]
\dfrac{2M^2 + 1}{3}, & m = i = M,
\end{cases}
\tag{5.6}
$$

where $c_0 = c_M = 2$ and $c_i = 1$, $i = 1, 2, \cdots, M-1$ (see [33]). Similarly, we have

$$
\begin{aligned}
\frac{\partial^2 u}{\partial x^2}(x_m) &= \sum_{j=0}^{M} D_{x,m,j} \frac{\partial u}{\partial x}(x_j) \\
&= \sum_{j=0}^{M} D_{x,m,j} \left[\sum_{i=0}^{M} D_{x,j,i} u(x_i) \right] \\
&= \sum_{i=0}^{M} \left(\sum_{j=0}^{M} D_{x,m,j} D_{x,j,i} \right) u(x_i) \\
&= \sum_{i=0}^{M} D_{xx,m,i} u(x_i),
\end{aligned}
\tag{5.7}
$$

where

$$
D_{xx,m,i} = \sum_{j=0}^{M} D_{x,m,j} D_{x,j,i}.
\tag{5.8}
$$

When the solution is very smooth, only a small M may be needed in order to get a satisfying result. In such a case, its performance could be better than the finite-difference approximations.

5.1.3 Approximate Integration

Trapezoidal rule. The approximation of the integral

$$
\int_a^b f(x)\,dx
$$

is needed in the numerical solution of integro-differential equations and sometimes in the numerical solution of partial differential equations. The simplest method for the approximation is called the **trapezoidal rule**. Let $h = (b - a)/M$, and $x_m = a + mh$, $m = 0, 1, \cdots, M$. In the subinterval $[x_m, x_{m+1}]$, we use the linear function

$$p_1(x) = \frac{x_{m+1} - x}{h} f(x_m) + \frac{x - x_m}{h} f(x_{m+1})$$

to approximate $f(x)$. Thus,

$$\int_{x_m}^{x_{m+1}} f(x)dx \approx \frac{1}{h} \int_{x_m}^{x_{m+1}} [(x_{m+1} - x)f(x_m) + (x - x_m)f(x_{m+1})]dx$$

$$= \frac{h}{2} [f(x_m) + f(x_{m+1})].$$

Using this for all subintervals, we obtain

$$\int_a^b f(x)dx = \sum_{m=0}^{M-1} \int_{x_m}^{x_{m+1}} f(x)dx$$

$$\approx \sum_{m=0}^{M-1} \frac{h}{2} [f(x_m) + f(x_{m+1})]$$

$$= \frac{h}{2} \left[f(a) + 2 \sum_{m=1}^{M-1} f(x_m) + f(b) \right].$$

The error of the trapezoidal rule is

$$-\frac{(b-a)h^2}{12} f''(\xi),$$

where $\xi \in (a, b)$.

Simpson's rule. Simpson's rule is a better approximation for the integral by using the quadratic interpolation polynomial. In the subinterval $[x_{m-1}, x_{m+1}]$, we use

$$p_2(x) = \frac{(x_m - x)(x_{m+1} - x)}{(x_m - x_{m-1})(x_{m+1} - x_{m-1})} f(x_{m-1})$$

$$+ \frac{(x - x_{m-1})(x_{m+1} - x)}{(x_m - x_{m-1})(x_{m+1} - x_m)} f(x_m)$$

$$+ \frac{(x - x_{m-1})(x - x_m)}{(x_{m+1} - x_{m-1})(x_{m+1} - x_m)} f(x_{m+1})$$

to approximate $f(x)$. Thus

$$\int_{x_{m-1}}^{x_{m+1}} f(x)dx \approx \frac{1}{2h^2} \int_{x_{m-1}}^{x_{m+1}} [(x_m - x)(x_{m+1} - x)f(x_{m-1})$$

$$+2(x - x_{m-1})(x_{m+1} - x)f(x_m)$$
$$+(x - x_{m-1})(x - x_m)f(x_{m+1})]dx$$

$$= \frac{h}{3}[f(x_{m-1}) + 4f(x_m) + f(x_{m+1})].$$

Using this for all subintervals, we obtain

$$\int_a^b f(x)dx \approx \frac{h}{3}\left[f(a) + 2\sum_{m=1}^{M/2-1} f(x_{2m}) + 4\sum_{m=1}^{M/2} f(x_{2m-1}) + f(b) \right].$$

The error of the Simpson's rule is

$$-\frac{(b-a)h^4}{180} f^{(4)}(\xi),$$

where $\xi \in (a,b)$. Obviously, M must be an even number.

5.1.4 Least Squares Approximation

In Subsection 5.1.1 we discussed various interpolations. In those cases, all the given points (x_m, f_m) are on the interpolation function. Here, we will discuss how to find an approximate function satisfying the following two conditions:

A) The number of parameters in the function is less than the number of given points.
B) Let the function have the "best fit" to those given points (x_m, f_m) in some sense.

Let x_m, $m = 0, 1, \cdots, M$, be distinct, and let $M+1$ points (x_m, f_m) be given. We want to find a product of a given function $g(x)$ and a polynomial of degree $N < M$

$$g(x)\sum_{n=0}^{N} a_n x^n$$

such that the value of the total least squares error

$$\sum_{m=0}^{M} b_m \left[f_m - g(x_m)\sum_{n=0}^{N} a_n x_m^n \right]^2$$

has a minimum, where b_m, $m = 0, 1, \cdots, M$, are given positive numbers called the weights. In order to minimize the least squares error, the necessary conditions are

$$\frac{\partial}{\partial a_i}\left\{\sum_{m=0}^{M} b_m \left[f_m - g(x_m)\sum_{n=0}^{N} a_n x_m^n\right]^2\right\}$$

$$= -2\sum_{m=0}^{M} b_m \left[f_m - g(x_m)\sum_{n=0}^{N} a_n x_m^n\right] g(x_m)x_m^i = 0,$$

$$i = 0, 1, \cdots, N.$$

This system can be written as

$$\sum_{n=0}^{N}\left[\sum_{m=0}^{M} b_m g^2(x_m)x_m^{n+i}\right] a_n = \sum_{m=0}^{M} b_m f_m g(x_m)x_m^i, \qquad (5.9)$$

$$i = 0, 1, \cdots, N.$$

It is a linear system for $a_0, a_1, \cdots, a_N$, which is usually known as the system of normal equations. When x_m are distinct, we can find $a_0, a_1, \cdots, a_N$ without any difficulty. This method is usually referred to as the least squares method with weights.

5.2 Solution of Systems and Eigenvalue Problems

5.2.1 LU Decomposition of Linear Systems

One efficient method for solving linear systems is the LU decomposition, which decomposes the matrix $\mathbf{A}$ into a product of a unit lower triangular matrix[2] $\mathbf{L}$ and an upper triangular matrix $\mathbf{U}$, i.e., $\mathbf{A} = \mathbf{L}\mathbf{U}$. In this subsection, we will give the details of the method for linear tridiagonal systems. For more complicated systems, the procedure is similar.

Consider a general linear tridiagonal system

$$\mathbf{A}\mathbf{x} = \mathbf{q}, \qquad (5.10)$$

where

$$\mathbf{A} = \begin{bmatrix} b_1 & c_1 & & & & \\ a_2 & b_2 & c_2 & & 0 & \\ & \ddots & \ddots & \ddots & & \\ & & \ddots & \ddots & \ddots & \\ & 0 & & a_{m-1} & b_{m-1} & c_{m-1} \\ & & & & a_m & b_m \end{bmatrix},$$

---

[2] A matrix $\mathbf{A}$ is called a unit lower triangular matrix if $a_{ii} = 1$ and $a_{ij} = 0$ for $i < j$. If $a_{ij} = 0$ for $i > j$, then the matrix $\mathbf{A}$ is called an upper triangular matrix.

$$\mathbf{x} = \begin{bmatrix} x_1 \\ x_2 \\ \vdots \\ x_m \end{bmatrix}, \qquad \mathbf{q} = \begin{bmatrix} q_1 \\ q_2 \\ \vdots \\ q_m \end{bmatrix}.$$

Let us discuss how to find the solution of this system. The first equation of the system is

$$b_1 x_1 + c_1 x_2 = q_1.$$

Let

$$u_1 = b_1, \quad y_1 = q_1,$$

the equation above can be written as

$$u_1 x_1 + c_1 x_2 = y_1. \tag{5.11}$$

Now suppose we have a relation in the form

$$u_{i-1} x_{i-1} + c_{i-1} x_i = y_{i-1}.$$

We put this relation and the i-th equation of the system together and obtain

$$\begin{cases} u_{i-1} x_{i-1} + c_{i-1} x_i = y_{i-1}, \\ a_i x_{i-1} + b_i x_i + c_i x_{i+1} = q_i. \end{cases}$$

Subtracting a_i / u_{i-1} times the first relation from the second equation, we can eliminate x_{i-1} and have another relation in the same form:

$$\left(b_i - c_{i-1} \frac{a_i}{u_{i-1}} \right) x_i + c_i x_{i+1} = q_i - y_{i-1} \frac{a_i}{u_{i-1}}$$

or

$$u_i x_i + c_i x_{i+1} = y_i,$$

where

$$u_i = b_i - \frac{c_{i-1} a_i}{u_{i-1}}, \quad y_i = q_i - \frac{y_{i-1} a_i}{u_{i-1}}.$$

Because we have (5.11) that is in this form, this procedure can be done for $i = 2, 3, \cdots, m$ successively and generates

$$u_i x_i + c_i x_{i+1} = y_i, \quad i = 2, 3, \cdots, m - 1 \tag{5.12}$$

and

$$u_m x_m = y_m. \tag{5.13}$$

Because in the last equation of the system x_{m+1} does not appear, which means $c_m = 0$, the last relation is in the form (5.13) instead of (5.12). This procedure can be called elimination or forward substitution.

When we obtain (5.13), we can have

$$x_m = \frac{y_m}{u_m}.$$

Furthermore, from (5.11)–(5.12) we can get

$$x_i = \frac{y_i - c_i x_{i+1}}{u_i}, \quad i = m - 1, \cdots, 1$$

successively. This procedure is called back substitution. Through these two procedures, we can obtain the solution of this system.

The elimination procedure can be written in matrix form

$$\mathbf{L}_{m-1} \ldots \mathbf{L}_2 \mathbf{L}_1 \mathbf{A} \mathbf{x} = \mathbf{U} \mathbf{x} = \mathbf{L}_{m-1} \cdots \mathbf{L}_1 \mathbf{q} = \mathbf{y}, \tag{5.14}$$

where

$$\mathbf{L}_i = \begin{bmatrix} 1 & 0 & & \cdots & & & 0 \\ 0 & 1 & \ddots & & & & \\ & \ddots & \ddots & \ddots & & & \\ \vdots & & 0 & 1 & \ddots & & \vdots \\ & & & -l_i & 1 & \ddots & \\ & & & & 0 & \ddots & \ddots \\ & & & & & \ddots & \ddots & 0 \\ 0 & & \cdots & & & & 0 & 1 \end{bmatrix},$$

$$\mathbf{U} = \begin{bmatrix} u_1 & c_1 & 0 & \cdots & & 0 \\ 0 & u_2 & c_2 & \ddots & & \vdots \\ \vdots & \ddots & \ddots & \ddots & 0 \\ \vdots & & \ddots & u_{m-1} & c_{m-1} \\ 0 & \cdots\cdots & 0 & u_m \end{bmatrix}, \quad \mathbf{y} = \begin{bmatrix} y_1 \\ y_2 \\ \vdots \\ y_m \end{bmatrix},$$

l_i equalling a_{i+1}/u_i and being in the i-th column and the $(i+1)$-th row of $\mathbf{L}_i$, $i = 1, 2, \cdots, m - 1$.

Let

$$\mathbf{L} = \begin{bmatrix} 1 & 0 & 0 & \cdots & 0 \\ l_1 & 1 & 0 & & \vdots \\ 0 & l_2 & 1 & \ddots & \vdots \\ \vdots & \ddots & \ddots & \ddots & 0 \\ 0 & \cdots & 0 & l_{m-1} & 1 \end{bmatrix}.$$

It is easy to see

$$\mathbf{L}_{m-1} \mathbf{L}_{m-2} \cdots \mathbf{L}_1 \mathbf{L} = \mathbf{I}.$$

Thus, we have

$$\mathbf{L}_{m-1}\mathbf{L}_{m-2}\cdots\mathbf{L}_1 = \mathbf{L}^{-1}$$

and (5.14) can be written as

$$\mathbf{L}^{-1}\mathbf{Ax} = \mathbf{Ux} = \mathbf{L}^{-1}\mathbf{q} = \mathbf{y}.$$

Consequently,

$$\mathbf{Ax} = \mathbf{LUx} = \mathbf{q}.$$

This means that $\mathbf{A}$ can be decomposed into a unit lower triangular matrix $\mathbf{L}$ multiplied by an upper triangular matrix $\mathbf{U}$. The procedure of solving the system is as follows. We first multiply the equation by $\mathbf{L}^{-1}$ so that the equation becomes $\mathbf{Ux} = \mathbf{L}^{-1}\mathbf{q} = \mathbf{y}$ and then solve $\mathbf{Ux} = \mathbf{y}$ to get $\mathbf{x} = \mathbf{U}^{-1}\mathbf{y}$. Because these two procedures are easy to perform, the method is quite popular.

5.2.2 Iteration Methods for Linear Systems

An alternative to LU decomposition is iteration. Iteration methods are especially effective for large systems with sparse coefficient matrices. Consider the linear system

$$\mathbf{Ax} = \mathbf{q}.$$

$\mathbf{A}$ may be decomposed as

$$\mathbf{A} = \mathbf{D} + \mathbf{L} + \mathbf{U}$$

where

$$\mathbf{L} = \begin{bmatrix} 0 & \cdots & & \cdots & 0 \\ a_{2,1} & 0 & & \cdots & 0 \\ \vdots & \ddots & \ddots & & \vdots \\ a_{m,1} & \cdots & a_{m,m-1} & 0 \end{bmatrix}, \quad \mathbf{U} = \begin{bmatrix} 0 & a_{1,2} & \cdots & & a_{1,m} \\ \vdots & \ddots & \ddots & & \vdots \\ 0 & \cdots & 0 & & a_{m-1,m} \\ 0 & \cdots & \cdots & & 0 \end{bmatrix},$$

and $\mathbf{D} = \mathrm{diag}\{a_{1,1}, a_{2,2}, \cdots, a_{m,m}\}$. Then, the linear system can be rewritten as the following system,

$$\mathbf{x} = \mathbf{D}^{-1}\left[\mathbf{q} - (\mathbf{L} + \mathbf{U})\mathbf{x}\right],$$

where we assume that $\mathbf{D}$ is invertible.

Jacobi iteration. A simple way to find the solution is to use the following iteration:

$$\mathbf{x}^{(k+1)} = \mathbf{D}^{-1}\left[\mathbf{q} - (\mathbf{L} + \mathbf{U})\mathbf{x}^{(k)}\right], \quad k = 0, 1, \cdots,$$

or in component form

$$x_1^{(k+1)} = \frac{1}{a_{1,1}}[q_1 - (a_{1,2}x_2^{(k)} + \cdots + a_{1,m}x_m^{(k)})],$$

$$x_2^{(k+1)} = \frac{1}{a_{2,2}}[q_2 - (a_{2,1}x_1^{(k)} + a_{2,3}x_3^{(k)} + \cdots + a_{2,m}x_m^{(k)})],$$

$$\vdots$$

$$x_m^{(k+1)} = \frac{1}{a_{m,m}}[q_m - (a_{m,1}x_1^{(k)} + \cdots + a_{m,m-1}x_{m-1}^{(k)})].$$

It is clear that in order to implement this iteration, an initial guess $\mathbf{x}^{(0)}$ should be given. This method is called the Jacobi iteration.

Gauss–Seidel iteration. In the Jacobi iteration, at the iteration step for $x_i^{(k+1)}$, all the solutions $x_1^{(k+1)}, \cdots, x_{i-1}^{(k+1)}$ have been obtained. Therefore, new variables can be used in the iteration, namely, we can have the following iteration:

$$x_1^{(k+1)} = \frac{1}{a_{1,1}}[q_1 - (a_{1,2}x_2^{(k)} + \cdots + a_{1,m}x_m^{(k)})],$$

$$x_2^{(k+1)} = \frac{1}{a_{2,2}}[q_2 - (a_{2,1}x_1^{(k+1)} + a_{2,3}x_3^{(k)} + \cdots + a_{2,m}x_m^{(k)})],$$

$$x_3^{(k+1)} = \frac{1}{a_{3,3}}[q_3 - (a_{3,1}x_1^{(k+1)} + a_{3,2}x_2^{(k+1)} + a_{3,4}x_4^{(k)} + \cdots + a_{3,m}x_m^{(k)})],$$

$$\vdots$$

$$x_m^{(k+1)} = \frac{1}{a_{m,m}}[q_m - (a_{m,1}x_1^{(k+1)} + \cdots + a_{m,m-1}x_{m-1}^{(k+1)})]$$

or in matrix form

$$\mathbf{x}^{(k+1)} = \mathbf{D}^{-1}\left[\mathbf{q} - \mathbf{L}\mathbf{x}^{(k+1)} - \mathbf{U}\mathbf{x}^{(k)}\right].$$

This method is called the Gauss–Seidel iteration.

SOR (successive over relaxation). The Gauss–Seidel iteration can be modified in the following way: Take a combination of the previous value of $\mathbf{x}$ and the current update (from the Gauss–Seidel method) as the next approximation:

$$\mathbf{x}^{(k+1)} = (1 - \omega)\mathbf{x}^{(k)} + \omega\mathbf{D}^{-1}\left[\mathbf{q} - \mathbf{L}\mathbf{x}^{(k+1)} - \mathbf{U}\mathbf{x}^{(k)}\right],$$

or in component form

$$x_1^{(k+1)} = (1 - \omega)x_1^{(k)} + \frac{\omega}{a_{1,1}}[q_1 - (a_{1,2}x_2^{(k)} + \cdots + a_{1,m}x_m^{(k)})],$$

$$x_2^{(k+1)} = (1 - \omega)x_2^{(k)} + \frac{\omega}{a_{2,2}}[q_2 - (a_{2,1}x_1^{(k+1)} + a_{2,3}x_3^{(k)} + \cdots + a_{2,m}x_m^{(k)})],$$

$$x_3^{(k+1)} = (1-\omega)x_3^{(k)} + \frac{\omega}{a_{3,3}}[q_3 - (a_{3,1}x_1^{(k+1)} + a_{3,2}x_2^{(k+1)}) + a_{3,4}x_4^{(k)} + \cdots +$$
$$a_{3,m}x_m^{(k)})],$$

$$\vdots$$

$$x_m^{(k+1)} = (1-\omega)x_m^{(k)} + \frac{\omega}{a_{m,m}}[q_m - (a_{m,1}x_1^{(k+1)} + \cdots + a_{m,m-1}x_{m-1}^{(k+1)})].$$

Here, ω is a real number. This method usually is called the method of successive over relaxation (SOR). When $\omega = 1$, it is the Gauss–Seidel iteration. The parameter ω should be chosen so that the method will converge and work better than the Gauss–Seidel iteration. The following result has been proven:

Theorem 5.2 *If* **A** *is a symmetric positive definite matrix and* $0 < \omega < 2$, *then the method of successive over relaxation will converge for any initial vector* **x**.

Practical computation shows that this method also works for some non-symmetric linear systems if ω is chosen properly. For many cases, this method gives faster convergence than the Gauss–Seidel iteration if $\omega \in (1,2)$. We would like to point out that in the books by Golub and Loan [32] and Saad [64], there are some other iteration methods that can also be used for solving linear systems in Chapters 6, 7, and 8. Interested readers are referred to these books.

5.2.3 Iteration Methods for Nonlinear Systems

In the numerical solution of partial differential equations, the resulting algebraic systems are sometimes nonlinear. In this section, we discuss three iteration methods for the nonlinear systems.

Newton's method. Consider the following nonlinear system,

$$f_1(x_1, x_2, \cdots, x_n) = 0,$$
$$f_2(x_1, x_2, \cdots, x_n) = 0,$$

$$\vdots$$

$$f_n(x_1, x_2, \cdots, x_n) = 0.$$

Let

$$\mathbf{x} = \begin{bmatrix} x_1 \\ x_2 \\ \vdots \\ x_n \end{bmatrix}, \quad \mathbf{f}(\mathbf{x}) = \begin{bmatrix} f_1(\mathbf{x}) \\ f_2(\mathbf{x}) \\ \vdots \\ f_n(\mathbf{x}) \end{bmatrix}.$$

Then, the nonlinear system has the form

$$\mathbf{f}(\mathbf{x}) = 0.$$

Suppose $\mathbf{x}^{(0)} = [x_1^0, x_2^0, \cdots, x_n^0]^T$ is a good initial guess to the true solution $\mathbf{x}^* = [x_1^*, x_2^*, \cdots, x_n^*]^T$, i.e.,

$$\delta\mathbf{x} = \mathbf{x}^* - \mathbf{x}^{(0)} = [\delta x_1, \delta x_2, \cdots, \delta x_n]^T$$

is small in norm. Then, for $i = 1, 2, \cdots, n$

$$0 = f_i(x_1^*, x_2^*, \cdots, x_n^*) = f_i(x_1^0 + \delta x_1, x_2^0 + \delta x_2, \cdots, x_n^0 + \delta x_n)$$

$$\approx f_i(x_1^0, x_2^0, \cdots, x_n^0) + \sum_{k=1}^{n} \frac{\partial f_i(x_1^0, x_2^0, \cdots, x_n^0)}{\partial x_k} \delta x_k.$$

In matrix form, we have

$$\begin{bmatrix} \dfrac{\partial f_1(\mathbf{x}^{(0)})}{\partial x_1} & \dfrac{\partial f_1(\mathbf{x}^{(0)})}{\partial x_2} & \cdots & \dfrac{\partial f_1(\mathbf{x}^{(0)})}{\partial x_n} \\ \dfrac{\partial f_2(\mathbf{x}^{(0)})}{\partial x_1} & \dfrac{\partial f_2(\mathbf{x}^{(0)})}{\partial x_2} & \cdots & \dfrac{\partial f_2(\mathbf{x}^{(0)})}{\partial x_n} \\ \cdots & \cdots & \cdots & \cdots \\ \dfrac{\partial f_n(\mathbf{x}^{(0)})}{\partial x_1} & \dfrac{\partial f_n(\mathbf{x}^{(0)})}{\partial x_2} & \cdots & \dfrac{\partial f_n(\mathbf{x}^{(0)})}{\partial x_n} \end{bmatrix} \begin{bmatrix} \delta x_1 \\ \delta x_2 \\ \vdots \\ \delta x_n \end{bmatrix} + \begin{bmatrix} f_1(\mathbf{x}^{(0)}) \\ f_2(\mathbf{x}^{(0)}) \\ \vdots \\ f_n(\mathbf{x}^{(0)}) \end{bmatrix} \approx 0,$$

or

$$\mathbf{J_f}(\mathbf{x}^{(0)}) \cdot \delta\mathbf{x} + \mathbf{f}(\mathbf{x}^{(0)}) \approx 0,$$

where $\mathbf{J_f}(\mathbf{x}^{(0)})$ denotes the above Jacobian matrix. Solving for $\delta\mathbf{x}$ we get

$$\delta\mathbf{x} \approx -[\mathbf{J_f}(\mathbf{x}^{(0)})]^{-1}\mathbf{f}(\mathbf{x}^{(0)})$$

or

$$\mathbf{x}^* \approx \mathbf{x}^{(0)} - [\mathbf{J_f}(\mathbf{x}^{(0)})]^{-1}\mathbf{f}(\mathbf{x}^{(0)}).$$

This means that the vector

$$\mathbf{x}^{(1)} = \mathbf{x}^{(0)} - [\mathbf{J_f}(\mathbf{x}^{(0)})]^{-1}\mathbf{f}(\mathbf{x}^{(0)})$$

will be a better approximation to the solution $\mathbf{x}^*$. In general, suppose $\mathbf{x}^{(k)}$ has been obtained, then

$$\mathbf{x}^{(k+1)} = \mathbf{x}^{(k)} - [\mathbf{J_f}(\mathbf{x}^{(k)})]^{-1}\mathbf{f}(\mathbf{x}^{(k)}). \tag{5.15}$$

This iteration method is called Newton's method. Because finding an inverse of a matrix is time consuming, in the real computation, Newton's method has the form

$$\begin{cases} \mathbf{J_f}(\mathbf{x}^{(k)})\mathbf{y} = -\mathbf{f}(\mathbf{x}^{(k)}), \\ \mathbf{x}^{(k+1)} = \mathbf{x}^{(k)} + \mathbf{y}. \end{cases}$$

Newton's method converges locally with second order. More precisely, it can be proven that the following result holds.

Theorem 5.3 *Let* $\mathbf{x}^*$ *be a solution of* $\mathbf{f}(\mathbf{x}) = 0$. *Assume that* $\mathbf{J_f}(\mathbf{x}^*)$ *is not singular, and that* $f_i(\mathbf{x})$ *has continuous second-order partial derivatives near* $\mathbf{x}^*$. *Then, if* $\mathbf{x}^{(0)}$ *is close enough to* $\mathbf{x}^*$, *Newton's method converges and*

$$\left\| \mathbf{x}^{(k+1)} - \mathbf{x}^* \right\|_\infty \leq C \left\| \mathbf{x}^{(k)} - \mathbf{x}^* \right\|_\infty^2.$$

Generalized secant method. One of weaknesses of Newton's method for solving nonlinear systems is that the Jacobian matrix must be computed at each iteration. The Jacobian matrix associated with a system $\mathbf{f}(\mathbf{x}) = 0$ requires n^2 partial derivatives to be evaluated. In many situations, the exact evaluation of the partial derivatives is inconvenient. This difficulty can be overcome by using finite-difference approximations to the partial derivatives. For example,

$$\frac{\partial f_i(x_1, x_2, \cdots, x_n)}{\partial x_k}$$

$$\approx \frac{1}{\Delta x_k}[f_i(x_1, \cdots, x_k + \Delta x_k, \cdots, x_n) - f_i(x_1, \cdots, x_k, \cdots, x_n)],$$

$$k = 1, 2, \cdots, n,$$

where Δx_k is small in absolute value. This approximation, however, still requires at least n^2 function evaluations to be performed in order to approximate the Jacobian and does not decrease the amount of calculations. Actually, if we have $\mathbf{f}(\mathbf{x})$ at $n+1$ points, then we usually can have an approximate Jacobian at some point. Suppose that we have $\mathbf{x}^{(l)}$ and $\mathbf{f}(\mathbf{x}^{(l)})$, $l = k-n, k-n+1, \cdots, k$. Because

$$\left[\mathbf{f}(\mathbf{x}^{(k-n)}) - \mathbf{f}(\mathbf{x}^{(k)}), \mathbf{f}(\mathbf{x}^{(k-n+1)}) - \mathbf{f}(\mathbf{x}^{(k)}), \cdots, \mathbf{f}(\mathbf{x}^{(k-1)}) - \mathbf{f}(\mathbf{x}^{(k)}) \right]$$

$$\approx \mathbf{J_f}(\mathbf{x}^{(k)}) \left[\mathbf{x}^{(k-n)} - \mathbf{x}^{(k)}, \mathbf{x}^{(k-n+1)} - \mathbf{x}^{(k)}, \cdots, \mathbf{x}^{(k-1)} - \mathbf{x}^{(k)} \right],$$

we have

$$\mathbf{J_f}(\mathbf{x}^{(k)})$$

$$\approx \left[\mathbf{f}(\mathbf{x}^{(k-n)}) - \mathbf{f}(\mathbf{x}^{(k)}), \mathbf{f}(\mathbf{x}^{(k-n+1)}) - \mathbf{f}(\mathbf{x}^{(k)}), \cdots, \mathbf{f}(\mathbf{x}^{(k-1)}) - \mathbf{f}(\mathbf{x}^{(k)}) \right]$$

$$\times \left[\mathbf{x}^{(k-n)} - \mathbf{x}^{(k)}, \mathbf{x}^{(k-n+1)} - \mathbf{x}^{(k)}, \cdots, \mathbf{x}^{(k-1)} - \mathbf{x}^{(k)} \right]^{-1}.$$

Therefore, Newton's method can be modified to

$$\mathbf{x}^{(k+1)} = \mathbf{x}^{(k)} - \left[\mathbf{x}^{(k-n)} - \mathbf{x}^{(k)}, \mathbf{x}^{(k-n+1)} - \mathbf{x}^{(k)}, \cdots, \mathbf{x}^{(k-1)} - \mathbf{x}^{(k)} \right]$$

$$\times \left[\mathbf{f}(\mathbf{x}^{(k-n)}) - \mathbf{f}(\mathbf{x}^{(k)}), \mathbf{f}(\mathbf{x}^{(k-n+1)}) - \mathbf{f}(\mathbf{x}^{(k)}), \right.$$

$$\left. \cdots, \mathbf{f}(\mathbf{x}^{(k-1)}) - \mathbf{f}(\mathbf{x}^{(k)}) \right]^{-1} \mathbf{f}(\mathbf{x}^{(k)}). \qquad (5.16)$$

Consequently, if we have $n+1$ guesses $\mathbf{x}^{(l)}$, $l = 0, 1, \cdots, n$, and the values of the function $\mathbf{f}(\mathbf{x})$ at these points, then we can do the iteration (5.16) for

$k = n, n + 1, \cdots$ and at each iteration we spend very little time to calculate a Jacobian. Of course, it needs to be guaranteed that the matrix

$$\left[\mathbf{f}(\mathbf{x}^{(k-n)}) - \mathbf{f}(\mathbf{x}^{(k)}), \cdots, \mathbf{f}(\mathbf{x}^{(k-1)}) - \mathbf{f}(\mathbf{x}^{(k)})\right]$$

is invertible. If during the iteration this matrix is not invertible, we need to find the guess $\mathbf{x}^{(k+1)}$ that is close to $\mathbf{x}^{(k)}$ in another way, for example, by changing a component of $\mathbf{x}^{(k)}$ a little bit. In practice, it happens very seldom.

If $n = 1$, then the vectors $\mathbf{x}$ and $\mathbf{f}$ becomes scalars x and f and (5.16) becomes

$$x^{(k+1)} = x^{(k)} - \frac{(x^{(k-1)} - x^{(k)})f(x^{(k)})}{f(x^{(k-1)}) - f(x^{(k)})}. \tag{5.17}$$

Thus, if we have two initial guesses $x^{(0)}$ and $x^{(1)}$, we can do this iteration starting from $k = 1$. This method is called the secant method, and (5.16) is referred to as the generalized secant method. Under some conditions, for (5.16) we can prove that the following relation holds:

$$\left\|\mathbf{x}^{(k+1)} - \mathbf{x}^*\right\|_\infty \leq C\left\|\mathbf{x}^{(k)} - \mathbf{x}^*\right\|_\infty^2 + C \sup_{1 \leq l \leq n} \left\|\mathbf{x}^{(k-l)} - \mathbf{x}^{(k)}\right\|_\infty \left\|\mathbf{f}(\mathbf{x}^{(k)})\right\|_\infty$$

for $k = n, n + 1, \cdots$, where C is a constant (see [88]).

Broyden's method. There are some other ways to avoid calculating the Jacobian for each iteration except for the first iteration. Another weakness of Newton's method is that an $n \times n$ linear system has to be solved at each iteration, which usually requires $O(n^3)$ arithmetic calculations. Here, we introduce Broyden's method, which avoids calculating the Jacobian at each iteration and reduces the number of arithmetic calculations to $O(n^2)$ at each iteration if we get the inverse of the matrix for the first iteration.

Suppose that an initial approximation $\mathbf{x}^{(0)}$ is given, and $\mathbf{x}^{(1)}$ is computed by Newton's method

$$\mathbf{x}^{(1)} = \mathbf{x}^{(0)} - [\mathbf{J_f}(\mathbf{x}^{(0)})]^{-1}\mathbf{f}(\mathbf{x}^{(0)}).$$

In order to get $\mathbf{x}^{(2)}$, we replace the matrix $\mathbf{J_f}(\mathbf{x}^{(1)})$ in Newton's method by a matrix $\mathbf{A}_1$ satisfying

$$\mathbf{A}_1(\mathbf{x}^{(1)} - \mathbf{x}^{(0)}) = \mathbf{f}(\mathbf{x}^{(1)}) - \mathbf{f}(\mathbf{x}^{(0)})$$

and

$$\mathbf{A}_1\mathbf{z} = \mathbf{J_f}(\mathbf{x}^{(0)})\mathbf{z} \quad \text{whenever} \quad (\mathbf{x}^{(1)} - \mathbf{x}^{(0)})^T\mathbf{z} = 0.$$

From these conditions, it can be proven that

$$\mathbf{A}_1 = \mathbf{J_f}(\mathbf{x}^{(0)}) + \frac{\mathbf{f}(\mathbf{x}^{(1)}) - \mathbf{f}(\mathbf{x}^{(0)}) - \mathbf{J_f}(\mathbf{x}^{(0)})(\mathbf{x}^{(1)} - \mathbf{x}^{(0)})}{\left\|\mathbf{x}^{(1)} - \mathbf{x}^{(0)}\right\|_2^2}(\mathbf{x}^{(1)} - \mathbf{x}^{(0)})^T.$$

Using this matrix in place of $\mathbf{J_f}(\mathbf{x}^{(1)})$, we have

$$\mathbf{x}^{(2)} = \mathbf{x}^{(1)} - \mathbf{A}_1^{-1}\mathbf{f}(\mathbf{x}^{(1)}).$$

In general, suppose we have $\mathbf{x}^{(i-1)}$, $\mathbf{x}^{(i)}$ and $\mathbf{A}_{i-1}$, then we can have $\mathbf{x}^{(i+1)}$ by

$$\mathbf{A}_i = \mathbf{A}_{i-1} + \frac{\mathbf{y}^{(i)} - \mathbf{A}_{i-1}\mathbf{s}^{(i)}}{\left\|\mathbf{s}^{(i)}\right\|_2^2}(\mathbf{s}^{(i)})^T,$$

and

$$\mathbf{x}^{(i+1)} = \mathbf{x}^{(i)} - \mathbf{A}_i^{-1}\mathbf{f}(\mathbf{x}^{(i)}),$$

where $\mathbf{y}^{(i)} = \mathbf{f}(\mathbf{x}^{(i)}) - \mathbf{f}(\mathbf{x}^{(i-1)})$ and $\mathbf{s}^{(i)} = \mathbf{x}^{(i)} - \mathbf{x}^{(i-1)}$. However, at each iteration step, the linear system

$$\mathbf{A}_i\mathbf{s}^{(i+1)} = -\mathbf{f}(\mathbf{x}^{(i)})$$

still needs to be solved. To further improve the method, we need the following theorem.

Theorem 5.4 *If $\mathbf{A} \in \mathbb{R}^{n\times n}$ is nonsingular, $\mathbf{x}, \mathbf{y} \in \mathbb{R}^n$, and $\mathbf{y}^T\mathbf{A}^{-1}\mathbf{x} \neq -1$, then $\mathbf{A} + \mathbf{x}\mathbf{y}^T$ is also nonsingular, moreover,*

$$(\mathbf{A} + \mathbf{x}\mathbf{y}^T)^{-1} = \mathbf{A}^{-1} - \frac{\mathbf{A}^{-1}\mathbf{x}\mathbf{y}^T\mathbf{A}^{-1}}{1 + \mathbf{y}^T\mathbf{A}^{-1}\mathbf{x}}.$$

This theorem suggests a simple way to find the inverse of $\mathbf{A}_i$. By setting

$$\mathbf{A} = \mathbf{A}_{i-1},$$
$$\mathbf{x} = \frac{\mathbf{y}^{(i)} - \mathbf{A}_{i-1}\mathbf{s}^{(i)}}{\left\|\mathbf{s}^{(i)}\right\|_2^2},$$
$$\mathbf{y} = \mathbf{s}^{(i)},$$
$$(\mathbf{A} + \mathbf{x}\mathbf{y}^T) = \mathbf{A}_i$$

in the above theorem, we have

$$\mathbf{A}_i^{-1} = \mathbf{A}_{i-1}^{-1} - \frac{\mathbf{A}_{i-1}^{-1}\left(\dfrac{\mathbf{y}^{(i)} - \mathbf{A}_{i-1}\mathbf{s}^{(i)}}{\left\|\mathbf{s}^{(i)}\right\|_2^2}(\mathbf{s}^{(i)})^T\right)\mathbf{A}_{i-1}^{-1}}{1 + (\mathbf{s}^{(i)})^T\mathbf{A}_{i-1}^{-1}\left(\dfrac{\mathbf{y}^{(i)} - \mathbf{A}_{i-1}\mathbf{s}^{(i)}}{\left\|\mathbf{s}^{(i)}\right\|_2^2}\right)}$$

$$= \mathbf{A}_{i-1}^{-1} - \frac{(\mathbf{A}_{i-1}^{-1}\mathbf{y}^{(i)} - \mathbf{s}^{(i)})(\mathbf{s}^{(i)})^T\mathbf{A}_{i-1}^{-1}}{\left\|\mathbf{s}^{(i)}\right\|_2^2 + (\mathbf{s}^{(i)})^T\mathbf{A}_{i-1}^{-1}\mathbf{y}^{(i)} - \left\|\mathbf{s}^{(i)}\right\|_2^2}$$

$$= \mathbf{A}_{i-1}^{-1} + \frac{(\mathbf{s}^{(i)} - \mathbf{A}_{i-1}^{-1}\mathbf{y}^{(i)})(\mathbf{s}^{(i)})^T\mathbf{A}_{i-1}^{-1}}{(\mathbf{s}^{(i)})^T\mathbf{A}_{i-1}^{-1}\mathbf{y}^{(i)}}.$$

This computation requires only $O(n^2)$ arithmetic calculations because it involves only matrix-vector multiplications. Therefore, we have the following Broyden's method:

- Given initial guess $\mathbf{x}^{(0)}$, compute $\mathbf{A}_0^{-1} = [\mathbf{J}_{\mathbf{f}}(\mathbf{x}^{(0)})]^{-1}$ and $\mathbf{x}^{(1)}$.
- For $i = 1, 2, \cdots$, do the following:

$$
\begin{cases}
\mathbf{y}^{(i)} = \mathbf{f}(\mathbf{x}^{(i)}) - \mathbf{f}(\mathbf{x}^{(i-1)}), \qquad \mathbf{s}^{(i)} = \mathbf{x}^{(i)} - \mathbf{x}^{(i-1)}, \\
\mathbf{A}_i^{-1} = \mathbf{A}_{i-1}^{-1} + \dfrac{(\mathbf{s}^{(i)} - \mathbf{A}_{i-1}^{-1}\mathbf{y}^{(i)})(\mathbf{s}^{(i)})^T \mathbf{A}_{i-1}^{-1}}{(\mathbf{s}^{(i)})^T \mathbf{A}_{i-1}^{-1}\mathbf{y}^{(i)}}, \\
\mathbf{x}^{(i+1)} = \mathbf{x}^{(i)} - \mathbf{A}_i^{-1}\mathbf{f}(\mathbf{x}^{(i)}).
\end{cases}
\tag{5.18}
$$

Broyden's method reduces a large amount of work from Newton's method. However, the quadratic convergence of Newton's method is lost. For Broyden's method, we have

$$
\lim_{i \to \infty} \frac{\left\|\mathbf{x}^{(i+1)} - \mathbf{x}^*\right\|}{\left\|\mathbf{x}^{(i)} - \mathbf{x}^*\right\|} = 0.
$$

This type of convergence is called **superlinear**.

5.2.4 Obtaining Eigenvalues and Eigenvectors

In this subsection, we will discuss how to get eigenvalues and eigenvectors of a square matrix, especially, a symmetric matrix. Before that, we introduce some basic tools we will need.

Consider an $m \times m$ matrix in the form

$$
\mathbf{H}_m = \mathbf{I}_m - \alpha \mathbf{v}\mathbf{v}^T,
$$

where $\mathbf{I}_m$ is an $m \times m$ identity matrix, $\mathbf{v}$ is an m-dimensional vector, and α is a number. Obviously, $\mathbf{H}_m$ is a symmetric matrix. We also want $\mathbf{H}_m$ to be orthogonal, namely,

$$
\begin{aligned}
\mathbf{H}_m^T \mathbf{H}_m &= \left(\mathbf{I}_m - \alpha \mathbf{v}\mathbf{v}^T\right)\left(\mathbf{I}_m - \alpha \mathbf{v}\mathbf{v}^T\right) \\
&= \mathbf{I}_m - 2\alpha \mathbf{v}\mathbf{v}^T + \alpha^2 \mathbf{v}\mathbf{v}^T \mathbf{v}\mathbf{v}^T \\
&= \mathbf{I}_m - \left(2\alpha - \alpha^2 \mathbf{v}^T\mathbf{v}\right)\mathbf{v}\mathbf{v}^T = \mathbf{I}_m.
\end{aligned}
$$

Therefore, we require

$$
\alpha = \frac{2}{\mathbf{v}^T\mathbf{v}}
$$

and

$$
\mathbf{H}_m = \mathbf{I}_m - \frac{2}{\mathbf{v}^T\mathbf{v}}\mathbf{v}\mathbf{v}^T.
\tag{5.19}
$$

The matrix defined by (5.19) is called a Householder matrix. We are especially interested in the Householder matrix satisfying

$$\mathbf{H}_m \mathbf{x} = \beta \mathbf{e}_1, \tag{5.20}$$

where $\mathbf{x} = [x_1, x_2, \cdots, x_m]^T$ is an m-dimensional vector, β is a number whose value may depend on the components of $\mathbf{x}$, and $\mathbf{e}_1 = [1, 0, \cdots, 0]^T$. Because

$$\mathbf{H}_m \mathbf{x} = \mathbf{x} - \frac{2}{\mathbf{v}^T \mathbf{v}} \mathbf{v} \mathbf{v}^T \mathbf{x} = \mathbf{x} - \frac{2 \mathbf{v}^T \mathbf{x}}{\mathbf{v}^T \mathbf{v}} \mathbf{v} = \beta \mathbf{e}_1,$$

we have

$$\mathbf{u} \equiv \frac{2 \mathbf{v}^T \mathbf{x}}{\mathbf{v}^T \mathbf{v}} \mathbf{v} = \mathbf{x} - \beta \mathbf{e}_1 \tag{5.21}$$

and

$$\mathbf{u}^T \mathbf{x} = \mathbf{x}^T \mathbf{x} - \beta x_1, \quad \mathbf{u}^T \mathbf{u} = \mathbf{x}^T \mathbf{x} - 2\beta x_1 + \beta^2.$$

Therefore, we further obtain

$$\mathbf{H}_m \mathbf{x} = \left(\mathbf{I}_m - \frac{2}{\mathbf{v}^T \mathbf{v}} \mathbf{v} \mathbf{v}^T \right) \mathbf{x} = \left(\mathbf{I}_m - \frac{2}{\mathbf{u}^T \mathbf{u}} \mathbf{u} \mathbf{u}^T \right) \mathbf{x}$$

$$= \mathbf{x} - \frac{2 \mathbf{u}^T \mathbf{x}}{\mathbf{u}^T \mathbf{u}} \mathbf{u} = \left(1 - \frac{2 \mathbf{u}^T \mathbf{x}}{\mathbf{u}^T \mathbf{u}} \right) \mathbf{x} + \frac{2 \mathbf{u}^T \mathbf{x}}{\mathbf{u}^T \mathbf{u}} \beta \mathbf{e}_1.$$

Because we want (5.20) to hold, we require

$$1 - \frac{2 \mathbf{u}^T \mathbf{x}}{\mathbf{u}^T \mathbf{u}} = 1 - \frac{2 \left(\mathbf{x}^T \mathbf{x} - \beta x_1 \right)}{\mathbf{x}^T \mathbf{x} - 2\beta x_1 + \beta^2} = 0$$

or

$$\beta = \pm \sqrt{\mathbf{x}^T \mathbf{x}}. \tag{5.22}$$

Usually, we take the $+$ sign so that the first component of the vector $\mathbf{H}_m \mathbf{x}$ is nonnegative. In this case

$$\mathbf{H}_m = \mathbf{I}_m - \frac{2}{\mathbf{u}^T \mathbf{u}} \mathbf{u} \mathbf{u}^T = \mathbf{I}_m - \frac{1}{\beta(\beta - x_1)} \mathbf{u} \mathbf{u}^T, \tag{5.23}$$

where $\mathbf{u}$ and β are given by (5.21) and (5.22).

An $n \times n$ matrix

$$\mathbf{A} = \begin{bmatrix} a_{1,1} & a_{1,2} & \cdots & a_{1,n} \\ a_{2,1} & a_{2,2} & \cdots & a_{2,n} \\ \vdots & \vdots & \ddots & \vdots \\ a_{n,1} & a_{n,2} & \cdots & a_{n,n} \end{bmatrix}$$

is called an upper triangular matrix if $a_{ij} = 0$ for $i > j$ and an upper Hessenberg matrix if $a_{ij} = 0$ for $i > j + 1$. Because a Householder matrix defined by (5.23) has the property (5.20), it can be used to reduce a matrix $\mathbf{A}$ to an upper triangular matrix or an upper Hessenberg matrix, which will be described below. Based on this fact, we can have the so-called $\mathbf{QR}$ algorithm for finding the eigenvalues of a matrix.

The first step of the $\mathbf{QR}$ algorithm for finding the eigenvalues of a matrix $\mathbf{A}$ is to reduce the matrix to an upper Hessenberg matrix. Let $\mathbf{P}_k$ be an $n \times n$ matrix in the form:

$$\mathbf{P}_k = \begin{bmatrix} \mathbf{I}_k & 0 \\ 0 & \mathbf{H}_{n-k} \end{bmatrix},$$

where k is equal to $0, 1, \cdots$, or $n-2$, $\mathbf{H}_{n-k}$ is an $(n-k) \times (n-k)$ matrix defined by (5.23). Clearly, $\mathbf{P}_k$ is a Householder matrix. Suppose that after using $k-1$ Householder transformations, $\mathbf{A}$ is changed to

$$\mathbf{A}_{k-1} = (\mathbf{P}_1 \cdots \mathbf{P}_{k-1})^T \mathbf{A} (\mathbf{P}_1 \cdots \mathbf{P}_{k-1}) = \begin{bmatrix} \mathbf{C}_{11} & \mathbf{C}_{12} & \mathbf{C}_{13} \\ \mathbf{C}_{21} & \mathbf{C}_{22} & \mathbf{C}_{23} \\ 0 & \mathbf{C}_{32} & \mathbf{C}_{33} \end{bmatrix},$$

where

$$\begin{bmatrix} \mathbf{C}_{11} & \mathbf{C}_{12} \\ \mathbf{C}_{21} & \mathbf{C}_{22} \end{bmatrix}$$

is a $k \times k$ upper Hessenberg matrix and $\mathbf{C}_{32}$ is a column vector. Now let us define $\mathbf{A}_k = \mathbf{P}_k^T \mathbf{A}_{k-1} \mathbf{P}_k$, and from the forms of $\mathbf{A}_{k-1}$ and $\mathbf{P}_k$ we have

$$\mathbf{A}_k = \mathbf{P}_k^T \mathbf{A}_{k-1} \mathbf{P}_k = \mathbf{P}_k \mathbf{A}_{k-1} \mathbf{P}_k = \begin{bmatrix} \mathbf{C}_{11} & \mathbf{C}_{12} & \mathbf{C}_{13} \mathbf{H}_{n-k} \\ \mathbf{C}_{21} & \mathbf{C}_{22} & \mathbf{C}_{23} \mathbf{H}_{n-k} \\ 0 & \mathbf{H}_{n-k} \mathbf{C}_{32} & \mathbf{H}_{n-k} \mathbf{C}_{33} \mathbf{H}_{n-k} \end{bmatrix}.$$

If we take $\mathbf{C}_{32}$ as $\mathbf{x}$ and determine $\mathbf{H}_{n-k}$, then we arrive at

$$\mathbf{H}_{n-k} \mathbf{C}_{32} = \begin{bmatrix} \beta, 0, \cdots, 0 \end{bmatrix}^T.$$

Therefore, the $(k+1) \times (k+1)$ submatrix at the upper-left corner of $\mathbf{A}_k$ is an upper Hessenberg matrix, and the procedure can continue for $k+1$. For $k = 1$, this procedure can be done. Consequently, we can do this procedure from $k = 1$ to $n - 2$, and finally obtain an upper Hessenberg matrix:

$$\mathbf{A}_{n-2} = (\mathbf{P}_1 \cdots \mathbf{P}_{n-2})^T \mathbf{A} (\mathbf{P}_1 \cdots \mathbf{P}_{n-2}). \tag{5.24}$$

Now let us discuss the second step. If the procedure above starts from $\mathbf{P}_0$ and a matrix is multiplied only by $\mathbf{P}_k^T$ from the left-hand side, then we will obtain an upper triangular matrix with nonnegative main diagonal entries. Therefore, for any matrix $\mathbf{B}$, we can find an orthogonal matrix $\mathbf{Q}^T$ such that $\mathbf{Q}^T \mathbf{B} = \mathbf{R}$ or $\mathbf{B} = \mathbf{QR}$, where $\mathbf{R}$ is an upper triangular matrix with nonnegative main diagonal entries. This procedure is called QR factorization. Using the QR factorization and letting $\mathbf{B}_1 = \mathbf{A}_{n-2}$, we have the following iteration:

$$\mathbf{B}_k = \mathbf{Q}_k \mathbf{R}_k,$$
$$\mathbf{B}_{k+1} = \mathbf{R}_k \mathbf{Q}_k = \mathbf{Q}_k^T \mathbf{B}_k \mathbf{Q}_k \tag{5.25}$$

for $k = 1, 2, \cdots$. That is, first get $\mathbf{Q}_k$ and $\mathbf{R}_k$ from $\mathbf{B}_k$ and then multiplying $\mathbf{R}_k$ by $\mathbf{Q}_k$ from the right-hand side yields $\mathbf{B}_{k+1}$. For this iteration, we have the following relation

$$\begin{aligned}
\mathbf{B}_{k+1} &= \mathbf{Q}_k^T \mathbf{B}_k \mathbf{Q}_k = \mathbf{Q}_k^T \cdots \mathbf{Q}_1^T \mathbf{A}_{n-2} \mathbf{Q}_1 \cdots \mathbf{Q}_k \\
&= (\mathbf{Q}_1 \cdots \mathbf{Q}_k)^T \mathbf{A}_{n-2} (\mathbf{Q}_1 \cdots \mathbf{Q}_k) \\
&= (\mathbf{P}_1 \cdots \mathbf{P}_{n-2} \mathbf{Q}_1 \cdots \mathbf{Q}_k)^T \mathbf{A} (\mathbf{P}_1 \cdots \mathbf{P}_{n-2} \mathbf{Q}_1 \cdots \mathbf{Q}_k),
\end{aligned}$$

or

$$\mathbf{B}_{k+1} = \mathbf{S}_k^T \mathbf{A} \mathbf{S}_k,$$

where

$$\mathbf{S}_k = \mathbf{P}_1 \cdots \mathbf{P}_{n-2} \mathbf{Q}_1 \cdots \mathbf{Q}_k.$$

Let $\mathbf{B}$ and $\mathbf{S}$ be the limits of $\mathbf{B}_{k+1}$ and $\mathbf{S}_k$ as $k \to \infty$ respectively, then we have

$$\mathbf{B} = \mathbf{S}^T \mathbf{A} \mathbf{S}.$$

The goal of the iteration is to find an upper triangular matrix that is similar to $\mathbf{A}$, so that we can have the eigenvalues of $\mathbf{A}$ from the main diagonal entries of the upper triangular matrix. From the relation (5.25), we can see as follows. First, we get an upper triangular matrix by multiplying an orthogonal matrix from the left-hand side, but in order to let the new matrix be similar to the old one, multiplying the same orthogonal matrix from the right-hand side is needed, which may destroy the goal of finding an upper triangular matrix. However, under certain conditions it will be proven that the limit $\mathbf{B}$ is an upper triangular matrix. Therefore, we may reach our goal at the end of the iteration.

In order to find the eigenvectors of $\mathbf{A}$, we first need to find the eigenvectors of $\mathbf{B}$. As soon as we find the eigenvectors of $\mathbf{B}$, the eigenvectors of $\mathbf{A}$ can be obtained through multiplying the eigenvectors of $\mathbf{B}$ from the left-hand side by $\mathbf{S}$. If $\mathbf{A}$ is symmetric, then $\mathbf{B}$ is diagonal and every column of $\mathbf{S}$ is an eigenvector of $\mathbf{A}$.

For the convergence of the iteration we have

Theorem 5.5 *Assume that the eigenvalues of $\mathbf{B}_1$ have distinct absolute values, and $\mathbf{X}^{-1}$ has an* **LU** *decomposition, where $\mathbf{X}$ is the matrix of eigenvectors. Then, $\mathbf{B}_k$ converges to an upper triangular matrix.*

Proof.   Suppose $\mathbf{X}$ has the decomposition

$$\mathbf{X} = \mathbf{Q}_x \mathbf{R}_x,$$

where $\mathbf{Q}_x$ is orthogonal and $\mathbf{R}_x$ is upper triangular with positive main diagonal entries. Then, we have

$$\begin{aligned}
\mathbf{B}_1^k &= \mathbf{X} \mathbf{\Lambda}^k \mathbf{X}^{-1} = \mathbf{X}(\mathbf{\Lambda}^k \mathbf{L} \mathbf{\Lambda}^{-k}) \mathbf{\Lambda}^k \mathbf{U} \\
&= \mathbf{Q}_x \mathbf{R}_x (\mathbf{I} + \mathbf{E}_k) \mathbf{\Lambda}^k \mathbf{U} \\
&= \mathbf{Q}_x (\mathbf{I} + \mathbf{R}_x \mathbf{E}_k \mathbf{R}_x^{-1}) \mathbf{R}_x \mathbf{\Lambda}^k \mathbf{U},
\end{aligned}$$

where $\mathbf{\Lambda}$ is the Jordan canonical matrix of $\mathbf{B}_1$ and $\mathbf{E}_k = \mathbf{\Lambda}^k \mathbf{L} \mathbf{\Lambda}^{-k} - \mathbf{I} \to 0$ as $k \to \infty$ because we assume $|\lambda_1| > |\lambda_2| > \cdots |\lambda_n|$, $|\lambda_i|$ being an eigenvalue of $\mathbf{B}_1$. Let

$$\mathbf{I} + \mathbf{R}_x \mathbf{E}_k \mathbf{R}_x^{-1} = \mathbf{Q}^{(k)} \mathbf{R}^{(k)},$$

where $\mathbf{Q}^{(k)}$ is orthogonal and $\mathbf{R}^{(k)}$ is upper triangular with positive main diagonal entries. Obviously,

$$\mathbf{Q}^{(k)} \to \mathbf{I}, \ \mathbf{R}^{(k)} \to \mathbf{I}.$$

Let $\mathbf{D}$ and $\mathbf{D}_u$ be diagonal matrices defined by

$$\mathbf{D} = \mathrm{diag}(\lambda_1/|\lambda_1|, \cdots, \lambda_n/|\lambda_n|),$$
$$\mathbf{D}_u = \mathrm{diag}(u_{11}/|u_{11}|, \cdots, u_{nn}/|u_{nn}|),$$

where u_{ii}, $i = 1, \cdots, n$, are the main diagonal entries of $\mathbf{U}$. Then, we have

$$\mathbf{B}_1^k = \mathbf{Q}_x \mathbf{Q}^{(k)} \mathbf{R}^{(k)} \mathbf{R}_x \mathbf{\Lambda}^k \mathbf{U}$$
$$= (\mathbf{Q}_x \mathbf{Q}^{(k)} \mathbf{D}_u \mathbf{D}^k)(\mathbf{D}^{-k} \mathbf{D}_u^{-1} \mathbf{R}^{(k)} \mathbf{R}_x \mathbf{\Lambda}^k \mathbf{U}).$$

Because a product of two upper triangular matrices is an upper triangular matrix, and because a main diagonal entry of the new matrix is the product of the corresponding main diagonal entries in each original matrix, this is a QR decomposition of $\mathbf{B}_1^k$ and the upper triangular matrix $\mathbf{D}^{-k} \mathbf{D}_u^{-1} \mathbf{R}^{(k)} \mathbf{R}_x \mathbf{\Lambda}^k \mathbf{U}$ has positive main diagonal entries. On the other hand, it can be shown that

$$\mathbf{B}_1^k = \hat{\mathbf{Q}}_k \hat{\mathbf{R}}_k,$$

where

$$\hat{\mathbf{Q}}_k = \mathbf{Q}_1 \cdots \mathbf{Q}_k, \ \hat{\mathbf{R}}_k = \mathbf{R}_k \cdots \mathbf{R}_1.$$

In fact

$$\mathbf{B}_k = \hat{\mathbf{Q}}_{k-1}^T \mathbf{B}_1 \hat{\mathbf{Q}}_{k-1}$$

or

$$\mathbf{B}_1 \hat{\mathbf{Q}}_{k-1} = \hat{\mathbf{Q}}_{k-1} \mathbf{B}_k = \hat{\mathbf{Q}}_{k-1} \mathbf{Q}_k \mathbf{R}_k = \hat{\mathbf{Q}}_k \mathbf{R}_k.$$

Multiplying $\hat{\mathbf{R}}_{k-1}$ from the right-hand side on both sides of the relation $\hat{\mathbf{Q}}_k \mathbf{R}_k = \mathbf{B}_1 \hat{\mathbf{Q}}_{k-1}$, we get

$$\hat{\mathbf{Q}}_k \hat{\mathbf{R}}_k = \mathbf{B}_1 \hat{\mathbf{Q}}_{k-1} \hat{\mathbf{R}}_{k-1}$$

and furthermore we obtain

$$\hat{\mathbf{Q}}_k \hat{\mathbf{R}}_k = \mathbf{B}_1 \hat{\mathbf{Q}}_{k-1} \hat{\mathbf{R}}_{k-1} = \mathbf{B}_1^2 \hat{\mathbf{Q}}_{k-2} \hat{\mathbf{R}}_{k-2} = \cdots = \mathbf{B}_1^k.$$

Therefore, we have another QR decomposition of $\mathbf{B}_1^k$. Because the QR decomposition is unique, we have

$$\hat{\mathbf{Q}}_k = \mathbf{Q}_x \mathbf{Q}^{(k)} \mathbf{D}_u \mathbf{D}^k, \ \hat{\mathbf{R}}_k = \mathbf{D}^{-k} \mathbf{D}_u^{-1} \mathbf{R}^{(k)} \mathbf{R}_x \boldsymbol{\Lambda}^k \mathbf{U}.$$

Therefore,

$$
\begin{aligned}
\mathbf{B}_{k+1} &= (\mathbf{D}^T)^k \mathbf{D}_u^T (\mathbf{Q}^{(k)})^T \mathbf{Q}_x^T \mathbf{B}_1 \mathbf{Q}_x \mathbf{Q}^{(k)} \mathbf{D}_u \mathbf{D}^k \\
&= (\mathbf{D}^T)^k \mathbf{D}_u^T (\mathbf{Q}^{(k)})^T \mathbf{Q}_x^T \mathbf{Q}_x \mathbf{R}_x \boldsymbol{\Lambda} \mathbf{R}_x^{-1} \mathbf{Q}_x^{-1} \mathbf{Q}_x \mathbf{Q}^{(k)} \mathbf{D}_u \mathbf{D}^k \\
&= (\mathbf{D}^T)^k \mathbf{D}_u^T (\mathbf{Q}^{(k)})^T \mathbf{R}_x \boldsymbol{\Lambda} \mathbf{R}_x^{-1} \mathbf{Q}^{(k)} \mathbf{D}_u \mathbf{D}^k.
\end{aligned}
$$

Because $\mathbf{Q}^{(k)} \to \mathbf{I}$ and an inverse of an upper triangular matrix is still an upper triangular matrix, $\mathbf{B}_{k+1}$ converges to an upper triangular matrix. ∎

From the proof, we can see that it is not necessary for $\mathbf{B}_1$ to be an upper Hessenberg matrix. Having a Hessenberg matrix at the first step is for the practical reason of reducing computational cost. If $\mathbf{B}_k$ is in upper Hessenberg form, then $\mathbf{B}_{k+1}$ is also in upper Hessenberg form. Thus, in the entire iteration process, we deal with upper Hessenberg matrices. For an upper Hessenberg matrix, the amount of computational work at each step of the QR factorization is $O(n^2)$, which is much smaller than $O(n^3)$ for a full matrix. In order to make computation faster, we can also speed up the convergence of the **QR** algorithm by combining the shifting technique. In addition, there are some other methods for finding eigenvalues of a matrix, for example, the Jacobi algorithm. The details of the shifting technique and other methods can be found, for example, from the book [32].

5.3 Finite-Difference Methods

In this section, we will discuss the finite-difference methods for parabolic partial differential equation problems (parabolic PDE problems). Usually, a parabolic partial differential equation problem is formulated as follows:

$$
\begin{cases}
\dfrac{\partial u}{\partial \tau} = a(x,\tau) \dfrac{\partial^2 u}{\partial x^2} + b(x,\tau) \dfrac{\partial u}{\partial x} + c(x,\tau)u + g(x,\tau), \\[2mm]
\qquad\qquad\qquad x_l \le x \le x_u, \quad 0 \le \tau \le T, \\[2mm]
u(x,0) = f(x), \qquad x_l \le x \le x_u, \\[2mm]
u(x_l,\tau) = f_l(\tau), \qquad 0 \le \tau \le T, \\[2mm]
u(x_u,\tau) = f_u(\tau), \qquad 0 \le \tau \le T,
\end{cases}
\tag{5.26}
$$

where $a(x,\tau) > 0$ on the domain $[x_l, x_u] \times [0, T]$. Sometimes, a problem of pricing a European option can be formulated in this way. In many cases, a European option problem is defined on an infinite domain and can be approximately formulated in such a way after giving some approximate boundary condition on each artificial boundary. However, such a problem can usually be transformed into the following degenerate parabolic partial differential equation problem:

$$
\begin{cases}
\dfrac{\partial u}{\partial \tau} = a(x,\tau)\dfrac{\partial^2 u}{\partial x^2} + b(x,\tau)\dfrac{\partial u}{\partial x} + c(x,\tau)u + g(x,\tau), \\[2mm]
\qquad\qquad x_l \le x \le x_u, \quad 0 \le \tau \le T, \\[2mm]
u(x,0) = f(x), \quad x_l \le x \le x_u,
\end{cases}
\tag{5.27}
$$

where $a(x,\tau) \ge 0$ on the domain $[x_l, x_u] \times [0,T]$,

$$
\begin{cases}
b(x_l,\tau) - \dfrac{\partial a}{\partial x}(x_l,\tau) \ge 0, \quad 0 \le \tau \le T, \\[2mm]
a(x_l,\tau) = 0, \quad 0 \le \tau \le T
\end{cases}
\tag{5.28}
$$

and

$$
\begin{cases}
b(x_u,\tau) - \dfrac{\partial a}{\partial x}(x_u,\tau) \le 0, \quad 0 \le \tau \le T, \\[2mm]
a(x_u,\tau) = 0, \quad 0 \le \tau \le T.
\end{cases}
\tag{5.29}
$$

For example, the prices of vanilla European call/put options are solutions of the problem

$$
\begin{cases}
\dfrac{\partial V}{\partial t} + \dfrac{1}{2}\sigma^2(S)S^2\dfrac{\partial^2 V}{\partial S^2} + (r - D_0)S\dfrac{\partial V}{\partial S} - rV = 0, \ 0 \le S, \ t \le T, \\[2mm]
V(S,t) = \max(\pm(S - E),0), \ 0 \le S.
\end{cases}
$$

Through the transformation

$$
\begin{cases}
\xi = \dfrac{S}{S + E}, \\[2mm]
\tau = T - t, \\[2mm]
V(S,t) = (S + E)\overline{V}(\xi,\tau),
\end{cases}
$$

the problem is converted into

$$
\begin{cases}
\dfrac{\partial \overline{V}}{\partial \tau} = \dfrac{1}{2}\bar{\sigma}^2(\xi)\xi^2(1 - \xi)^2\dfrac{\partial^2 \overline{V}}{\partial \xi^2} + (r - D_0)\xi(1 - \xi)\dfrac{\partial \overline{V}}{\partial \xi} - [r(1 - \xi) + D_0\xi]\overline{V}, \\[2mm]
\qquad\qquad\qquad\qquad\qquad\qquad\qquad\qquad 0 \le \xi \le 1, \quad 0 \le \tau, \\[2mm]
\overline{V}(\xi,0) = \max(\pm(2\xi - 1),0), \qquad\qquad\qquad 0 \le \xi \le 1,
\end{cases}
$$

where $\bar{\sigma}(\xi) = \sigma(E\xi/(1 - \xi))$. (For details, see Subsection 2.3.2.) Clearly, this problem is in the form of (5.27). Moreover, if a stochastic model

$$
dS = udt + wdX
$$

is defined on $[S_l, S_u]$, and the conditions

$$\begin{cases} u\left(S_l,t\right) - w(S_l,t)\dfrac{\partial}{\partial S}w(S_l,t) \geq 0, \\[2mm] w\left(S_l,t\right) = 0 \end{cases}$$

and

$$\begin{cases} u\left(S_u,t\right) - w(S_u,t)\dfrac{\partial}{\partial S}w(S_u,t) \leq 0, \\[2mm] w\left(S_u,t\right) = 0 \end{cases}$$

hold, then prices of European-style derivatives on this random variable also are solutions of (5.27). (For details, see Section 2.9.)

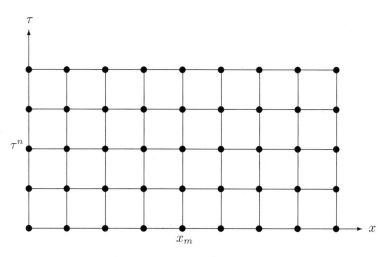

Fig. 5.1. A mesh for finite-difference methods

To find an approximate solution of a partial differential equation problem by finite-difference methods, we first divide the domain $[x_l, x_u] \times [0, T]$ into small subdomains using lines $x_m = x_l + m\Delta x$ and $\tau^n = n\Delta\tau$, where $\Delta x = (x_u - x_l)/M$, $\Delta\tau = T/N$ and M, N are positive integers. These lines form a grid, and these points (x_m, τ^n) are called grid points (see Fig. 5.1). We want to find the approximate values of the solution on these grid points.

Let us look at the problem (5.27). First consider the case[3]

$$b(x_l, \tau) = 0, \quad 0 \leq \tau \leq T$$

---

[3] As pointed out in Section 4.7, because $a(x, \tau) \geq 0$ on $[x_l, x_u]$ and $a(x_l, \tau) = a(x_u, \tau) = 0$, from the inequality conditions in (5.28) and (5.29), we have $b(x_l, \tau) \geq \dfrac{\partial a}{\partial x}(x_l, \tau) \geq 0$ and $b(x_u, \tau) \leq \dfrac{\partial a}{\partial x}(x_u, \tau) \leq 0$. Therefore, the two conditions below imply $\dfrac{\partial a}{\partial x}(x_l, \tau) = \dfrac{\partial a}{\partial x}(x_u, \tau) = 0$.

and
$$b(x_u, \tau) = 0, \quad 0 \le \tau \le T.$$

In this case, the partial differential equation in (5.27) degenerates into an ordinary differential equation at each boundary, and the degenerate parabolic problem (5.27) can be discretized in the following way.

Using forward difference for $\frac{\partial u}{\partial \tau}(x_m, \tau^n)$, second-order central difference for $\frac{\partial u}{\partial x}(x_m, \tau^n)$ and $\frac{\partial^2 u}{\partial x^2}(x_m, \tau^n)$ in equation (5.27) at the point (x_m, τ^n), we have

$$\frac{u(x_m, \tau^{n+1}) - u(x_m, \tau^n)}{\Delta\tau} - \frac{\Delta\tau}{2}\frac{\partial^2 u}{\partial\tau^2}(x_m, \eta)$$
$$= a_m^n \left[\frac{u(x_{m+1}, \tau^n) - 2u(x_m, \tau^n) + u(x_{m-1}, \tau^n)}{\Delta x^2} - \frac{\Delta x^2}{12}\frac{\partial^4 u}{\partial x^4}(\xi, \tau^n) \right]$$
$$+ b_m^n \left[\frac{u(x_{m+1}, \tau^n) - u(x_{m-1}, \tau^n)}{2\Delta x} - \frac{\Delta x^2}{6}\frac{\partial^3 u}{\partial x^3}(\bar\xi, \tau^n) \right]$$
$$+ c_m^n u(x_m, \tau^n) + g_m^n,$$

where
$$\eta \in (\tau^n, \tau^{n+1}), \quad \xi \in (x_{m-1}, x_{m+1}), \quad \bar\xi \in (x_{m-1}, x_{m+1}),$$

and a_m^n, b_m^n, c_m^n, and g_m^n denote $a(x_m, \tau^n), b(x_m, \tau^n), c(x_m, \tau^n)$, and $g(x_m, \tau^n)$, respectively. Dropping the term $-\frac{\Delta\tau}{2}\frac{\partial^2 u}{\partial\tau^2}(x_m, \eta)$ from the left-hand side and the two terms $-a_m^n \frac{\Delta x^2}{12}\frac{\partial^4 u}{\partial x^4}(\xi, \tau^n)$ and $-b_m^n \frac{\Delta x^2}{6}\frac{\partial^3 u}{\partial x^3}(\bar\xi, \tau^n)$ from the right-hand side, and denoting the approximate solution of $u(x_m, \tau^n)$ by u_m^n, we obtain the following approximation to the partial differential equation in (5.27):

$$\frac{u_m^{n+1} - u_m^n}{\Delta\tau} = a_m^n \frac{u_{m+1}^n - 2u_m^n + u_{m-1}^n}{\Delta x^2} + b_m^n \frac{u_{m+1}^n - u_{m-1}^n}{2\Delta x} + c_m^n u_m^n + g_m^n,$$
$$m = 0, 1, \cdots, M, \quad n = 0, 1, \cdots, N-1.$$

From the initial condition in (5.27), we have $u_m^0 = f(x_m)$, $m = 0, 1, \cdots, M$. Therefore, the degenerate parabolic problem (5.27) can be discretized by

$$\begin{cases} u_m^{n+1} = \left(\frac{a_m^n \Delta\tau}{\Delta x^2} + \frac{b_m^n \Delta\tau}{2\Delta x} \right) u_{m+1}^n + \left(1 - 2\frac{a_m^n \Delta\tau}{\Delta x^2} + c_m^n \Delta\tau \right) u_m^n \\ \qquad + \left(\frac{a_m^n \Delta\tau}{\Delta x^2} - \frac{b_m^n \Delta\tau}{2\Delta x} \right) u_{m-1}^n + g_m^n \Delta\tau, \\ \qquad m = 0, 1, \cdots, M, \quad n = 0, 1, \cdots, N-1, \\ u_m^0 = f(x_m), \quad m = 0, 1, \cdots, M. \end{cases} \tag{5.30}$$

Here, we need to point out that because we discretize ordinary differential equations at the boundaries, only u_0^n appears in the equation for $m = 0$ and only u_M^n for $m = M$. That is, because $a_0^n = b_0^n = a_M^n = b_M^n = 0$, u_{-1}^n and u_{M+1}^n actually do not appear in the equations above.

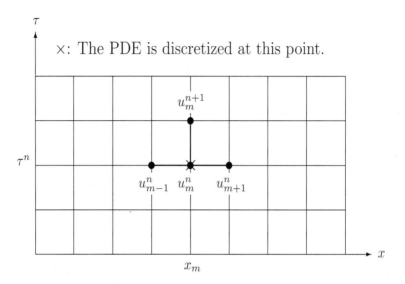

Fig. 5.2. An explicit finite-difference discretization

When u_m^n, $m = 0, 1, \cdots, M$ are known, we can find u_m^{n+1}, $m = 0, 1, \cdots, M$ by (5.30). Because u_m^0, $m = 0, 1, \cdots, M$ are given in (5.30), this procedure can be done for $n = 0, 1, \cdots, N - 1$ successively, and the approximate solution on all the grid points can be obtained. This method is called an **explicit finite-difference method**. This is because when u_m^n has been obtained, one equation involves only one unknown, so the unknown u_m^{n+1} can be computed from u_{m-1}^n, u_m^n and u_{m+1}^n explicitly. Figure 5.2 gives a diagram for this procedure. When we have the approximation (5.30), we have dropped the terms

$$\frac{\Delta \tau}{2} \frac{\partial^2 u}{\partial \tau^2}(x_m, \eta) - a_m^n \frac{\Delta x^2}{12} \frac{\partial^4 u}{\partial x^4}(\xi, \tau^n) - b_m^n \frac{\Delta x^2}{6} \frac{\partial^3 u}{\partial x^3}(\bar{\xi}, \tau^n)$$

from the equations. These terms as a whole are called the **truncation error** for scheme (5.30). Because the truncation error can be rewritten as $O(\Delta x^2, \Delta \tau)$, we say that for scheme (5.30), the truncation error is second order in Δx and first order in $\Delta \tau$.

Now let us discretize the problem (5.27) at the point $(x_m, \tau^{n+1/2})$. For $\frac{\partial u}{\partial \tau}(x_m, \tau^{n+1/2})$, we use the central scheme. The derivative $\frac{\partial u}{\partial x}(x_m, \tau^{n+1/2})$ is approximated first by the average of the values at the points (x_m, τ^n) and

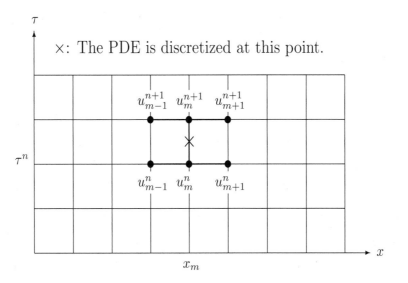

Fig. 5.3. An implicit finite-difference discretization

(x_m, τ^{n+1}), and then the derivatives at these two points are discretized by the central difference. The second derivative $\dfrac{\partial^2 u}{\partial x^2}(x_m, \tau^{n+1/2})$ is dealt with similarly. Using this way, the degenerate parabolic problem (5.27) can be approximated by the implicit finite-difference method:

$$
\begin{cases}
\dfrac{u_m^{n+1} - u_m^n}{\Delta\tau} = \dfrac{a_m^{n+1/2}}{2}\left(\dfrac{u_{m+1}^{n+1} - 2u_m^{n+1} + u_{m-1}^{n+1}}{\Delta x^2}\right. \\[3mm]
\qquad\qquad\qquad\left. + \dfrac{u_{m+1}^n - 2u_m^n + u_{m-1}^n}{\Delta x^2}\right) \\[3mm]
\qquad + \dfrac{b_m^{n+1/2}}{2}\left(\dfrac{u_{m+1}^{n+1} - u_{m-1}^{n+1}}{2\Delta x} + \dfrac{u_{m+1}^n - u_{m-1}^n}{2\Delta x}\right) \\[3mm]
\qquad + \dfrac{c_m^{n+1/2}}{2}(u_m^{n+1} + u_m^n) + g_m^{n+1/2}, \\[3mm]
\qquad\qquad m = 0, 1, \cdots, M, \quad n = 0, 1, \cdots, N-1, \\[3mm]
u_m^0 = f(x_m), \qquad m = 0, 1, \cdots, M.
\end{cases}
\tag{5.31}
$$

From here, we see that each equation involves six grid points (see Fig. 5.3) and that there are three unknowns. As we know, the error of a central difference is second order. For a function, the average of the values at the points (x_m, τ^n) and (x_m, τ^{n+1}) is an approximate value at the point $(x_m, \tau^{n+1/2})$

with an error of $O(\Delta\tau^2)$ because it actually is the result obtained by the linear interpolation. Therefore, the truncation error of this scheme is $O(\Delta x^2, \Delta\tau^2)$.

Similar to the scheme (5.30), because we actually discretize ordinary differential equations at the boundaries, the equations for $m = 0$ and $m = M$ can be written as

$$\frac{u_m^{n+1} - u_m^n}{\Delta\tau} = \frac{c_m^{n+1/2}}{2}(u_m^{n+1} + u_m^n) + g_m^{n+1/2},$$
$$m = 0, M, \quad n = 0, 1, \cdots, N - 1.$$

Consequently, these equations actually do not involve u_{-1}^n and u_{M+1}^n. Furthermore, the equations for $m = 0$ alone can determine u_0^n, $n = 1, 2, \cdots, N$ from u_0^0. For u_M^n, the situation is similar. However, for u_m^n, $m \neq 0$ and M, the situation is different. We cannot determine u_m^{n+1} only from a few equations. In order to obtain u_m^{n+1}, $m = 1, 2, \cdots, M - 1$, we have to solve a tridiagonal system of linear equations, and each of u_m^{n+1} is determined by all the u_m^n. Consequently, this method is called an **implicit finite-difference method**.

The problem (5.26) can be discretized similarly. The only difference is that the partial differential equation should not be discretized for $m = 0$ and $m = M$ because the boundary conditions $u(x_l, \tau) = f_l(\tau)$ and $u(x_u, \tau) = f_u(\tau)$ provide the equations we need. When $a(x, \tau)$ is equal to a positive constant a, $b(x, \tau) = 0$, $c(x, \tau) = 0$, and $g(x, \tau) = 0$, i.e., for the heat conductivity problem

$$\begin{cases} \dfrac{\partial u}{\partial \tau} = a\dfrac{\partial^2 u}{\partial x^2}, & x_l \leq x \leq x_u, \quad 0 \leq \tau \leq T, \\[2mm] u(x, 0) = f(x), & x_l \leq x \leq x_u, \\[2mm] u(x_l, \tau) = f_l(\tau), & 0 \leq \tau \leq T, \\[2mm] u(x_u, \tau) = f_u(\tau), & 0 \leq \tau \leq T, \end{cases} \qquad (5.32)$$

corresponding to the explicit scheme (5.30), (5.32) can be approximated by

$$\begin{cases} u_m^{n+1} = \alpha u_{m+1}^n + (1 - 2\alpha)u_m^n + \alpha u_{m-1}^n, \\[2mm] \quad m = 1, 2, \cdots, M - 1, \quad n = 0, 1, \cdots, N - 1, \\[2mm] u_0^{n+1} = f_l(\tau^{n+1}), & n = 0, 1, \cdots, N - 1, \\[2mm] u_M^{n+1} = f_u(\tau^{n+1}), & n = 0, 1, \cdots, N - 1, \\[2mm] u_m^0 = f(x_m), & m = 0, 1, \cdots, M, \end{cases} \qquad (5.33)$$

where

$$\alpha = \frac{a\Delta\tau}{\Delta x^2}.$$

Similar to the implicit scheme (5.31), (5.32) can also be approximated by

$$\begin{cases} \dfrac{u_m^{n+1} - u_m^n}{\Delta\tau} = \dfrac{a}{2}\left(\dfrac{u_{m+1}^{n+1} - 2u_m^{n+1} + u_{m-1}^{n+1}}{\Delta x^2}\right. \\ \qquad\qquad\qquad \left. +\dfrac{u_{m+1}^n - 2u_m^n + u_{m-1}^n}{\Delta x^2}\right), \\ \quad m = 1, 2, \cdots, M-1, \qquad n = 0, 1, \cdots, N-1, \\ u_0^{n+1} = f_l(\tau^{n+1}), \qquad\qquad n = 0, 1, \cdots, N-1, \\ u_M^{n+1} = f_u(\tau^{n+1}), \qquad\qquad n = 0, 1, \cdots, N-1, \\ u_m^0 = f(x_m), \qquad\qquad\quad m = 0, 1, \cdots, M, \end{cases} \tag{5.34}$$

which is called the Crank–Nicolson scheme.

Since $u(x_l, \tau)$ and $u(x_u, \tau)$ are given, there are only $M-1$ unknowns for each time level, and the $M-1$ equations in (5.34) can be written together in matrix form:

$$\mathbf{A}\mathbf{u}^{n+1} = \mathbf{B}\mathbf{u}^n + \mathbf{b}^n, \tag{5.35}$$

where

$$\mathbf{A} = \begin{bmatrix} 1+\alpha & -\alpha/2 & 0 & \cdots & 0 \\ -\alpha/2 & 1+\alpha & -\alpha/2 & \ddots & \vdots \\ 0 & -\alpha/2 & \ddots & \ddots & 0 \\ \vdots & \ddots & \ddots & \ddots & -\alpha/2 \\ 0 & \cdots & 0 & -\alpha/2 & 1+\alpha \end{bmatrix},$$

$$\mathbf{B} = \begin{bmatrix} 1-\alpha & \alpha/2 & 0 & \cdots & 0 \\ \alpha/2 & 1-\alpha & \alpha/2 & \ddots & \vdots \\ 0 & \alpha/2 & \ddots & \ddots & 0 \\ \vdots & \ddots & \ddots & \ddots & \alpha/2 \\ 0 & \cdots & 0 & \alpha/2 & 1-\alpha \end{bmatrix},$$

$$\mathbf{u}^n = \begin{bmatrix} u_1^n \\ u_2^n \\ \vdots \\ u_{M-2}^n \\ u_{M-1}^n \end{bmatrix} \quad \text{and} \quad \mathbf{b}^n = \begin{bmatrix} \alpha u_0^n/2 + \alpha u_0^{n+1}/2 \\ 0 \\ \vdots \\ 0 \\ \alpha u_M^n/2 + \alpha u_M^{n+1}/2 \end{bmatrix}.$$

Now we consider the problem (5.27) for the case

$$b(x_l, \tau) > 0, \quad 0 \le \tau \le T$$

and

$$b(x_u, \tau) < 0, \quad 0 \le \tau \le T.$$

In this case, the partial differential equation degenerates into hyperbolic differential equations at the boundaries, and the first derivative in the equation has to be discretized by a one-sided difference. For example, if in the scheme (5.30) or (5.31), we use a one-sided difference for the first derivative in the equations for $m = 0$ and $m = M$, we can have the approximation we need. However, here the way of discretizing the first derivative at $m = 0$ is different from that at $m = 1$, namely, the discretization "jumps" from $m = 0$ to $m = 1$, so from the finite-difference equation at $m = 0$ to $m = 1$, the coefficients do not satisfy the Lipschitz condition. This causes some problems when doing stability analysis. A similar situation occurs from $m = M - 1$ to $m = M$. In order to avoid the "jump," we can approximate the degenerate parabolic problem (5.27) by the explicit finite-difference method:

$$
\begin{cases}
\dfrac{u_m^{n+1} - u_m^n}{\Delta \tau} = a_m^n \dfrac{u_{m+1}^n - 2u_m^n + u_{m-1}^n}{\Delta x^2} + \Phi_m^n + c_m^n u_m^n + g_m^n, \\
\qquad\qquad m = 0, 1, \cdots, M, \quad n = 0, 1, \cdots, N - 1, \\
u_m^0 = f(x_m), \qquad m = 0, 1, \cdots, M,
\end{cases}
\tag{5.36}
$$

where

$$
\Phi_m^n = \begin{cases}
b_m^n \dfrac{-u_{m+2}^n + 4u_{m+1}^n - 3u_m^n}{2\Delta x}, & \text{if} \quad b_m^n > 0, \\
0, & \text{if} \quad b_m^n = 0, \\
b_m^n \dfrac{3u_m^n - 4u_{m-1}^n + u_{m-2}^n}{2\Delta x}, & \text{if} \quad b_m^n < 0
\end{cases}
$$

or by the implicit finite-difference method:

$$
\begin{cases}
\dfrac{u_m^{n+1} - u_m^n}{\Delta \tau} = \dfrac{a_m^{n+1/2}}{2}\left(\dfrac{u_{m+1}^{n+1} - 2u_m^{n+1} + u_{m-1}^{n+1}}{\Delta x^2} \right. \\
\qquad\qquad\qquad \left. + \dfrac{u_{m+1}^n - 2u_m^n + u_{m-1}^n}{\Delta x^2} \right) \\
\qquad\qquad + \Phi_m^{n+1/2} + \dfrac{c_m^{n+1/2}}{2}(u_m^{n+1} + u_m^n) + g_m^{n+1/2}, \\
\qquad\qquad m = 0, 1, \cdots, M, \quad n = 0, 1, \cdots, N - 1, \\
u_m^0 = f(x_m), \qquad m = 0, 1, \cdots, M,
\end{cases}
\tag{5.37}
$$

where

$$\Phi_m^{n+1/2} = \begin{cases} \dfrac{b_m^{n+1/2}}{2}\left(\dfrac{-u_{m+2}^{n+1} + 4u_{m+1}^{n+1} - 3u_m^{n+1}}{2\Delta x}\right. \\ \qquad \left. +\dfrac{-u_{m+2}^n + 4u_{m+1}^n - 3u_m^n}{2\Delta x}\right), \text{ if } \quad b_m^{n+1/2} > 0, \\[4pt] 0, \qquad\qquad\qquad\qquad\qquad\quad\ \text{ if } \quad b_m^{n+1/2} = 0, \\[4pt] \dfrac{b_m^{n+1/2}}{2}\left(\dfrac{3u_m^{n+1} - 4u_{m-1}^{n+1} + u_{m-2}^{n+1}}{2\Delta x}\right. \\ \qquad \left. +\dfrac{3u_m^n - 4u_{m-1}^n + u_{m-2}^n}{2\Delta x}\right), \text{ if } \quad b_m^{n+1/2} < 0. \end{cases}$$

Scheme (5.37) usually involves eight points, among them there is four unknowns (see Fig. 5.4). However, at boundaries there are three unknowns because $a_0 = a_M = 0$. When the partial differential equation is discretized in this way, the stability analysis can be done much easier. In the paper [68] by Sun, Yang, and Zhu, the stability problem of scheme (5.37) has been carefully studied. Clearly, the truncation error of the scheme (5.36) is $O(\Delta x^2, \Delta\tau)$ and that of the scheme (5.37) is $O(\Delta x^2, \Delta\tau^2)$.

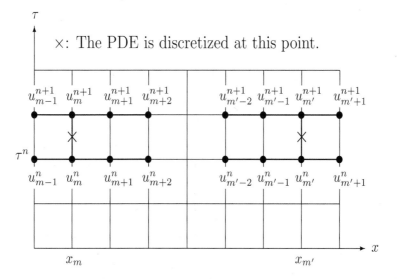

Fig. 5.4. Implicit eight-point finite-difference discretizations

Therefore, in order to find a solution, we can use either an explicit finite-difference method or an implicit finite-difference method. From the next section, we will see that for an explicit method, the step size $\Delta\tau$ must be less than a constant times Δx^2 for a stable computation. Thus, if a small Δx must be adopted in order to have satisfying results, the computation could take quite a long time. However, there is no restriction on the step size $\Delta \tau$ for implicit finite-difference methods. This is the main advantage of implicit methods over explicit methods.

A European-style derivative could involve several random state variables. In this case, we need to discretize a multi-dimensional problem, which will be dealt with in Chapters 6 and 8. Usually, an American-style derivative problem can be formulated as a free boundary problem. Discretization of such a problem will be discussed in Chapter 7.

5.4 Stability and Convergence Analysis

5.4.1 Stability

Stability is concerned with the propagation of errors. During the computation, truncation errors are brought into approximate solutions at each step. Also rounding errors are introduced into solutions all the time because any computer has a finite number of digits for numbers. If for a given finite-difference method, the errors are not magnified at each step in some norm, then we say that the finite-difference method is stable. There are two different norms that are often used in studying stability. Suppose

$$\mathbf{x} = (x_1, x_2, \cdots, x_{M-1})^T$$

is a vector with $M-1$ components. The L_∞ and L_2 norms of the vector $\mathbf{x}$ are defined as follows:

$$||\mathbf{x}||_{L_\infty} = \max_{1 \leq m \leq M-1} |x_m|$$

and

$$||\mathbf{x}||_{L_2} = \left(\frac{1}{M-1} \sum_{m=1}^{M-1} x_m^2 \right)^{1/2}.$$

Here, $M-1$ could be any positive integer and is allowed to go to infinity.

Stability of explicit finite-difference methods for the heat equation.
Consider the explicit finite-difference method (5.33) for the heat conductivity problem. Suppose an initial error e_m^0 appears in computing $f(x_m)$ for $m = 1, 2, \cdots, M-1$. That is, instead of $f(x_m)$, $f(x_m) + e_m^0$ is given as the initial value. We assume that there is no error from boundary conditions, that is, $e_0^0 = e_M^0 = 0$. Let $\tilde{u}_m^n, m = 0, 1, \cdots, M, n = 0, 1, \cdots, N$, be the computed solution. We want to study how $\tilde{u}_m^n$ is affected by e_m^0. This is usually referred to as studying the stability of schemes with respect to initial values. Clearly, $\tilde{u}_m^n$ satisfies

$$\begin{cases} \tilde{u}_m^{n+1} = \alpha\tilde{u}_{m+1}^n + (1-2\alpha)\tilde{u}_m^n + \alpha\tilde{u}_{m-1}^n, \\ \qquad m = 1, 2, \cdots, M-1, \qquad n = 0, 1, \cdots, N-1, \\ \tilde{u}_0^{n+1} = f_l(\tau^{n+1}), \qquad\qquad n = 0, 1, \cdots, N-1, \\ \tilde{u}_M^{n+1} = f_u(\tau^{n+1}), \qquad\qquad n = 0, 1, \cdots, N-1, \\ \tilde{u}_m^0 = f(x_m) + e_m^0, \qquad\quad m = 0, 1, \cdots, M. \end{cases}$$

Let
$$e_m^n = \tilde{u}_m^n - u_m^n, \quad m = 0, 1, \cdots, M, \quad n = 0, 1, \cdots, N.$$

Taking the difference of (5.33) and this system, we get

$$\begin{cases} e_m^{n+1} = \alpha e_{m+1}^n + (1-2\alpha)e_m^n + \alpha e_{m-1}^n, \\ \qquad m = 1, 2, \cdots, M-1, \qquad n = 0, 1, \cdots, N-1, \\ e_0^{n+1} = 0, \qquad\qquad\qquad n = 0, 1, \cdots, N-1, \qquad (5.38) \\ e_M^{n+1} = 0, \qquad\qquad\qquad n = 0, 1, \cdots, N-1, \\ e_m^0 = e_m^0, \qquad\qquad\qquad m = 0, 1, \cdots, M. \end{cases}$$

For this scheme, we can analyze its stability in two ways. First, we show that this scheme is stable in the maximum norm if $\alpha \leq 1/2$. In this case, all the coefficients in the right-hand side of the finite-difference equation, α, $1-2\alpha$, α, are nonnegative, so

$$\begin{aligned} |e_m^{n+1}| &= |\alpha e_{m+1}^n + (1-2\alpha)e_m^n + \alpha e_{m-1}^n| \\ &\leq \alpha|e_{m+1}^n| + (1-2\alpha)|e_m^n| + \alpha|e_{m-1}^n| \\ &\leq \max_{1\leq m\leq M-1} |e_m^n|, \quad m = 1, 2, \cdots, M-1, \end{aligned}$$

or
$$\max_{1\leq m\leq M-1} |e_m^{n+1}| \leq \max_{1\leq m\leq M-1} |e_m^n|,$$

where we have used the fact $e_0^n = e_M^n = 0$, $n = 0, 1, \cdots, N$. This is true for any n. Therefore,

$$\max_{1\leq m\leq M-1} |e_m^n| \leq \max_{1\leq m\leq M-1} |e_m^0|$$

or
$$||\mathbf{e}^n||_{L_\infty} \leq ||\mathbf{e}^0||_{L_\infty}.$$

Consequently, (5.33) is stable with respect to initial value in the maximum norm. This method of analyzing stability is very simple. Unfortunately, it seems that this method works only for explicit schemes with positive coefficients on the right-hand side.

Now let us study the stability of scheme (5.33) in another way. Suppose

$$e_m^0 = \sin m\omega_k, \quad m = 0, 1, \cdots, M,$$

where
$$\omega_k = k\pi/M, \quad k = 1, 2, \cdots, \text{ or } M - 1.$$

Such a function satisfies the conditions $e_0^0 = e_M^0 = 0$. In this case, for $m = 1, 2, \cdots, M - 1$, we have

$$
\begin{aligned}
e_m^1 &= \alpha \sin\left(m+1\right)\omega_k + (1 - 2\alpha)\sin m\omega_k + \alpha\sin\left(m-1\right)\omega_k \\
&= \sin m\omega_k + \alpha\left[\sin\left(m+1\right)\omega_k - 2\sin m\omega_k + \sin\left(m-1\right)\omega_k\right] \\
&= \left(1 - 4\alpha\sin^2\frac{\omega_k}{2}\right)\sin m\omega_k \\
&= \lambda_{\omega_k}\sin m\omega_k, \quad m = 1, 2, \cdots, M - 1,
\end{aligned}
\tag{5.39}
$$

where $\lambda_{\omega_k} = 1 - 4\alpha\sin^2(\omega_k/2)$ is independent of m. Therefore, e_m^1 has the same form as e_m^0 for $m = 1, 2, \cdots, M - 1$, and the only change is that the coefficient varies from one in e_m^0 to λ_{ω_k} in e_m^1. Because $e_0^n = e_M^n = 0$, for $n = 1, 2, \cdots, N$, using the same way, we can further show

$$e_m^2 = \lambda_{\omega_k}\lambda_{\omega_k}\sin m\omega_k = \lambda_{\omega_k}^2\sin m\omega_k = \lambda_{\omega_k}^2 e_m^0, \quad m = 1, 2, \cdots, M - 1$$

and in general, we obtain

$$e_m^n = \lambda_{\omega_k}\lambda_{\omega_k}^{n-1}\sin m\omega_k = \lambda_{\omega_k}^n\sin m\omega_k = \lambda_{\omega_k}^n e_m^0, \quad m = 1, 2, \cdots, M - 1.$$

Here and in the rest of this section, $\lambda_{\omega_k}^n$ stands for the n-th power of λ_{ω_k}. If $0 \le \alpha \le 1/2$, then

$$-1 \le 1 - 4\alpha \le \lambda_{\omega_k} = 1 - 4\alpha\sin^2(\omega_k/2) \le 1$$

for any ω_k. Thus, e_m^n does not grow as n increases, and the computation is stable. If we choose $\alpha > 1/2$, then when M is large enough, we can find at least one ω_k with $\lambda_{\omega_k} < -1$. For such an ω_k and some m,

$$|e_m^n| = |\lambda_{\omega_k}^n\sin m\omega_k| \to \infty \text{ as } n \to \infty.$$

Thus, the computation is unstable. This happens for a special initial error. What will happen for a general case?

Let

$$
\mathbf{A}_1 =
\begin{bmatrix}
1 - 2\alpha & \alpha & 0 & \cdots & 0 \\
\alpha & 1 - 2\alpha & \alpha & \ddots & \vdots \\
0 & \alpha & \ddots & \ddots & 0 \\
\vdots & \ddots & \ddots & \ddots & \alpha \\
0 & \cdots & 0 & \alpha & 1 - 2\alpha
\end{bmatrix},
\quad
\mathbf{e}^n =
\begin{bmatrix}
e_1^n \\
e_2^n \\
\vdots \\
\vdots \\
e_{M-1}^n
\end{bmatrix}.
\tag{5.40}
$$

From (5.38), we see that between $\mathbf{e}^{n+1}$ and $\mathbf{e}^n$ there is the following relation:

$$\mathbf{e}^{n+1} = \mathbf{A}_1 \mathbf{e}^n.$$

From (5.39), we also know that λ_{ω_k} is an eigenvalue of A_1 and

$$\mathbf{e}_{\omega_k} = \begin{bmatrix} \sin \omega_k \\ \sin 2\omega_k \\ \vdots \\ \vdots \\ \sin(M-1)\omega_k \end{bmatrix}$$

is the associated eigenvector. λ_{ω_k}, $k = 1, 2, \cdots, M-1$ are distinct eigenvalues of $\mathbf{A}_1$, so the $M-1$ associated eigenvectors, $\mathbf{e}_{\omega_k}$, $k = 1, 2, \cdots, M-1$, are linearly independent. Thus, any vector with $M-1$ components, especially $\mathbf{e}^0$, can be expressed as linear combination of $\mathbf{e}_{\omega_k}$. That is, any error $\mathbf{e}^0$ can be expressed as

$$\mathbf{e}^0 = \sum_{k=1}^{M-1} \varepsilon_{\omega_k} \mathbf{e}_{\omega_k}.$$

Substituting this expression into $\mathbf{e}^{n+1} = \mathbf{A}_1 \mathbf{e}^n$, we have

$$\mathbf{e}^1 = \mathbf{A}_1 \mathbf{e}^0 = \sum_{k=1}^{M-1} \varepsilon_{\omega_k} \lambda_{\omega_k} \mathbf{e}_{\omega_k}$$

and furthermore

$$\mathbf{e}^n = \sum_{k=1}^{M-1} \varepsilon_{\omega_k} \lambda_{\omega_k}^n \mathbf{e}_{\omega_k}$$

or in component form

$$e_m^n = \sum_{k=1}^{M-1} \varepsilon_{\omega_k} \lambda_{\omega_k}^n \sin m\omega_k, \quad m = 1, 2, \cdots, M-1.$$

As eigenvectors of a symmetric matrix $\mathbf{A}_1$, $\mathbf{e}_{\omega_k}$, $k = 1, 2, \cdots, M-1$ are orthogonal. Thus, from the expressions of $\mathbf{e}^0$ and $\mathbf{e}^n$ above, we have

$$\|\mathbf{e}^0\|_{L_2} = \left(\frac{1}{M-1} \sum_{m=1}^{M-1} \varepsilon_{\omega_k}^2 \|\mathbf{e}_{\omega_k}\|_{L_2}^2 \right)^{1/2}$$

and

$$||\mathbf{e}^n||_{L_2} = \left(\frac{1}{M-1} \sum_{m=1}^{M-1} \varepsilon_{\omega_k}^2 \lambda_{\omega_k}^{2n} ||\mathbf{e}_{\omega_k}||_{L_2}^2 \right)^{1/2}.$$

Consequently, we obtain

$$||\mathbf{e}^n||_{L_2} \leq ||\mathbf{e}^0||_{L_2}$$

if all the eigenvalues of $\mathbf{A}_1$ are in $[-1,1]$. Hence, if $0 \leq \alpha \leq 1/2$, then the computation is stable with respect to the initial value. If $\alpha > 1/2$, then when M is large enough, some of the eigenvalues of $\mathbf{A}_1$ must be less than -1. Hence, if a component of $\mathbf{e}^0$ associated with such an eigenvalue is not zero, then the corresponding component of $\mathbf{e}^n$ will be greater than the component of $\mathbf{e}^0$ and go to infinity as n goes to infinity. Because the errors are random variables, the ε_{ω_k} corresponding to such an eigenvalue λ_{ω_k} might not be zero. Thus, the computation is unstable. Consequently, we have the following conclusion: Scheme (5.33) is stable if

$$\alpha = \frac{a\Delta\tau}{\Delta x^2} \leq 1/2;$$

whereas the scheme is unstable if

$$\alpha = \frac{a\Delta\tau}{\Delta x^2} > 1/2.$$

Stability of implicit finite-difference methods for the heat equation. The second method used above to analyze stability can be applied to other cases, for example, implicit finite-difference methods. For an implicit finite-difference scheme, suppose $\mathbf{e}^n$ satisfies

$$\mathbf{A}\mathbf{e}^{n+1} = \mathbf{B}\mathbf{e}^n,$$

where $\mathbf{A}$ and $\mathbf{B}$ are two matrices, and $\mathbf{A}$ is invertible. Also, assume that the following relation holds:

$$\lambda_{\omega_k}\mathbf{A}\mathbf{e}_{\omega_k} = \mathbf{B}\mathbf{e}_{\omega_k}. \tag{5.41}$$

In this case, this method still works: if all the $\lambda_{\omega_k} \in [-1,1]$, then the scheme is stable; if certain λ_{ω_k} does not belong to $[-1,1]$, then the scheme is unstable. In fact, any initial error can be expressed as

$$\mathbf{e}^0 = \sum_{k=1}^{M-1} \varepsilon_{\omega_k}\mathbf{e}_{\omega_k}$$

and because of (5.41), we have

$$\mathbf{e}^n = \sum_{k=1}^{M-1} \varepsilon_{\omega_k}\lambda_{\omega_k}^n\mathbf{e}_{\omega_k}$$

for any n. Therefore, the scheme is stable if and only if

$$|\lambda_{\omega_k}| \leq 1$$

for all the ω_k.

For the Crank–Nicolson scheme (5.34), we have

$$\lambda_{\omega_k}\left[(1+\alpha)\sin m\omega_k - \frac{\alpha}{2}(\sin(m+1)\omega_k + \sin(m-1)\omega_k)\right]$$
$$= \left[(1-\alpha)\sin m\omega_k + \frac{\alpha}{2}(\sin(m+1)\omega_k + \sin(m-1)\omega_k)\right].$$

Thus,

$$\lambda_{\omega_k} = \frac{(1-\alpha)\sin m\omega_k + \alpha\sin m\omega_k \cos\omega_k}{(1+\alpha)\sin m\omega_k - \alpha\sin m\omega_k \cos\omega_k} = \frac{1 - 2\alpha\sin^2\frac{\omega_k}{2}}{1 + 2\alpha\sin^2\frac{\omega_k}{2}}$$

and $|\lambda_{\omega_k}| \leq 1$ for any ω_k. Consequently, (5.34) is stable in the L_2 norm.

Stability for periodic problems. In schemes (5.33) and (5.34), the values are given at both boundaries, and during stability analysis, we assume that there is no error at the boundaries. It is clear that this is not always the case. Consider problems satisfying periodic conditions and assume $u_m^n = u_{m+M}^n$. In this case, we only need to find u_m^n, $m = 0, 1, \cdots, M - 1$ for each time level. If the coefficients of the problem are constant, then we can analyze the stability in a similar way. Let us further assume that the solution satisfies the system:

$$a_1 u_{m+1}^{n+1} + a_0 u_m^{n+1} + a_{-1} u_{m-1}^{n+1} = b_1 u_{m+1}^n + b_0 u_m^n + b_{-1} u_{m-1}^n, \quad m = 0, 1, \cdots, M-1.$$

If e_m^n is the error of u_m^n, then e_m^n satisfy the same system. Thus, the system for e_m^n can be written as

$$\mathbf{A}_2 \mathbf{e}^{n+1} = \mathbf{B}_2 \mathbf{e}^n,$$

where we have used the conditions

$$e_{-1}^n = e_{M-1}^n, \quad e_M^n = e_0^n$$

and adopted the following notation:

$$\mathbf{A}_2 = \begin{bmatrix} a_0 & a_1 & 0 & \cdots & a_{-1} \\ a_{-1} & a_0 & a_1 & \ddots & \vdots \\ 0 & a_{-1} & a_0 & \ddots & 0 \\ \vdots & \ddots & \ddots & \ddots & a_1 \\ a_1 & \cdots & 0 & a_{-1} & a_0 \end{bmatrix}, \quad \mathbf{e}^n = \begin{bmatrix} e_0^n \\ e_1^n \\ \vdots \\ \vdots \\ e_{M-1}^n \end{bmatrix}$$

and

$$
\mathbf{B}_2 = \begin{bmatrix} b_0 & b_1 & 0 & \cdots & b_{-1} \\ b_{-1} & b_0 & b_1 & \ddots & \vdots \\ 0 & b_{-1} & b_0 & \ddots & 0 \\ \vdots & \ddots & \ddots & \ddots & b_1 \\ b_1 & \cdots & 0 & b_{-1} & b_0 \end{bmatrix}.
$$

For this system, the eigenvectors are

$$
\mathbf{e}_{\theta_k} = \begin{bmatrix} 1 \\ \mathrm{e}^{\mathrm{i}\theta_k} \\ \vdots \\ \vdots \\ \mathrm{e}^{\mathrm{i}(M-1)\theta_k} \end{bmatrix}, \quad k = 0, 1, \cdots, M-1,
$$

where $\theta_k = 2k\pi/M$ and the eigenvalues are

$$
\lambda_{\theta_k} = \frac{b_1 \mathrm{e}^{\mathrm{i}\theta_k} + b_0 + b_{-1}\mathrm{e}^{-\mathrm{i}\theta_k}}{a_1 \mathrm{e}^{\mathrm{i}\theta_k} + a_0 + a_{-1}\mathrm{e}^{-\mathrm{i}\theta_k}}, \quad k = 0, 1, \cdots, M-1.
$$

By using the relations $\mathrm{e}^{-\mathrm{i}\theta_k} = \mathrm{e}^{\mathrm{i}(M-1)\theta_k}$ and $\mathrm{e}^{\mathrm{i}M\theta_k} = 1$, this can be shown by a straightforward calculation. If $|\lambda_{\theta_k}| \leq 1$, $k = 0, 1, \cdots, M-1$, then the method is stable. If $|\lambda_{\theta_k}| > 1$ for some k, then the method is unstable. Because M can go to infinity, θ_k indeed can be any number in the interval $[0, 2\pi]$. Therefore, if for any $\theta \in [0, 2\pi]$,

$$
|\lambda_\theta| = \left| \frac{b_1 \mathrm{e}^{\mathrm{i}\theta} + b_0 + b_{-1}\mathrm{e}^{-\mathrm{i}\theta}}{a_1 \mathrm{e}^{\mathrm{i}\theta} + a_0 + a_{-1}\mathrm{e}^{-\mathrm{i}\theta}} \right| \leq 1, \tag{5.42}
$$

then the scheme is stable. Otherwise, the method is unstable. Such a method of analyzing stability is usually called the von Neumann method and λ_θ is called the amplification factor. This method gives a complete stability analysis for periodic initial value problems with constant coefficients. This method can be performed in the following way. Assume

$$
e_m^n = \lambda_\theta^n \mathrm{e}^{\mathrm{i}m\theta}, \tag{5.43}
$$

where θ can be any real number in the interval $[0, 2\pi]$. Substituting this expression into the finite-difference equation, we can find λ_θ. If all $|\lambda_\theta| \leq 1$, then the scheme is stable; if some $|\lambda_\theta| > 1$, then the scheme is unstable.

Stability analysis in practice. In practice, most problems have variable coefficients. Therefore, the von Neumann method does not give a complete stability analysis. However, it is still very useful. The von Neumann method can be applied in practice in the following way.

Consider the following scheme with variable coefficients:

$$a_{1,m}^n u_{m+1}^{n+1} + a_{0,m}^n u_m^{n+1} + a_{-1,m}^n u_{m-1}^{n+1} = b_{1,m}^n u_{m+1}^n + b_{0,m}^n u_m^n + b_{-1,m}^n u_{m-1}^n,$$
$$(5.44)$$

where for simplicity, we assume that only three points in the x direction are involved. If more points are involved, the procedure is still the same. Suppose

$$|f_{m+1}^n - f_m^n| < c\Delta x, \quad |f_{m+1}^n - 2f_m^n + f_{m-1}^n| < c\Delta x^2,$$

and

$$|f_m^{n+1} - f_m^n| < c\Delta \tau$$

for $f = a_1, a_0, a_{-1}, b_1, b_0$, and b_{-1}. Assume that e_m^n has the form (5.43). Substituting this expression into the finite-difference equation (5.44) yields

$$\lambda_\theta(x_m, \tau^n) = \frac{b_{1,m}^n e^{i(m+1)\theta} + b_{0,m}^n e^{im\theta} + b_{-1,m}^n e^{i(m-1)\theta}}{a_{1,m}^n e^{i(m+1)\theta} + a_{0,m}^n e^{im\theta} + a_{-1,m}^n e^{i(m-1)\theta}}.$$

If for the amplification factor, we have

$$|\lambda_\theta(x_m, \tau^n)| \le 1$$

for every point and the treatment of boundary conditions is reasonable, then we can expect the scheme to be stable. Clearly, the condition $|\lambda_\theta(x_m, \tau^n)| \le 1$ is equivalent to

$$|b_{1,m}^n e^{i\theta} + b_{0,m}^n + b_{-1,m}^n e^{-i\theta}|^2 - |a_{1,m}^n e^{i\theta} + a_{0,m}^n + a_{-1,m}^n e^{-i\theta}|^2 \le 0. \quad (5.45)$$

The latter is easier to use in practice than the former.

Let us analyze the stability of scheme (5.31) in this way. This scheme has the form (5.44) with

$$a_{1,m}^n = -\left(\frac{a_m^{n+1/2}}{2\Delta x^2} + \frac{b_m^{n+1/2}}{4\Delta x}\right)\Delta\tau,$$

$$a_{0,m}^n = 1 + \frac{a_m^{n+1/2}}{\Delta x^2}\Delta\tau,$$

$$a_{-1,m}^n = -\left(\frac{a_m^{n+1/2}}{2\Delta x^2} - \frac{b_m^{n+1/2}}{4\Delta x}\right)\Delta\tau,$$

$$b_{1,m}^n = -a_{1,m}^n,$$

$$b_{0,m}^n = 2 - a_{0,m}^n,$$

$$b_{-1,m}^n = -a_{-1,m}^n.$$

Here, we assume $g_m^{n+1/2} = c_m^{n+1/2} = 0$ because we analyze the stability with respect to initial values only and ignoring a term of $O(\Delta\tau)$ in coefficients will have no effect on the conclusion on stability. The left-hand side of the condition (5.45) for this scheme is

$$
\begin{aligned}
&\left[-a_{1,m}^n e^{i\theta} + (2 - a_{0,m}^n) - a_{-1,m}^n e^{-i\theta}\right]\left[-a_{1,m}^n e^{-i\theta} + (2 - a_{0,m}^n) - a_{-1,m}^n e^{i\theta}\right] \\
&- (a_{1,m}^n e^{i\theta} + a_{0,m}^n + a_{-1,m}^n e^{-i\theta})(a_{1,m}^n e^{-i\theta} + a_{0,m}^n + a_{-1,m}^n e^{i\theta}) \\
&= (a_{1,m}^n)^2 + (a_{0,m}^n - 2)^2 + (a_{-1,m}^n)^2 + 2a_{1,m}^n(a_{0,m}^n - 2)\cos\theta \\
&\quad + 2(a_{0,m}^n - 2)a_{-1,m}^n \cos\theta + 2a_{1,m}^n a_{-1,m}^n \cos 2\theta \\
&\quad - \left[(a_{1,m}^n)^2 + (a_{0,m}^n)^2 + (a_{-1,m}^n)^2 + 2a_{1,m}^n a_{0,m}^n \cos\theta + 2a_{0,m}^n a_{-1,m}^n \cos\theta \right. \\
&\quad \left. + 2a_{1,m}^n a_{-1,m}^n \cos 2\theta\right] \\
&= (a_{0,m}^n - 2)^2 - (a_{0,m}^n)^2 - 4a_{1,m}^n \cos\theta - 4a_{-1,m}^n \cos\theta \\
&= -\frac{4a_m^{n+1/2}}{\Delta x^2}\Delta\tau + \frac{4a_m^{n+1/2}}{\Delta x^2}\Delta\tau\cos\theta \\
&= \frac{4a_m^{n+1/2}}{\Delta x^2}\Delta\tau(\cos\theta - 1).
\end{aligned}
$$

This expression is always nonpositive. Therefore, the condition (5.45) is satisfied at every grid point. For scheme (5.31), there is no other boundary condition. Consequently, the scheme is expected to be stable.

So far, we say that a scheme is stable with respect to initial values if the error of the solution caused by the error in the initial condition is less than or equal to the error in the initial condition. However, generally speaking, we say that a scheme is stable with respect to initial values if the error of the solution caused by the error in the initial condition is less than c times the error in the initial condition. c is a constant independent of Δx and $\Delta\tau$, but is allowed to be greater than one. That is, the error is allowed to increase by a certain factor, but the factor must be bounded and independent of Δx and $\Delta\tau$. Therefore, we can take

$$|\lambda_\theta(x_m, \tau^n)| \leq 1 + \bar{c}\Delta\tau \tag{5.46}$$

as a criterion for stability. In fact, if (5.46) holds for any θ, then usually we can have

$$\|e^n\|_{L_2} \leq (1 + \bar{c}\Delta\tau)\|e^{n-1}\|_{L_2} \leq (1 + \bar{c}\Delta\tau)^n\|e^0\|_{L_2} \leq e^{\bar{c}nT/N}\|e^0\|_{L_2}$$

for any $n \leq N$, so the error increases at most by a factor $e^{\bar{c}T}$.

Now let us study the stability of (5.30) by using (5.46). We consider the stability with respect to initial values only, so we can set $g_m^n = 0$. In this case, the scheme has the form (5.44) with $a_{1,m}^n = 0$, $a_{0,m}^n = 1$, $a_{-1,m}^n = 0$ and

$$b_{1,m}^n = \frac{a_m^n \Delta\tau}{\Delta x^2} + \frac{b_m^n \Delta\tau}{2\Delta x},$$

$$b_{0,m}^n = 1 - 2\frac{a_m^n \Delta\tau}{\Delta x^2} + c_m^n \Delta\tau,$$

$$b_{-1,m}^n = \frac{a_m^n \Delta\tau}{\Delta x^2} - \frac{b_m^n \Delta\tau}{2\Delta x}.$$

Therefore,

$$
\begin{aligned}
\lambda_\theta(x_m, \tau^n) &= b_{1,m}^n e^{i\theta} + b_{0,m}^n + b_{-1,m}^n e^{-i\theta} \\
&= b_{0,m}^n + \left(b_{1,m}^n + b_{-1,m}^n\right)\cos\theta + i\left(b_{1,m}^n - b_{-1,m}^n\right)\sin\theta \\
&= 1 - 2\frac{a_m^n \Delta\tau}{\Delta x^2} + c_m^n \Delta\tau + 2\frac{a_m^n \Delta\tau}{\Delta x^2}\cos\theta + i\frac{b_m^n \Delta\tau}{\Delta x}\sin\theta \\
&= 1 - 4\frac{a_m^n \Delta\tau}{\Delta x^2}\sin^2\frac{\theta}{2} + c_m^n \Delta\tau + i\frac{b_m^n \Delta\tau}{\Delta x}\sin\theta.
\end{aligned}
$$

If

$$\max \frac{a_m^n \Delta\tau}{\Delta x^2} \le \frac{1}{2} \quad \text{or} \quad \frac{\Delta\tau}{\Delta x^2} \le \frac{1}{2\max a_m^n}, \tag{5.47}$$

then

$$
\begin{aligned}
|\lambda_\theta(x_m, \tau^n)|^2 &\le (1 + |c_m^n|\,\Delta\tau)^2 + \left(\frac{b_m^n \Delta\tau}{\Delta x}\right)^2 \\
&\le (1 + |c_m^n|\,\Delta\tau)^2 + \frac{(b_m^n)^2}{2\max a_m^n}\Delta\tau.
\end{aligned}
$$

Thus, let $\bar{c} = |c_m^n| + (b_m^n)^2/(4\max a_m^n)$, we have

$$|\lambda_\theta(x_m, \tau^n)| \le 1 + \bar{c}\Delta\tau$$

and we can expect this scheme to be stable if (5.47) holds.

In fact, the stability of scheme (5.31) with variable coefficients has been proven rigorously in the paper [68] by Sun, Yan, and Zhu. By a similar method, the stability of scheme (5.30) with variable coefficients can also be shown when (5.47) holds. If the reader is interested in such a subject, please see that paper and the book [88] by Zhu, Zhong, Chen, and Zhang.

5.4.2 Convergence

If a scheme is stable with respect to initial values, and the truncation error of the scheme goes to zero as Δx and $\Delta\tau$ tend to zero, then the approximate solution will usually go to the exact solution. Such a result is usually referred to as the Lax equivalence theorem. We are not going to prove this conclusion for general cases but explain this result intuitively through proving this result for special cases.

Consider the explicit finite-difference method (5.33). We know that the exact solution $u(x, \tau)$ satisfies the equation

$$u(x_m, \tau^{n+1})$$
$$= \alpha u(x_{m+1}, \tau^n) + (1 - 2\alpha)u(x_m, \tau^n) + \alpha u(x_{m-1}, \tau^n) + \Delta\tau R_m^n(\Delta x^2, \Delta\tau),$$
$$m = 1, 2, \cdots, M, \quad n = 0, 1, \cdots, N - 1,$$

where

$$R_m^n(\Delta x^2, \Delta\tau) = \frac{\Delta\tau}{2} \frac{\partial^2 u}{\partial \tau^2}(x_m, \eta) - a \frac{\Delta x^2}{12} \frac{\partial^4 u}{\partial x^4}(\xi, \tau^n).$$

Let e_m^n be the error of the approximate solution on the point (x_m, τ^n), that is,

$$e_m^n = u(x_m, \tau^n) - u_m^n, \quad m = 0, 1, \cdots, M, \; n = 0, 1, \cdots, N.$$

Then, e_m^n is the solution of the problem

$$\begin{cases} e_m^{n+1} = \alpha e_{m+1}^n + (1 - 2\alpha)e_m^n + \alpha e_{m-1}^n + \Delta\tau R_m^n(\Delta x^2, \Delta\tau), \\ \qquad m = 1, 2, \cdots, M - 1, \qquad n = 0, 1, \cdots, N - 1, \\ e_0^{n+1} = 0, \qquad\qquad\qquad\qquad\qquad n = 0, 1, \cdots, N - 1, \\ e_M^{n+1} = 0, \qquad\qquad\qquad\qquad\qquad n = 0, 1, \cdots, N - 1, \\ e_m^0 = 0, \qquad\qquad\qquad\qquad\qquad m = 0, 1, \cdots, M. \end{cases}$$

Because $e_0^n = e_M^n = 0$ for any n, the system can be written as

$$\begin{cases} \mathbf{e}^{n+1} = \mathbf{A}_1 \mathbf{e}^n + \Delta\tau \mathbf{R}^n(\Delta x^2, \Delta\tau), \quad n = 0, 1, \cdots, N - 1, \\ \mathbf{e}^0 = 0, \end{cases}$$

where $\mathbf{e}^n$ is a vector with $M - 1$ components e_m^n, $m = 1, 2, \cdots, M - 1$ and

$$\mathbf{R}^n(\Delta x^2, \Delta\tau) = \begin{bmatrix} R_1^n(\Delta x^2, \Delta\tau) \\ R_2^n(\Delta x^2, \Delta\tau) \\ \vdots \\ R_{M-1}^n(\Delta x^2, \Delta\tau) \end{bmatrix}.$$

Actually, $\mathbf{e}^n$ can be written as $\sum_{k=1}^n \mathbf{e}_{(k)}^n$. Here, for $k = n$,

$$\mathbf{e}_{(n)}^n = \Delta\tau \mathbf{R}^{n-1}(\Delta x^2, \Delta\tau)$$

and for $k = 1, 2, \cdots, n - 1$, $\mathbf{e}_{(k)}^n$ is the solution of the following problem

$$\begin{cases} \mathbf{e}_{(k)}^{\bar{n}+1} = \mathbf{A}_1 \mathbf{e}_{(k)}^{\bar{n}}, \quad \bar{n} = k, k + 1, \cdots, n - 1, \\ \mathbf{e}_{(k)}^k = \Delta\tau \mathbf{R}^{k-1}(\Delta x^2, \Delta\tau). \end{cases}$$

Because the error does not increase for the scheme (5.33) if $\alpha \leq 1/2$, $||\mathbf{e}^n||_{L_2}$ should not be greater than $\sum_{k=1}^n \Delta\tau ||\mathbf{R}^{k-1}(\Delta x^2, \Delta\tau)||_{L_2}$. Noticing $n \leq T/\Delta\tau$, we see that e_m^n goes to zero as $R_m^{k-1}(\Delta x^2, \Delta \tau)$ tends to zero for $k = 1, 2, \cdots, n$ and $m = 1, 2, \cdots, M - 1$. Hence, the approximate solution converges to the exact solution as Δx and $\Delta \tau$ tend to zero and α stays less than $1/2$ and $||e^n||_{L_2}$ has an order of $O(\Delta x^2, \Delta \tau)$. Usually, $\alpha = a\Delta \tau/\Delta x^2$ stays constant as Δx and $\Delta \tau$ tend to zero. Therefore, $||e^n||_{L_2} = O(\Delta \tau)$, and we say that the scheme (5.33) converges with order of $\Delta \tau$.

For implicit schemes, the situation is similar. Consider the Crank–Nicolson scheme (5.34). The exact solution satisfies

$$\frac{u(x_m, \tau^{n+1}) - u(x_m, \tau^n)}{\Delta \tau}$$

$$= \frac{a}{2} \left[\frac{u(x_{m+1}, \tau^{n+1}) - 2u(x_m, \tau^{n+1}) + u(x_{m-1}, \tau^{n+1})}{\Delta x^2} \right.$$

$$\left. + \frac{u(x_{m+1}, \tau^n) - 2u(x_m, \tau^n) + u(x_{m-1}, \tau^n)}{\Delta x^2} \right] + R_m^n(\Delta x^2, \Delta \tau^2),$$

$$m = 1, 2, \cdots, M - 1,$$

where

$$R_m^n(\Delta x^2, \Delta \tau^2)$$

$$= \Delta \tau^2 \left[\frac{1}{24} \frac{\partial^3 u}{\partial \tau^3}(x_m, \eta^{(1)}) - \frac{a}{8} \frac{\partial^4 u}{\partial x^2 \tau^2}(x_m, \eta^{(2)}) \right] - \frac{\Delta x^2 a}{12} \frac{\partial^4 u}{\partial x^4}(\xi, \eta^{(3)}).$$

In this case, the error satisfies

$$\mathbf{A}e^{n+1} = \mathbf{B}e^n + \Delta \tau \mathbf{R}^n(\Delta x^2, \Delta \tau^2),$$

where e^n and $\mathbf{R}^n(\Delta x^2, \Delta \tau^2)$ are two $(M-1)$-dimensional vectors with e_m^n and $R_m^n(\Delta x^2, \Delta \tau^2)$ as components, respectively, and $\mathbf{A}$ and $\mathbf{B}$ are given in (5.35). Just like in the case of the scheme (5.33), e^n can also be written as $\sum_{k=1}^n e_{(k)}^n$. Here, for $k = n$,

$$e_{(n)}^n = \Delta \tau \mathbf{A}^{-1} \mathbf{R}^{n-1}(\Delta x^2, \Delta \tau^2)$$

and for $k = 1, 2, \cdots, n-1$, $e_{(k)}^n$ is the solution of the following problem:

$$\begin{cases} \mathbf{A}e_{(k)}^{\bar{n}+1} = \mathbf{B}e_{(k)}^{\bar{n}}, & \bar{n} = k, k+1, \cdots, n-1, \\ e_{(k)}^k = \Delta \tau \mathbf{A}^{-1} \mathbf{R}^{k-1}(\Delta x^2, \Delta \tau^2). \end{cases}$$

The Crank–Nicolson scheme is stable with respect to the initial value. Thus, $||e^n||_{L_2}$ does not exceed $\sum_{k=1}^n \Delta \tau ||\mathbf{A}^{-1} \mathbf{R}^{k-1}(\Delta x^2, \Delta \tau^2)||_{L_2}$. Because

$$\mathbf{A}e_{\omega_k} = \left(1 + 2\alpha \sin^2 \frac{\omega_k}{2}\right) e_{\omega_k},$$

we see that $1+2\alpha \sin^2(\omega_k/2)$ is an eigenvalue of $\mathbf{A}$. Thus, $1/[1+2\alpha \sin^2(\omega_k/2)]$ is an eigenvalue of $\mathbf{A}^{-1}$. This means that $\mathbf{A}^{-1}$ always exists and that its norm is bounded for any case. Consequently, $||\mathbf{e}^n||_{\mathrm{L}_2}$ goes to zero as Δx and $\Delta \tau$ tend to zero. In this case, we say that this scheme is convergent. Furthermore, because $||\mathbf{e}^n||_{\mathrm{L}_2}$ is of the order $O(\Delta x^2, \Delta \tau^2)$, we say that the scheme has a second-order convergence or possesses a second-order accuracy.

For schemes with variable coefficients, from the stability with respect to initial values and the consistency of a scheme, we also can have its convergence. Here, we say that a scheme is consistent with the partial differential equation if the truncation error of the scheme goes to zero as Δx and $\Delta \tau$ tends to zero. In the paper [68] by Sun, Yan, and Zhu, some results on this issue are given.

5.5 Extrapolation of Numerical Solutions

When a partial differential equation problem is discretized, a truncation error is introduced that causes the numerical solution to have an error. What is the relation between the truncation error and the error of the numerical solution? Intuitively, the answer should be that a term of $O(\Delta x^{k_1}, \Delta \tau^{k_2})$ in the truncation error causes an error of $O(\Delta x^{k_1}, \Delta \tau^{k_2})$ in the numerical solution. Let us illustrate this fact.

Consider the following problem

$$
\begin{cases}
\dfrac{\partial u}{\partial \tau} = a(x,\tau)\dfrac{\partial^2 u}{\partial x^2} + b(x,\tau)\dfrac{\partial u}{\partial x} + c(x,\tau)u + g(x,\tau), \\[2mm]
\qquad\qquad 0 \le x \le 1, \quad 0 \le \tau \le T, \\[2mm]
u(x,0) = f(x), \quad 0 \le x \le 1,
\end{cases}
$$

where $a(x,\tau) \ge 0$ and $b(0,\tau) = a_x(0,\tau) = b(1,\tau) = a_x(1,\tau) = 0$. This problem can be approximated by

$$
\begin{cases}
\delta_\tau u_m^{n+1/2} = a_m^{n+1/2}\delta_x^2 u_m^{n+1/2} + b_m^{n+1/2}\delta_{0x} u_m^{n+1/2} + c_m^{n+1/2} u_m^{n+1/2} + g_m^{n+1/2}, \\[2mm]
\qquad\qquad\qquad\qquad\qquad\qquad 0 \le m \le M, \quad 0 \le n \le N-1, \\[2mm]
u_m^0 = f(x_m), \qquad\qquad\qquad\qquad 0 \le m \le M.
\end{cases}
$$

$$(5.48)$$

Here,

$$
\delta_\tau u_m^{n+1/2} = \frac{u_m^{n+1} - u_m^n}{\Delta \tau},
$$

$$
\delta_x^2 u_m^{n+1/2} = \frac{1}{2}\left(\frac{u_{m+1}^{n+1} - 2u_m^{n+1} + u_{m-1}^{n+1}}{\Delta x^2} + \frac{u_{m+1}^n - 2u_m^n + u_{m-1}^n}{\Delta x^2}\right),
$$

$$
\delta_{0x} u_m^{n+1/2} = \frac{1}{2}\left(\frac{u_{m+1}^{n+1} - u_{m-1}^{n+1}}{2\Delta x} + \frac{u_{m+1}^n - u_{m-1}^n}{2\Delta x}\right),
$$

$$f_m^{n+1/2} = \frac{1}{2}\left(f_m^{n+1} + f_m^n\right), \quad f \text{ being } u, a, b, c, g,$$

and the same notation will be used for other functions in what follows. The truncation error of this scheme is $O(\Delta x^2) + O(\Delta \tau^2)$ everywhere; more accurately, it is in the form

$$P_m^{n+1/2}\Delta x^2 + R_m^{n+1/2}\Delta \tau^2 + O(\Delta x^4 + \Delta \tau^4),$$

where $P_m^{n+1/2}$ and $R_m^{n+1/2}$ denote the values of two functions $P(x,\tau)$ and $R(x,\tau)$ at $x = x_m$ and $\tau = \tau^{n+1/2}$. That is, the exact solution satisfies the following equation:

$$\begin{cases} \delta_\tau U_m^{n+1/2} = a_m^{n+1/2}\delta_x^2 U_m^{n+1/2} + b_m^{n+1/2}\delta_{0x} U_m^{n+1/2} + c_m^{n+1/2} U_m^{n+1/2} + g_m^{n+1/2} \\ \qquad + P_m^{n+1/2}\Delta x^2 + R_m^{n+1/2}\Delta \tau^2 + O(\Delta x^4 + \Delta \tau^4), \\ \qquad\qquad\qquad\qquad\qquad\qquad 0 \le m \le M, \quad 0 \le n \le N-1, \\ u_m^0 = f(x_m), \qquad\qquad\qquad\qquad\quad 0 \le m \le M, \end{cases}$$

where U_m^n stands for $u(x_m, \tau^n)$. Suppose v_1 and v_2 are the solutions of the problems

$$\begin{cases} \dfrac{\partial v_1}{\partial \tau} = a(x,\tau)\dfrac{\partial^2 v_1}{\partial x^2} + b(x,\tau)\dfrac{\partial v_1}{\partial x} + c(x,\tau)v_1 + P(x,\tau), \\ \qquad\qquad\qquad\qquad\qquad 0 \le x \le 1, \quad 0 \le \tau \le T, \\ v_1(x,0) = 0, \qquad\qquad\qquad 0 \le x \le 1 \end{cases}$$

and

$$\begin{cases} \dfrac{\partial v_2}{\partial \tau} = a(x,\tau)\dfrac{\partial^2 v_2}{\partial x^2} + b(x,\tau)\dfrac{\partial v_2}{\partial x} + c(x,\tau)v_2 + R(x,\tau), \\ \qquad\qquad\qquad\qquad\qquad 0 \le x \le 1, \quad 0 \le \tau \le T, \\ v_2(x,0) = 0, \qquad\qquad\qquad 0 \le x \le 1 \end{cases}$$

respectively. Let $V_{1,m}^n$ and $V_{2,m}^n$ denote $v_1(x_m, \tau^n)$ and $v_2(x_m, \tau^n)$. Then,

$$\begin{cases} \delta_\tau V_{1,m}^{n+1/2} = a_m^{n+1/2}\delta_x^2 V_{1,m}^{n+1/2} + b_m^{n+1/2}\delta_{0x} V_{1,m}^{n+1/2} + c_m^{n+1/2} V_{1,m}^{n+1/2} + P_m^{n+1/2} \\ \qquad + O(\Delta x^2 + \Delta \tau^2), \qquad\qquad 0 \le m \le M, \quad 0 \le n \le N-1, \\ V_{1,m}^0 = 0, \qquad\qquad\qquad\qquad 0 \le m \le M, \end{cases}$$

and

$$\begin{cases} \delta_\tau V_{2,m}^{n+1/2} = a_m^{n+1/2}\delta_x^2 V_{2,m}^{n+1/2} + b_m^{n+1/2}\delta_{0x} V_{2,m}^{n+1/2} + c_m^{n+1/2} V_{2,m}^{n+1/2} + R_m^{n+1/2} \\ \qquad + O(\Delta x^2 + \Delta \tau^2), \qquad\qquad 0 \le m \le M, \quad 0 \le n \le N-1, \\ V_{2,m}^0 = 0, \qquad\qquad\qquad\qquad 0 \le m \le M. \end{cases}$$

Let us define

$$W_m^n = U_m^n - u_m^n - V_{1,m}^n \Delta x^2 - V_{2,m}^n \Delta \tau^2.$$

It is clear that W_m^n satisfies

$$\begin{cases} \delta_\tau W_m^{n+1/2} = a_m^{n+1/2} \delta_x^2 W_m^{n+1/2} + b_m^{n+1/2} \delta_{0x} W_m^{n+1/2} + c_m^{n+1/2} W_m^{n+1/2} \\ \qquad + O(\Delta x^4 + \Delta x^2 \Delta \tau^2 + \Delta \tau^4), \quad 0 \le m \le M, \quad 0 \le n \le N-1, \\ W_m^0 = 0, \qquad\qquad\qquad\qquad\qquad\qquad 0 \le m \le M. \end{cases}$$

Because the scheme is stable with respect to the initial value and the nonhomogeneous term (see the paper [68] by Sun, Yang, and Zhu for the details of the proof) and $O(\Delta x^2 \Delta \tau^2)$ can be expressed as $O(\Delta x^4 + \Delta \tau^4)$, we have

$$\|U_m^n - u_m^n - V_{1,m}^n \Delta x^2 - V_{2,m}^n \Delta \tau^2\| \le O(\Delta x^4 + \Delta \tau^4),$$

or we can write this relation as

$$u(x_m, \tau^n) - u_m^n(\Delta x, \Delta \tau) = v_1(x_m, \tau^n) \Delta x^2 + v_2(x_m, \tau^n) \Delta \tau^2 + O(\Delta x^4 + \Delta \tau^4),$$

that is,

$$u_m^n(\Delta x, \Delta \tau) = u(x_m, \tau^n) - v_1(x_m, \tau^n) \Delta x^2 - v_2(x_m, \tau^n) \Delta \tau^2 \\ + O(\Delta x^4 + \Delta \tau^4). \tag{5.49}$$

Here, we write u_m^n as $u_m^n(\Delta x, \Delta \tau)$ in order to indicate that the approximate solution is obtained on a mesh with mesh sizes Δx and $\Delta \tau$. For this case, the error of a numerical solution is in the form

$$-v_1(x_m, \tau^n) \Delta x^2 - v_2(x_m, \tau^n) \Delta \tau^2 + O(\Delta x^4 + \Delta \tau^4),$$

which has the same form as the truncation error given above. Similarly, if the truncation error of a numerical scheme, including the algorithms for boundary conditions, is

$$P \Delta x^2 + Q \Delta x \Delta \tau + R \Delta \tau^2 + O(\Delta \tau^3),$$

i.e., the scheme is second order and stable, then the numerical solution can be expressed as

$$u_m^n(\Delta x, \Delta \tau) = u(x_m, \tau^n) - v_1(x_m, \tau^n) \Delta x^2 - v_{12}(x_m, \tau^n) \Delta x \Delta \tau \\ - v_2(x_m, \tau^n) \Delta x^2 + O(\Delta \tau^3), \tag{5.50}$$

where $O(\Delta \tau^3)$ means $O(\Delta x^3 + \Delta x^2 \Delta \tau + \Delta x \Delta \tau^2 + \Delta \tau^3)$ for simplicity.

Here, the approximate value is given only at the nodes. Now let us generate a function defined on the domain $[0,1] \times [0,T]$ by some type of interpolation. We assume that the interpolation function generated from the values on the nodes by an interpolation method is an approximation to $f(x, \tau)$

with an error of $O(\Delta\tau^3)$ for any smooth enough function $f(x,\tau)$. For example, if we use quadratic interpolation, then the interpolation function generated has such a property. Let $u(x,\tau;\Delta x,\Delta\tau)$ denote such a function generated by $u(x_m,\tau^n;\Delta x,\Delta\tau)$. Because $u(x_m,\tau^n;\Delta x,\Delta\tau)$ consists of $u(x_m,\tau^n) - v_1(x_m,\tau^n)\Delta x^2 - v_{12}(x_m,\tau^n)\Delta x\Delta\tau - v_2(x_m,\tau^n)\Delta\tau^2$ and $O(\Delta\tau^3)$, the interpolation function also has two parts. One part is the interpolation function generated by $u(x_m,\tau^n) - v_1(x_m,\tau^n)\Delta x^2 - v_{12}(x_m,\tau^n)\Delta x\Delta\tau - v_2(x_m,\tau^n)\Delta\tau^2$, which we call $u_1(x,\tau;\Delta x,\Delta\tau)$. The other part is generated by the term $O(\Delta\tau^3)$, which is denoted by $u_2(x,\tau;\Delta x,\Delta\tau)$. Clearly,

$$u_1(x,\tau;\Delta x,\Delta\tau) - u(x,\tau) + v_1(x,\tau)\Delta x^2 + v_{12}(x,\tau)\Delta x\Delta\tau + v_2(x,\tau)\Delta\tau^2$$

is a term of $O(\Delta\tau^3)$. The function $u_2(x,\tau;\Delta x,\Delta\tau)$ is also a term of $O(\Delta\tau^3)$. Consequently, we have

$$u(x,\tau;\Delta x,\Delta\tau) = u_1(x,\tau;\Delta x,\Delta\tau) + u_2(x,\tau;\Delta x,\Delta\tau)$$

$$= u(x,\tau) - v_1(x,\tau)\Delta x^2 - v_{12}(x,\tau)\Delta x\Delta\tau - v_2(x,\tau)\Delta\tau^2$$

$$+ O(\Delta\tau^3).$$

In this case, we can use the following technique to eliminate the error of $O(\Delta x^2 + \Delta x\Delta\tau + \Delta\tau^2)$ if we have numerical solutions on a mesh with mesh sizes Δx and $\Delta\tau$ and on a mesh with mesh sizes $2\Delta x$ and $2\Delta\tau$. Let us consider a linear combination of the solutions on the two different meshes, which are denoted by $u(x,\tau;\Delta x,\Delta\tau)$ and $u(x,\tau;2\Delta x,2\Delta\tau)$:

$$(1-d) \times u(x,\tau;\Delta x,\Delta\tau) + d \times u(x,\tau;2\Delta x,2\Delta\tau)$$

$$= u(x,\tau) - v_1(x,\tau)(1-d+4d)\Delta x^2 - v_{12}(x,\tau)(1-d+4d)\Delta x\Delta\tau$$

$$- v_2(x,\tau)(1-d+4d)\Delta\tau^2 + O(\Delta\tau^3).$$

If we choose d such that $1 - d + 4d = 0$, that is, $d = -\dfrac{1}{3}$, then

$$(1-d) \times u(x,\tau;\Delta x,\Delta\tau) + d \times u(x,\tau;2\Delta x,2\Delta\tau) = u(x,\tau) + O(\Delta\tau^3).$$

Therefore,

$$\frac{1}{3}[4u(x,\tau;\Delta x,\Delta\tau) - u(x,\tau;2\Delta x,2\Delta\tau)] \tag{5.51}$$

is an approximate to $u(x,y)$ with an error of $O(\Delta\tau^3)$.

However, for the approximation (5.48), the expression of the numerical solution is in the form (5.49), and the extrapolation formula of numerical solutions (5.51) gives an approximation to $u(x,y)$ with an error of $O(\Delta\tau^4)$. This is a special case. Generally speaking, if for a second-order scheme we have three solutions $u_m^n(\Delta x,\Delta\tau)$, $u_m^n(2\Delta x,2\Delta\tau)$, and $u_m^n(4\Delta x,4\Delta\tau)$, then we can have an approximation with an error of $O(\Delta\tau^4)$. In order to do that, we first generate an interpolation function from the values at these nodes and require the interpolation with an error of $O(\Delta\tau^4)$. This can be done, for example, by cubic interpolation. Let $u(x,\tau;\Delta x,\Delta\tau)$, $u(x,\tau;2\Delta x,2\Delta\tau)$, and $u(x,\tau;4\Delta x,4\Delta\tau)$ represent these functions. Then, consider a linear combination of them:

$$(1-d_1-d_2)u(x,\tau;\Delta x,\Delta\tau)+d_1 u(x,\tau;2\Delta x,2\Delta\tau)+d_2 u(x,\tau;4\Delta x,4\Delta\tau).$$

If we choose d_1 and d_2 such that

$$\begin{cases} 1-d_1-d_2+2^2 d_1+4^2 d_2=0, \\ 1-d_1-d_2+2^3 d_1+4^3 d_2=0, \end{cases}$$

which gives

$$\begin{cases} d_1=\dfrac{-12}{21}, \\ d_2=\dfrac{1}{21}, \end{cases}$$

then all the terms of $O(\Delta\tau^2)$ and the terms of $O(\Delta\tau^3)$ in

$$(1-d_1-d_2)u(x,\tau;\Delta x,\Delta\tau)+d_1 u(x,\tau;2\Delta x,2\Delta\tau)+d_2 u(x,\tau;4\Delta x,4\Delta\tau)$$

are eliminated. Therefore

$$\frac{1}{21}[32u(x,\tau;\Delta x,\Delta\tau)-12u(x,\tau;2\Delta x,2\Delta\tau)+u(x,\tau;4\Delta x,4\Delta\tau)] \quad (5.52)$$

gives an approximation to $u(x,\tau)$ with an error of $O(\Delta\tau^4)$ for any second-order scheme.

Here, we need to point out that in order to obtain an approximate solution with an error of $O(\Delta\tau^3)$, it is not necessary for both $\Delta x_1/\Delta x_2$ and $\Delta\tau_1/\Delta\tau_2$ to equal two, where $\Delta x_1,\Delta\tau_1$ are mesh sizes for one mesh and $\Delta x_2,\Delta\tau_2$ for the other. For example, if we have a solution on a 12×16 mesh and a solution on a 9×12 mesh, then we still can obtain an approximate solution with an error of $O(\Delta\tau^3)$ by using extrapolation. Furthermore, if there exist solutions on 15×20, 12×16, and 9×12 meshes, then we can have an approximate solution with an error of $O(\Delta\tau^4)$ by using extrapolation. These are left as a problem for the reader to prove.

The technique of generating more accurate results by combining several numerical results, which is similar to Richardson's extrapolation in numerical methods for ordinary differential equations, is referred to as the extrapolation technique of numerical solutions in next few chapters.

5.6 Determination of Parameters in Models

In order to price an option on a specified underlying asset, we must have a model for the asset. We can have various models, and we have to determine the parameters in the model before pricing. In this section, we will discuss how to determine the parameters in models from the market data.

5.6.1 Constant Variances and Covariances

Assume that the stochastic process of an asset price S can be described by

$$dS = a\,dt + b\,dX,$$

where a and b are constants and dX is a Wiener process. Because we assume that the parameters in the stochastic process do not depend on time, we can determine a and b according to the historical data. Clearly,

$$\mathrm{E}\,[dS] = a\,dt$$

and

$$\mathrm{Var}\,[dS] = \mathrm{E}\left[(dS - a\,dt)^2\right] = \mathrm{E}\left[(b\,dX)^2\right] = b^2\,dt,$$

that is,

$$a = \frac{1}{dt}\mathrm{E}\,[dS]$$

and

$$b^2 = \frac{1}{dt}\mathrm{Var}\,[dS].$$

Suppose that from the market, we have the values of the asset price S at time $t^i = T_1 + (i-1)dt$, $i = 1, 2, \cdots, I+1$. From any statistics textbook, we know that the mean and variance of dS can be approximated by

$$\mathrm{E}\,[dS] \approx \frac{1}{I}\sum_{i=1}^{I} dS_i = \frac{1}{I}\sum_{i=1}^{I}(S_{i+1} - S_i)$$

and

$$\mathrm{Var}\,[dS] \approx \frac{1}{I-1}\sum_{i=1}^{I}\left[S_{i+1} - S_i - \frac{1}{I}\sum_{i=1}^{I}(S_{i+1} - S_i)\right]^2.$$

Thus, we have the estimates for a and b^2 as follows:

$$a \approx \frac{1}{I\,dt}\sum_{i=1}^{I}(S_{i+1} - S_i) \tag{5.53}$$

and

$$b^2 \approx \frac{1}{(I-1)dt}\sum_{i=1}^{I}\left[S_{i+1} - S_i - \frac{1}{I}\sum_{i=1}^{I}(S_{i+1} - S_i)\right]^2$$

$$= \frac{1}{(I-1)dt}\left[\sum_{i=1}^{I}(S_{i+1} - S_i)^2 - \frac{1}{I}\left(\sum_{i=1}^{I}(S_{i+1} - S_i)\right)^2\right]. \tag{5.54}$$

Now suppose

$$dS = \mu S dt + \sigma S dX$$

and let us discuss how to find μ and σ from the market data. Because $dS = \mu S dt + \sigma S dX$ can be written as

$$d \ln S = (\mu - \sigma^2/2)dt + \sigma dX,$$

then we can estimate μ and σ^2 by

$$\sigma^2 \approx \frac{1}{(I-1)dt}\left[\sum_{i=1}^{I}(\ln S_{i+1} - \ln S_i)^2 - \frac{1}{I}\left(\sum_{i=1}^{I}(\ln S_{i+1} - \ln S_i)\right)^2\right]$$

$$\approx \frac{1}{(I-1)dt}\left[\sum_{i=1}^{I}\left(\frac{S_{i+1}-S_i}{S_i}\right)^2 - \frac{1}{I}\left(\sum_{i=1}^{I}\frac{S_{i+1}-S_i}{S_i}\right)^2\right] \qquad (5.55)$$

and

$$\mu - \sigma^2/2 \approx \frac{1}{Idt}\sum_{i=1}^{I}(\ln S_{i+1} - \ln S_i) \approx \frac{1}{Idt}\sum_{i=1}^{I}\frac{S_{i+1}-S_i}{S_i}$$

or

$$\mu \approx \frac{1}{Idt}\sum_{i=1}^{I}\frac{S_{i+1}-S_i}{S_i} + \sigma^2/2. \qquad (5.56)$$

Here, we have used the approximate relation

$$\ln S_{i+1} - \ln S_i \approx \frac{S_{i+1}-S_i}{S_i}.$$

Suppose that there are two stochastic processes:

$$dS_1 = a_1 dt + b_1 dX_1$$

and

$$dS_2 = a_2 dt + b_2 dX_2,$$

where a_1, b_1, a_2, and b_2 are constants, dX_1, dX_2 are two Wiener processes correlated with $\mathrm{E}[dX_1 dX_2] = \rho dt$. Assume that we have the values of the asset prices S_1 and S_2 at time $t^i = T_1 + (i-1)dt$, which are denoted by $S_{1,i}$ and $S_{2,i}$, $i = 1, 2, \cdots, I+1$. We can have estimates for a_1, b_1, a_2, and b_2 by (5.53) and (5.54). Now let us discuss how to estimate ρ from $S_{1,i}$ and $S_{2,i}$, $i = 1, 2, \cdots, I+1$. Because

$$\mathrm{E}[dX_1 dX_2] = \mathrm{E}\left[\frac{dS_1 - a_1 dt}{b_1} \times \frac{dS_2 - a_2 dt}{b_2}\right] = \frac{1}{b_1 b_2}\left\{\mathrm{E}[dS_1 dS_2] - a_1 a_2 dt^2\right\},$$

we have

$$\rho = \frac{1}{b_1 b_2 dt}\left\{\mathrm{E}[dS_1 dS_2] - a_1 a_2 dt^2\right\}.$$

From statistics, we know

$$E\left[dS_1 dS_2\right] \approx \frac{1}{I-1}\sum_{i=1}^{I}\left(S_{1,i+1}-S_{1,i}\right)\left(S_{2,i+1}-S_{2,i}\right),$$

so we have

$$\rho \approx \frac{1}{b_1 b_2 dt}\left[\frac{1}{I-1}\sum_{i=1}^{I}\left(S_{1,i+1}-S_{1,i}\right)\left(S_{2,i+1}-S_{2,i}\right)-a_1 a_2 dt^2\right]. \quad (5.57)$$

On the market, the data are given hourly, daily, and so forth, and only on workdays. Suppose we use the data given daily and the adopted time unit is year. When doing the computation, we should think that dt between two successive workdays is always equal to $1/I_w$, where I_w is the number of workdays per year.

5.6.2 Variable Parameters

From Figs. 1.1–1.7, we can see that the assumption of the volatility being constant might not be a good assumption. For example, Figs. 1.1 and 1.2 show that the prices of IBM and GE stocks have less volatilities if the price is lower. Therefore, we assume that volatilities are functions of stock prices S. That is, the stochastic process of S is described by

$$dS = a\left(S\right)dt + b\left(S\right)dX,$$

where $a\left(S\right)$ and $b\left(S\right)$ are functions of S to be determined. Because we do not assume the dependence of the parameters on time t, we can still determine $a\left(S\right)$ and $b\left(S\right)$ from the historical data.

Again, suppose that we have $I+1$ prices of an asset from the market: S_i, $i = 1, 2, \cdots, I+1$. Let $S_{\max}$ and $S_{\min}$ be the maximum and minimum values among them. Set $S_{(m)} = S_{\min} - \varepsilon + m\left(S_{\max} - S_{\min} + \varepsilon\right)/(M+1)$, $m = 0, 1, \cdots, M+1$, where ε is a small positive number. Clearly, $S_{(0)} = S_{\min} - \varepsilon$ and $S_{(M+1)} = S_{\max}$. The entire interval $\left(S_{(0)}, S_{(M+1)}\right]$ is divided into $M+1$ subintervals $\left(S_{(m-1)}, S_{(m)}\right], m = 1, 2, \cdots, M+1$. Every S_i belongs to one of these subintervals. Consider S_i, $i = 1, 2, \cdots, I$. If $S_i \in \left(S_{(m-1)}, S_{(m)}\right]$, then we say that S_i belongs to the set $\mathcal{S}_{(m)}$. Let I_m be the number of elements in the set $\mathcal{S}_{(m)}$. It is clear that $\sum_{m=1}^{M+1} I_m = I$. For each set $\mathcal{S}_{(m)}$, we can have a mean $a_{(m)}$ and a variance $b_{(m)}^2$ by (5.53) and (5.54).

$b_{(m)}^2$ is an approximate variance of the random variable S at $S = (S_{(m-1)} + S_{(m)})/2$, $m = 1, 2, \cdots, M+1$. We define $S_{(m-1/2)} = \left(S_{(m-1)} + S_{(m)}\right)/2$, so $b(S_{(m-1/2)}) \approx b_{(m)}$. Because S is defined on $[0, \infty)$, $b(S)$ is a function on $[0, \infty)$. However, it is not convenient to approximate the function $b(S)$ defined on an infinite interval. Hence we introduce a transformation

$$\xi = \frac{S}{S + P_m},$$

where P_m is a positive number. This transformation maps $[0, \infty)$ to $[0, 1)$. Therefore, we assume that $b(S)$ is in the form $\bar{b}(\xi)$ and find $\bar{b}(\xi)$ on the interval $[0, 1)$. It is clear that $b_{(m)}$ should be an approximation to $\bar{b}\left(\xi_{(m-1/2)}\right)$, where $\xi_{(m-1/2)} = \dfrac{S_{(m-1/2)}}{S_{(m-1/2)} + P_m}$. Now the problem is reduced to finding a function $\bar{b}(\xi)$ such that the points $\left(\xi_{(m-1/2)}, b_{(m)}\right)$, $m = 1, 2, \cdots, M+1$, are as close to $\bar{b}(\xi)$ as possible. Assume

$$\bar{b}(\xi) = g(\xi) \sum_{n=0}^{N} a_n \xi^n,$$

where $N < M$ and $g(\xi)$ is a given function, for example, $g(\xi) = 1$ or $\dfrac{P_m \xi}{1 - \xi}$.

Under this assumption, using the points $\left(\xi_{\left(m-\frac{1}{2}\right)}, b_{(m)}\right)$, $m = 1, 2, \cdots, M+1$ and taking the weights $b_m = I_m/I$, we can find $a_0, a_1, \cdots, a_N$ by the least squares method with weights in Subsection 5.1.4. As soon as we find $\bar{b}(\xi)$, we have $b(S)$ by

$$b(S) = g\left(\frac{S}{S + P_m}\right) \sum_{n=0}^{N} a_n \left(\frac{S}{S + P_m}\right)^n.$$

If $b(S) < 0$ in some small regions, then a local modification is needed in order to guarantee $b(S) \geq 0$ for all $S \in [0, \infty)$. For $a(S)$, the method is similar.

Now let us discuss the case involving several stochastic processes. For simplicity, suppose we have two stochastic processes governed by

$$dS_1 = a_1(S_1)dt + b_1(S_1)dX_1$$

and

$$dS_2 = a_2(S_2)dt + b_2(S_2)dX_2$$

with $\mathrm{E}\left[dX_1 dX_2\right] = \rho dt$, ρ being a constant. Using the method given above, we can find $a_1(S_1)$, $b_1(S_1)$, $a_2(S_2)$, and $b_2(S_2)$. Because we assume that ρ is a constant, it can be determined by

$$\begin{aligned}
\rho &= \frac{1}{dt}\mathrm{E}\left[dX_1 dX_2\right] \\
&= \frac{1}{dt}\mathrm{E}\left[\frac{dS_1 - a_1(S_1)dt}{b_1(S_1)} \times \frac{dS_2 - a_2(S_2)dt}{b_2(S_2)}\right] \\
&\approx \frac{1}{(I-1)dt}\sum_{i=1}^{I}\left[\frac{S_{1,i+1} - S_{1,i} - a_1(S_{1,i})dt}{b_1(S_{1,i})} \times \frac{S_{2,i+1} - S_{2,i} - a_2(S_{2,i})dt}{b_2(S_{2,i})}\right].
\end{aligned}$$

Problems

1. Suppose $x_m = m\Delta x$.
   a) Find the order of the error of the following approximate function

   $$u(x) \approx \frac{x_{m+1} - x}{\Delta x} u(x_m) + \frac{x - x_m}{\Delta x} u(x_{m+1})$$

   without using Theorem 5.1. Here, $x \in [x_m, x_{m+1}]$.
   b) Find the order of the error of the following approximate function

   $$u(x) \approx \frac{(x - x_m)(x - x_{m+1})}{2\Delta x^2} u(x_{m-1})$$
   $$- \frac{(x - x_{m-1})(x - x_{m+1})}{\Delta x^2} u(x_m)$$
   $$+ \frac{(x - x_{m-1})(x - x_m)}{2\Delta x^2} u(x_{m+1})$$

   without using Theorem 5.1. Here, $x \in [x_{m-1}, x_{m+1}]$.

2. Show that from

   $$\begin{cases} a_{m+1} = a_m + b_m h_m + c_m h_m^2 + d_m h_m^3, \\ b_{m+1} = b_m + 2c_m h_m + 3d_m h_m^2, \\ c_{m+1} = c_m + 3d_m h_m, \\ \qquad m = 0, 1, \cdots, M - 2, \end{cases}$$

   and

   $$\begin{cases} a_M = a_{M-1} + b_{M-1} h_{M-1} + c_{M-1} h_{M-1}^2 + d_{M-1} h_{M-1}^3, \\ c_M = c_{M-1} + 3d_{M-1} h_{M-1}, \end{cases}$$

   the following relation can be derived:

   $$h_{m-1} c_{m-1} + 2(h_{m-1} + h_m) c_m + h_m c_{m+1}$$
   $$= \frac{3(a_{m+1} - a_m)}{h_m} - \frac{3(a_m - a_{m-1})}{h_{m-1}},$$
   $$m = 1, 2, \cdots, M - 1.$$

3. Suppose $x_m = m\Delta x$ and $\tau^n = n\Delta\tau$. Find the error of each of the following approximations:
   a)

   $$u(x_m, \tau^{n+1/2}) \approx \frac{u(x_m, \tau^{n+1}) + u(x_m, \tau^n)}{2};$$

   b)

   $$\frac{\partial u}{\partial \tau}(x_m, \tau^n) \approx \frac{u(x_m, \tau^{n+1}) - u(x_m, \tau^n)}{\Delta\tau};$$

c)

$$\frac{\partial u}{\partial \tau}(x_m, \tau^{n+1/2}) \approx \frac{u(x_m, \tau^{n+1}) - u(x_m, \tau^n)}{\Delta \tau};$$

d)

$$\frac{\partial u}{\partial x}(x_m, \tau^n) \approx \frac{u(x_{m+1}, \tau^n) - u(x_m, \tau^n)}{\Delta x};$$

e)

$$\frac{\partial u}{\partial x}(x_m, \tau^n) \approx \frac{u(x_{m+1}, \tau^n) - u(x_{m-1}, \tau^n)}{2\Delta x};$$

f)

$$\frac{\partial u}{\partial x}(x_m, \tau^n) \approx \frac{3u(x_m, \tau^n) - 4u(x_{m-1}, \tau^n) + u(x_{m-2}, \tau^n)}{2\Delta x};$$

g)

$$\frac{\partial^2 u}{\partial x^2}(x_m, \tau^n) \approx \frac{u(x_{m+1}, \tau^n) - 2u(x_m, \tau^n) + u(x_{m-1}, \tau^n)}{\Delta x^2};$$

h)

$$\frac{\partial^2 u}{\partial x \partial y}(x_m, y_l, \tau^n) \approx \frac{1}{2\Delta x}\left[\frac{u(x_{m+1}, y_{l+1}, \tau^n) - u(x_{m+1}, y_{l-1}, \tau^n)}{2\Delta y}\right.$$
$$\left. - \frac{u(x_{m-1}, y_{l+1}, \tau^n) + u(x_{m-1}, y_{l-1}, \tau^n)}{2\Delta y}\right].$$

4. Derive the formulae of the LU decomposition method for the following almost tridiagonal system

$$\mathbf{A}\mathbf{x} = \mathbf{q},$$

where

$$\mathbf{A} = \begin{bmatrix} b_1 & c_1 & & & & d_1 \\ a_2 & b_2 & c_2 & & 0 & d_2 \\ & & \ddots & \ddots & \ddots & & \vdots \\ & & & \ddots & \ddots & \ddots & \vdots \\ & 0 & & & a_{m-1} & b_{m-1} & d_{m-1} \\ & & & & & a_m & d_m \end{bmatrix},$$

$$\mathbf{x} = \begin{bmatrix} x_1 \\ x_2 \\ \vdots \\ x_m \end{bmatrix}, \qquad \mathbf{q} = \begin{bmatrix} q_1 \\ q_2 \\ \vdots \\ q_m \end{bmatrix}.$$

5. *Describe the Jacobi iteration, the Gauss–Seidel iteration, and the method of successive over relaxation for an $n \times n$ system of linear equations.

6. *Suppose $f(x) = 0$ is a nonlinear equation. Derive the iteration formulae of Newton's method and the secant method for solving the nonlinear equation.

7. Suppose

$$\mathbf{A}_1 = \mathbf{J_f}(\mathbf{x}^{(0)}) + \frac{\mathbf{f}(\mathbf{x}^{(1)}) - \mathbf{f}(\mathbf{x}^{(0)}) - \mathbf{J_f}(\mathbf{x}^{(0)})(\mathbf{x}^{(1)} - \mathbf{x}^{(0)})}{\left\| \mathbf{x}^{(1)} - \mathbf{x}^{(0)} \right\|_2^2}(\mathbf{x}^{(1)} - \mathbf{x}^{(0)})^T.$$

Show that the following relations hold:

$$\mathbf{A}_1(\mathbf{x}^{(1)} - \mathbf{x}^{(0)}) = \mathbf{f}(\mathbf{x}^{(1)}) - \mathbf{f}(\mathbf{x}^{(0)})$$

and

$$\mathbf{A}_1 \mathbf{z} = \mathbf{J_f}(\mathbf{x}^{(0)})\mathbf{z} \quad \text{whenever} \quad (\mathbf{x}^{(1)} - \mathbf{x}^{(0)})^T \mathbf{z} = 0.$$

8. Prove that if $\mathbf{A} \in \mathbb{R}^{n \times n}$ is nonsingular, $\mathbf{x}, \mathbf{y} \in \mathbb{R}^n$, and $\mathbf{y}^T \mathbf{A}^{-1} \mathbf{x} \neq -1$, then $\mathbf{A} + \mathbf{x}\mathbf{y}^T$ is also nonsingular, moreover,

$$(\mathbf{A} + \mathbf{x}\mathbf{y}^T)^{-1} = \mathbf{A}^{-1} - \frac{\mathbf{A}^{-1}\mathbf{x}\mathbf{y}^T \mathbf{A}^{-1}}{1 + \mathbf{y}^T \mathbf{A}^{-1} \mathbf{x}}.$$

9. a) *Show

$$\mathbf{H}_m \mathbf{x} = \beta \mathbf{e}_1,$$

where

$$\mathbf{x} = [x_1, x_2, \cdots, x_m]^T,$$
$$\beta = \sqrt{\mathbf{x}^T \mathbf{x}},$$
$$\mathbf{H}_m = \mathbf{I}_m - \frac{1}{\beta(\beta - x_1)}\mathbf{u}\mathbf{u}^T,$$
$$\mathbf{u} \text{ being } \mathbf{x} - \beta \mathbf{e}_1.$$

   b) *Using the result in (a), design a method to obtain an orthogonal matrix $\mathbf{Q}$ from $\mathbf{A}$ such that $\mathbf{A} = \mathbf{Q}\mathbf{R}$, where $\mathbf{R}$ is an upper triangular matrix with nonnegative diagonals.

10. *Let f_m^n denote $f(m\Delta x, n\Delta\tau)$. Find the truncation error of the explicit difference scheme

$$\frac{u_m^{n+1} - u_m^n}{\Delta\tau} = a_m^n \frac{u_{m+1}^n - 2u_m^n + u_{m-1}^n}{\Delta x^2}$$
$$+ b_m^n \frac{u_{m+1}^n - u_{m-1}^n}{2\Delta x} + c_m^n u_m^n$$

to the parabolic partial differential equation

$$\frac{\partial u}{\partial \tau} = a(x, \tau)\frac{\partial^2 u}{\partial x^2} + b(x, \tau)\frac{\partial u}{\partial x} + c(x, \tau)u.$$

11. Show that the truncation error of the Crank–Nicloson scheme for the heat equation at the point $(x_m, \tau^{n+1/2})$ is in the following form:

$$\Delta\tau^2 \left[\frac{1}{24} \frac{\partial^3 u}{\partial \tau^3}(x_m, \eta^{(1)}) - \frac{a}{8} \frac{\partial^4 u}{\partial x^2 \partial \tau^2}(x_m, \eta^{(2)}) \right] - \frac{\Delta x^2 a}{12} \frac{\partial^4 u}{\partial x^4}(\xi, \eta^{(3)}),$$

where $\xi \in (x_{m-1}, x_{m+1})$ and $\eta^{(k)} \in (\tau^n, \tau^{n+1})$, $k = 1, 2, 3$.

12. *Let f_m^n denote $f(m\Delta x, n\Delta\tau)$. Find the truncation error of the implicit difference scheme

$$\frac{u_m^{n+1} - u_m^n}{\Delta\tau}$$

$$= \frac{a_m^{n+1/2}}{2} \left(\frac{u_{m+1}^{n+1} - 2u_m^{n+1} + u_{m-1}^{n+1}}{\Delta x^2} + \frac{u_{m+1}^n - 2u_m^n + u_{m-1}^n}{\Delta x^2} \right)$$

$$+ \frac{b_m^{n+1/2}}{2} \left(\frac{u_{m+1}^{n+1} - u_{m-1}^{n+1}}{2\Delta x} + \frac{u_{m+1}^n - u_{m-1}^n}{2\Delta x} \right) + \frac{c_m^{n+1/2}}{2} (u_m^{n+1} + u_m^n)$$

to the parabolic partial differential equation

$$\frac{\partial u}{\partial \tau} = a(x, \tau) \frac{\partial^2 u}{\partial x^2} + b(x, \tau) \frac{\partial u}{\partial x} + c(x, \tau) u.$$

13. The heat equation

$$\frac{\partial u}{\partial \tau} = \frac{\partial^2 u}{\partial x^2}$$

can also be discretized by

$$\frac{u_m^{n+1} - u_m^n}{\Delta\tau} = \theta \left(\frac{u_{m+1}^{n+1} - 2u_m^{n+1} + u_{m-1}^{n+1}}{\Delta x^2} \right) + (1-\theta) \left(\frac{u_{m+1}^n - 2u_m^n + u_{m-1}^n}{\Delta x^2} \right)$$

or

$$u_m^{n+1} - \theta\alpha(u_{m+1}^{n+1} - 2u_m^{n+1} + u_{m-1}^{n+1}) = u_m^n + (1-\theta)\alpha(u_{m+1}^n - 2u_m^n + u_{m-1}^n),$$

where $0 \le \theta \le 1$ and $\alpha = \Delta\tau/\Delta x^2$. This scheme is called the θ–scheme. It is clear that when $\theta = 0$, the scheme reduces to the explicit scheme and when $\theta = 1/2$, the scheme becomes the Crank–Nicolson scheme. Show that the truncation error of the θ–scheme is

$$O\left((1 - 2\theta)\Delta\tau + \Delta\tau^2 + \Delta x^2\right).$$

(Hint: Discretize the partial differential equation at $x = x_m$ and $\tau = \tau^{n+\theta}$.)

14. *Consider the three-point explicit finite-difference scheme:

$$u_m^{n+1} = a_m u_{m-1}^n + b_m u_m^n + c_m u_{m+1}^n, \quad m = 0, 1, \cdots, M,$$

where $a_m \geq 0$, $b_m = 1 - a_m - c_m \geq 0$, $c_m \geq 0$ and $a_0 = c_M = 0$. Show

$$\max_{0 \leq m \leq M} |u_m^{n+1}| \leq \max_{0 \leq m \leq M} |u_m^n|.$$

This means that the numerical procedure is stable under the maximum norm.

15. *Consider the explicit scheme

$$\frac{u_m^{n+1} - u_m^n}{\Delta \tau} = a \frac{u_{m+1}^n - 2u_m^n + u_{m-1}^n}{\Delta x^2}, \quad m = 1, 2, \cdots, M - 1$$

with $u_0^{n+1} = f_l(\tau^{n+1})$ and $u_M^{n+1} = f_u(\tau^{n+1})$. Determine when it is stable with respect to initial values in L_2 norm and when it is unstable. (Suppose $a > 0$.)

16. *Consider the implicit scheme

$$\frac{u_m^{n+1} - u_m^n}{\Delta \tau} = \frac{a}{2} \left(\frac{u_{m+1}^{n+1} - 2u_m^{n+1} + u_{m-1}^{n+1}}{\Delta x^2} + \frac{u_{m+1}^n - 2u_m^n + u_{m-1}^n}{\Delta x^2} \right),$$

$$m = 1, 2, \cdots, M - 1$$

with $u_0^{n+1} = f_l(\tau^{n+1})$ and $u_M^{n+1} = f_u(\tau^{n+1})$. Show that it is always stable with respect to initial values in L_2 norm. (Suppose $a > 0$.)

17. By using the von Neumann method, show that for periodic problems, the θ–scheme (see Problem 13) is stable for all $\alpha > 0$ if $\frac{1}{2} \leq \theta \leq 1$ and that it is stable for $0 < \alpha \leq \frac{1}{2(1 - 2\theta)}$ if $0 < \theta < \frac{1}{2}$.

18. Show that if

$$\max_{0 \leq m \leq M} \frac{x_m^2 (1 - x_m)^2 \bar{\sigma}_m^2}{2} \frac{\Delta \tau}{\Delta x^2} \leq \frac{1}{2},$$

then for the scheme with variable coefficients

$$\frac{u_m^{n+1} - u_m^n}{\Delta \tau} = \frac{1}{2} [x_m (1 - x_m) \bar{\sigma}_m]^2 \frac{u_{m+1}^n - 2u_m^n + u_{m-1}^n}{\Delta x^2}$$
$$+ (r - D_0) x_m (1 - x_m) \frac{u_{m+1}^n - u_{m-1}^n}{2\Delta x}$$
$$- [r (1 - x_m) + D_0 x_m] u_m^n,$$

the condition $|\lambda_\theta (x_m, \tau^n)| \leq 1 + O(\Delta \tau)$ is satisfied for any $x_m = m/M \in [0, 1]$. (When you prove this result, you should derive the condition (5.47) by yourself.)

19. For the scheme with variable coefficients

$$\frac{u_m^{n+1} - u_m^n}{\Delta \tau}$$
$$= \frac{1}{4} (x_m (1 - x_m) \bar{\sigma}_m)^2 \left(\frac{u_{m+1}^{n+1} - 2u_m^{n+1} + u_{m-1}^{n+1}}{\Delta x^2} + \frac{u_{m+1}^n - 2u_m^n + u_{m-1}^n}{\Delta x^2} \right),$$

$$+ \frac{1}{2}(r - D_0)x_m(1 - x_m)\left(\frac{u_{m+1}^{n+1} - u_{m-1}^{n+1}}{2\Delta x} + \frac{u_{m+1}^n - u_{m-1}^n}{2\Delta x}\right),$$

show that the condition $|\lambda_\theta(x_m, \tau^n)| \leq 1$ is satisfied for any $x \in [0, 1]$.

20. a) Suppose we have two approximate solutions with errors of $O(\Delta\tau^2)$:

$$u\left(x, \tau; \frac{1}{12}, \frac{T}{16}\right) \text{ and } u\left(x, \tau; \frac{1}{9}, \frac{T}{12}\right). \text{ Find a linear combination}$$

$$(1 - d) \times u\left(x, \tau; \frac{1}{12}, \frac{T}{16}\right) + d \times u\left(x, \tau; \frac{1}{9}, \frac{T}{12}\right)$$

such that it is an approximate solution with an error of $O(\Delta\tau^3)$.

   b) Suppose we further have another approximate solution with errors of $O(\Delta\tau^2)$: $u\left(x, \tau; \frac{1}{15}, \frac{T}{20}\right)$. Find a linear combination

$$d_0 \times u\left(x, \tau; \frac{1}{15}, \frac{T}{20}\right) + d_1 \times u\left(x, \tau; \frac{1}{12}, \frac{T}{16}\right) + d_2 \times u\left(x, \tau; \frac{1}{9}, \frac{T}{12}\right)$$

such that it is an approximate solution with an error of $O(\Delta\tau^4)$, where $d_0 = 1 - d_1 - d_2$.

21. *Assume that the volatility of a stock is a function of the stock price. Describe a method determining the function from the market data.

Projects

General Requirements

A) *Submit a code or codes in C or C++ that will work on a computer the instructor can get access to. At the beginning of the code, write down the name of the student and indicate on which computer it works and the procedure to make it work.*

B) *Each code should use an input file to specify all the problem parameters and the computational parameters and an output file to store all the results. In an output file, the name of the problem, all the problem parameters, and the computational parameters should be given, so that one can know what the results are and how they were obtained. The input file should be submitted with the code.*

C) *Submit results in form of tables. When a result is given, always provide the problem parameters and the computational parameters.*

1. **Cumulative distribution function and Black–Scholes formulae.** Write a function

   double N(double z)

for computing approximate values of the cumulative distribution function for the standardized normal variable, where z is the independent variable (see Subsection 2.4.2). Also write a function

double BS(double S, double E, double T, double r, double D0, double sigma, int option),

which gives prices of the European options by using Black–Scholes formulae (see Subsection 2.4.2). When the value of the integer 'option' is equal to 0, the value of the European call needs to be evaluated. Otherwise, the value of the European put needs to be evaluated.

- Give the values of $N(z)$ for $z = -1, -2, 0, 1, 2$.
- For European call and put options, give the results for the cases: $S = 100$, $E = 95, 100, 105$, $T = 1$, $r = 0.1$, $D_0 = 0.05$, $\sigma = 0.2$.
- For European call and put options, give the results for the cases: $S = 100$, $E = 95, 100, 105$, $T = 1$, $r = 0.05$, $D_0 = 0.1$, $\sigma = 0.2$.

2. **Quadratic interpolation and LU decomposition of a tridiagonal system.**   Write a function

double Qint(double x, double x1, double x2, double x3, double f1, double f2, double f3)

for the the quadratic interpolation method (see Subsection 5.1.1). Also write a function

int LUT(int m, double *a, double *b, double *c, double *q).

Suppose that we have a tridiagonal system (5.10). The number of unknowns is given in the integer 'm.' The nonhomogeneous term q_i is given in q[i-1] (the i-th component of the array 'q'). The coefficients a_i, b_i, and c_i are given in the i-th component of the arrays 'a,' 'b,' and 'c,' respectively. Write a function to solve the system by using the method described in Subsection 5.2.1. If all the u_i are not equal to zero, then the code should return an integer number 0 and gives the value of the i-th unknown in the i-th component of the array q. If one of u_i is equal to zero, then the solution(s) of the system cannot be found by the method (or the system has no solution), and the code should return an integer number 1. The values in the arrays 'a', 'b' and 'c' are required unchanged.

- Let $x_1 = 0.1$, $x_2 = 0.2$, $x_3 = 0.4$, $f_1 = 5$, $f_2 = 10$, and $f_3 = 15$. For $x = 0.11, 0.19, 0.35$, find the values of f.
- Let $m = 4$, $a_2 = a_3 = a_4 = -0.48$, $b_1 = b_2 = b_3 = b_4 = 1$, $c_1 = c_2 = c_3 = -0.49$, $q_1 = 0.02$, $q_2 = 0.05$, $q_3 = 0.08$, and $q_4 = 2.56$. Find the solution of the system (5.10).

6

Initial-Boundary Value and LC Problems

Evaluation of European-style derivatives can be reduced to solving initial value or initial-boundary value problems of parabolic partial differential equations. This chapter discusses numerical methods for such problems. If an American option problem is formulated as a linear complementarity problem, then the only difference between solving a European option and an American option is that if the solution obtained by the partial differential equation does not satisfy the constraint at some point, then the solution of the PDE at the point should be replaced by the value determined from the constraint condition. Such methods are usually referred to as projected methods for American-style derivatives. Therefore, the two methods are very close, and we also study the projected methods in this chapter.

In this chapter, there are four sections. The first two sections are devoted to explicit and implicit schemes, respectively. As we know, the derivative of the function representing the payoff of an option usually is discontinuous. This fact makes numerical methods inefficient. In many cases, an option problem can be reduced to another problem that has either a smooth solution or a solution with a weaker singularity than the solution of the option problem itself, and the numerical solution of the new problem can be obtained efficiently. We call such a method the singularity-separating method. In Section 6.3, we give several examples to illustrate how such a method works. In the final section, we discuss the pseudo-spectral method, which is very efficient if the solution is smooth. Examples are given to explain this fact.

6.1 Explicit Methods

6.1.1 Pricing European Options by Using $\overline{V}$, ξ, τ or u, x, $\bar{\tau}$ Variables

In Section 2.3, we obtained the formulation of the problem satisfied by a call/put option on a finite domain:

$$
\begin{cases}
\dfrac{\partial \overline{V}}{\partial \tau} = \dfrac{1}{2}\bar{\sigma}^2(\xi)\xi^2(1-\xi)^2\dfrac{\partial^2 \overline{V}}{\partial \xi^2} + (r-D_0)\xi(1-\xi)\dfrac{\partial \overline{V}}{\partial \xi} \\[2mm]
\qquad\qquad -[r(1-\xi)+D_0\xi]\overline{V}, \qquad 0\le\xi\le1, \quad 0\le\tau\le T, \\[2mm]
\overline{V}(\xi,0) = \max(\pm(2\xi-1),0), \qquad 0\le\xi\le1,
\end{cases}
\tag{6.1}
$$

where the sign "+" in $\pm$ corresponds to the call option and the sign "$-$" in $\pm$ the put option. Here, we assume that the volatility depends on S, so $\bar{\sigma}$ is a function of ξ in the equation. Let

$$
\begin{cases}
\xi_m = m\Delta\xi, \ m = 0,1,\cdots,M, \\[2mm]
\tau^n = n\Delta\tau, \ n = 0,1,\cdots,N,
\end{cases}
\tag{6.2}
$$

where M and N are given integers, and $\Delta\xi = 1/M$ and $\Delta\tau = T/N$. This means that we use an $M \times N$ equidistant mesh on the domain $[0,1] \times [0,T]$. Let v_m^n denote the approximate value of $\overline{V}(\xi,\tau)$ at $\xi = \xi_m$ and $\tau = \tau^n$, and $\{v_m^n\}$ represent the set v_m^n, $m = 0,1,\cdots,M$. Discretizing the partial differential equation in (6.1) at the point (ξ_m, τ^n) by scheme (5.30), i.e., by using the forward difference for $\dfrac{\partial \overline{V}}{\partial \tau}$ and the central difference for $\dfrac{\partial^2 \overline{V}}{\partial \xi^2}$ and $\dfrac{\partial \overline{V}}{\partial \xi}$, we get

$$
\begin{aligned}
\frac{v_m^{n+1}-v_m^n}{\Delta\tau} &= \frac{1}{2}\bar{\sigma}_m^2\xi_m^2(1-\xi_m)^2\frac{v_{m+1}^n-2v_m^n+v_{m-1}^n}{\Delta\xi^2} \\[2mm]
&\quad +(r-D_0)\xi_m(1-\xi_m)\frac{v_{m+1}^n-v_{m-1}^n}{2\Delta\xi} \\[2mm]
&\quad -[r(1-\xi_m)+D_0\xi_m]v_m^n
\end{aligned}
$$

or

$$
\begin{aligned}
v_m^{n+1} &= \frac{1}{2}\left[\bar{\sigma}_m^2\xi_m^2(1-\xi_m)^2 - (r-D_0)\xi_m(1-\xi_m)\Delta\xi\right]\alpha v_{m-1}^n \\[2mm]
&\quad +[1-\bar{\sigma}_m^2\xi_m^2(1-\xi_m)^2\alpha - (r(1-\xi_m)+D_0\xi_m)\Delta\tau]v_m^n \\[2mm]
&\quad +\frac{1}{2}\left[\bar{\sigma}_m^2\xi_m^2(1-\xi_m)^2 + (r-D_0)\xi_m(1-\xi_m)\Delta\xi\right]\alpha v_{m+1}^n, \\[2mm]
&\qquad m = 0,1,\cdots,M, \quad n = 0,1,\cdots,N-1,
\end{aligned}
\tag{6.3}
$$

where

$$
\alpha = \frac{\Delta\tau}{\Delta\xi^2}.
$$

In order for scheme (6.3) to be stable, we require

$$
\max_{0\le m\le M}[\bar{\sigma}_m^2\xi_m^2(1-\xi_m)^2]\frac{\Delta\tau}{2\Delta\xi^2} \le \frac{1}{2}
$$

because if this is true, then $|\lambda_\theta| \le 1 + O(\Delta\tau)$ (see Problem 18 in Chapter 5). In practice, we can replace this condition by a slightly stronger condition as follows:

$$\Delta\tau \le \frac{16\Delta\xi^2}{\max_{0 \le m \le M} \bar{\sigma}_m^2}. \tag{6.4}$$

From (6.3), we know that when the values v_{m-1}^n, v_m^n, and v_{m+1}^n are given, v_m^{n+1} can be obtained immediately. At a glance, it appears that v_{-1}^n and v_{M+1}^n are needed when v_0^{n+1} and v_M^{n+1} are calculated. As pointed out in Section 5.3, because the coefficients of v_{-1}^n and v_{M+1}^n equal zero, the values of v_{-1}^n and v_{M+1}^n will not be used. Therefore, if $\{v_m^n\}$ is given, then $\{v_m^{n+1}\}$ can be obtained by (6.3). According to the initial condition given in (6.1), we have

$$v_m^0 = \max(\pm(2\xi_m - 1), 0).$$

Therefore, from $\{v_m^0\}$, we can get $\{v_m^n\}$, $n = 1, 2, \cdots, N$ successively. Usually, we need the value of V at a certain point S^* at time zero. After $\{v_m^N\}$ have been obtained, $V(S^*, 0)$ can be found in the following way. First, we need to find $v(\xi^*, T)$ by using the quadratic interpolation given in Section 5.1, where $\xi^* = \dfrac{S^*}{S^* + E}$. Then, we can obtain $V(S^*, 0)$ from $v(\xi^*, T)$ by

$$V(S^*, 0) = (S^* + E)v(\xi^*, T).$$

This method works not only for a constant σ but also for a variable σ, namely, $\sigma = \sigma(S)$, even $\sigma = \sigma(S, t)$. In what follows, this scheme is referred to as the explicit finite-difference method I, and its abbreviation is EFDI.

If σ is a constant, then an alternative way to find the approximate solution of the European options is to use u, x, $\bar{\tau}$ variables. From Section 2.3, we know that if $E = 1$, i.e., if the stock price and the option price has been divided by the exercise price, then pricing a call/put option can be reduced to finding $u(x, \bar{\tau})$, which is the solution of the problem:

$$\begin{cases} \dfrac{\partial u}{\partial \bar{\tau}} = \dfrac{\partial^2 u}{\partial x^2}, & -\infty < x < \infty, \quad 0 \le \bar{\tau} \le \frac{1}{2}\sigma^2 T, \\ u(x, 0) = \max(\pm(e^x - 1), 0), & -\infty < x < \infty. \end{cases} \tag{6.5}$$

Here,

$$x = \ln S + (r - D_0 - \sigma^2/2)(T - t), \quad \bar{\tau} = \sigma^2(T - t)/2$$

and

$$u(x, \bar{\tau}) = e^{r(T-t)}V(S, t).$$

Let $x_m = a + m\Delta x$, a being a given number and $\bar{\tau}^n = n\Delta\bar{\tau}$, and let u_m^n denote the approximate value of $u(x_m, \bar{\tau}^n)$. Then, the partial differential equation can be discretized by (5.33):

$$u_m^{n+1} = \bar{\alpha}u_{m+1}^n + (1 - 2\bar{\alpha})u_m^n + \bar{\alpha}u_{m-1}^n, \tag{6.6}$$

where

$$\bar{\alpha} = \frac{\Delta \bar{\tau}}{\Delta x^2}.$$

From Subsection 5.4.1, we know that in order for the scheme to be stable, we need to require

$$\bar{\alpha} = \frac{\Delta \bar{\tau}}{\Delta x^2} \le \frac{1}{2}. \tag{6.7}$$

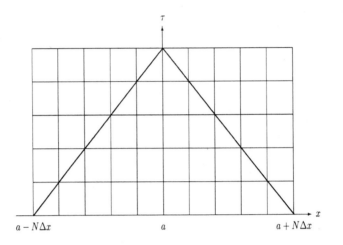

Fig. 6.1. A triangle mesh ($N = 5$)

Suppose again that we need to find $V(S^*, 0)$, i.e., we need to know $u \left(\ln S^* + (r - D_0 - \sigma^2/2)T, \sigma^2 T/2 \right)$. Assume that we will use N steps in $\bar{\tau}$ direction, i.e., $\Delta \bar{\tau} = \dfrac{\sigma^2 T}{2N}$. In order to find $u \left(\ln S^* + (r - D_0 - \sigma^2/2)T, \sigma^2 T/2 \right)$, we can use a triangle mesh (see Fig. 6.1): $\bar{\tau}^n = n\Delta\bar{\tau}, n = 0, 1, \cdots, N$ and for each n, $x_m = \ln S^* + (r - D_0 - \sigma^2/2)T + m\Delta x$, $m = -N + n, -N + n + 1, \cdots, N - n - 1, N - n$. From the initial condition at $\bar{\tau} = 0$, we have

$$u_m^0 = \max\left(\pm(e^{x_m} - 1), 0\right), \quad m = -N, -N + 1, \cdots, N - 1, N.$$

It is clear that when $u_m^n, m = -N+n, -N+n+1, \cdots, N-n-1, N-n$ are given, we can obtain $u_m^{n+1}, m = -N + n + 1, -N + n + 2, \cdots, N - n - 2, N - n - 1$. Therefore, starting from $u_m^0, m = -N, -N + 1, \cdots, N - 1, N$, we can find $u_m^n, m = -N + n, -N + n + 1, \cdots, N - n - 1, N - n$ for $n = 1, 2, \cdots, N$ successively. When we get u_0^N, $V(S^*, 0)$ can be calculated by

$$V(S^*, 0) = e^{-rT} u_0^N$$

because $V(S, t) = e^{-r(T-t)} u(\ln S + (r - D_0 - \sigma^2/2)(T - t), \sigma^2(T - t)/2)$.

Table 6.1. Values of European put options (EFDI)

$(E = 50, S = 48, r = 0.05, \sigma = 0.20, \text{ and } D_0 = 0)$

$\Delta\tau$	$T = 0.25$	$T = 0.50$	$T = 0.75$	$T = 1.00$
0.01	2.7220	3.1163	3.4045	3.5852
0.001	2.7087	3.1275	3.3989	3.5910
0.0001	2.7083	3.1272	3.3986	3.5907
Exact	$2.708349\cdots$	$3.127199\cdots$	$3.398586\cdots$	$3.590738\cdots$

Assume that we want to calculate the value of an option on a stock when the stock price is \$100 and the exercise price is \$90. In this method above, the stock price and the option price has been divided by E, so S^* should be 100/90, and the real option price should be obtained by $90 \times V(S^*, 0)$. This method is referred to as the explicit finite-difference method II, and its abbreviation is EFDII.

Example 1: Using EFDI with

$$\Delta\xi = \sqrt{\max_{0 \le m \le M} \bar{\sigma}_m^2 \Delta\tau / 4},$$

we have solved European put problems using different $\Delta\tau$. Numerical results for $T = 0.25, 0.5, 0.75,$ and 1.00 are listed in Table 6.1, and the other problem parameters are also shown there. From the table we see that for $\Delta\tau = 0.01, 0.001,$ and 0.0001, the error is about on the second, third, and fourth decimal places.

6.1.2 Projected Methods for LC Problems

In Section 2.5, we saw that an American option problem could be formulated as a linear complementarity problem. When the variables $\overline{V}, \xi, \tau$ are adopted, the linear complementarity problem is

$$\begin{cases} \left(\dfrac{\partial \overline{V}}{\partial \tau} - \mathbf{L}_\xi \overline{V} \right) \left[\overline{V}(\xi, \tau) - \max(\pm(2\xi - 1), 0) \right] = 0, \\[2mm] \dfrac{\partial \overline{V}}{\partial \tau} - \mathbf{L}_\xi \overline{V} \ge 0, \\[2mm] \overline{V}(\xi, \tau) - \max(\pm(2\xi - 1), 0) \ge 0, \\[2mm] \overline{V}(\xi, 0) = \max(\pm(2\xi - 1), 0), \end{cases} \qquad (6.8)$$

where

$$\mathbf{L}_\xi = \frac{1}{2}\bar{\sigma}^2(\xi)\xi^2(1 - \xi)^2 \frac{\partial^2}{\partial \xi^2} + (r - D_0)\xi(1 - \xi)\frac{\partial}{\partial \xi} - [r(1 - \xi) + D_0\xi];$$

whereas if the variables $u, x, \bar{\tau}$ are used, the linear complementarity problem is

$$
\begin{cases}
\left(\dfrac{\partial u}{\partial \bar{\tau}} - \dfrac{\partial^2 u}{\partial x^2}\right)[u(x,\bar{\tau}) - g(x,\bar{\tau})] = 0, \\[3mm]
\dfrac{\partial u}{\partial \bar{\tau}} - \dfrac{\partial^2 u}{\partial x^2} \geq 0, \\[3mm]
u(x,\bar{\tau}) - g(x,\bar{\tau}) \geq 0, \\[3mm]
u(x,0) = g(x,0),
\end{cases}
\tag{6.9}
$$

where

$$
g(x,\bar{\tau}) = \max\left(\pm(e^{x+(2D_0/\sigma^2+1)\bar{\tau}} - e^{2r\bar{\tau}/\sigma^2}), 0\right).
$$

Such a formulation can be described in another way. Let us take (6.9) as an example in order to explain it. Suppose that we have obtained the solution at $\bar{\tau} = \bar{\tau}^*$, $u(x,\bar{\tau}^*)$. Starting from $u(x,\bar{\tau}^*)$, we can find the solution $u(x,\bar{\tau}^* + \Delta\bar{\tau})$ in the following way. Let $\tilde{u}(x,\bar{\tau}^* + \Delta\bar{\tau})$ be the solution determined by an approximation to the equation

$$
\frac{\partial \tilde{u}}{\partial \bar{\tau}} - \frac{\partial^2 \tilde{u}}{\partial x^2} = 0.
$$

If

$$
\tilde{u}(x,\bar{\tau}^* + \Delta\bar{\tau}) \geq g(x,\bar{\tau}^* + \Delta\bar{\tau})
$$

at a point, then

$$
u(x,\bar{\tau}^* + \Delta\bar{\tau}) = \tilde{u}(x,\bar{\tau}^* + \Delta\bar{\tau});
$$

otherwise

$$
u(x,\bar{\tau}^* + \Delta\bar{\tau}) = g(x,\bar{\tau}^* + \Delta\bar{\tau}).
$$

That is, for each x,

$$
u(x,\bar{\tau}^* + \Delta\bar{\tau}) = \max\left(\tilde{u}(x,\bar{\tau}^* + \Delta\bar{\tau}),\ g(x,\bar{\tau}^* + \Delta\bar{\tau})\right).
$$

Does the solution determined in this way satisfy all the requirements in (6.9)? The first equality in (6.9) holds because at any point, either $\dfrac{\partial u}{\partial \bar{\tau}} - \dfrac{\partial^2 u}{\partial x^2} = 0$ or $u(x,\bar{\tau}) - g(x,\bar{\tau}) = 0$. Because $u(x,\bar{\tau}^* + \Delta\bar{\tau})$ takes the maximum between $\tilde{u}(x,\bar{\tau}^*+\Delta\bar{\tau})$ and $g(x,\bar{\tau}^*+\Delta\bar{\tau})$, the third inequality must also hold. $\tilde{u}(x,\bar{\tau}^* + \Delta\bar{\tau})$ satisfies $\dfrac{\partial \tilde{u}}{\partial \bar{\tau}} - \dfrac{\partial^2 \tilde{u}}{\partial x^2} = 0$ and $u(x,\bar{\tau}^*+\Delta\bar{\tau}) \geq \tilde{u}(x,\bar{\tau}^*+\Delta\bar{\tau})$, so $\dfrac{\partial u}{\partial \bar{\tau}} - \dfrac{\partial^2 u}{\partial x^2} \geq 0$ must be true because $u(x,\bar{\tau}^*) = \tilde{u}(x,\bar{\tau}^*)$. If the problem is formulated in the form (6.8), the situation is the same.

Therefore, if an American option is formulated as a linear complementarity problem, the difference between the numerical methods for European options and American options is not big. In fact, if the formulation (6.8) is used, then we can compute the value of American options by

$$v_m^{n+1} = \max(\tilde{v}_m^{n+1}, \pm(2\xi - 1), 0),\qquad (6.10)$$

where

$$\begin{aligned}
\tilde{v}_m^{n+1} = &\frac{1}{2}\left[\bar{\sigma}_m^2\xi_m^2(1-\xi_m)^2 - (r-D_0)\xi_m(1-\xi_m)\Delta\xi\right]\alpha v_{m-1}^n\\
&+\left[1 - \bar{\sigma}_m^2\xi_m^2(1-\xi_m)^2\alpha - (r(1-\xi_m)+D_0\xi)\Delta\tau\right]v_m^n\\
&+\frac{1}{2}\left[\bar{\sigma}_m^2\xi_m^2(1-\xi_m)^2 + (r-D_0)\xi_m(1-\xi_m)\Delta\xi\right]\alpha v_{m+1}^n.
\end{aligned}$$

If the formulation (6.9) is adopted, then the computation is done by

$$u_m^{n+1} = \max\left(\bar{\alpha}u_{m+1}^n + (1-2\bar{\alpha})u_m^n + \bar{\alpha}u_{m-1}^n,\ g(x_m, \bar{\tau}^{n+1})\right).\qquad (6.11)$$

Table 6.2. American call option (PEFDII)

$(r = 0.1, \sigma = 0.2, D_0 = 0.05, S = E = 100, T = 1,$
and the exact solution is $C = 9.94092345\cdots$)

Numbers of time steps	Results	|Errors|	CPU(sec.)
50	9.902768	0.038156	0.0003
100	9.921822	0.019102	0.0013
200	9.931367	0.009557	0.0053
400	9.936144	0.004780	0.0220
800	9.938533	0.002390	0.0880

Table 6.3. American put option (PEFDII)

$(r = 0.1, \sigma = 0.2, D_0 = 0.05, S = E = 100, T = 1,$
and the exact solution is $P = 5.92827717\cdots$)

Numbers of time steps	Results	|Errors|	CPU(sec.)
50	5.911829	0.016448	0.0003
100	5.920472	0.007805	0.0013
200	5.924476	0.003801	0.0054
400	5.926424	0.001853	0.0220
800	5.927360	0.000917	0.0880

Finding the prices of American options in such a way is referred to as a projected method in the book [71] by Wilmott, Dewynne, and Howison. We call (6.10) and (6.11) projected explicit finite-difference methods I and II, respectively, and their abbreviations are PEFDI and PEFDII. Clearly, PEFDI can be applied to the cases with both a constant σ and a variable σ, and PEFDII is suitable only for the case that σ is a constant. In Tables 6.2

and 6.3, the results of call and put options on several meshes are given. The method used is PEFDII. The error and the CPU time needed are also shown. In order to have an error, we must have the exact solutions. The exact solution for the American call and put option problems with these parameters are $C = 9.94092345\cdots$ and $P = 5.92827717\cdots$, which are obtained by the SSM given in Chapter 7. Here, the first nine digits are given, and it is enough to determine the first few digits of the errors given in these tables. Computation is done on a Space Ultra 10 computer. In this book, when a CPU time is mentioned, the computation is done on such a computer if no other explanation is given.

6.1.3 Binomial and Trinomial Methods

This subsection is devoted to the binomial and trinomial methods. In these methods, there is a lattice of possible asset prices. Thus, such methods are also called lattice methods.

Binomial methods. The binomial method is a simple and very effective method for computing the option prices. To see how the method works, we consider the following call option. If at the expiry date T the AT&T share price will be \$60 and \$40 with equal probabilities and the exercise price is \$50, then the value of expected return at the expiry date will be $\frac{1}{2}\max(40 - 50, 0) + \frac{1}{2}\max(60 - 50, 0)$, and the value of the call option at time $t < T$ is

$$e^{-r(T-t)}\left[\frac{1}{2}\max(40 - 50, 0) + \frac{1}{2}\max(60 - 50, 0)\right] = 5\cdot e^{-r(T-t)}. \quad (6.12)$$

Now the question is how we determine the possible prices and the probability for each price. In the example given, we have

$$\mathrm{E}\,[S] = \frac{1}{2}\cdot 40 + \frac{1}{2}\cdot 60 = 50,$$

$$\mathrm{Var}\,[S] = \frac{1}{2}(40 - 50)^2 + \frac{1}{2}(60 - 50)^2 = 100,$$

where S is the price at time $t = T$. If $\mathrm{E}\,[S] \neq 50$ or $\mathrm{Var}\,[S] \neq 100$, the value $5\cdot e^{-r(T-t)}$ is meaningless. Therefore, in order for such a way to give an acceptable result, we need to require both $\mathrm{E}\,[S]$ and $\mathrm{Var}\,[S]$ to have the correct values.

Suppose that the asset price at time $t + n\Delta t$ is S_n and assume that at time $t + (n + 1)\Delta t$, the price will be $S_{n+1,1}$ or $S_{n+1,0}$ and the corresponding probabilities are p_n and $1 - p_n$, i.e., the random variable for the price at time $t + (n + 1)\Delta t$ is

$$S_{n+1} = \begin{cases} S_{n+1,1}, \text{ with probability } p_n, \\ S_{n+1,0}, \text{ with probability } 1 - p_n. \end{cases}$$

In this case, we have three parameters, $S_{n+1,1}$, $S_{n+1,0}$, p_n, to be determined, but we have only two conditions: correct $\mathrm{E}\,[S_{n+1}]$ and $\mathrm{Var}\,[S_{n+1}]$. In order for these parameters to be determined uniquely, we need to give another condition. Let us require $S_{n+1,1}S_{n+1,0} = S_n^2$, that is, we let

$$S_{n+1,1} = S_n u_n$$

and

$$S_{n+1,0} = S_n d_n,$$

where

$$d_n = \frac{1}{u_n}.$$

Now there are only two independent parameters p_n, u_n, which can be determined by the two conditions. Under the assumption here, we have

$$
\begin{aligned}
\mathrm{E}\,[S_{n+1}] &= p_n S_n u_n + (1 - p_n) S_n d_n \\
&= S_n(p_n u_n + (1 - p_n) d_n), \\
\mathrm{Var}\,[S_{n+1}] &= \mathrm{E}\,\big[S_{n+1}^2\big] - \mathrm{E}^2\,[S_{n+1}] \\
&= S_n^2\,\big[p_n u_n^2 + (1 - p_n) d_n^2\big] - \mathrm{E}^2\,[S_{n+1}] \\
&= S_n^2\,\Big\{p_n u_n^2 + (1 - p_n) d_n^2 - [p_n u_n + (1 - p_n) d_n]^2\Big\}.
\end{aligned}
$$

As we know, if r, D_0, and σ are constants, then for a European option the solution is

$$V(S,t) = \mathrm{e}^{-r(T-t)} \int_0^\infty V_T(S') G(S', T; S, t)\,dS',$$

where

$$G(S', T; S, t) = \frac{1}{S'\sigma\sqrt{2\pi(T - t)}} \mathrm{e}^{-\left[\ln(S'/S) - (r - D_0 - \sigma^2/2)(T - t)\right]^2 / 2\sigma^2(T - t)}.$$

$G(S', T; S, t)$ is a probability density function for a lognormal distribution. From (2.6), the mean and variance for this distribution are

$$\mathrm{E}\,[S'] = S\mathrm{e}^{(r - D_0)(T - t)}$$

and

$$\mathrm{Var}\,[S'] = S^2 \mathrm{e}^{2(r - D_0)(T - t)}\left(\mathrm{e}^{\sigma^2(T - t)} - 1\right).$$

Let S' and S be the prices S_{n+1} and S_n, respectively, then we have

$$\mathrm{E}\,[S_{n+1}] = S_n \mathrm{e}^{(r - D_0)\Delta t} \tag{6.13}$$

and

$$\mathrm{Var}\,[S_{n+1}] = S_n^2 \mathrm{e}^{2(r - D_0)\Delta t}\left(\mathrm{e}^{\sigma^2 \Delta t} - 1\right) \tag{6.14}$$

and furthermore obtain

$$\begin{cases} p_n u_n + (1 - p_n)d_n = e^{(r-D_0)\Delta t}, \\ p_n u_n^2 + (1 - p_n)d_n^2 = e^{(2(r-D_0)+\sigma^2)\Delta t} \end{cases} \tag{6.15}$$

with $d_n = \dfrac{1}{u_n}$. When r, D_0, and σ are constants, p_n, u_n, and d_n do not depend on n, so we can omit the subscript n, and (6.15) can be simplified into

$$\begin{cases} pu + (1 - p)d = e^{(r-D_0)\Delta t}, \\ pu^2 + (1 - p)d^2 = e^{(2(r-D_0)+\sigma^2)\Delta t}. \end{cases}$$

From these two equations we have

$$p = \frac{e^{(r-D_0)\Delta t} - d}{u - d}$$

and

$$p = \frac{e^{(2(r-D_0)+\sigma^2)\Delta t} - d^2}{u^2 - d^2}.$$

Dividing one by the other of the two relations above yields

$$u + d = \frac{e^{(2(r-D_0)+\sigma^2)\Delta t} - d^2}{e^{(r-D_0)\Delta t} - d}$$

or

$$u\left(e^{(r-D_0)\Delta t} - d\right) + e^{(r-D_0)\Delta t}d = e^{(2(r-D_0)+\sigma^2)\Delta t}.$$

Noticing $d = \dfrac{1}{u}$, we have

$$e^{(r-D_0)\Delta t}u^2 - \left(1 + e^{(2(r-D_0)+\sigma^2)\Delta t}\right)u + e^{(r-D_0)\Delta t} = 0$$

or

$$u^2 - \frac{1 + e^{(2(r-D_0)+\sigma^2)\Delta t}}{e^{(r-D_0)\Delta t}}u + 1 = 0.$$

Assume $S_{n+1,1} > S_{n+1,0}$, in other words, $u > d$. Consequently

$$u = \frac{1}{2}\left[\frac{1 + e^{(2(r-D_0)+\sigma^2)\Delta t}}{e^{(r-D_0)\Delta t}} + \sqrt{\left(\frac{1 + e^{(2(r-D_0)+\sigma^2)\Delta t}}{e^{(r-D_0)\Delta t}}\right)^2 - 4}\right]. \tag{6.16}$$

As soon as we find u, we can have d and p by

$$d = \frac{1}{u}, \tag{6.17}$$

$$p = \frac{e^{(r-D_0)\Delta t} - d}{u - d}. \tag{6.18}$$

Because

$$\frac{1 + e^{(2(r-D_0)+\sigma^2)\Delta t}}{e^{(r-D_0)\Delta t}}$$

$$= \left[2 + \sigma^2 \Delta t + 2(r - D_0)\Delta t + O\left(\Delta t^2\right)\right]\left[1 - (r - D_0)\Delta t + O\left(\Delta t^2\right)\right]$$

$$= 2 + \sigma^2 \Delta t + O\left(\Delta t^2\right),$$

we have

$$u = \frac{1}{2}\left[2 + \sigma^2 \Delta t + O\left(\Delta t^2\right) + \sqrt{(2 + \sigma^2 \Delta t + O\left(\Delta t^2\right))^2 - 4}\right]$$

$$= 1 + \frac{1}{2}\sigma^2 \Delta t + \sqrt{\sigma^2 \Delta t + O\left(\Delta t^2\right)} + O\left(\Delta t^2\right)$$

$$= 1 + \frac{1}{2}\sigma^2 \Delta t + \sigma\sqrt{\Delta t}\left[1 + O(\Delta t)\right] + O\left(\Delta t^2\right)$$

$$= e^{\sigma\sqrt{\Delta t}} + O\left(\Delta t^{3/2}\right).$$

Therefore, if Δt is very small, then we can ignore the term $O\left(\Delta t^{3/2}\right)$ and let

$$u = e^{\sigma\sqrt{\Delta t}} \tag{6.19}$$

instead of (6.16).

As soon as u, d, and p are calculated, we can have $S_{n+1,1} = S_n u$ and $S_{n+1,0} = S_n d$. If $V\left(S_{n+1,1}, (n+1)\Delta t\right)$ and $V\left(S_{n+1,0}, (n+1)\Delta t\right)$ are given, then the option value $V(S_n, n\Delta t)$ can be determined by

$$V(S_n, n\Delta t)$$

$$= e^{-r\Delta t}\left[pV\left(S_{n+1,1}, (n+1)\Delta t\right) + (1 - p)V\left(S_{n+1,0}, (n+1)\Delta t\right)\right], \tag{6.20}$$

which is similar to (6.12). Cox, Ross, and Rubinstein [21] proposed such a method. In their paper, the formulae for u, d, p are (6.19), (6.17), and (6.18).

Now let us describe the complete method. Suppose the asset price at the current time t to be S, and we divide the remaining life of the derivative security into N equal time subintervals with time step $\Delta t = (T-t)/N$. At the first time level $t + \Delta t$, there are two possible asset prices Su and $Sd = Su^{-1}$. At the second time level $t + 2\Delta t$, there are three possible asset prices, Su^2, $Sud = Sdu = S$, and $Sd^2 = Su^{-2}$. At the third time level $t + 3\Delta t$, there are four possible asset prices, Su^3, $Su^2d = Su$, $Sud^2 = Su^{-1}$, and $Sd^3 = Su^{-3}$. In general, at the n-th time level $t + n\Delta t$, there are $n + 1$ possible values of the asset price. Originally, at the n-th time level, there should be 2^n possible values of the asset price. However since $d = 1/u$ is required, many points are the same. For example, S, Su^2d^2, Su^4d^4, $\cdots$ are the same point. Hence the number of possible values is greatly reduced. This is why $d = 1/u$ is chosen as the third condition when the method is derived. Let $S_{n,m}, m = 0, 1, \cdots, n$,

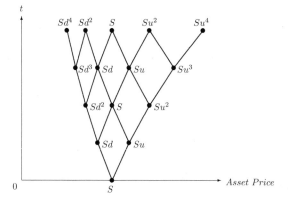

Fig. 6.2. Tree of asset prices for a binomial model

denote the $n+1$ possible values of the asset price at the n-th time level from the smallest to the largest. Then

$$S_{n,m} = Su^{2m-n}, \quad m = 0, 1, \cdots, n. \tag{6.21}$$

For $N = 4$, all the possible prices for each n are given in Fig. 6.2. This plot is usually referred to as a tree or lattice of possible asset prices.

Assuming that we know the payoff function for our derivative security and that it depends only on the values of the underlying asset at expiry, this enables us to value it at expiry, the N-th time level. If we are considering a call, for example, we find

$$c_{N,m} = \max(S_{N,m} - E, 0), \quad m = 0, 1, \cdots, N, \tag{6.22}$$

where E is the exercise price and $c_{N,m}$ denotes the value of the call for the m-th possible asset value $S_{N,m}$ at time-step N. For a put, we know that

$$p_{N,m} = \max(E - S_{N,m}, 0), \quad m = 0, 1, \cdots, N, \tag{6.23}$$

where $p_{N,m}$ denotes the value of the put for the m-th possible asset value $S_{N,m}$ at expiry.

We can now find the expected value of the derivative security at the (N-1)-th time level and for possible asset prices $S_{N-1,m}$, $m = 0, 1, ..., N - 1$ because we know that the probability of an asset price moving from $S_{N-1,m}$ to $S_{N,m+1}$ during a time step is p and that the probability of it moving to $S_{N,m}$ is $(1-p)$. Using the discounting factor $e^{-r\Delta t}$, we can obtain the value of the security at each possible asset price for the $(N - 1)$-th time level. This procedure can be applied to the n-th time level if the values of the option for the $(n + 1)$-th time level have been obtained, and the computational formula is (6.20) or, in a general form,

$$V_{n,m} = e^{-r\Delta t}(pV_{n+1,m+1} + (1-p)V_{n+1,m}), \quad m = 0, 1, \cdots, n. \qquad (6.24)$$

Here, $V_{n,m}$ denotes the value of a European option at the n-th time level and corresponding to asset price $S_{n,m}$. According to this formula, starting from the payoff function, $V_{N,m}$, $m = 0, 1, \cdots, N$, we can recursively determine $V_{n,m}$, $m = 0, 1, \cdots, n$ for $n = N-1, N-2, \cdots, 0$, and the final value $V_{0,0}$ is the current value of the option.

For American options, we can easily incorporate the possibility of early exercise of an option into the binomial model. Because the price of an American call option must be greater than or equal to

$$\max(S_{n,m} - E, 0), \qquad (6.25)$$

when calculating the price of an American call option, we need to replace (6.24) by

$$C_{n,m} = \max\left(S_{n,m} - E, \ 0, \ e^{-r\Delta t}\left[pC_{n+1,m+1} + (1-p)C_{n+1,m}\right]\right) \qquad (6.26)$$

at each point. Similarly, for an American put option, the formula is

$$P_{n,m} = \max\left(E - S_{n,m}, \ 0, \ e^{-r\Delta t}\left[pP_{n+1,m+1} + (1-p)P_{n+1,m}\right]\right) \qquad (6.27)$$

because the price of an American put option has to be at least

$$\max(E - S_{n,m}, 0). \qquad (6.28)$$

From what has been described, we see that the entire computation can be done in two steps. In the first step, we calculate all the $S_{n,m}$ to be used. Then, we find $V_{N,m}$, $m = 0, 1, \cdots, N$ and calculate $V_{n,m}$, $m = 0, 1, \cdots, n$ for $n = N-1, N-2, \cdots, 0$ successively. When a European option is calculated, only the $S_{N,m}$, $m = 0, 1, \cdots, N$, are used in order to find $V_{N,m}$. When an American option is evaluated, all the $S_{n,m}$ are needed. However, because $S_{n,m} = Su^{2m-n} = Su^{2(m-1)-(n-2)} = S_{n-2,m-1}$, we indeed only need to calculate $S_{N,m}$, $m = 0, 1, \cdots, N$ and $S_{N-1,m}$, $m = 0, 1, \cdots, N-1$, i.e., Su^m, $m = -N, -N+1, \ldots, N$. For this method, the total number of nodes is $(N+2)(N+1)/2$, so the execution time for computing all the $V_{n,m}$ is $O(N^2)$.

Trinomial methods. If σ depends on S, then u is not a constant. In this case, generally speaking, at the n-th time level, there are 2^n possible values of the asset prices that need to be considered, and the total nodes and the execution time will be very large if a binomial method is used. In order to reduce the nodes for a problem with variable σ, we can use trinomial methods. In a trinomial method, given a current asset value S, the asset value after a time-step Δt can take any of the three values

$$Su, \ Sq, \ Sd,$$

where $0 \leq d < q < u$. Let p_u be the probability of the value of the asset after a time-step Δt being Su, p_q be the probability of the value being Sq, and p_d

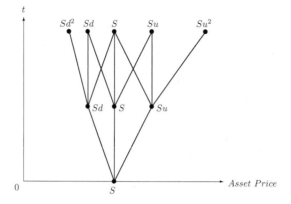

Fig. 6.3. Lattice generated by a trinomial model

be the probability of the value being Sd. Because there are only three possible cases, we must have

$$p_u + p_q + p_d = 1, \quad 0 \le p_u \le 1, \quad 0 \le p_q \le 1, \quad 0 \le p_u \le 1$$

and according to (6.15), the following relations are true:

$$p_u u + p_q q + p_d d = e^{(r-D_0)\Delta t},$$
$$p_u u^2 + p_q q^2 + p_d d^2 = e^{(2(r-D_0)+\sigma^2)\Delta t}.$$

Because there are three equations above for six unknowns, u, q, d, p_u, p_q, p_d, we can choose three parameters. In order for the number of the possible asset prices to not be 3^n at the n-th time level, we can choose

$$d = 1/u \quad \text{and} \quad q = 1. \tag{6.29}$$

In this case, the corresponding lattice is illustrated in Fig. 6.3. Now there are only four parameters u, p_u, p_q, p_d left. They should satisfy the three conditions above. If u is given, then this is a linear system for p_u, p_q, p_d and can be solved for them easily. Its solution is

$$\begin{cases} p_u = \dfrac{e^{(2(r-D_0)+\sigma^2)\Delta t} - e^{(r-D_0)\Delta t}(q+d) + qd}{(u-q)(u-d)}, \\[3mm] p_q = \dfrac{e^{(2(r-D_0)+\sigma^2)\Delta t} - e^{(r-D_0)\Delta t}(d+u) + du}{(q-d)(q-u)}, \\[3mm] p_d = \dfrac{e^{(2(r-D_0)+\sigma^2)\Delta t} - e^{(r-D_0)\Delta t}(u+q) + uq}{(d-u)(d-q)}. \end{cases} \tag{6.30}$$

Because they represent probabilities, we need to choose such a u that p_u, p_q and p_d all are nonnegative. If σ depends on S and t, then p_u, p_q and p_d will be different at different points. In this case, we need to choose such a u that at all the points p_u, p_q and p_d are nonnegative and the trinomial method described above can still be used.

Table 6.4. American call option (binomial method)

$(r = 0.1, \sigma = 0.2, D_0 = 0.05, T = 1$ year, $S = E = 100$, and the exact solution is $C = 9.94092345\cdots)$

Numbers of time steps	Results	|Errors|	CPU(sec.)
50	9.902969	0.037955	0.0004
100	9.921921	0.019002	0.0013
200	9.931416	0.009507	0.0053
400	9.936168	0.004755	0.0220
800	9.938546	0.002378	0.0890

Table 6.5. American put option (binomial method)

$(r = 0.1, \sigma = 0.2, D_0 = 0.05, T = 1$ year, $S = E = 100$, and the exact solution is $P = 5.92827717\cdots)$

Numbers of time steps	Results	|Errors|	CPU(sec.)
50	5.911020	0.017257	0.0004
100	5.920066	0.008211	0.0014
200	5.924273	0.004005	0.0053
400	5.926323	0.001955	0.0210
800	5.927309	0.000968	0.0880

The details for evaluating derivative securities using a trinomial method are nearly identical to the binomial method. The only major difference is that the expected value of the security at the n-th time level depends on the three possible values at the $(n+1)$-th time level, and that at the n-th time level, there are $2n+1$ possible asset prices, $S_{n,m} = Su^m$, $m = -n, -n+1, \cdots, n$, and $2n+1$ security prices rather than $n+1$. Let $V_{n,m}$ be the security price at $S_{n,m}$. Then, the formula for finding the expected value of a security at time level $n+1$ is

$$E[V_{n+1,m}] = (p_u V_{n+1,m+1} + p_q V_{n+1,m} + p_d V_{n+1,m-1})$$

and the value of a European derivative security for $S_{n,m}$ is

$$V_{n,m} = e^{-r\Delta t}(p_u V_{n+1,m+1} + p_q V_{n+1,m} + p_d V_{n+1,m-1}),$$

and for American puts and calls we have

$$P_{n,m} = \max\left(E - S_{n,m},\ 0,\ e^{-r\Delta t}\left[p_u P_{n+1,m+1} + p_q P_{n+1,m} + p_d P_{n+1,m-1}\right]\right),$$
(6.31)

$$C_{n,m} = \max\left(S_{n,m} - E,\ 0,\ e^{-r\Delta t}\left[p_u C_{n+1,m+1} + p_q C_{n+1,m} + p_d C_{n+1,m-1}\right]\right).$$
(6.32)

In Tables 6.4 and 6.5, we give binomial lattice approximations to American call and put options. The errors and the CPU times are also shown.

6.1.4 Relations Between the Lattice Methods and the Explicit Finite-Difference Methods

Let $y = \ln S$. In this case, the Black–Scholes equation becomes

$$\frac{\partial V}{\partial t} + \frac{1}{2}\sigma^2 \frac{\partial^2 V}{\partial y^2} + \left(r - D_0 - \frac{1}{2}\sigma^2\right)\frac{\partial V}{\partial y} - rV = 0.$$
(6.33)

This can be approximated by

$$\frac{V_m^{n+1} - V_m^n}{\Delta t} + \frac{1}{2}\sigma^2 \frac{V_{m+1}^{n+1} - 2V_m^{n+1} + V_{m-1}^{n+1}}{\Delta y^2}$$

$$+ \left(r - D_0 - \frac{1}{2}\sigma^2\right)\frac{V_{m+1}^{n+1} - V_{m-1}^{n+1}}{2\Delta y} - rV_m^n = 0,$$

or

$$V_m^n = \frac{1}{1 + r\Delta t}\left(a V_{m+1}^{n+1} + b V_m^{n+1} + c V_{m-1}^{n+1}\right),$$
(6.34)

where

$$a = \frac{1}{2}\left[\frac{\Delta t}{\Delta y^2}\sigma^2 + \frac{\Delta t}{\Delta y}\left(r - D_0 - \frac{1}{2}\sigma^2\right)\right],$$

$$b = 1 - \frac{\Delta t}{\Delta y^2}\sigma^2,$$

$$c = \frac{1}{2}\left[\frac{\Delta t}{\Delta y^2}\sigma^2 - \frac{\Delta t}{\Delta y}\left(r - D_0 - \frac{1}{2}\sigma^2\right)\right].$$

If

$$\sigma\sqrt{\Delta t} \le \Delta y \le \frac{\sigma^2}{|r - D_0 - \sigma^2/2|},$$
(6.35)

then a, b, and c are all nonnegative. In this case, this is an explicit finite-difference scheme with nonnegative coefficients. This scheme has a natural interpretation as a discrete random walk. Imagine a random walk taking place on the (y, t) mesh with equally spaced y-nodes, a distance Δy apart, and equally spaced t-nodes, a distance Δt apart, as in Fig. 6.4. Noticing $a + b + c = 1$, then a, b, and c could be seen as probabilities of three possible events if a, b, and c are all nonnegative. The coefficient a could be understood as the

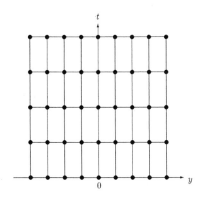

Fig. 6.4. A uniform mesh on (y, t)-plane

probability of the event that from $t = t^n$ to $t = t^{n+1}$, the price moves from y_m to y_{m+1}, b corresponds to the event that the price stays at y_m, and c corresponds to the event that the price moves from y_m to y_{m-1}. Therefore,

$$aV_{m+1}^{n+1} + bV_m^{n+1} + cV_{m-1}^{n+1}$$

could be understood as the expected value of a security at time $(n+1)\Delta t$. The term $1/(1 + r\Delta t)$ is a discounting factor that is an approximation of $e^{-r\Delta t}$. Therefore, the expression above times $1/(1 + r\Delta t)$ gives V_m^n. This method actually is a trinomial method. If we take

$$\frac{\Delta t}{\Delta y^2} = \frac{1}{\sigma^2}, \tag{6.36}$$

then $b = 0$ and the method becomes the binomial method.

Let $y_m = m\Delta y + \bar{y}$. Then, from $y = \ln S$, we have

$$S(y_m) = e^{m\Delta y + \bar{y}} = \left(e^{\Delta y}\right)^m e^{\bar{y}}.$$

Therefore, the corresponding mesh on (S, t)-plane is not a uniform mesh (see Fig. 6.5) and is similar to the mesh in Fig. 6.3 with $u = e^{\Delta y}$. Consequently, this explicit difference method can be understood as a trinomial method with a lattice in Fig. 6.3 and as a binomial method with a lattice in Fig. 6.2 if (6.36) holds. If (6.36) holds, then

$$\Delta y = \sigma\sqrt{\Delta t},$$

$$a = \frac{1}{2}\left[1 + \frac{\sqrt{\Delta t}}{\sigma}\left(r - D_0 - \frac{1}{2}\sigma^2\right)\right],$$

$$c = 1 - a$$

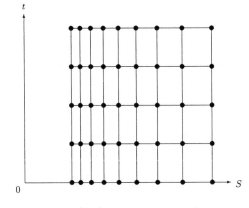

Fig. 6.5. The mesh on (S, t)-plane corresponding to a uniform mesh on (y, t)-plane

and

$$u = e^{\Delta y} = e^{\sigma\sqrt{\Delta t}}.$$

From (6.18), we have

$$p = \frac{e^{(r-D_0)\Delta t} - e^{-\sigma\sqrt{\Delta t}}}{e^{\sigma\sqrt{\Delta t}} - e^{-\sigma\sqrt{\Delta t}}}$$

$$= \frac{(r - D_0)\Delta t + \sigma\sqrt{\Delta t} - \dfrac{1}{2}\sigma^2 \Delta t + \dfrac{1}{6}\sigma^3 \Delta t^{3/2} + O\left(\Delta t^2\right)}{\sigma\sqrt{\Delta t} + \dfrac{1}{2}\sigma^2 \Delta t + \dfrac{1}{6}\sigma^3 \Delta t^{3/2} + \sigma\sqrt{\Delta t} - \dfrac{1}{2}\sigma^2 \Delta t + \dfrac{1}{6}\sigma^3 \Delta t^{3/2} + O\left(\Delta t^2\right)}$$

$$= \frac{\sigma\sqrt{\Delta t} + \left(r - D_0 - \dfrac{1}{2}\sigma^2\right)\Delta t + \dfrac{1}{6}\sigma^3 \Delta t^{3/2} + O\left(\Delta t^2\right)}{2\sigma\sqrt{\Delta t} + \dfrac{1}{3}\sigma^3 \Delta t^{3/2} + O\left(\Delta t^2\right)}$$

$$= \frac{1}{2} \cdot \frac{1 + \left(r - D_0 - \dfrac{1}{2}\sigma^2\right)\sqrt{\Delta t}/\sigma + \dfrac{1}{6}\sigma^2 \Delta t + O\left(\Delta t^{3/2}\right)}{1 + \dfrac{1}{6}\sigma^2 \Delta t + O\left(\Delta t^{3/2}\right)}$$

$$= \frac{1}{2}\left[1 + \frac{\sqrt{\Delta t}}{\sigma}\left(r - D_0 - \frac{1}{2}\sigma^2\right)\right] + O\left(\Delta t^{3/2}\right).$$

Thus, the difference between p and a for $\Delta y = \sigma\sqrt{\Delta t}$ is $O(\Delta t^{3/2})$. Consequently, the explicit method with $\Delta y = \sigma\sqrt{\Delta t}$ is almost the binomial method. Because the convergence of the explicit scheme here with $\Delta t/\Delta y^2 = \sigma^{-2}$ can be easily proved and the difference between a and p is $O\left(\Delta t^{3/2}\right)$, the convergence of the binomial method can also be proved.

Therefore, the binomial method and the trinomial method can be understood as explicit finite-difference methods in some sense.

6.1.5 Examples of Unstable Schemes

As has been pointed out in Subsection 6.1.1, when the scheme (6.3) or (6.6) is used, stability condition (6.4) or (6.7) is required. What will happen if these conditions are violated?

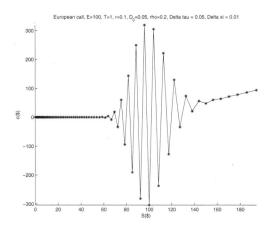

Fig. 6.6. A unstable solution of EFDI
(The solution appears when (6.4) is violated. $E = 100$, $T = 1$,
$r = 0.1$, $D_0 = 0.05$, $\sigma = 0.2$, $\Delta\tau = 0.05$, and $\Delta\xi = 0.01$.)

Let us try scheme (6.3) for a European call option with parameters $E = 100$, $T = 1$, $r = 0.1$, $D_0 = 0.05$, and $\sigma = 0.2$. Take $\Delta\tau = 0.05$ and $\Delta\xi = 0.01$. The solution at $t = 0$ is shown in Fig. 6.6, where we see that rather large oscillations develop. In this case, $\Delta\tau = 0.05$ and $16\Delta\xi^2/\sigma^2 = 0.04$, so (6.4) does not hold, and the scheme is unstable. We cannot get a useful solution if such a set of $\Delta\tau$ and $\Delta\xi$ is adopted.

6.2 Implicit Methods

6.2.1 Pricing European Options by Using $\overline{V}$, ξ, τ Variables

The problem (6.1) can also be discretized by an implicit scheme, for example, by scheme (5.31). In this case, the difference equations are

$$\frac{v_m^{n+1} - v_m^n}{\Delta\tau}$$

$$= \frac{1}{4}\bar{\sigma}_m^2 \xi_m^2 (1 - \xi_m)^2 \left(\frac{v_{m+1}^{n+1} - 2v_m^{n+1} + v_{m-1}^{n+1}}{\Delta\xi^2} + \frac{v_{m+1}^n - 2v_m^n + v_{m-1}^n}{\Delta\xi^2} \right)$$

$$+\frac{1}{2}(r - D_0)\xi_m(1 - \xi_m)\left(\frac{v_{m+1}^{n+1} - v_{m-1}^{n+1}}{2\Delta\xi} + \frac{v_{m+1}^{n} - v_{m-1}^{n}}{2\Delta\xi}\right)$$

$$-\frac{1}{2}[r(1 - \xi_m) + D_0\xi_m](v_m^{n+1} + v_m^{n}),$$

$$m = 0, 1, \cdots, M, \quad n = 0, 1, \cdots, N - 1. \tag{6.37}$$

Like the explicit scheme case, v_{-1}^{n} and v_{M+1}^{n} do not appear in the equation. For a fixed n, there are $M + 1$ equations in the following form:

$$a_m v_{m-1}^{n+1} + b_m v_m^{n+1} + c_m v_{m+1}^{n+1} = -a_m v_{m-1}^{n} + (2 - b_m)v_m^{n} - c_m v_{m+1}^{n},$$

$$m = 0, 1, \cdots, M,$$

where

$$a_m = \left[(r - D_0)m(1 - \xi_m) - \bar{\sigma}_m^2 m^2(1 - \xi_m)^2\right]\Delta\tau/4,$$

$$b_m = 1 + [\bar{\sigma}_m^2 m^2(1 - \xi_m)^2 + r(1 - \xi_m) + D_0\xi_m]\Delta\tau/2,$$

$$c_m = \left[-(r - D_0)m(1 - \xi_m) - \bar{\sigma}_m^2 m^2(1 - \xi_m)^2\right]\Delta\tau/4,$$

$$m = 0, 1, \cdots, M.$$

In matrix form, we have

$$A\mathbf{v}^{n+1} = \mathbf{q}^{n},$$

where

$$A = \begin{bmatrix} b_0 & c_0 & & & & \\ a_1 & b_1 & c_1 & & 0 & \\ & \ddots & \ddots & \ddots & & \\ & & \ddots & \ddots & & \ddots \\ & 0 & & a_{M-1} & b_{M-1} & c_{M-1} \\ & & & & a_M & b_M \end{bmatrix},$$

$$\mathbf{v}^{n} = \begin{bmatrix} v_0^{n} \\ v_1^{n} \\ \vdots \\ v_M^{n} \end{bmatrix}, \quad \mathbf{q}^{n} = \begin{bmatrix} (2 - b_0)v_0^{n} - c_0 v_1^{n} \\ -a_1 v_0^{n} + (2 - b_1)v_1^{n} - c_1 v_2^{n} \\ \vdots \\ -a_M v_{M-1}^{n} + (2 - b_M)v_M^{n} \end{bmatrix}.$$

The difference between an implicit method and an explicit method is that for an implicit method, a linear system needs to be solved in order to get $\mathbf{v}^{n+1}$ from $\mathbf{v}^{n}$. This can be done by the LU decomposition or an iteration method given in Subsections 5.2.1 and 5.2.2. The linear system here has a variable coefficient matrix, however, it does not depend on time if σ does not depend on t. Thus, the linear system can be solved with only slightly more cost compared to a linear system with a constant coefficient matrix. It is clear that scheme (6.37) can even be applied to the case when σ depends on S and t. We will refer to this scheme as the implicit finite-difference scheme. From

Problem 19 in Chapter 5, we can expect this scheme to be stable without any condition on the ratio $\Delta\tau/\Delta\xi$. In fact, in the paper by Sun, Yan, and Zhu [68], it is rigorously proven that this scheme with variable coefficients is unconditionally stable.

When σ is a constant, we can also use the variables u, x and $\bar{\tau}$. In this case, (5.34) can be applied to the equation in problem (6.5) in x, $\bar{\tau}$ variables. However, when (5.34) is used for problem (6.5), we have to modify the problem formulation slightly. Let the problem be defined on a finite domain $[x_l, x_u]$ and give an artificial boundary condition on each boundary. From (2.27) and (2.31) in Subsection 2.3.2, we know at $S = 0, V(0, t) = V(0, T)e^{-r(T-t)}$ and for $S \approx \infty$, $V(S, t) \approx V(S, T)e^{-D_0(T-t)}$. Therefore, noticing $u(x, \bar{\tau}) = e^{r(T-t)}V(S, t)$, for $S \approx 0$, i.e., $x \approx -\infty$ we have

$$u(x, \bar{\tau}) \approx V(S, T)$$

and for $S \approx \infty$, i.e., $x \approx \infty$,

$$u(x, \bar{\tau}) \approx V(S, T)e^{(r-D_0)(T-t)},$$

where $x = \ln S + (r - D_0 - \sigma^2/2)(T - t)$ and $\bar{\tau} = \sigma^2(T - t)/2$. These two relations can be taken as artificial boundary conditions at $x = x_l$ and $x = x_u$, respectively, if x_l is small enough and x_u is large enough. For example, in order to calculate a call option,

$$u(x_l, \bar{\tau}) = 0 \quad \text{and} \quad u(x_u, \bar{\tau}) = (e^{x_u - (2(r-D_0)/\sigma^2 - 1)\bar{\tau}} - E)e^{2(r-D_0)\bar{\tau}/\sigma^2}$$

can be adopted as artificial boundary conditions. If the call option has parameters $r = 0.1$, $D_0 = 0.05$, $\sigma = 0.2$, $E = 1$, and $T = 1$, we can let $x_l = \ln 0.2$ and $x_u = \ln 2.3$.

6.2.2 European Options with Discrete Dividends and Asian and Lookback Options with Discrete Sampling

A holder of a stock usually obtains dividends at certain days, not continuously. Thus, in practice, it is important to know how to price options on stocks with discrete dividends. This subsection is devoted to discuss how to evaluate such a European option. For European-style Asian and lookback options with discrete sampling, the prices can be obtained in a similar way. Some results on such options are also given here.

Let $V(S, t)$ be the price of an option on stocks with discrete dividends. From Subsection 2.2.2, we know that $V(S, t)$ is the solution of the following problem:

$$\begin{cases} \dfrac{\partial V}{\partial t} + \dfrac{1}{2}\sigma^2(S)S^2\dfrac{\partial^2 V}{\partial S^2} + [rS - D(S,t)]\dfrac{\partial V}{\partial S} - rV = 0, \\ \qquad\qquad 0 \le S, \quad t \le T, \\ V(S, T) = V_T(S), \quad 0 \le S, \end{cases} \qquad (6.38)$$

where

$$D(S,t) = \sum_{i=1}^{I} D_i(S)\delta(t - t_i) \quad \text{and} \quad D_i(S) \le S \quad \text{for any } S.$$

Here, the meaning of the condition $D_i(S) \le S$ is that the price of a stock at any time should be greater than or equal to the dividend paid at that time. From (6.38), we know the following: at $t \ne t_i, i = 1, 2, \cdots, I$, V satisfies

$$\frac{\partial V}{\partial t} + \frac{1}{2}\sigma^2(S)S^2\frac{\partial^2 V}{\partial S^2} + rS\frac{\partial V}{\partial S} - rV = 0, \quad 0 \le S \tag{6.39}$$

and at $t = t_i, i = 1, 2, \cdots,$ or I, the equation

$$\frac{\partial V}{\partial t} - D_i(S)\delta(t - t_i)\frac{\partial V}{\partial S} = 0, \quad 0 \le S$$

holds. From Subsection 2.10.2, we see that this equation gives

$$V(S, t_i^-) = V\left(S - D_i(S), t_i^+\right). \tag{6.40}$$

As we know from Subsection 2.3.2, through the transformation

$$\begin{cases} \xi = \dfrac{S}{S + P_m}, \\[2mm] \tau = T - t, \\[2mm] \overline{V}(\xi, \tau) = \dfrac{V(S, t)}{S + P_m}, \end{cases} \tag{6.41}$$

(6.39) becomes

$$\frac{\partial \overline{V}}{\partial \tau} = \frac{1}{2}\bar{\sigma}^2(\xi)\xi^2(1-\xi)^2\frac{\partial^2 \overline{V}}{\partial \xi^2} + r\xi(1-\xi)\frac{\partial \overline{V}}{\partial \xi} - r(1-\xi)\overline{V}, \quad 0 \le \xi \le 1, \tag{6.42}$$

where $\bar{\sigma}(\xi) = \sigma\left(\dfrac{P_m\xi}{1-\xi}\right)$, the final condition in (6.38) is converted into an initial condition of the form:

$$\overline{V}(\xi, 0) = \frac{1-\xi}{P_m}V_T\left(\frac{P_m\xi}{1-\xi}\right), \quad 0 \le \xi \le 1 \tag{6.43}$$

and the condition (6.40) is transferred to

$$\overline{V}(\xi, \tau_i^+) = \left[1 - D_i\left(\frac{\xi P_m}{1-\xi}\right)\frac{1-\xi}{P_m}\right]\overline{V}\left(\frac{P_m\xi - D_i\left(\dfrac{\xi P_m}{1-\xi}\right)(1-\xi)}{P_m - D_i\left(\dfrac{\xi P_m}{1-\xi}\right)(1-\xi)}, \tau_i^-\right).$$

$$\tag{6.44}$$

We solve the problem here on a mesh given by (6.2). Again, let v_m^n be an approximate value of $\overline{V}$ at $\xi = \xi_m$ and $\tau = \tau_n$ and $\{v_m^n\}$ denote v_m^n, $m = 0, 1, \cdots, M$. For simplicity, we assume that $\tau_0 = 0$, $\tau_{I+1} = T$ and that $\tau_i = T - t_i = n_i \Delta\tau$ with an integer n_i, $i = 1, 2, \cdots, I$. Furthermore, we suppose that $\tau_0 < \tau_1 < \cdots < \tau_I < \tau_{I+1}$. The problem can be solved in the following way. When $\{v_m^{n_i^+}\}$ are known at τ_i^+, we can obtain $\{v_m^{n_{i+1}^-}\}$ at τ_{i+1}^- by a scheme approximating (6.42), for example, the scheme (6.37). Then, we use condition (6.44) to interpolate $\{v_m^{n_{i+1}^+}\}$ from $\{v_m^{n_{i+1}^-}\}$. At $t = 0$, the option values are the same for $t = 0^-$ and $t = 0^+$. Thus, from (6.43) we can have $\{v_m^{n_0^+}\}$. Consequently, we can do the procedure of getting $\{v_m^{n_{i+1}^+}\}$ from $\{v_m^{n_i^+}\}$ for $i = 0, 1, \cdots, I - 1$ successively. As soon as we have $\{v_m^{n_I^+}\}$, we can find $\{v_m^{n_{I+1}^-}\}$, i.e., $\{v_m^N\}$ by scheme (6.37).

Table 6.6. European options on stocks with discrete dividends

$(r = 0.1, \sigma = 0.2, T = 1$ year and $S = 100$.

There are two dividend payments, and each pays 1.25.)

		t=1/12 & 7/12	t=2/12 & 8/12	t=3/12 & 9/12	t=4/12 & 10/12	t=5/12 & 11/12	$D_0 = 0.025$
	E						
Call	95	14.59	14.61	14.64	14.67	14.70	14.50
	100	11.62	11.65	11.68	11.71	11.73	11.53
	105	9.09	9.11	9.14	9.17	9.19	9.00
Put	95	2.96	2.97	2.98	2.99	2.99	2.92
	100	4.52	4.53	4.54	4.55	4.56	4.48
	105	6.51	6.52	6.53	6.54	6.54	6.47

In Table 6.6, we give some values of one-year European options with two dividend payments. Each time, the dividend payment is 1.25 if the price of stock is greater than or equal to 1.25. If $S < 1.25$, we let $D_i(S) = S$ in the computation. The payments are given at $t = 1/12$ & 7/12, 2/12 & 8/12, $\cdots$, or 5/12 & 11/12. By the way, the results on the options with $D_0 = 0.025$ are also listed. From there, we can see that the higher the exercise price, the lower the call option price and the higher the put option price. This is clear from finance. We also see that for these cases, the option prices do not depend on the date of the dividend payment very much. The option prices on a stock paying the "same" dividend continuously is little lower than and close to these corresponding prices.

For European-style Asian and lookback options with discrete sampling, the method is similar. That is, the problem is solved by numerical schemes for partial differential equations and interpolation alternately. In Section 3.3, some results on European-style Asian options with discrete sampling have been given. Here, we give some results again to show the effect of the extrapolation

Table 6.7. Double average call option prices on four meshes ($D_0 < r$)

$(T = 1, S = 100, r = 0.05, D_0 = 0, \sigma = 0.2,$
$T_{1s} = 0.1, T_{1e} = 0.5, K_1 = 5, T_s = 0.6, T_e = 1.0, K = 5, P_m = 0.4,$
the payoff $= \max\left(\dfrac{I}{K} - \dfrac{I_1}{K_1}, 0\right)$, and the exact solution $= 5.872133\cdots)$

Mesh sizes	Results	|Errors|	CPU times	Results without extrapolation	|Errors|	CPU times
200 × 20	5.870320	0.001813	0.0042	5.869883	0.002250	0.0020
400 × 40	5.871861	0.000272	0.0094	5.871367	0.000766	0.0077
800 × 80	5.872133	0.000000	0.0282	5.871942	0.000191	0.0203
1600 × 160	5.872126	0.000007	0.0928	5.872080	0.000053	0.0745

Table 6.8. Double average call option prices on four meshes ($D_0 > r$)

$(T = 1, S = 100, r = 0.05, D_0 = 0.1, \sigma = 0.2,$
$T_{1s} = 0.1, T_{1e} = 0.5, K_1 = 5, T_s = 0.6, T_e = 1.0, K = 5, P_m = 0.2,$
the payoff $= \max\left(\dfrac{I}{K} - \dfrac{I_1}{K_1}, 0\right)$, and the exact solution $= 3.244201\cdots)$

Mesh sizes	Results	|Errors|	CPU times	Results without extrapolation	|Errors|	CPU times
200 × 20	3.241122	0.003079	0.0052	3.235091	0.009110	0.0030
400 × 40	3.244162	0.000039	0.0116	3.241894	0.002307	0.0084
800 × 80	3.244263	0.000062	0.0321	3.243671	0.000530	0.0217
1600 × 160	3.244196	0.000005	0.1009	3.244064	0.000137	0.0813

technique and how the approximate solutions converge to exact solutions in
Tables 6.7 and 6.8. There the results on two double average call options are
given. In Table 6.7 $D_0 = 0.1 > r = 0.05$, and in Table 6.8 $D_0 = 0 < r = 0.05$.
There are ten samplings at $t = 0.1, 0.2, \cdots, 1.0$. From these tables, we can see
that the extrapolation technique greatly improves the rate of convergence and
the accuracy with about 25% extra CPU time. For details of the methods and
more results, see the papers by Andreasen[2] and Zhu[80].

6.2.3 Projected Direct Methods for the LC Problem

As seen in Subsection 6.2.1, using implicit finite-difference methods for European options is straightforward. From Subsection 6.1.2, if an American option
is formulated as a linear complementarity problem, then there is not a big difference between explicit finite-difference methods for European and American
options. The implicit methods for American options are also only a little more
complicated than the methods for European options.

Suppose we use a direct method to solve the system related to an American
call option, which is formulated as (6.8). Assuming that the partial differential
equation holds everywhere and using scheme (6.37), we have a system in the

form:

$$a_m \bar{v}_{m-1}^{n+1} + b_m \bar{v}_m^{n+1} + c_m \bar{v}_{m+1}^{n+1} = q_m^n, \quad m = 0, 1, 2, \cdots, M. \tag{6.45}$$

Actually, $\bar{v}_{-1}^{n+1}$ and $\bar{v}_{M+1}^{n+1}$ do not appear in the system because

$$a_0 = c_M = 0.$$

It is clear that the solution of (6.45) may not be the solution of the American option. However, we can find the solution of the American option with the aid of (6.45).

Similar to what we did in Subsection 5.2.1, if we let

$$u_0 = b_0, \quad y_0 = q_0^n, \tag{6.46}$$

and

$$u_m = b_m - \frac{c_{m-1} a_m}{u_{m-1}}, \quad y_m = q_m^n - \frac{y_{m-1} a_m}{u_{m-1}}, \quad m = 1, 2, \cdots, M, \tag{6.47}$$

then the equations in (6.45) can be rewritten as

$$\bar{v}_m^{n+1} = \frac{y_m - c_m \bar{v}_{m+1}^{n+1}}{u_m}, \quad m = M, M-1, \cdots, 0, \tag{6.48}$$

where the relation with $m = M$ actually is

$$\bar{v}_M^{n+1} = \frac{y_M}{u_M}$$

because $c_M = 0$. From the derivation, we know that the relations in (6.48) with $m = 0, 1, \cdots, M_f$ are equivalent to the equations in (6.45) with $m = 0, 1, \cdots, M_f$, where M_f is any positive integer less than or equal to M. Obviously, $\bar{v}_m^{n+1}$ may not be greater than or equal to $\max(2\xi_m - 1, 0)$. Therefore, we need to find the value of the American option by

$$v_m^{n+1} = \max\left(\bar{v}_m^{n+1}, 2\xi_m - 1, 0\right), \quad m = 0, 1, \cdots, M \tag{6.49}$$

or by

$$v_m^{n+1} = \max\left(\frac{y_m - c_m v_{m+1}^{n+1}}{u_m}, 2\xi_m - 1, 0\right), \quad m = M, M-1, \cdots, 0, \tag{6.50}$$

successively. This method is referred to as the projected implicit finite-difference method I (PIFDI).

Is there any difference between (6.49) and (6.50)? The answer is yes. Let us explain this. As we know from Subsection 2.6.1, there is only one free boundary for a call option. It is natural to expect that when (6.49) is used, there exists an M_f so that $v_m^{n+1} = \bar{v}_m^{n+1}$ for $m = 0, 1, \cdots, M_f$ and $v_m^{n+1} = \max(2\xi_m - 1, 0)$

Table 6.9. American call option (PIFDI)

$(r = 0.05, \sigma = 0.2, D_0 = 0.1, S = E = 100, T = 1,$
and the exact solution is $C = 5.92827717\cdots)$

Meshes	Results by (6.49)	|Errors|	Results by (6.50)	|Errors|
50×50	5.752424	0.175853	5.760096	0.168181
100×100	5.878708	0.049569	5.884210	0.044067
200×200	5.914582	0.013695	5.917403	0.010874
400×400	5.924045	0.004132	5.925541	0.002736
800×800	5.926810	0.001467	5.927574	0.000703
1600×1600	5.927706	0.000571	5.928097	0.000180
3200×3200	5.928032	0.000245	5.928230	0.000047

for $m = M_f+1, M_f+2, \cdots, M$. When $\bar{v}_m^{n+1}$ are determined, we assume all the equations in (6.45) to hold. Even though for $m = M_f+1, M_f+2, \cdots, M$ we do not take $\bar{v}_m^{n+1}$ as solutions so that the constraint condition is satisfied, v_m^{n+1}, $m = 0, 1, \cdots, M_f$ are determined under the assumption of all the equations in (6.45) holding. For (6.50), the situation is different. We assume that for $m = M, M-1, \cdots, M_f+1$, $v_m^{n+1} = \max(2\xi_m - 1, 0)$ and for $m = M_f, M_f-1, \cdots, 0$,

$$v_m^{n+1} = \frac{y_m - c_m v_{m+1}^{n+1}}{u_m}.$$

In this case, we only use the relations in (6.48) with $m = M_f, M_f - 1, \cdots, 0$, which are equivalent to the equations in (6.45) with $m = M_f, M_f - 1, \cdots, 0$. Therefore, we only assume that the equations in (6.45) hold for $m = M_f, M_f - 1, \cdots, 0$. Consequently, this is closer to what the situation should be. In Table 6.9, results obtained by (6.49) and (6.50) and their errors are listed. You can see that on the same mesh, the error of the results obtained by (6.49) is greater than (6.50) and that the smaller the mesh size, the greater the difference. Even though (6.49) can be used to obtain the price of American options, it brings some error that can be avoided if (6.50) is used. However, if the free boundary is far away from $S = E$, then in the region $S \approx E$, the difference of the solutions obtained by the two direct methods is very small.

When an implicit scheme is used to solve problem (6.9), we need to choose the lower and upper bounds of the computational domain and give some artificial boundary conditions at these two boundaries because we cannot do computation on an infinite domain. Let the lower and upper bounds be x_l and x_u. For a call option, we assume $u(x_l, \bar{\tau}) = 0$ and $u(x_u, \bar{\tau}) = g(x_u, \bar{\tau})$, and for a put option, $u(x_l, \bar{\tau}) = g(x_l, \bar{\tau})$ and $u(x_u, \bar{\tau}) = 0$. As soon as we set these conditions, the problem (6.9) can be discretized and solved in the same way as described above for the problem (6.8). This method is referred to as the projected implicit finite-difference method II (PIFDII).

In Tables 6.10 and 6.11, the values of American call and put options obtained by PIFDII are given. When we do computation, we take $x_l =$

Table 6.10. American call option (PIFDII)

($r = 0.1$, $\sigma = 0.2$, $D_0 = 0.05$, $S = E = 100$, $T = 1$,
and the exact solution is $C = 9.94092345\cdots$)

Meshes	Results	|Errors|	CPU(sec.)
100×25	9.928528	0.012396	0.0025
200×50	9.937831	0.003093	0.0096
400×100	9.940151	0.000773	0.0400
800×200	9.940729	0.000194	0.1700
1600×400	9.940875	0.000048	0.6700

Table 6.11. American put option (PIFDII)

($r = 0.1$, $\sigma = 0.2$, $D_0 = 0.05$, $S = E = 100$, $T = 1$,
and the exact solution is $P = 5.92827717\cdots$)

Meshes	Results	|Errors|	CPU(sec.)
100×25	5.922275	0.006002	0.0025
200×50	5.926394	0.001883	0.0094
400×100	5.927654	0.000623	0.0400
800×200	5.928050	0.000227	0.1700
1600×400	5.928188	0.000089	0.6700

$\ln(S_l/E) - |(r - D_0 - \sigma^2/2)T|$ and $x_u = \ln(S_u/E) + |(r - D_0 - \sigma^2/2)T|$.
For the call option, $S_l = 20$ and $S_u = 230$, and for the put, $S_l = 80$ and
$S_u = 350$. There, we also give a solution with an error less than 10^{-8} in each
table, which is obtained by the SSM given in Chapter 7. Therefore, we can
have the errors of the solutions on different meshes. The CPU time used is
also given, so you can have a notion about the performance of the method.

6.2.4 Projected Iteration Methods for the LC Problem

As we know, there are two types of methods to solve a linear system: iteration
methods and direct methods. Similarly, there are two ways to solve the system
related to American options. We discussed direct methods in the last subsection. Now let us study an iteration method. We still consider call options and
use the system (6.45). This problem can be solved by a method similar to the
SOR method for a system of linear equations given in Subsection 5.2.2. Any
equation in the system (6.45) can be rewritten as

$$\bar{v}_m^{n+1} = (1 - \omega)\bar{v}_m^{n+1} + \frac{\omega}{b_m}\left(q_m^n - a_m\bar{v}_{m-1}^{n+1} - c_m\bar{v}_{m+1}^{n+1}\right),$$

where ω is a constant. The value of the American option v_m^{n+1} satisfies the
relation above if $\bar{v}_m^{n+1} > \max(2\xi_m - 1, 0)$ or equal to $\max(2\xi_m - 1, 0)$ otherwise.
Therefore, for v_m^{n+1} we have the following relations:

$$v_m^{n+1} = \max\left((1-\omega)v_m^{n+1} + \frac{\omega}{b_m}\left(q_m^n - a_m v_{m-1}^{n+1} - c_m v_{m+1}^{n+1}\right), 2\xi_m - 1, 0\right),$$

$$m = 0, 1, \cdots, M.$$

We use an iteration method for finding its solution. Let $v_m^{(k)}$ be the k-th iteration of v_m^{n+1}, and the relation above can be rewritten in the following iteration form:

$$v_m^{(k+1)} = \max\left((1-\omega)v_m^{(k)} + \frac{\omega}{b_m}\left(q_m^n - a_m v_{m-1}^{(k+1)} - c_m v_{m+1}^{(k)}\right), 2\xi_m - 1, 0\right),$$

$$(6.51)$$

where $\omega \in [0, 2]$. Let $v_m^{(0)} = v_m^n$ for $m = 0, 1, \cdots, M$. As soon as we have $v_m^{(k)}$ for all m, the $(k+1)$-th iterative value of v_m^{n+1} can be obtained by (6.51) for $m = 0, 1, \cdots, M$ successively, starting from $k = 0$. When

$$\frac{1}{M+1}\sum_{m=0}^{M}\left(v_m^{(k)} - v_m^{(k+1)}\right)^2 \le \epsilon^2,$$

where ϵ^2 is a small number given according to the required accuracy, we can stop the iteration because for any m, $v_m^{(k)}$ and $v_m^{(k+1)}$ are very close to each other. This method is referred to as the projected successive over relaxation method I (PSORI). If the formulation (6.9) is adopted, after setting the values of x_l, x_u and the artificial boundary conditions, we can have a similar method and the corresponding method is referred to as PSORII. The details of the PSORII are left for the reader to write as Problem 11.

Table 6.12. American call option (PSORII)

$(r = 0.1, \sigma = 0.2, D_0 = 0.05, S = E = 100,$
and the exact solution is $C = 9.94092345\cdots)$

Meshes	Results	|Errors|	CPU(sec.)	ϵ^2
100×25	9.929351	0.011573	0.0240	10^{-8}
200×50	9.938037	0.002887	0.1100	$0.5 \cdot 10^{-9}$
400×100	9.940202	0.000721	0.5300	$0.25 \cdot 10^{-10}$
800×200	9.940743	0.000181	2.7500	$0.125 \cdot 10^{-11}$
1600×400	9.940878	0.000046	20.000	$0.6125 \cdot 10^{-13}$

In Tables 6.12 and 6.13, the prices of American call and put options on several meshes obtained by PSORII are given. The corresponding errors, CPU times, and ϵ^2 are also listed. All the parameters are the same as those given in Tables 6.10 and 6.11. The only difference between the results here and there is the way we solved the system.

Comparing Tables 6.10 and 6.11 with Tables 6.12 and 6.13 shows that the CPU time here is longer. This implies that the cost of PSORII method

Table 6.13. American put option (PSORII)

$(r = 0.1, \sigma = 0.2, D_0 = 0.05, S = E = 100,$
and the exact solution is $P = 5.92827717\cdots)$

Meshes	Results	|Errors|	CPU(sec.)	ϵ^2
100×25	5.922349	0.005928	0.0180	10^{-8}
200×50	5.926410	0.001867	0.0960	$0.5 \cdot 10^{-9}$
400×100	5.927651	0.000626	0.6100	$0.25 \cdot 10^{-10}$
800×200	5.928048	0.000230	5.2200	$0.125 \cdot 10^{-11}$
1600×400	5.928188	0.000089	46.300	$0.6125 \cdot 10^{-13}$

is greater than the PIFDII method for this case. However, we need to point out that for most of multi-dimensional problems, the iteration methods may be better than the direct methods even though here we show that the direct method is better than the iteration method for one-dimensional problems.

6.2.5 Comparison with Explicit Methods

Explicit methods are usually very simple and very easy to use. The main problem of explicit methods is the stability requirement. For the explicit method (6.6), the stability requirement is

$$\bar{\alpha} \le \frac{1}{2} \quad \text{or} \quad \Delta\bar{\tau} \le \frac{1}{2}\Delta x^2.$$

Thus, if the accuracy of the solution requires a small Δx, then a much smaller $\Delta\bar{\tau}$ must be taken in order to satisfy the stability condition, which slows down the computation. For implicit methods, no such restrictions are needed, and we can let $\Delta\bar{\tau}/\Delta x = constant$. Therefore, if we require higher accuracy, an implicit scheme will give a better performance. This can be seen in the following way.

Suppose we solve the problem (6.5) by the explicit scheme (6.6) and the implicit scheme (5.34). Assume that for (6.6) $\Delta\bar{\tau} = \alpha\Delta x^2$, where α is a constant not greater than $1/2$ and that for the scheme (5.34), $\Delta\bar{\tau} = \beta\Delta x$, where β is a constant. For the explicit scheme (6.6), the amount of computational work is

$$W_e = \frac{a_e}{\Delta\bar{\tau}\Delta x} = \frac{a_e}{\alpha\Delta x^3},$$

and the error is

$$E = b_{e\bar{\tau}}\Delta\bar{\tau} + b_{ex}\Delta x^2 = (b_{e\bar{\tau}}\alpha + b_{ex})\Delta x^2,$$

where a_e, $b_{e\bar{\tau}}$, and b_{ex} are three parameters related to scheme (6.6) and the solution. From these two relations for the scheme (6.6), we have the relation between the amount of work and the error required:

$$W_e = \frac{a_e[b_{e\bar{\tau}}\alpha + b_{ex}]^{3/2}}{\alpha} E^{-3/2}.$$

For the scheme (5.34),

$$W_i = \frac{a_i}{\Delta\bar{\tau}\Delta x} = \frac{a_i}{\beta\Delta x^2}$$

and

$$E = b_{i\bar{\tau}}\Delta\bar{\tau}^2 + b_{ix}\Delta x^2 = (b_{i\bar{\tau}}\beta^2 + b_{ix})\Delta x^2,$$

where a_i, $b_{i\bar{\tau}}$, and b_{ix} are three parameters related to scheme (5.34) and the solution. Here, we assume that a direct method is used for solving the linear system. Therefore, the relation between the amount of work and the error required is

$$W_i = \frac{a_i(b_{i\bar{\tau}}\beta^2 + b_{ix})}{\beta} E^{-1}.$$

Usually, a_i is greater than a_e because for (5.34) a linear system needs to be solved at each time step. Consequently, when E is not too small, it is possible that W_i is greater than W_e for the same E, which means that (6.6) is better than (5.34). When the solution is much smoother in the $\bar{\tau}$-direction than in the x-direction, the scheme (5.34) might be better than (6.6) even if E is not very small. This is because in this case for (5.34) we can choose a big β such that $b_{i\bar{\tau}}\beta^2$ is close to b_{ix}, which makes W_i smaller, but for (6.6) we cannot take this advantage because of the stability requirement. However, when E is small enough, then W_i must be less than W_e. This can be seen from comparing Tables 6.2 and 6.3 with Tables 6.10 and 6.11. The tables show that for the American call problem with the parameters given there, in order to reach an error about 0.003, the CPU time for (6.6) is about 0.06 and the CPU time for (5.34) is about 0.01.

6.3 Singularity-Separating Method

In this section, we will discuss how to make numerical methods more efficient. Generally speaking, the smoother the solution, the smaller the truncation error. Therefore, if the solution is smooth, even on a coarse mesh, the numerical result is still quite good. Suppose that the solution we need to find is not very smooth but has a certain type of singularity caused by the final condition. Also, we assume that there is an analytic expression that satisfies the same final condition and the same equation or a similar equation. If both the final conditions and the equations are the same, their singularities caused by the final conditions are the same, and the difference between them is a smooth function; if only the final conditions are the same, they possess similar singularities, and the difference between them is usually smoother than the solution we need to find. In both cases, we can first compute the difference using numerical methods and then have our solution by adding the analytic expression and the difference together. Such a method or technique will be referred to as singularity-separating method (SSM), or singularity-separating technique, in this book. Because computing the difference is quite efficient, we can have the solution quite efficiently. Of course, there is some extra work in order to compute the difference. However, from the examples we are going to show, such a way can truly make numerical methods more efficient. In this section, we will give some details of the method for European double moving barrier options, European vanilla option with variable volatilities, Bermudan options, European Parisian options, European average price options, two-factor vanilla options, and two-factor convertible bonds with $D_0 = 0$. Indeed, the method can be used for many more cases, including multi-factor derivative securities.

6.3.1 Barrier Options

If the option has a fixed barrier and σ, r, and D_0 are constants, we can find analytic solutions of barrier options (see Section 3.2). However, if the option has two moving barriers, analytic solutions may not exist even if σ, r, and D_0 are constants, and we may need to rely on numerical methods for pricing such an option. Here, we discuss how to make numerical methods more efficient.

The price $V(S, t)$ of a double moving barrier call option with rebates satisfies the equation

$$
\begin{cases}
\dfrac{\partial V}{\partial t} + \dfrac{1}{2}\sigma^2 S^2 \dfrac{\partial^2 V}{\partial S^2} + (r - D_0)S\dfrac{\partial V}{\partial S} - rV = 0, \\[2mm]
\qquad\qquad\qquad\qquad\qquad f(t) \le S \le g(t), \quad 0 \le t \le T, \\[2mm]
V(S, T) = \max(S - E, 0), \quad f(T) \le S \le g(T), \\[2mm]
V(f(t), t) = 0, \qquad\qquad\quad 0 \le t \le T, \\[2mm]
V(g(t), t) = g(t) - E, \qquad\; 0 \le t \le T,
\end{cases}
\tag{6.52}
$$

where $f(t)$ and $g(t)$ are the locations of the lower and upper barriers with

$$f(t) < E$$

and

$$g(t) > E,$$

and we assume that at the lower barrier, there is no rebate and at the upper barrier, the rebate is

$$g(t) - E.$$

Because the derivative of the payoff function

$$\max(S - E, 0)$$

is discontinuous at $S = E$, the solution $V(S, t)$ at $t \approx T$ and $S \approx E$ is not very smooth. Therefore, the error of numerical solutions in the region around

$t = T$ and $S = E$ is relatively large compared with that in the region far away from this point. In order to make the numerical solution better, we introduce a new function

$$\overline{V}(S,t) = V(S,t) - c(S,t),$$

where $c(S,t)$ is the price of the vanilla call option. Because $c(S,t)$ also satisfies the partial differential equation and the final condition in (6.52), $\overline{V}(S,t)$ satisfies

$$
\begin{cases}
\dfrac{\partial \overline{V}}{\partial t} + \dfrac{1}{2}\sigma^2 S^2 \dfrac{\partial^2 \overline{V}}{\partial S^2} + (r - D_0)S\dfrac{\partial \overline{V}}{\partial S} - r\overline{V} = 0, & \\
& f(t) \le S \le g(t), \quad 0 \le t \le T, \\
\overline{V}(S,T) = 0, & f(T) \le S \le g(T), \\
\overline{V}(f(t),t) = -c(f(t),t), & 0 \le t \le T, \\
\overline{V}(g(t),t) = g(t) - E - c(g(t),t), & 0 \le t \le T.
\end{cases}
\tag{6.53}
$$

The derivative of $\overline{V}(S,t)$ at $t \approx T$ and $S \approx E$ is very smooth, so the error of the numerical solution of $\overline{V}(S,t)$ is usually smaller than that of $V(S,t)$. Therefore, in order to get a better $V(S,t)$, we can first obtain the numerical solution of $\overline{V}(S,t)$ and then have $V(S,t)$ by adding $\overline{V}(S,t)$ and $c(S,t)$ together. We refer to this procedure as the singularity-separating method (SSM) or the singularity-separating technique for European barrier options. The reason is as follows. The derivative of $V(S,t)$ is discontinuous at $t = T$ and $S = E$. Thus, we say that $V(S,t)$ has some weak singularity. The function $\overline{V}(S,t)$, which will be determined numerically, is smooth. Therefore, the weak singularity has been "separated" from the numerical computation. The CPU time of getting $\overline{V}(S,t)$ is slightly longer than that of getting $V(S,t)$ directly because $c(f(t),t)$ and $c(g(t),t)$ need to be computed in order to get $\overline{V}(S,t)$. Because the error is smaller, we can usually expect better performance, i.e., we can usually expect to have the same accuracy by spending less CPU time or to spend the same CPU time for a better accuracy. Consequently, the singularity-separating technique can usually improve the performance.

Both $V(S,t)$ and $\overline{V}(S,t)$ are solutions of the following problem

$$
\begin{cases}
\dfrac{\partial \bar{u}}{\partial t} + \dfrac{1}{2}\sigma^2 S^2 \dfrac{\partial^2 \bar{u}}{\partial S^2} + (r - D_0)S\dfrac{\partial \bar{u}}{\partial S} - r\bar{u} = 0, & \\
& f(t) \le S \le g(t), \quad 0 \le t \le T, \\
\bar{u}(S,T) = \bar{f}_1(S), & f(T) \le S \le g(T), \\
\bar{u}(f(t),t) = \bar{b}_l(t), & 0 \le t \le T, \\
\bar{u}(g(t),t) = \bar{b}_u(t), & 0 \le t \le T.
\end{cases}
\tag{6.54}
$$

The only difference between the two cases is the functions in the final condition and in the boundary conditions. Thus, no matter whether the singularity-separating technique is used, we need a numerical method for problem (6.54) in order to have $V(S,t)$.

Problem (6.54) is a typical moving boundary problem. In order to convert it into a problem with fixed boundaries and transfer the final condition to an initial condition, we use the following transformation:

$$\begin{cases} \eta = \dfrac{S - f(t)}{g(t) - f(t)}, \\ \tau = T - t. \end{cases} \tag{6.55}$$

Let

$$u(\eta,\ \tau) = u(\eta(S,t),\ T - t) = \bar{u}(S,\ t),$$
$$F(\tau) = F(T - t) = f(t),$$
$$G(\tau) = G(T - t) = g(t).$$

Because

$$\frac{\partial \bar{u}}{\partial t} = \frac{\partial u}{\partial \eta}\frac{\partial \eta}{\partial t} + \frac{\partial u}{\partial \tau}\frac{\partial \tau}{\partial t}$$

$$= -\frac{1}{g - f}\left[\frac{df}{dt} + \eta\left(\frac{dg}{dt} - \frac{df}{dt}\right)\right]\frac{\partial u}{\partial \eta} - \frac{\partial u}{\partial \tau}$$

$$= \frac{1}{G - F}\left[\frac{dF}{d\tau} + \eta\left(\frac{dG}{d\tau} - \frac{dF}{d\tau}\right)\right]\frac{\partial u}{\partial \eta} - \frac{\partial u}{\partial \tau},$$

$$\frac{\partial \bar{u}}{\partial S} = \frac{1}{G(\tau) - F(\tau)}\frac{\partial u}{\partial \eta},$$

$$\frac{\partial^2 \bar{u}}{\partial S^2} = \frac{1}{[G(\tau) - F(\tau)]^2}\frac{\partial^2 u}{\partial \eta^2},$$

$u(\eta,\tau)$ is the solution of the problem

$$\begin{cases} \dfrac{\partial u}{\partial \tau} = \mathbf{L}_{\eta 1} u, & 0 \le \eta \le 1 \quad 0 \le \tau \le T, \\ u(\eta, 0) = f_1(\eta), & 0 \le \eta \le 1, \\ u(0, \tau) = b_l(\tau), & 0 \le \tau \le T, \\ u(1, \tau) = b_u(\tau), & 0 \le \tau \le T, \end{cases} \tag{6.56}$$

where

$$\mathbf{L}_{\eta 1} = \frac{1}{2}\left(\frac{S\sigma}{G - F}\right)^2\frac{\partial^2}{\partial \eta^2} + \left\{\frac{S}{G - F}(r - D_0)\right.$$
$$\left. + \frac{1}{G - F}\left[\frac{dF}{d\tau} + \eta\left(\frac{dG}{d\tau} - \frac{dF}{d\tau}\right)\right]\right\}\frac{\partial}{\partial \eta} - r,$$

$$f_1(\eta) = \bar{f}_1\left(F(0) + \eta[G(0) - F(0)]\right),$$
$$b_l(\tau) = \bar{b}_l(T - \tau),$$
$$b_u(\tau) = \bar{b}_u(T - \tau).$$

The problem (6.56) can be solved by explicit finite-difference schemes or implicit finite-difference schemes and even by pseudo-spectral methods. Here, we give some results to explain the effect of this technique if implicit finite-difference methods are used.

Table 6.14. Implicit methods with and without the SSM

$(S = 95,\ T = 1,\ E = 100,\ \sigma = 0.25,\ r = 0.1,\ D_0 = 0,$
$f(t) = 0.9Ee^{-0.1t},\ g(t) = 1.6Ee^{0.1t},$ the rebate $= g(t) - E,$
and the exact solution is $6.8441468\cdots)$

Meshes	Without SSM			With SSM		
	Solution	|Errors|	CPU	Solution	|Errors|	CPU
12×48	6.845973	0.001826	0.00039	6.843292	0.000855	0.00049
25×100	6.844623	0.000476	0.0019	6.844205	0.000058	0.0019
50×200	6.844187	0.000040	0.0062	6.844163	0.000016	0.0063
100×400	6.844167	0.000020	0.0221	6.844150	0.000003	0.0221

We have solved an identical problem by scheme (5.31) in two different ways: with and without SSM. In Table 6.14, the results, the errors, and the CPU time in seconds for four meshes are given. There, $N \times M$ in the column "Meshes" stands for a mesh that has $N + 1$ nodes in the t-direction (the τ-direction) and $M + 1$ nodes in the S-direction (the η-direction). The lower and upper knock-out boundaries are

$$f(t) = 0.9Ee^{-0.1t} \quad \text{and} \quad g(t) = 1.6Ee^{0.1t}.$$

There, the results both with and without SSM are given. In order to give errors, we have to find the exact solution. To our knowledge, no analytic solution for such a problem has been found. Therefore, we take a very accurate approximate solution as an exact solution. For this case, the exact solution is $6.8441468\cdots$ (here the eight digits are correct). From there, we can see that the result with SSM is clearly better than without SSM on the same mesh whereas the CPU time difference between the two cases is very small. Therefore, the advantage of the singularity-separating technique is obvious for this case. As we know, if the error $\approx a\Delta\tau^{\alpha} = a(T/N)^{\alpha}$ (suppose $\Delta\tau/\Delta\eta = constant$), then we say that the convergence rate is $O(\Delta\tau^{\alpha})$. From Table 6.14, we can see that when N is doubled, the error of the implicit finite-difference method with the singularity-separating technique decreases by a factor of about 4. This implies that the convergence rate of this method is $O(\Delta\tau^2)$.

In what follows, we give an intuitive explanation on why the singularity-separating method can improve the numerical results. The functions computed

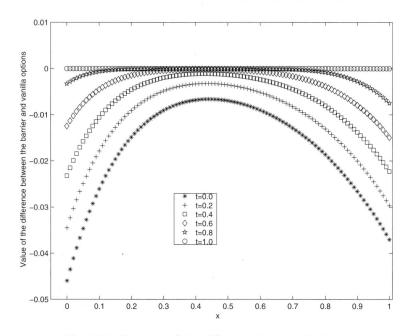

Fig. 6.7. Variation of the difference between the barrier
and vanilla option values

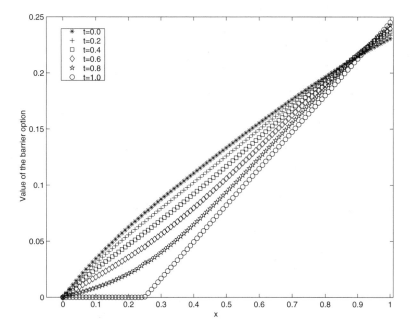

Fig. 6.8. Variation of the barrier option value

numerically for the methods with and without the singularity-separating technique are plotted in Figures 6.7 and 6.8 respectively. In each figure, there are six curves, which correspond to $t = 0, 0.2, 0.4, 0.6, 0.8, 1.0$. In Fig. 6.8, the functions are not as smooth as those in Fig. 6.7, especially, the derivative of the function for $t = 1$ in Fig. 6.8 is discontinuous. Therefore, when the singularity-separating technique is used, the truncation is smaller.

When there is no rebate at the upper barrier, such a method can still improve the performance. This is left for the reader to study (see Problem 13). For the case discussed in this subsection, the singularity is removed completely. For the European options with discrete dividends and some other cases, the singularity can also be completely removed in the same way. In many other cases, the singularity cannot be completely separated but can be made much weaker. In the next several subsections, we will discuss how the SSM works for other cases.

6.3.2 European Vanilla Options with Variable Volatilities

When σ is a constant, for European vanilla options we can get their prices by the Black–Scholes formulae. However, it seems that the assumption of σ being a constant needs to be modified. One of the modifications is to let σ be a function of S. In this case, in order to evaluate an option, we usually need to solve a partial differential equation problem numerically. In order to overcome the problem caused by the discontinuous derivative in the payoff, we can do the following.

Let us consider call options. Their prices $c(S,t)$ are solutions of the problem:

$$
\begin{cases}
\dfrac{\partial c}{\partial t} + \dfrac{1}{2}\sigma^2(S)S^2\dfrac{\partial^2 c}{\partial S^2} + (r - D_0)S\dfrac{\partial c}{\partial S} - rc = 0, & 0 \le S, \ t \le T, \\
c(S,T) = \max(S - E, 0), & 0 \le S.
\end{cases}
$$

Suppose that $c_E(S,t;\sigma(E))$ is the price of the option with the volatility at $S = E$, $\sigma(E)$, i.e., $c_E(S,t;\sigma(E))$ satisfies

$$
\begin{cases}
\dfrac{\partial c_E}{\partial t} + \dfrac{1}{2}\sigma^2(E)S^2\dfrac{\partial^2 c_E}{\partial S^2} + (r - D_0)S\dfrac{\partial c_E}{\partial S} - rc_E = 0, & 0 \le S, \ t \le T, \\
c_E(S,T) = \max(S - E, 0), & 0 \le S.
\end{cases}
$$

Let $\bar{c}(S,t) = c(S,t) - c_E(S,t;\sigma(E))$. Then, $\bar{c}(S,t)$ is the solution of the problem

$$
\begin{cases}
\dfrac{\partial \bar{c}}{\partial t} + \dfrac{1}{2}\sigma^2(S)S^2\dfrac{\partial^2 \bar{c}}{\partial S^2} + (r - D_0)S\dfrac{\partial \bar{c}}{\partial S} - r\bar{c} = f(S,t), \\
\qquad\qquad\qquad\qquad\qquad\qquad 0 \le S, \quad t \le T, \\
\bar{c}(S,T) = 0, \qquad\qquad\qquad\qquad 0 \le S,
\end{cases}
\tag{6.57}
$$

where

$$f(S,t) = \frac{1}{2}\left[\sigma^2(E) - \sigma^2(S)\right] S^2 \frac{\partial^2 c_E}{\partial S^2}$$

$$= \frac{1}{2\sigma(E)\sqrt{2\pi(T-t)}}\left[\sigma^2(E) - \sigma^2(S)\right] S e^{-(D_0(T-t)+d_1^2/2)} \quad (6.58)$$

and

$$d_1 = \left\{\ln(S/E) + \left[r - D_0 + \frac{1}{2}\sigma^2(E)\right](T-t)\right\} \Big/ \left[\sigma(E)\sqrt{T-t}\right].$$

This problem is defined on an infinite domain. In order to convert it into a problem on a finite domain with a bounded solution, we use the following transformation:

$$\begin{cases} \xi = \dfrac{S}{S+E}, \\ \tau = \dot{T} - t, \\ \bar{c}(S,t) = (S+E)\overline{V}(\xi,\tau). \end{cases}$$

Finally, we have

$$\begin{cases} \dfrac{\partial \overline{V}}{\partial \tau} = \dfrac{1}{2}\bar{\sigma}^2(\xi)\xi^2(1-\xi)^2\dfrac{\partial^2 \overline{V}}{\partial \xi^2} + (r - D_0)\xi(1-\xi)\dfrac{\partial \overline{V}}{\partial \xi} \\ \qquad\quad - [r(1-\xi) + D_0\xi]\overline{V} + \bar{f}(\xi,\tau), \quad 0 \le \xi \le 1, \quad 0 \le \tau, \\ \overline{V}(\xi,0) = 0, \qquad\qquad\qquad\qquad\qquad\qquad 0 \le \xi \le 1, \end{cases} \quad (6.59)$$

where

$$\bar{\sigma}(\xi) = \sigma(\xi E/(1-\xi)),$$

$$\bar{f}(\xi,\tau) = \frac{-f(S,t)}{S+E} = \frac{1}{2\sigma(E)\sqrt{2\pi\tau}}\left[\bar{\sigma}^2(\xi) - \sigma^2(E)\right]\xi e^{-(D_0\tau+d_1^2/2)}$$

and

$$d_1 = \left\{\ln\frac{\xi}{(1-\xi)} + \left[r - D_0 + \frac{1}{2}\sigma^2(E)\right]\tau\right\} \Big/ \left[\sigma(E)\sqrt{\tau}\right].$$

In order to do some computation, we need the function $\sigma(S)$ or $\bar{\sigma}(\xi)$. For the Japanese yen–U.S. dollar exchange rate, we determine the function by the method in Subsection 5.6.2. In order to avoid approximating a function on an infinite domain, a new variable $\xi = S/(S + P_m)$ is introduced. Because the exchange rate is around 0.01, we set $P_m = 0.01$. Using the data of 1990–2000 from the market (see the curve in Fig. 1.5), we find the maximum and minimum values, $S_{max} = 0.01232741616$ and $S_{min} = 0.00625390870$. The corresponding values of ξ are

$$\xi_l = \frac{S_{min}}{S_{min} + P_m} = 0.384763371, \quad \xi_u = \frac{S_{max}}{S_{max} + P_m} = 0.552120141.$$

Assume that the function $\bar{\sigma}(\xi)$ is in the form:

$$\bar{\sigma}(\xi) = \begin{cases} c_l + a_l \left[1 - \left(\dfrac{\xi}{\xi_l}\right)^{200}\right], & 0 \leq \xi < \xi_l, \\[3ex] a_0 + a_1\xi + a_2\xi^2 + a_3\xi^3, & \xi_l \leq \xi \leq \xi_u, \\[3ex] c_u + a_u \left[1 - \left(\dfrac{1-\xi}{1-\xi_u}\right)^{200}\right], & \xi_u \leq \xi \leq 1, \end{cases}$$

where c_l, a_l, a_0, a_1, a_2, a_3, c_u, a_u are eight parameters to be determined. Taking the data of 1990–2000 from the market, using the method described in Subsection 5.6.2 with $g(\xi) \equiv 1$ and setting $M = 7$, we find the values of a_0, a_1, a_2, a_3:

$$a_0 = -10.7848, \quad a_1 = 72.8005, \quad a_2 = -161.134, \quad a_3 = 118.208.$$

Then, requiring the continuity of the function at $\xi = \xi_l$ and $\xi = \xi_u$ up to the first derivative yields

$$c_l = 0.104667, \quad a_l = -0.00250664, \quad c_u = 0.185335, \quad a_u = 0.00665520.$$

In Fig. 6.9, this function is plotted as a solid line, and the circles are the volatilities for different S obtained by statistics.

As soon as we have this function, we can evaluate the price of options on the Japanese yen–U.S. dollar exchange rate. Discretizing problem (6.59) by (5.31) and solving the linear system by the LU decomposition, we can find the price. In Fig. 6.10, the solid line gives the value of the European call option. There, we also compare different models. Another model is to let the volatility be a constant. Using the same data, we find $\sigma = 0.1165$. The dashed line in Fig. 6.10 gives the option price for this model obtained by the Black–Scholes formula. The maximum difference of the results between the two models is more than 30% if $S \in [0.0115, 0.0125]$. If we assume σ to take the value of $\sigma(E)$ (the result for this case is given by the dotted line in Fig. 6.10), the maximum difference is more than 8% for $S \in [0.0115, 0.0125]$. Therefore, among the results obtained by using different models, there is quite a big difference. In our computation for the model with variable volatility, the numerical method is quite efficient because we are calculating the difference numerically. For this example problem, on a 60×4 mesh for the option price at $S = E$, the error is $6 \times 10^{-5}E$ when the SSM is used and $1 \times 10^{-3}E$ when the SSM is not used.

Finally, we would like to point out that unlike the barrier options, in this case the weak singularity is not removed completely. However, the singularity is weakened so the SSM still succeeds as shown above. Let us explain this

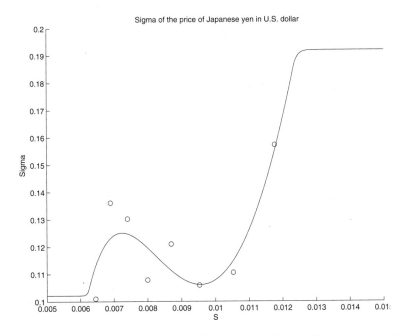

Fig. 6.9. The volatility function for Japanese yen–U.S. dollar exchange rate

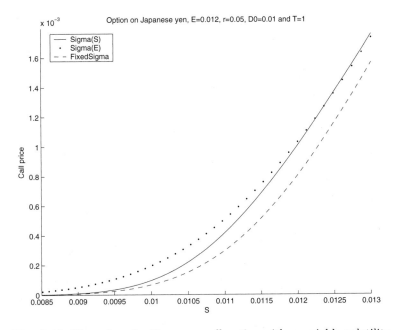

Fig. 6.10. The value of a European call option with a variable volatility
with $E = 0.012$, $r = 0.05$, $D_0 = 0.01$, and $T = 1$ year

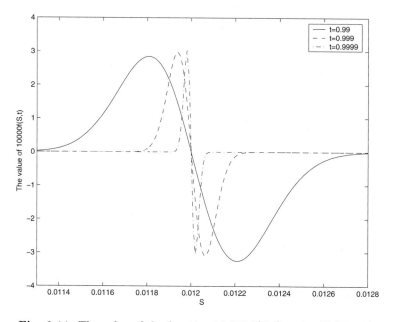

Fig. 6.11. The value of the function $10,000f(S,t)$ at $t \approx T$ $(T = 1)$

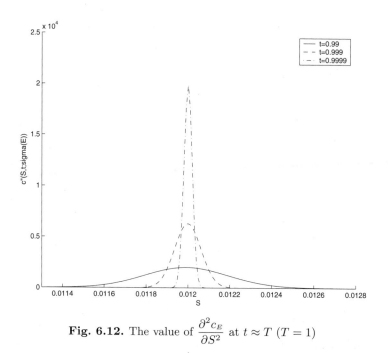

Fig. 6.12. The value of $\dfrac{\partial^2 c_E}{\partial S^2}$ at $t \approx T$ $(T = 1)$

matter as follows. Because $\dfrac{\partial^2 c_E}{\partial S^2}$ has some singularity at the point $T = t$

and $S = E$, the function $f(S,t) = \dfrac{1}{2}\left[\sigma^2(E) - \sigma^2(S)\right] S^2 \dfrac{\partial^2 c_E}{\partial S^2}$ also has some

singularity. However, because the term $\sigma^2(E) - \sigma^2(S)$ is equal to zero at $S = E$,

the singularity of $f(S,t)$ at that point is much weaker than that of $\dfrac{\partial^2 c_E}{\partial S^2}$. In

Figs. 6.11 and 6.12, $f(S,t)$ used in this example and $\dfrac{\partial^2 c_E}{\partial S^2}$ for $t = T - 0.01$,

$T - 0.001$, $T - 0.0001$ are plotted, respectively. Noticing the maximum value

of $|f(S,t)|$ is about 3.5×10^{-4} and the value of $\dfrac{\partial^2 c_E}{\partial S^2}$ could be very large,

reaching 2×10^4 at $t = T - 0.0001$, we can see that the singularity of $f(S,t)$

at that point is truly weaker than that of $\dfrac{\partial^2 c_E}{\partial S^2}$. Because the singularity of

$f(S,t)$ is quite weak and the singularity of $\bar{c}(S,t)$ is weaker than $f(S,t)$, the
function $\bar{c}(S,t)$ is quite smooth. This is an important reason to guarantee the
success of the SSM.

6.3.3 Bermudan Options

A Bermudan option is an option that can be exercised early, but only on
predetermined dates. It is clear that the holder of a Bermudan option has more
rights than the holder of a European option and less rights than the holder
of an American option, just like the fact that Bermuda is situated between
America and Europe. This is how the option got its name. If we use projected
methods, it is easy to price. Here, we suggest some more efficient methods.
Assume the expiry of the option to be T and suppose the option can be
exercised at time $t = T_1, T_2, \cdots, T_K = T$, where $T_k = kT/K$, $k = 1, 2, \cdots, K$.

Let us consider a Bermudan call option with $D_0 > 0$ and a variable $\sigma(S)$,
and denote its value by $C_b(S,t)$. Define $T_0 = 0$ and assume $T_0 < T_1 < \cdots < T_K$. Then, $C_b(S,t)$ is a solution of K successive problems:

$$
\begin{cases}
\dfrac{\partial C_b}{\partial t} + \dfrac{1}{2}\sigma^2(S)S^2\dfrac{\partial^2 C_b}{\partial S^2} + (r - D_0)S\dfrac{\partial C_b}{\partial S} - rC_b = 0, \\[2mm]
\qquad\qquad 0 \le S, \;\; T_{k-1} < t < T_k, \\[2mm]
C_b(S, T_k^-) = \max(C_b(S, T_k^+), \max(S - E, 0)), \quad 0 \le S,
\end{cases}
\tag{6.60}
$$

$$k = K, K - 1, \cdots, 1$$

with $C_b(S, T_K^+) = \max(S - E, 0)$. Clearly, at $t = T_K$, $C_b(S, T_K^-) = \max(S - E, 0)$, $0 \le S$. At $t = T_k$, $k = K - 1, K - 2, \cdots, 1$, the whole interval $[0, \infty)$ is
divided into two parts $[0, S_k^*]$ and (S_k^*, ∞). On $[0, S_k^*]$, $C_b(S, T_k^-) \ge \max(S - E, 0)$ and on (S_k^*, ∞), $C_b(S, T_k^+) < \max(S - E, 0)$. Because these functions are
nonnegative and continuous, $S_k^* \ge E$ and $C_b(S_k^*, T_k^+) = S_k^* - E$. Therefore,
the final condition of each problem above can be written as

$$C_b(S, T_k^-) = \begin{cases} C_b(S, T_k^+), & \text{if } 0 \le S \le S_k^*, \\ S - E, & \text{if } S_k^* < S. \end{cases}$$

Because a European call option with a constant volatility has a closed-form solution, just like what we did in the last subsection, we consider the difference between the Bermudan call option and the European call option with a constant volatility $\sigma(E)$ and denote the difference by

$$\tilde{C}_b = C_b - c_E(S, t; \sigma(E)).$$

It is clear that $\tilde{C}_b$ satisfies the partial differential equation in (6.57). At $t = T_k$, we have

$$\tilde{C}_b(S, T_k^-) = \begin{cases} \tilde{C}_b(S, T_k^+), & \text{if } 0 \le S \le S_k^*, \\ S - E - c_E(S, T_k; \sigma(E)), & \text{if } S_k^* < S. \end{cases}$$

Therefore, $\tilde{C}_b$ is the solution of the following K successive problems:

$$\begin{cases} \dfrac{\partial \tilde{C}_b}{\partial t} + \dfrac{1}{2}\sigma^2(S)S^2\dfrac{\partial^2 \tilde{C}_b}{\partial S^2} + (r - D_0)S\dfrac{\partial \tilde{C}_b}{\partial S} - r\tilde{C}_b = f(S,t), \\ \qquad\qquad 0 \le S, \ T_{k-1} < t < T_k, \\ \tilde{C}_b(S, T_k^-) = \max\left(\tilde{C}_b(S, T_k^+), \max(S - E, 0) - c_E(S, T_k; \sigma(E))\right), \\ \qquad\qquad 0 \le S, \\ \qquad\qquad k = K, K-1, \cdots, 1 \end{cases} \tag{6.61}$$

with $\tilde{C}_b(S, T_K^+) = 0$. This problem can be solved in a way similar to what we have used to find the solution of a European option with discrete dividends in Subsection 6.2.2. The only difference is that using jump conditions should be replaced by taking the maximum between $\tilde{C}_b(S, T_k^+)$ and $\max(S - E, 0) - c_E(S, t; \sigma(E))$ at these specified times.

In many cases, this method can be further improved by doing the following. For $k = K - 1, K - 2, \cdots, 1$, let us define $K - 1$ polynomials of degree J: $f_k(S) = a_{0,k} + a_{1,k}S + \cdots + a_{J,k}S^J$ on $[S_k^{**}, S_k^*]$, which satisfies the conditions $f_k(S_k^*) = S_k^* - E$ and $f_k(S_k^{**}) = 0$. Besides satisfying these two conditions, we choose these coefficients $a_{0,k}, a_{1,k}, \cdots, a_{J,k}$ and $S_k^{**} \in [0, S_k^*]$ such that the norm of the function

$$\begin{cases} C_b(S, T_k^-), & \text{if } 0 \le S < S_k^{**}, \\ C_b(S, T_k^-) - f_k(S), & \text{if } S_k^{**} \le S < S_k^* \end{cases}$$

is as small as possible. It is clear that the function

$$\begin{cases} 0, & \text{if } 0 \le S < S_k^{**}, \\ f_k(S), & \text{if } S_k^{**} \le S < S_k^*, \\ S - E, & \text{if } S_k^* \le S \end{cases}$$

is a good approximation to $C_b(S, T_k^-)$. For $k = K$, if we define $S_k^* = S_k^{**} = E$, then the function defined above is equal to $C_b(S, T_K^-)$. Therefore, we assume the function above to be defined for $k = K, K - 1, \cdots, 1$.

Consider the problems

$$
\begin{cases}
\dfrac{\partial c_b}{\partial t} + \dfrac{1}{2}\sigma^2(S_k^*)S^2\dfrac{\partial^2 c_b}{\partial S^2} + (r - D_0)S\dfrac{\partial c_b}{\partial S} - rc_b = 0, \\[2mm]
\qquad 0 \le S, \ T_{k-1} < t < T_k, \\[2mm]
c_b(S, T_k^-) = \begin{cases} 0, & \text{if } 0 \le S < S_k^{**}, \\ f_k(S), & \text{if } S_k^{**} \le S < S_k^*, \\ S - E, & \text{if } S_k^* \le S. \end{cases}
\end{cases}
\tag{6.62}
$$

Noticing that for any integer n, we have (see Problem 14 in Chapter 2)

$$
\frac{1}{\sqrt{2\pi b}}\int_c^d S^n e^{-(\ln(S/a)+b^2/2)^2/2b^2}\frac{dS}{S}
$$
$$
= a^n e^{(n^2-n)b^2/2}\left[N\left(\frac{\ln(d/a) + (1/2 - n)b^2}{b}\right) - N\left(\frac{\ln(c/a) + (1/2 - n)b^2}{b}\right)\right],
$$

we can find a closed-form solution of (6.62) (see Problem 18 in Chapter 2)

$$
c_b(S, t) = \sum_{n=0}^J \left\{ a_{n,k} S^n e^{[(n-1)r - nD_0 + (n-1)n\sigma^2(S_k^*)/2](T_k - t)} \right.
$$
$$
\left. \times \left[N\left(d_k^* - n\sigma(S_k^*)\sqrt{T_k - t}\right) - N\left(d_k^{**} - n\sigma(S_k^*)\sqrt{T_k - t}\right)\right]\right\}
$$
$$
+ Se^{-D_0(T_k-t)}\left[1 - N\left(d_k^* - \sigma(S_k^*)\sqrt{T_k - t}\right)\right] - Ee^{-r(T_k-t)}[1 - N(d_k^*)],
\tag{6.63}
$$

where $t \in (T_{k-1}, T_k)$ and

$$
d_k^* = \left[\ln(S_k^*/S) - \left(r - D_0 - \frac{1}{2}\sigma^2(S_k^*)\right)(T_k - t)\right] \Big/ \left(\sigma(S_k^*)\sqrt{T_k - t}\right),
$$
$$
d_k^{**} = \left[\ln(S_k^{**}/S) - \left(r - D_0 - \frac{1}{2}\sigma^2(S_k^*)\right)(T_k - t)\right] \Big/ \left(\sigma(S_k^*)\sqrt{T_k - t}\right).
$$

It is easy to see that for $t \in (T_{K-1}, T_K]$, c_b represents the price of the European option with a constant volatility $\sigma(S^*) = \sigma(E)$ because $S^* = E$ at time $t = T$, that is, $c_b(S, t)$ is equal to $c_E(S, t; \sigma(E))$ for this period.

At the point $S = S_k^*$ and $t = T_k$, the singularity of the solution of (6.62) is very close to that of (6.60). Therefore, the function

$$
\overline{C}_b = C_b - c_b
$$

is smooth near this point for $t \in (T_{k-1}, T_k)$ and its value is quite small if $T_k - T_{k-1}$ is not big. This function satisfies the following equation and condition:

$$
\begin{cases}
\dfrac{\partial \overline{C}_b}{\partial t} + \dfrac{1}{2}\sigma^2(S)S^2\dfrac{\partial^2 \overline{C}_b}{\partial S^2} + (r - D_0)S\dfrac{\partial \overline{C}_b}{\partial S} - r\overline{C}_b = \\[2mm]
\qquad \dfrac{1}{2}(\sigma^2(S_k^*) - \sigma^2(S))S^2\dfrac{\partial^2 c_b}{\partial S^2}, \quad 0 \le S, \quad T_{k-1} < t < T_k, \\[3mm]
\overline{C}_b(S, T_k^-) = \begin{cases}
C_b(S, T_k^-), & \text{if } 0 \le S < S_k^{**}, \\
C_b(S, T_k^-) - f_k(S), & \text{if } S_k^{**} \le S < S_k^*, \\
0, & \text{if } S_k^* \le S.
\end{cases}
\end{cases}
\tag{6.64}
$$

Therefore, in order to have $C_b(S, T_{k-1}^+)$, we can first find $\overline{C}_b(S, T_{k-1}^+)$ by solving (6.64) from $t = T_k$ to T_{k-1} and then obtain $C_b(S, T_{k-1}^+)$ by

$$
C_b(S, T_{k-1}^+) = \overline{C}_b(S, T_{k-1}^+) + c_b(S, T_{k-1}^+).
$$

Because for a variable σ the partial differential equation in (6.64) is nonhomogeneous and the right-hand side is quite complicated, the amount of computation of solving (6.64) is greater than solving (6.60) on the same mesh. However, in order to have a solution with the same accuracy, the number of mesh points needed for (6.64) is much smaller than (6.60). It is expected that in order to reach the same accuracy, solving (6.64) is better. If $\sigma = constant$, then (6.64) becomes

$$
\begin{cases}
\dfrac{\partial \overline{C}_b}{\partial t} + \dfrac{1}{2}\sigma^2 S^2\dfrac{\partial^2 \overline{C}_b}{\partial S^2} + (r - D_0)S\dfrac{\partial \overline{C}_b}{\partial S} - r\overline{C}_b = 0, \\[3mm]
\qquad 0 \le S, \quad T_{k-1} < t < T_k, \\[3mm]
\overline{C}_b(S, T_k^-) = \begin{cases}
C_b(S, T_k^-), & \text{if } 0 \le S < S_k^{**}, \\
C_b(S, T_k^-) - f_k(S), & \text{if } S_k^{**} \le S < S_k^*, \\
0, & \text{if } S_k^* \le S.
\end{cases}
\end{cases}
\tag{6.65}
$$

The partial differential equation in (6.65) is a homogeneous equation. Hence, the amount of computation of solving (6.65) is very close to that of solving the original problem (6.60).

Sometimes, the singularities at the points $S = S_k^*$ and $t = T_k$, $k = K - 1, K - 2, \cdots, 1$, are quite weak and far away from the region $S \approx E$. Therefore, these singularities only cause small errors in the region $S \approx E$. Also, $[S_k^{**}, S_k^*]$ is not a small interval, so $f_k(S)$ may not be a good approximation to $C_b(S, T_k^-)$. In this case, using the method described at the beginning of this subsection might be better.

In what follows, we give some examples. Consider a Bermudan call option with $r = 0.1$, $D_0 = 0.05$, and $T_k = k/4$, $k = 1, 2, 3, 4$. The price of the option is evaluated by two different ways. One is to solve problem (6.60) by the implicit method (5.31) and the other is to take $J = 6$ and solve problem (6.65) by (5.31). For $S = 100$, the results obtained by the two ways, the errors, and CPU times needed on a Pentium III 800 MHz computer are given in Table 6.15. From there, we can see that for this case in order to have a result with an error about 10^{-2} (the corresponding relative error to E is 10^{-4}), CPU time needed is about 0.003 seconds if the singularity-separating method described here is used, and CPU time needed is about 0.01 seconds if the singularity-separating method is not used. Therefore, even though on an identical mesh, the CPU time needed for the SSM is much longer, overall the SSM is still fast for a fixed accuracy.

Table 6.15. Bermudan call option prices ($r < D_0$)

($S = 100$, $E = 100$, $T = 1$, $r = 0.05$, $D_0 = 0.1$, $\sigma = 0.2$, $K = 4$, and the exact solution $= 5.77654\cdots$)

	Implicit method			SSM		
Mesh sizes	Results	‖Errors‖	CPU(seconds)	Results	‖Errors‖	CPU(seconds)
24 × 24	5.0474	0.7291	0.0002	5.8564	0.0799	0.0014
36 × 36	5.4507	0.3258	0.0005	5.7788	0.0023	0.0019
48 × 48	5.6143	0.1622	0.0008	5.7881	0.0116	0.0028
60 × 60	5.6732	0.1033	0.0013	5.7845	0.0080	0.0037
72 × 72	5.7069	0.0696	0.0018	5.7833	0.0068	0.0048
84 × 84	5.7332	0.0433	0.0024	5.7833	0.0068	0.0061
96 × 96	5.7362	0.0403	0.0032	5.7809	0.0044	0.0073
108 × 108	5.7479	0.0286	0.0039	5.7807	0.0042	0.0086
120 × 120	5.7543	0.0222	0.0049	5.7797	0.0032	0.0101
132 × 132	5.7599	0.0166	0.0059	5.7804	0.0039	0.0119
144 × 144	5.7592	0:0173	0.0073	5.7800	0.0035	0.0134
156 × 156	5.7649	0.0116	0.0082	5.7799	0.0034	0.0152
168 × 168	5.7674	0.0091	0.0096	5.7790	0.0025	0.0172
180 × 180	5.7659	0.0106	0.0109	5.7784	0.0019	0.0190

The next example is a Bermudan option with $r = 0.1$, $D_0 = 0.05$, and $T_k = k/12$, $k = 1, 2, \cdots, 12$. The other parameters are the same as those for the first example. In this case, the singularities at the points $S = S_k^*$ and $t = T_k$, $k = K - 1, K - 2, \cdots, 1$, are weak and we choose $c_E(S, t; \sigma(E))$ as c_b and solve problem (6.61) by (5.31). The results for $S = 100$ are given in Table 6.16. When the SSM is not used, the errors are close to those in the first example. However, when the SSM is used, the errors are even much less than those in the first example due to the very small value of $\overline{C}_b$.

In Fig. 6.13, the price of the first Bermudan call option is given as a function of S. The prices of the American and European call options are also given there. The figure shows that the price of the Bermudan option

Table 6.16. Bermudan call option prices $(r > D_0)$

$(S = 100, E = 100, T = 1, r = 0.1, D_0 = 0.05, \sigma = 0.2, K = 12,$
and the exact solution $= 9.940918 \cdots)$

Mesh sizes	Implicit method			SSM		
	Results	‖Errors‖	CPU(seconds)	Results	‖Errors‖	CPU(seconds)
24×24	9.1488	0.7922	0.0003	9.9411	0.0002	0.0017
36×36	9.6261	0.3148	0.0006	9.9410	0.0001	0.0026
48×48	9.7704	0.1705	0.0011	9.9410	0.0001	0.0037
60×60	9.8333	0.1076	0.0015	9.9409	0.0000	0.0049
72×72	9.8667	0.0742	0.0020	9.9409	0.0000	0.0062
84×84	9.8866	0.0543	0.0027	9.9409	0.0000	0.0075
96×96	9.8995	0.0414	0.0034	9.9409	0.0000	0.0090
108×108	9.9082	0.0327	0.0043	9.9409	0.0000	0.0105
120×120	9.9145	0.0264	0.0052	9.9409	0.0000	0.0121

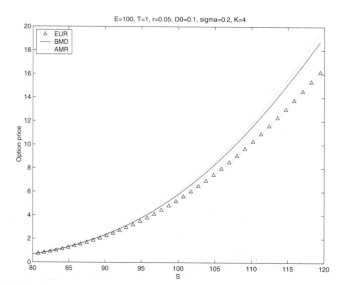

Fig. 6.13. Prices of American, Bermudan, and European call options
$(E = 100, T = 1, r = 0.05, D_0 = 0.1, \sigma = 0.2, \text{ and } K = 4)$

is less than the price of the American option and greater than the price of
the European option, and it is quite close to the price of the corresponding
American option. The financial reason of this fact is as follows. As has been
pointed out at the beginning of this subsection, the holder of a Bermudan
option has more rights than a holder of a European option and less rights
than a holder of an American option. Thus, the money paid by the holder of
the Bermudan option should be greater than the price of a European option
and less than the price of an American option.

The symmetry relations also hold for Bermudan options, which is left for the reader to prove. Therefore, we only need to study numerical methods for Bermudan call options. In order to obtain the price of a put option, we first solve a corresponding call option problem and then find the price of the put option by the symmetry relation.

6.3.4 European Parisian Options

Let us take a European Parisian up-and-out call option with continuous sampling as an example to show how the singularity-separating method works for Parisian options.

Suppose c_p is the price of the Parisian up-and-out call option. From Subsection 3.2.4, we see that $c_p(S, t_d, t)$ is the solution of problem (3.6):

$$
\begin{cases}
\dfrac{\partial c_p}{\partial t} + \dfrac{1}{2}\sigma^2 S^2 \dfrac{\partial^2 c_p}{\partial S^2} + (r - D_0) S \dfrac{\partial c_p}{\partial S} + H(S - B_u) \dfrac{\partial c_p}{\partial t_d} - r c_p = 0, \\[2mm]
\quad 0 \le S, \quad t_d = 0 \quad \text{and} \quad B_u \le S, \quad 0 < t_d \le T_d, \quad 0 \le t \le T, \\[2mm]
c_p(S, t_d, T) = \begin{cases} \max(S - E, 0), & 0 \le S < B_u, \quad t_d = 0, \\ S - E, & B_u \le S, \quad 0 \le t_d < T_d, \\ 0, & B_u \le S, \quad t_d = T_d, \end{cases} \\[6mm]
c_p(B_u, t_d, t) = c_p(B_u, 0, t), \quad t_d \in (0, T_d), \quad 0 \le t \le T, \\[2mm]
c_p(S, T_d, t) = 0, \quad B_u \le S, \quad 0 \le t \le T.
\end{cases}
$$

Let $c(S, t)$ be the price of the European vanilla call option and define

$$
\bar{c}_p(S, t_d, t) = c_p(S, t_d, t) - c(S, t).
$$

Because $c(S, t)$ does not depend on t_d, it is clear that $c(S, t)$ also satisfies the partial differential equation in (3.6). Therefore, $\bar{c}_p(S, t_d, t)$ is the solution of the following problem:

$$
\begin{cases}
\dfrac{\partial \bar{c}_p}{\partial t} + \dfrac{1}{2}\sigma^2 S^2 \dfrac{\partial^2 \bar{c}_p}{\partial S^2} + (r - D_0) S \dfrac{\partial \bar{c}_p}{\partial S} + H(S - B_u) \dfrac{\partial \bar{c}_p}{\partial t_d} - r \bar{c}_p = 0, \\[2mm]
\quad 0 \le S, \quad t_d = 0 \quad \text{and} \quad B_u \le S, \quad 0 < t_d \le T_d, \quad 0 \le t \le T, \\[2mm]
\bar{c}_p(S, t_d, T) = \begin{cases} 0, & 0 \le S < B_u, \quad t_d = 0, \\ 0, & B_u \le S, \quad 0 \le t_d < T_d, \\ E - S, & B_u \le S, \quad t_d = T_d, \end{cases} \\[6mm]
\bar{c}_p(B_u, t_d, t) = \bar{c}_p(B_u, 0, t), \quad t_d \in (0, T_d), \quad 0 \le t \le T, \\[2mm]
\bar{c}_p(S, T_d, t) = -c(S, t), \quad B_u \le S, \quad 0 \le t \le T.
\end{cases}
\tag{6.66}
$$

Table 6.17. Numerical solutions for Parisian up-and-out call options
$(r = 0.1, D_0 = 0.05, \sigma = 0.25, E = 100, T = 0.5, B_u = 150, \text{ and } T_d = 0.02)$

Meshes	$S = 100$ Solutions	|Errors|	$S = 120$ Solution	|Errors|	$S = 150$ Solution	|Errors|
200×100	7.4139	$1.08 \cdot 10^{-3}$	15.3107	$7.79 \cdot 10^{-3}$	5.0574	$3.73 \cdot 10^{-2}$
300×150	7.4067	$1.08 \cdot 10^{-4}$	15.2886	$6.33 \cdot 10^{-3}$	4.9389	$1.30 \cdot 10^{-2}$
400×200	7.4059	—	15.1924	—	4.8754	—

Table 6.18. Numerical solutions for Parisian up-and-out call options (with SSM)
$(r = 0.1, D_0 = 0.05, \sigma = 0.25, E = 100, T = 0.5, B_u = 150, \text{ and } T_d = 0.02)$

Meshes	$S = 100$ Solutions	|Errors|	$S = 120$ Solution	|Errors|	$S = 150$ Solution	|Errors|
200×100	7.3943	$1.76 \cdot 10^{-4}$	15.2016	$5.13 \cdot 10^{-4}$	4.9232	$2.09 \cdot 10^{-2}$
300×150	7.3936	$8.16 \cdot 10^{-5}$	15.1947	$5.92 \cdot 10^{-5}$	4.8251	$5.18 \cdot 10^{-4}$
400×200	7.3930	—	15.1938	—	4.8226	—

Because $c_p(S, t_d, t)$ and $c(S, t)$ have the same singularity at the point $S = E$ and $t = T$, $\bar{c}_p$ is quite smooth near $S = E$ and $t = T$, that is, the singularity has been separated. Therefore, it is expected that on the same mesh, the error of the numerical results obtained by solving (6.66) is smaller than that obtained by solving (3.6). Tables 6.17 and 6.18 (see [53]) give the results and the relative errors when the SSM is not used and when it is used, respectively. From there, we can see that the results with the SSM are much better than the results without the SSM.

Problem (6.66) is a two-dimensional problem. However, it can be solved by a modified one-dimensional method. Let us explain why this problem can be solved like a one-dimensional problem. Because there is no second derivative in the t_d-direction, the coefficient of $\dfrac{\partial \bar{c}_p}{\partial t_d}$ is positive or zero, and the boundary condition is given at $t_d = T_d$, for a fixed time t^* the solution of the problem can be obtained from $t_d = T_d$ to $t_d = 0$ successively. Suppose the value of $\bar{c}_p$ for $t = t^*$ and $t_d \geq t_d^*$ has been obtained. We want to find the value of $\bar{c}_p$ for $t = t^*$ and $t_d = t_d^* - \Delta t_d$ with a positive Δt_d. Because the value at $t = t^*$ and $t_d = t_d^*$ is known, the value at $t = t^*$ and $t_d = t_d^* - \Delta t_d$ can be found by solving a one-dimensional problem on an (S, t)-plane. This can be done by various methods. After transforming the problem to one defined on a finite domain by the transformation (2.25), the partial differential equation can be discretized by scheme (5.31) at interior points, and the right boundary point and the solution can be found from these finite difference equations. The results given in this subsection are obtained by using a method that is a little different from what we have described here. For details, see the paper [53] by Luo and Wu.

When σ is a function of S, the SSM method can still be used. However, a European vanilla call option has a closed-form solution only when σ is a constant. Therefore, we do not have a closed-form solution for the corresponding European vanilla call option. In this case, we can consider the difference between the Parisian call option and the vanilla call option with a constant volatility $\sigma(E)$. This difference satisfies a nonhomogeneous equation (for details, see Subsection 6.3.2), but we still can expect that the SSM will make the computation more efficient.

6.3.5 European Average Price Options

In the last few subsections, we always computed the difference between an option and the corresponding vanilla option with a constant volatility. However, other functions can also be used as long as they have a similar singularity, and even they may be better. In this subsection, we give such an example.

From (3.20), we know that if sampling is done continuously, then the European-style Asian option may be modeled by the following partial differential equation

$$\frac{\partial V}{\partial t} + \frac{1}{2}\sigma^2 S^2 \frac{\partial^2 V}{\partial S^2} + (r - D_0)S\frac{\partial V}{\partial S} + \frac{S}{T}\frac{\partial V}{\partial I} - rV = 0, \qquad (6.67)$$

where

$$I = \frac{1}{T}\int_0^t S(\tau)d\tau.$$

Let us consider an average price call option whose final condition is

$$V(S, I, T) = \max(I - E, 0). \qquad (6.68)$$

Zhang[1] in his paper [79] proposed to solve the problem in the following way. By letting (see Section 3.3.4)

$$\eta = \frac{I - E}{S} \quad \text{and} \quad W(\eta, t) = \frac{V(S, I, t)}{S},$$

the two-dimensional equation (6.67) can be converted into a one-dimensional equation:

$$\frac{\partial W}{\partial t} + \frac{1}{2}\sigma^2\eta^2\frac{\partial^2 W}{\partial \eta^2} + \left[(D_0 - r)\eta + \frac{1}{T}\right]\frac{\partial W}{\partial \eta} - D_0 W = 0$$

and the final condition becomes

$$W(\eta, T) = \max(\eta, 0).$$

---

[1] In his paper, he assumes $D_0 = 0$. However, it is not difficult to generalize that result to the case with $D_0 \neq 0$.

Because the equation

$$\frac{d\eta}{dt} = (D_0 - r)\eta + \frac{1}{T}$$

has solutions in the form

$$\eta e^{-(r-D_0)(T-t)} + \frac{1}{(r-D_0)T}\left(1 - e^{-(r-D_0)(T-t)}\right) = constant,$$

introducing the transformation

$$\begin{cases} \xi = \eta e^{-(r-D_0)(T-t)} + \dfrac{1}{(r-D_0)T}\left(1 - e^{-(r-D_0)(T-t)}\right), \\[2mm] \tau = T - t, \\[2mm] W(\eta,t) = e^{-D_0\tau}f(\xi,\tau), \end{cases} \tag{6.69}$$

we can get rid of the first derivative of W and the function W, and we arrive at an initial value problem of a heat equation

$$\begin{cases} \dfrac{\partial f}{\partial \tau} - \dfrac{1}{2}\sigma^2\left[\xi - \dfrac{1}{(r-D_0)T}\left(1 - e^{-(r-D_0)\tau}\right)\right]^2 \dfrac{\partial^2 f}{\partial \xi^2} = 0, \\[4mm] \hspace{4cm} -\infty < \xi < \infty, \quad 0 \leq \tau \leq T, \\[2mm] f(\xi,0) = \max(\xi,0), \hspace{2cm} -\infty < \xi < \infty. \end{cases} \tag{6.70}$$

The initial condition $f(\xi,0) = \max(\xi,0)$ is not smooth at the point $\xi = 0$. To separate the singularity, the problem that is obtained by setting $\xi = 0$ in the above equation

$$\begin{cases} \dfrac{\partial \tilde{f}_0}{\partial \tau} - \dfrac{\sigma^2}{2(r-D_0)^2T^2}\left(1 - e^{-(r-D_0)\tau}\right)^2 \dfrac{\partial^2 \tilde{f}_0}{\partial \xi^2} = 0, \\[4mm] \hspace{4cm} -\infty < \xi < \infty, \quad 0 \leq \tau \leq T, \\[2mm] \tilde{f}_0(\xi,0) = \max(\xi,0), \hspace{2cm} -\infty < \xi < \infty \end{cases} \tag{6.71}$$

is considered. Introducing a new variable $\tau_1(\tau)$ by

$$d\tau_1 = \frac{\sigma^2}{2(r-D_0)^2T^2}\left(1 - e^{-(r-D_0)\tau}\right)^2 d\tau \quad \text{with} \quad \tau_1(0) = 0,$$

which gives

$$\tau_1(\tau) = \int_0^\tau \frac{\sigma^2}{2(r-D_0)^2T^2}\left(1 - e^{-(r-D_0)\tau}\right)^2 d\tau$$

$$= \frac{\sigma^2}{4(r-D_0)^3T^2}\left[2(r-D_0)\tau + 4e^{-(r-D_0)\tau} - e^{-2(r-D_0)\tau} - 3\right],$$

$$\tag{6.72}$$

and letting $f_0(\xi, \tau_1) = \tilde{f}_0(\xi, \tau(\tau_1))$, we obtain the following parabolic problem

$$
\begin{cases}
\dfrac{\partial f_0}{\partial \tau_1} - \dfrac{\partial^2 f_0}{\partial \xi^2} = 0, & -\infty < \xi < \infty, \quad 0 \leq \tau_1 \leq \tau_1(T), \\[2mm]
f_0(\xi, 0) = \max(\xi, 0), & -\infty < \xi < \infty.
\end{cases}
\tag{6.73}
$$

The solution of this problem is given by

$$
f_0(\xi, \tau_1) = \int_0^\infty \frac{\xi_T}{2\sqrt{\pi \tau_1}} e^{-(\xi_T - \xi)^2/4\tau_1} d\xi_T = \xi N\left(\frac{\xi}{\sqrt{2\tau_1}}\right) + \sqrt{\frac{\tau_1}{\pi}} e^{-\xi^2/4\tau_1}.
\tag{6.74}
$$

This analytic formula gives quite a good approximation to the prices of European average price call options. That is, the value of the difference between $f(\xi, \tau)$ and $f_0(\xi, \tau_1(\tau))$,

$$
f_1(\xi, \tau) = f(\xi, \tau) - f_0(\xi, \tau_1(\tau)),
\tag{6.75}
$$

is quite small. If we want to have more accurate results, we need to find $f_1(\xi, \tau)$. This function satisfies the following equation and initial condition:

$$
\begin{cases}
\dfrac{\partial f_1}{\partial \tau} - \dfrac{1}{2}\sigma^2 \left[\xi - \dfrac{1}{(r-D_0)T}\left(1 - e^{-(r-D_0)\tau}\right)\right]^2 \dfrac{\partial^2 f_1}{\partial \xi^2} = \dfrac{\sigma^2 \xi e^{-\xi^2/4\tau_1}}{4\sqrt{\pi \tau_1}} \\[4mm]
\quad \times \left[\xi - \dfrac{2}{(r-D_0)T}\left(1 - e^{-(r-D_0)\tau}\right)\right], \quad -\infty < \xi < \infty, \quad 0 \leq \tau \leq T, \\[4mm]
f_1(\xi, 0) = 0, \quad -\infty < \xi < \infty.
\end{cases}
\tag{6.76}
$$

The function $f_1(\xi, \tau)$ is smooth, and its value is quite small compared with $f(\xi, \tau)$, so in order to get a very good numerical solution, we need only a very coarse mesh. In this way, we can find quite accurate solutions very fast. The problem (6.76) is defined on an infinite domain. In order to convert the infinite domain into a finite domain, we can introduce the following transformation:

$$
\xi_1 = \frac{1}{2}\left(\frac{\xi}{|\xi| + P_m} + 1\right) \quad \text{and} \quad u(\xi_1, \tau) = \frac{f_1(\xi, \tau)}{|\xi| + P_m}.
$$

After this transformation, the problem for $u(\xi_1, \tau)$ is defined on $[0, 1] \times [0, T]$ in the (ξ_1, τ)-space and can be solved by scheme (5.31).

We can also take the difference between the price of a European-style Asian option and the price of a European vanilla option and do the numerical computation. However, the performance might not be as good as the method here. The reason is that the difference in the method given here is smaller than the difference between the price of a European-style Asian option and the price of a European vanilla option. This can be roughly explained as follows. Consider the following linear parabolic problem:

$$\begin{cases} \dfrac{\partial u}{\partial \tau} = a_2 \dfrac{\partial^2 u}{\partial \xi^2} + a_1 \dfrac{\partial u}{\partial \xi} + a_0 u + g(\xi, \tau), & -\infty < \xi < \infty, \quad 0 \le \tau \le T, \\ u(\xi, 0) = f(\xi), & -\infty < \xi < \infty. \end{cases}$$

Suppose that $\tilde{u}$ is an approximate solution by a numerical method on a certain mesh. It is clear that $v = u/10$ is the solution of the problem:

$$\begin{cases} \dfrac{\partial v}{\partial \tau} = a_2 \dfrac{\partial^2 v}{\partial \xi^2} + a_1 \dfrac{\partial v}{\partial \xi} + a_0 v + g(\xi, \tau)/10, & -\infty < \xi < \infty, \quad 0 \le \tau \le T, \\ v(\xi, 0) = f(\xi)/10, & -\infty < \xi < \infty. \end{cases}$$

Let $\tilde{v}$ be the approximate solution of this problem by using the same method on the same mesh. Just like the relation between u and v, we have $\tilde{v} = \tilde{u}/10$. Thus, $v - \tilde{v} = (u - \tilde{u})/10$, which means that the smaller the solution, the smaller the error of approximate solutions. Therefore, when we choose an analytic solution, we should let the analytic solution be as close to the desired solution as possible. In this way, we can have a better performance.

6.3.6 European Two-Factor Options

In Subsection 6.3.2, we pointed out that the assumption of the volatility being constant might need to be modified. One possible modification is to let the volatility be a given function of S. In Subsection 6.3.2, we discussed how to solve such a problem. Another possible modification is to allow the volatility to be a random variable, i.e., the volatility is stochastic. This subsection is devoted to studying how to solve this problem. In this case, option prices depend on two random variables. In what follows, such an option will be referred to as a two-factor option, and we will call an option a one-factor option if only the stock price is considered as a random variable.

Now let us discuss how to evaluate quickly such a European vanilla option or American vanilla option with $D_0 = 0$. We assume that the asset price S and the stochastic volatility are governed by the following two stochastic processes

$$\begin{cases} dS = \mu S dt + \sigma S dX_1, & 0 \le S, \\ d\sigma = p(\sigma, t)dt + q(\sigma, t)dX_2, & \sigma_l \le \sigma \le \sigma_u, \end{cases} \tag{6.77}$$

where dX_1 and dX_2 are two Wiener processes. These two random variables could be correlated and $E[dX_1 dX_2] = \rho dt$.

As we have seen in Section 2.9, in order to guarantee $\sigma \in [\sigma_l, \sigma_u]$, p and q in the model for the volatility need to satisfy the following reversion conditions:

$$\begin{cases} p(\sigma_l, t) - q(\sigma_l, t) \dfrac{\partial q(\sigma_l, t)}{\partial \sigma} \ge 0, \\ q(\sigma_l, t) = 0 \end{cases} \tag{6.78}$$

and

$$\begin{cases} p(\sigma_u,t) - q(\sigma_u,t)\dfrac{\partial q(\sigma_u,t)}{\partial \sigma} \le 0, \\[4mm] q(\sigma_u,t) = 0. \end{cases} \tag{6.79}$$

It is clear that if $\dfrac{\partial q(\sigma_l,t)}{\partial \sigma}$ and $\dfrac{\partial q(\sigma_u,t)}{\partial \sigma}$ are bounded, then (6.78) and (6.79) are simplified into

$$\begin{cases} p(\sigma_l,t) \ge 0, \\[2mm] q(\sigma_l,t) = 0 \end{cases} \tag{6.80}$$

and

$$\begin{cases} p(\sigma_u,t) \le 0, \\[2mm] q(\sigma_u,t) = 0. \end{cases} \tag{6.81}$$

Suppose $V(S,\sigma,t)$ is the value of an option depending on two random variables S and σ. From Section 2.9, such an option satisfies the following equation:

$$\frac{\partial V}{\partial t} + \mathbf{L}_{\mathbf{S},\sigma} V = 0, \tag{6.82}$$

where $\mathbf{L}_{\mathbf{S},\sigma}$ is an operator defined by

$$\begin{aligned} \mathbf{L}_{\mathbf{S},\sigma} = {} & \frac{1}{2}\sigma^2 S^2 \frac{\partial^2}{\partial S^2} + \rho\sigma Sq\frac{\partial^2}{\partial S\partial\sigma} + \frac{1}{2}q^2\frac{\partial^2}{\partial\sigma^2} \\ & + (r - D_0)S\frac{\partial}{\partial S} + (p - \lambda q)\frac{\partial}{\partial\sigma} - r. \end{aligned} \tag{6.83}$$

Consider a two-factor European vanilla call option problem, and let its value be $c(S,\sigma,t)$. Because the volatility model satisfies the reversion conditions, no boundary conditions need to be given at the boundaries $\sigma = \sigma_l$ and $\sigma = \sigma_u$. Therefore, the value of the two-factor European vanilla call option is the solution of the following final-value problem:

$$\begin{cases} \dfrac{\partial c}{\partial t} + \mathbf{L}_{\mathbf{S},\sigma} c = 0, & 0 \le S, \quad \sigma_l \le \sigma \le \sigma_u, \quad t \le T, \\[3mm] c(S,\sigma,T) = \max(S - E,0), & 0 \le S, \quad \sigma_l \le \sigma \le \sigma_u. \end{cases} \tag{6.84}$$

In order to make the computed solution smoother, which will make numerical methods more efficient, we let

$$\overline{c}(S,\sigma,t) = c(S,\sigma,t) - c_1(S,\sigma,t) \tag{6.85}$$

on the entire computational domain. $c_1(S,\sigma,t)$ is the price of the one-factor European vanilla call option, that is, the price of the European vanilla call option with a parameter σ. Here, we denote the value of this option by $c_1(S,\sigma,t)$ instead of $c(S,t)$ in order to indicate explicitly its dependence on σ and to explain that it is the price of the one-factor model. From Subsection 2.4.4, we know that its expression is given by

$$c_1(S, \sigma, t) = Se^{-D_0(T-t)} N(d_1) - Ee^{-r(T-t)} N(d_2),$$

where

$$N(z) = \frac{1}{\sqrt{2\pi}} \int_{-\infty}^{z} e^{-\xi^2/2} d\xi,$$

$$d_1 = \left[\ln \frac{Se^{-D_0(T-t)}}{Ee^{-r(T-t)}} + \frac{1}{2}\sigma^2(T-t) \right] \Big/ \left(\sigma\sqrt{T-t} \right),$$

$$d_2 = d_1 - \sigma\sqrt{T-t}.$$

Because $c_1(S, \sigma, t)$ satisfies the Black–Scholes equation, the difference $\bar{c}$ is the solution of the following final-value problem:

$$
\begin{cases}
\dfrac{\partial \bar{c}}{\partial t} + \mathbf{L}_{s,\sigma}\bar{c} = f(S, \sigma, t), & 0 \le S, \quad \sigma_l \le \sigma \le \sigma_u, \quad 0 \le t \le T, \\
\bar{c}(S, \sigma, T) = 0, & 0 \le S, \quad \sigma_l \le \sigma \le \sigma_u,
\end{cases}
\tag{6.86}
$$

where

$$f(S, \sigma, t) = -\rho\sigma Sq\frac{\partial^2 c_1}{\partial S \partial \sigma} - \frac{1}{2}q^2\frac{\partial^2 c_1}{\partial \sigma^2} - (p - \lambda q)\frac{\partial c_1}{\partial \sigma}.$$

From the expressions of $c_1(S, \sigma, t)$, noticing

$$\frac{\partial c_1}{\partial S} = e^{-D_0(T-t)} N(d_1),$$

$$\frac{\partial d_1}{\partial \sigma} = \sqrt{T-t} - \left[\ln \frac{Se^{-D_0(T-t)}}{Ee^{-r(T-t)}} + \frac{1}{2}\sigma^2(T-t) \right] \Big/ (\sigma^2\sqrt{T-t})$$

$$= \sqrt{T-t} - \frac{d_1}{\sigma},$$

$$\frac{\partial d_2}{\partial \sigma} = \frac{\partial d_1}{\partial \sigma} - \sqrt{T-t} = -\frac{d_1}{\sigma},$$

$$N'(z) = \frac{1}{\sqrt{2\pi}}e^{-z^2/2},$$

we can easily find

$$
\begin{cases}
\dfrac{\partial c_1}{\partial \sigma} = Se^{-D_0(T-t)} N'(d_1)\dfrac{\partial d_1}{\partial \sigma} - Ee^{-r(T-t)} N'(d_2)\dfrac{\partial d_2}{\partial \sigma} \\
\qquad = S\sqrt{T-t}\,e^{-D_0(T-t)} N'(d_1), \\
\dfrac{\partial^2 c_1}{\partial \sigma^2} = S\sqrt{T-t}\,e^{-D_0(T-t)} N''(d_1)\dfrac{\partial d_1}{\partial \sigma} \\
\qquad = -S\sqrt{T-t}\,e^{-D_0(T-t)} d_1 N'(d_1)\dfrac{\partial d_1}{\partial \sigma}, \\
\dfrac{\partial^2 c_1}{\partial S \partial \sigma} = e^{-D_0(T-t)} N'(d_1)\dfrac{\partial d_1}{\partial \sigma}.
\end{cases}
\tag{6.87}
$$

As we see from (6.84), the derivative of $c(S, \sigma, t)$ with respect to S is discontinuous at $t = T$ and $S = E$. However, (6.86) shows the derivative of $\bar{c}(S, \sigma, t)$ with respect to S to be identically equal to zero at $t = T$. Therefore, when a numerical method is used, the truncation error for the problem (6.86) will be much smaller than the problem (6.84). This is why we consider the formulation (6.86) instead of (6.84).

The final-value problem (6.86) is defined on an infinite domain. In order to convert it into a problem on a finite domain, we introduce the following transformation

$$
\begin{cases}
\xi = \dfrac{S}{S + P_m}, \\[2mm]
\sigma = \sigma, \\[2mm]
\tau = T - t, \\[2mm]
u(\xi, \sigma, \tau) = \dfrac{\bar{c}(S, \sigma, t)}{S + P_m}.
\end{cases}
\tag{6.88}
$$

In the $\{\xi, \sigma, \tau\}$-space, we need to solve a problem on the domain $[0, 1] \times [\sigma_l, \sigma_u] \times [0, T]$. This is a finite domain, and it is easy to construct numerical methods to solve the problem on this domain. From (6.88), we have

$$
\bar{c}(S, \sigma, t) = (S + P_m) u(\xi, \sigma, \tau) = \frac{P_m}{1 - \xi} u(\xi, \sigma, \tau) \quad \text{and} \quad \frac{\partial \xi}{\partial S} = \frac{(1 - \xi)^2}{P_m}.
$$

Therefore, among the derivatives of $\bar{c}$ and u, there are the following relations:

$$
\frac{\partial \bar{c}}{\partial t} = -\frac{P_m}{1 - \xi} \frac{\partial u}{\partial \tau},
$$

$$
\frac{\partial \bar{c}}{\partial S} = (1 - \xi) \frac{\partial u}{\partial \xi} + u,
$$

$$
\frac{\partial \bar{c}}{\partial \sigma} = \frac{P_m}{1 - \xi} \frac{\partial u}{\partial \sigma},
$$

$$
\frac{\partial^2 \bar{c}}{\partial S^2} = \frac{(1 - \xi)^3}{P_m} \frac{\partial^2 u}{\partial \xi^2},
$$

$$
\frac{\partial^2 \bar{c}}{\partial S \partial \sigma} = (1 - \xi) \frac{\partial^2 u}{\partial \xi \partial \sigma} + \frac{\partial u}{\partial \sigma},
$$

$$
\frac{\partial^2 \bar{c}}{\partial \sigma^2} = \frac{P_m}{1 - \xi} \frac{\partial^2 u}{\partial \sigma^2}.
$$

Substituting them into the partial differential equation in (6.86) yields

$$
\frac{\partial u}{\partial \tau} = a_1 \frac{\partial^2 u}{\partial \xi^2} + a_2 \frac{\partial^2 u}{\partial \xi \partial \sigma} + a_3 \frac{\partial^2 u}{\partial \sigma^2} + a_4 \frac{\partial u}{\partial \xi} + a_5 \frac{\partial u}{\partial \sigma} + a_6 u + a_7,
$$

where

$$a_1 = \frac{1}{2}\sigma^2\xi^2(1-\xi)^2,$$

$$a_2 = \rho\sigma\xi(1-\xi)q,$$

$$a_3 = \frac{1}{2}q^2,$$

$$a_4 = (r - D_0)\xi(1-\xi),$$

$$a_5 = p - (\lambda - \rho\sigma\xi)q,$$

$$a_6 = -[r(1-\xi) + D_0\xi],$$

$$\begin{aligned}
a_7 &= -f(\xi P_m/(1-\xi), \sigma, T - \tau)(1-\xi)/P_m \\
&= \rho\sigma\xi q\, e^{-D_0(T-t)}N'(d_1)\frac{\partial d_1}{\partial\sigma} - \frac{1}{2}q^2\xi\sqrt{T-t}\,e^{-D_0(T-t)}d_1 N'(d_1)\frac{\partial d_1}{\partial\sigma} \\
&\quad + (p - \lambda q)\xi\sqrt{T-t}\,e^{-D_0(T-t)}N'(d_1) \\
&= \frac{1}{\sqrt{2\pi}}\xi e^{-D_0\tau - d_1^2/2}\left[q(\sqrt{\tau} - d_1/\sigma)(\rho\sigma - q\sqrt{\tau}d_1/2) + (p - \lambda q)\sqrt{\tau}\right].
\end{aligned}$$

Therefore, the problem (6.86) becomes

$$\begin{cases}
\dfrac{\partial u}{\partial \tau} = a_1\dfrac{\partial^2 u}{\partial \xi^2} + a_2\dfrac{\partial^2 u}{\partial \xi \partial \sigma} + a_3\dfrac{\partial^2 u}{\partial \sigma^2} + a_4\dfrac{\partial u}{\partial \xi} + a_5\dfrac{\partial u}{\partial \sigma} + a_6 u + a_7, \\
\qquad\qquad 0 \le \xi \le 1, \quad \sigma_l \le \sigma \le \sigma_u, \quad 0 \le \tau \le T, \\
u(\xi, \sigma, 0) = 0, \qquad 0 \le \xi \le 1, \quad \sigma_l \le \sigma \le \sigma_u.
\end{cases} \qquad (6.89)$$

Once we have $u(\xi, \sigma, \tau)$, we can get the value of the two-factor European call option by

$$\begin{aligned}
c(S, \sigma, t) &= \bar{c}(S,\ \sigma,\ t) + c_1(S,\ \sigma,\ t) \\
&= (S + P_m)u\left(\frac{S}{S + P_m},\ \sigma,\ T - t\right) + c_1(S,\ \sigma,\ t).
\end{aligned} \qquad (6.90)$$

The solution of problem (6.89) can be obtained numerically. Let $u_{m,i}^n$ be the approximate value of u at $\xi = m\Delta\xi$, $\sigma = \sigma_l + i\Delta\sigma$, and $\tau = n\Delta\tau$, where $\Delta\xi = 1/M$, $\Delta\sigma = (\sigma_u - \sigma_l)/I$, and $\Delta\tau = 1/N$, M, I, N being integers. This partial differential equation can be discretized by the following scheme. If $\sigma \ne \sigma_l$ and $\sigma \ne \sigma_u$, at a point $(\xi_m, \sigma_i, \tau^{n+1/2})$ the partial differential equation in (6.89) can be discretized by the following second-order approximation:

$$\begin{aligned}
&\frac{u_{m,i}^{n+1} - u_{m,i}^n}{\Delta\tau} \\
&= \frac{a_1}{2\Delta\xi^2}\left(u_{m+1,i}^{n+1} - 2u_{m,i}^{n+1} + u_{m-1,i}^{n+1} + u_{m+1,i}^n - 2u_{m,i}^n + u_{m-1,i}^n\right) \\
&\quad + \frac{a_2}{8\Delta\sigma\Delta\xi}\big(u_{m+1,i+1}^{n+1} - u_{m+1,i-1}^{n+1} - u_{m-1,i+1}^{n+1} + u_{m-1,i-1}^{n+1} \\
&\qquad\qquad + u_{m+1,i+1}^n - u_{m+1,i-1}^n - u_{m-1,i+1}^n + u_{m-1,i-1}^n\big)
\end{aligned}$$

$$+\frac{a_3}{2\Delta\sigma^2}(u_{m,i+1}^{n+1} - 2u_{m,i}^{n+1} + u_{m,i-1}^{n+1}$$

$$+u_{m,i+1}^n - 2u_{m,i}^n + u_{m,i-1}^n) \tag{6.91}$$

$$+\frac{a_4}{4\Delta\xi}(u_{m+1,i}^{n+1} - u_{m-1,i}^{n+1} + u_{m+1,i}^n - u_{m-1,i}^n)$$

$$+\frac{a_5}{4\Delta\sigma}(u_{m,i+1}^{n+1} - u_{m,i-1}^{n+1} + u_{m,i+1}^n - u_{m,i-1}^n)$$

$$+\frac{a_6}{2}(u_{m,i}^{n+1} + u_{m,i}^n) + a_7,$$

$$m = 0, 1, \cdots, M, \quad i = 1, 2, \cdots, I - 1.$$

Here, all the coefficients a_1–a_7 should be evaluated at the point $(\xi_m, \sigma_i, \tau^{n+1/2})$ in order to guarantee second-order accuracy.

At the boundaries $\sigma = \sigma_l$ and $\sigma = \sigma_u$, due to $q = 0$ the partial differential equation in (6.89) becomes

$$\frac{\partial u}{\partial \tau} = a_1 \frac{\partial^2 u}{\partial \xi^2} + a_4 \frac{\partial u}{\partial \xi} + a_5 \frac{\partial u}{\partial \sigma} + a_6 u + a_7,$$

which possesses hyperbolic properties in the σ-direction. From the reversion conditions, we see $a_5 = p - (\lambda - \rho\sigma\xi)q = p \geq p - q\dfrac{\partial q}{\partial \sigma} \geq 0$ at the boundary $\sigma = \sigma_l$ and $a_5 = p - (\lambda - \rho\sigma\xi)q = p \leq p - q\dfrac{\partial q}{\partial \sigma} \leq 0$ at $\sigma = \sigma_u$. These facts tell us that the value of u on the boundaries $\sigma = \sigma_l$ and $\sigma = \sigma_u$ can be determined by the value of u inside the domain. Hence, we can approximate the partial differential equation in (6.89) at the boundaries $\sigma = \sigma_l$ and $\sigma = \sigma_u$ by

$$\frac{u_{m,0}^{n+1} - u_{m,0}^n}{\Delta\tau}$$

$$= \frac{a_1}{2\Delta\xi^2}(u_{m+1,0}^{n+1} - 2u_{m,0}^{n+1} + u_{m-1,0}^{n+1} + u_{m+1,0}^n - 2u_{m,0}^n + u_{m-1,0}^n)$$

$$+\frac{a_4}{4\Delta\xi}(u_{m+1,0}^{n+1} - u_{m-1,0}^{n+1} + u_{m+1,0}^n - u_{m-1,0}^n) \tag{6.92}$$

$$+\frac{a_5}{4\Delta\sigma}(-u_{m,2}^{n+1} + 4u_{m,1}^{n+1} - 3u_{m,0}^{n+1} - u_{m,2}^n + 4u_{m,1}^n - 3u_{m,0}^n)$$

$$+\frac{a_6}{2}(u_{m,0}^{n+1} + u_{m,0}^n) + a_7,$$

$$m = 0, 1, \cdots, M$$

and

$$\frac{u_{m,I}^{n+1} - u_{m,I}^n}{\Delta\tau}$$

$$= \frac{a_1}{2\Delta\xi^2}(u_{m+1,I}^{n+1} - 2u_{m,I}^{n+1} + u_{m-1,I}^{n+1} + u_{m+1,I}^n - 2u_{m,I}^n + u_{m-1,I}^n)$$

$$+\frac{a_4}{4\Delta\xi}(u_{m+1,I}^{n+1} - u_{m-1,I}^{n+1} + u_{m+1,I}^n - u_{m-1,I}^n) \tag{6.93}$$

$$+\frac{a_5}{4\Delta\sigma}(3u_{m,I}^{n+1} - 4u_{m,I-1}^{n+1} + u_{m,I-2}^{n+1} + 3u_{m,I}^{n} - 4u_{m,I-1}^{n} + u_{m,I-2}^{n})$$

$$+\frac{a_6}{2}(u_{m,I}^{n+1} + u_{m,I}^{n}) + a_7,$$

$$m = 0, 1, \cdots, M$$

respectively. Here, $\dfrac{\partial u}{\partial \sigma}$ is discretized by one-sided second-order scheme in order for all the node points involved to be in the computational domain. a_1 and a_4–a_7 are also evaluated at the point $(\xi_m, \sigma_i, \tau^{n+1/2})$, $i = 0$ or I. When $u_{m,i}^{n}$, $m = 0, 1, \cdots, M$, $i = 0, 1, \cdots, I$ are known, from (6.91)–(6.93) we can determine $u_{m,i}^{n+1}$, $m = 0, 1, \cdots, M$, $i = 0, 1, \cdots, I$. The initial condition gives $u_{m,i}^{0}$, $m = 0, 1, \cdots, M$, $i = 0, 1, \cdots, I$. Therefore, we can do this procedure for $n = 0, 1, \cdots, N-1$ successively and finally find $u_{m,i}^{N}$, $m = 0, 1, \cdots, M$, $i = 0, 1, \cdots, I$.

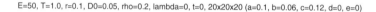

E=50, T=1.0, r=0.1, D0=0.05, rho=0.2, lambda=0, t=0, 20x20x20 (a=0.1, b=0.06, c=0.12, d=0, e=0)

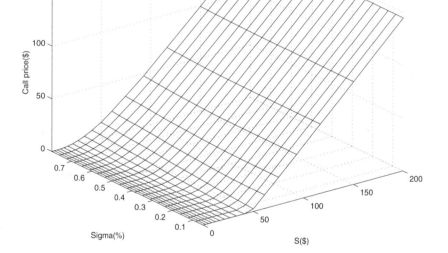

Fig. 6.14. The price of a two-factor European call option

In Fig. 6.14, the price of a European call option obtained in this way is given. The mesh used is $20\times20\times20$, where the first, second, and third numbers are M, I, and N, respectively. The parameters of the problem are given in the figure and the parameter functions are

$$\begin{cases} p = a(b - \sigma), & \sigma_l \leq \sigma \leq \sigma_u, \\[2mm] q = c\dfrac{1 - \left(1 - 2\dfrac{\sigma - \sigma_l}{\sigma_u - \sigma_l}\right)^2}{1 - 0.975\left(1 - 2\dfrac{\sigma - \sigma_l}{\sigma_u - \sigma_l}\right)^2}\,\sigma, & \sigma_l \leq \sigma \leq \sigma_u, \\[2mm] \rho = 0.2, \\[1mm] \lambda = d + e\sigma, & \sigma_l \leq \sigma \leq \sigma_u, \end{cases}$$

where $a = 0.1$, $b = 0.06$, $c = 0.12$, $d = 0$, $e = 0$, $\sigma_l = 0.05$, and $\sigma_u = 0.8$.

When the singularity-separating technique is not adopted, the scheme above can also be used. In that case,

$$a_7 = 0 \quad \text{and} \quad u(\xi, \sigma, 0) = \max(2\xi - 1, 0).$$

Table 6.19. SSM with and without extrapolation technique

($S = 50$, $E = 50$, $T = 1$, $\sigma = 0.2$, $r = 0.1$, $D_0 = 0.05$, $a = 0.1$, $b = 0.06$, $c = 0.12$, $d = 0$, and $e = 0$. The exact solution is $4.848069\cdots$.)

Meshes	Without extrapolation		With extrapolation	
	Solution	|Errors|	Solution	|Errors|
$10 \times 10 \times 10$	4.8143085	0.033761	—	—
$20 \times 20 \times 20$	4.8361039	0.011966	4.8433691	0.004700
$40 \times 40 \times 40$	4.8460151	0.002054	4.8493188	0.001249
$80 \times 80 \times 80$	4.8476154	0.000454	4.8481488	0.000079
$160 \times 160 \times 160$	4.8479592	0.000110	4.8480738	0.000004
$320 \times 320 \times 320$	4.8480421	0.000027	4.8480697	less than 10^{-6}
$640 \times 640 \times 640$	4.8480626	0.000007	4.8480694	less than 10^{-6}
$960 \times 960 \times 960$	4.8480664	0.000003	4.8480694	less than 10^{-6}

In order to give some idea about the performance of the method described in this subsection, we list the values of the option obtained by the method here and by the same difference method without using singularity-separating in Tables 6.19 and 6.20 for $S = 50$ and $\sigma = 0.2$. When these results were computed, for the first five coarser meshes, the linear systems were solved by the LU decomposition method and for the last three finer meshes, the Gauss-Seidel iteration was used in order to solve the linear systems. From the tables, we see that the exact solution up to the sixth decimal place is 4.848069, which we obtained by a very fine mesh. Therefore, we can find out the errors of the results up to the sixth decimal place, which are also listed there. Table 6.19 shows that for a $20 \times 20 \times 20$ mesh with extrapolation, the error relative to E is $0.0047/50 \approx 10^{-4}$ and that the error relative to the option value is $0.0047/4.848069 \approx 10^{-3}$. In practice, requiring such accuracy is reasonable. The CPU time on a Pentium III 800 MHz computer is 0.07 second. If the singularity-separating technique is not used, in order to reach a similar accuracy, the mesh is between $40 \times 40 \times 40$ and $80 \times 80 \times 80$ and the CPU time is between 1 to 8 seconds and close to 8 seconds, respectively.

Table 6.20. Implicit method with and without extrapolation technique

$(S = 50, E = 50, T = 1, \sigma = 0.2, r = 0.1, D_0 = 0.05,$
$a = 0.1, b = 0.06, c = 0.12, d = 0, \text{ and } e = 0.$
The exact solution is $4.848069\cdots.$)

Meshes	Without extrapolation		With extrapolation	
	Solution	|Errors|	Solution	|Errors|
$10 \times 10 \times 10$	3.1774889	1.670580	—	—
$20 \times 20 \times 20$	4.2406270	0.607442	4.5950063	0.253063
$40 \times 40 \times 40$	4.7179697	0.130100	4.8770840	0.029015
$80 \times 80 \times 80$	4.8171183	0.030951	4.8501678	0.002098
$160 \times 160 \times 160$	4.8404088	0.007661	4.8481722	0.000103
$320 \times 320 \times 320$	4.8461590	0.001910	4.8480758	0.000006
$640 \times 640 \times 640$	4.8475923	0.000477	4.8480700	0.000001
$960 \times 960 \times 960$	4.8478575	0.000212	4.8480697	less than 10^{-6}

Noticing

$$\frac{\partial p_1}{\partial \sigma} = \frac{\partial c_1}{\partial \sigma}, \quad \frac{\partial^2 p_1}{\partial \sigma^2} = \frac{\partial^2 c_1}{\partial \sigma^2}, \quad \frac{\partial^2 p_1}{\partial S \partial \sigma} = \frac{\partial^2 c_1}{\partial S \partial \sigma},$$

where p_1 is the price of the one-factor put option, we see that the difference between the two-factor and one-factor put options is also the solution of (6.89). Therefore, in order to have the price of a European put option, we proceed as follows. First solving (6.89), then we can have the put price by

$$p(S, \sigma, t) = (S + P_m)u\left(\frac{S}{S + P_m}, \sigma, T - t\right) + p_0(S, \sigma, t).$$

In Fig. 6.15, the price of a two-factor European put option obtained by this way is shown. The parameters of the problem and the parameter functions are the same as these for the two-factor European call option. Also, for European vanilla options, both the put–call parity relation and the put–call symmetric relation exist. The put–call parity relation still is

$$p(S, \sigma, t) = c(S, \sigma, t) - Se^{-D_0(T-t)} + Ee^{-r(T-t)}. \tag{6.94}$$

When we calculate put option prices without using SSM, this relation can be used to check the correctness of the code to some extent. First, we compute the prices of a call option and a put option with the same parameters. Then,

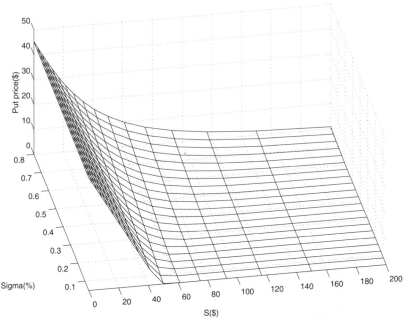

Fig. 6.15. The price of a two-factor European put option

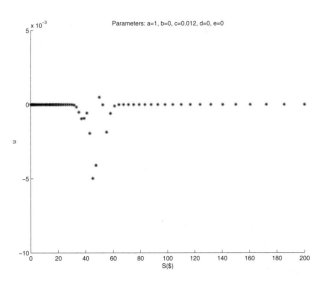

Fig. 6.16. An unstable solution of implicit schemes
(Variation of u with respect to S on the line $\sigma = 0.05$ at $t = 0$. $E = 50$,
$T = 2$, $r = 0.1$, and $D_0 = 0.05$. The solution is on a $80 \times 40 \times 80$ mesh.)

the results are substituted into the put–call parity relation to see if it holds. If it holds with a small error, then the code most likely gives correct results; if the relation does not hold, then the code must have some problems.

Finally, we give an example to explain that if the reversion conditions are not satisfied, then the final-value problem (6.89) is not well-posed and we cannot determine the solution using only the partial differential equation and the final condition in (6.89). Consider a problem with $a = 1$, $b = 0$, $c = 0.012$, $d = 0$, $e = 0$, and $T = 2$. The other parameters are the same as before. We still use the numerical method above to find the numerical solution. In Fig. 6.16, we give the variation of u with S on the line $\sigma = \sigma_l$ at time $t = 0$. From there, we can see some "nonphysical" oscillations, which means that the computation is unstable even though an implicit scheme is used. This indicates that for this case, the solution is not determined only by the partial differential equation and final condition. The reason is that a proper boundary condition is needed at the boundary $\sigma = \sigma_l$ because the inequality condition in (6.80) is not satisfied at $\sigma = \sigma_l$ due to $b = 0 < \sigma_l = 0.05$. If a reasonable condition cannot be given, then an artificial boundary condition has to be added. If the artificial boundary condition is not proper, then one will encounter some difficulty during computation.

6.3.7 Two-Factor Convertible Bonds with $D_0 = 0$

If $D_0 = 0$, then the convertible bond problem has no free boundary, and the problem has the same form as a European-style two-factor derivative problem does. The only difference is that the another random variable is the spot interest rate instead of the volatility. In order to make numerical methods more efficient, there are also two things we need to deal with. The first thing is the weak singularity generated by a discontinuous derivative of the payoff function. In order to separate this singularity, we can calculate numerically the difference between the values of two-factor and one-factor convertible bonds for the case $D_0 = 0$. We will not give the method here because it is similar to the method for two-factor options and the method for two-factor convertible bonds with $D_0 \neq 0$, which will be given in Subsection 7.1.2. The second thing is that the problem is defined on an infinite domain. Through a transformation similar to (6.88), the problem can be converted into a problem similar to (6.89) and the solution can be obtained by numerical methods efficiently. The details are similar to what we have done for two-factor options and left for the reader to complete (Problem 20).

6.4 Pseudo-Spectral Methods

After the singularity-separating method is used, the solution to be computed numerically (the difference between the original unknown solution and a closed-form solution) is quite smooth. In this case, the pseudo-spectral method might be another good choice for computing the difference numerically. The basic principle of the method was discussed in Chapter 5. In this subsection, we give some details when the pseudo-spectral method is applied to problems (5.26) and (5.27).

Let us take $M + 1$ grid points x_m, $m = 0, 1, \cdots, M$, on $[0, 1]$ and assume that the values of a function $u(x)$ for any x_m are given. Then, the values of the derivatives of $u(x)$ can be expressed as linear combinations of $u(x_m)$. Especially, if x_m is given by (5.4), then the first derivative is approximated by (5.5):

$$\frac{\partial u}{\partial x}(x_m) = \sum_{i=0}^{M} D_{x,m,i} u(x_i)$$

and the second derivative by (5.7):

$$\frac{\partial^2 u}{\partial x^2}(x_m) = \sum_{i=0}^{M} D_{xx,m,i} u(x_i),$$

where $D_{x,m,i}$ and $D_{xx,m,i}$ are given by (5.6) and (5.8), respectively. Consequently, (5.27) can be approximated by

$$\frac{u^{n+1}(x_m) - u^n(x_m)}{\Delta \tau}$$
$$= \frac{1}{2} \left[a_m^{n+\frac{1}{2}} \sum_{i=0}^{M} D_{xx,m,i} u^{n+1}(x_i) + b_m^{n+\frac{1}{2}} \sum_{i=0}^{M} D_{x,m,i} u^{n+1}(x_i) + c_m^{n+\frac{1}{2}} u^{n+1}(x_m) \right]$$
$$+ \frac{1}{2} \left[a_m^{n+\frac{1}{2}} \sum_{i=0}^{M} D_{xx,m,i} u^n(x_i) + b_m^{n+\frac{1}{2}} \sum_{i=0}^{M} D_{x,m,i} u^n(x_i) + c_m^{n+\frac{1}{2}} u^n(x_m) \right]$$
$$+ g_m^{n+\frac{1}{2}},$$
$$m = 0, 1, \cdots, M, \qquad (6.95)$$

where $u^{n+1}(x_m) = u(x_m, (n+1)\Delta\tau)$. Just like the implicit finite-difference method, if $u^n(x_m)$, $m = 0, 1, \cdots, M$ are given, we can determine $u^{n+1}(x_m)$, $m = 0, 1, \cdots, M$ by the linear system (6.95). However, the matrix of the current system is a full matrix, and the CPU time needed for solving this system is longer than the implicit finite-difference method if M is the same. When the solution is very smooth, only a small M might be needed in order to get a satisfying result. In such a case, its performance could be better than the implicit finite-difference method. This numerical method is referred to as the implicit pseudo-spectral method for one-dimensional problems.

If we consider problem (5.26), the only difference is that instead of the partial differential equation being discretized at x_m, $m = 0, 1, \cdots, M$, now it is discretized at x_m, $m = 1, 2, \cdots, M - 1$, and these equations and the boundary conditions given in (5.26) form the entire system we need.

Table 6.21. Pseudo-spectral methods

$(S = 95,\ T = 1,\ E = 100,\ \sigma = 0.25,\ r = 0.1,\ D_0 = 0,$
$f(t) = (0.9 - 0.05t)E,\ g(t) = (1.6 + 0.05t)E,$ and
the rebate $= g(t) - E$. The exact solution is $6.43129316\cdots$.)

	Without SSM			With SSM		
Meshes	Solutions	|Errors|	CPU	Solution	|Errors|	CPU
7×50	6.454922	0.023629	0.0007	6.431842	0.000549	0.0014
7×100	6.454789	0.023596	0.0015	6.431438	0.000145	0.0022
7×200	6.454755	0.023462	0.0028	6.431426	0.000133	0.0043
8×50	6.438364	0.007071	0.0010	6.431351	0.000058	0.0014
8×100	6.438227	0.006934	0.0019	6.431305	0.000012	0.0028
8×200	6.438193	0.006900	0.0038	6.431293	0.0000005	0.0058
9×50	6.404701	0.026592	0.0013	6.431350	0.000057	0.0021
9×100	6.404555	0.026738	0.0024	6.431304	0.000011	0.0036
9×200	6.404518	0.002678	0.0044	6.431292	0.000001	0.0065

Table 6.21 gives some results obtained by the implicit pseudo-spectral method described above with $M = 7, 8, 9$. The corresponding time steps used are $\Delta\tau = 1/N$, $N = 50, 100, 200$, respectively. In the column "Meshes," $M \times N$ is given. The problem is a double barrier call option whose lower and upper knock-out boundaries are $f(t) = (0.9 - 0.05t)E$ and $g(t) = (1.6 + 0.05t)E$. The other parameters are given in the table. When the computation is done, the independent variable x is defined by

$$x = \frac{\dfrac{S}{E+S} - \dfrac{f(t)}{E+f(t)}}{\dfrac{g(t)}{E+g(t)} - \dfrac{f(t)}{E+f(t)}}.$$

The exact solution for this case is $6.43129316\cdots$, where the nine digits given are correct. When we have the exact solution, we can have the error of the solution, which is also given. The CPU time in seconds is also shown in order to see the performance.

In Table 6.21, both the results with and without the SSM are listed. From there, we can see that if the SSM is not used, the result obtained by using higher order polynomials might be worse than the results obtained by using lower order polynomials. However, it shows that when the pseudo-spectral method is combined with the singularity-separating technique, the higher the polynomial order, the better the result. Hence, the result of the pseudo-spectral method with the singularity-separating technique is much better than without it. Consequently, if the pseudo-spectral method is adopted, then combining it with SSM is essential. In Figs. 6.7 and 6.8, the functions computed when SSM is used and not used are shown, respectively. As pointed out, the functions in Fig. 6.8 are not as smooth as those in Fig. 6.7, especially, the derivative of the function for $t = 1$ in Fig. 6.8 is discontinuous. Therefore, the pseudo-spectral method does not provide a good performance for this case. However, if the singularity-separating technique is used, then the functions determined numerically are always very smooth, which can be seen from Fig. 6.7. In this case, the performance of the pseudo-spectral method is very good, and in certain cases it may even be better than the second-order implicit finite-difference methods because a pseudo-spectral method can be understood as a high-order difference method. In Fig. 6.17, the performances of the implicit finite-difference method and the implicit pseudo-spectral method with the singularity-separating technique are compared, which confirms this conclusion.

The idea described here also works for double moving barrier put options with rebates and many other cases. For details, see the paper [82] by Zhu and Abifaker.

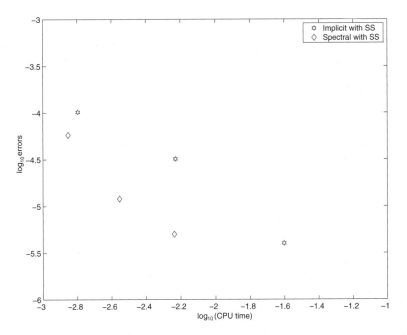

Fig. 6.17. Comparison between an implicit scheme and a pseudo-spectral method

In Section 6.3, we pointed out that two-dimensional European-style derivative problems and American-style derivative problems that do not have free boundaries could be written in the form (6.89). The pseudo-spectral method can also be applied to such a problem. When this method is combined with the singularity-separating method, a good performance can be expected. For details of this method for two-dimensional case, see Chapter 7.

Problems

1. *Suppose that we determine the price of an American vanilla call/put option through solving the following problem:

$$
\begin{cases}
\left(\dfrac{\partial u}{\partial \bar{\tau}} - \dfrac{\partial^2 u}{\partial x^2}\right)[u(x,\bar{\tau}) - g(x,\bar{\tau})] = 0, \\[2ex]
\dfrac{\partial u}{\partial \bar{\tau}} - \dfrac{\partial^2 u}{\partial x^2} \geq 0, \\[2ex]
u(x,\bar{\tau}) - g(x,\bar{\tau}) \geq 0, \\[2ex]
u(x,0) = g(x,0),
\end{cases}
$$

where $-\infty < x < \infty, \bar{\tau} \geq 0$ and

$$
g(x,\bar{\tau}) = \max\left(\pm(e^{x+(2D_0/\sigma^2+1)\bar{\tau}} - e^{2r\bar{\tau}/\sigma^2}), 0\right).
$$

Describe a numerical method for solving this problem by using an explicit scheme.

2. As we know from Section 3.4, an American lookback strike put option is the solution of the following linear complementarity problem:

$$
\begin{cases}
\left(\dfrac{\partial W}{\partial t} + \mathbf{L}_\eta W\right)[W - \max(\eta - \beta, 0)] = 0, & 1 \leq \eta, \quad t \leq T, \\[2ex]
\dfrac{\partial W}{\partial t} + \mathbf{L}_\eta W \leq 0, & 1 \leq \eta, \quad t \leq T, \\[2ex]
W - \max(\eta - \beta, 0) \geq 0, & 1 \leq \eta, \quad t \leq T, \\[2ex]
W(\eta, T) = \max(\eta - \beta, 0), & 1 \leq \eta, \\[2ex]
\dfrac{\partial W}{\partial \eta}(1,t) = 0, & t \leq T,
\end{cases}
$$

where we assume $\beta \geq 1$ and the operator $\mathbf{L}_\eta$ to be defined by

$$
\mathbf{L}_\eta \equiv \frac{1}{2}\sigma^2 \eta^2 \frac{\partial^2}{\partial \eta^2} + (D_0 - r)\eta \frac{\partial}{\partial \eta} - D_0.
$$

Convert this problem into a problem on $[0,1]$, and design an explicit method for solving the new problem.

3. In a binomial method, u and p can be determined by

$$
\begin{cases}
pu + (1-p)u^{-1} = e^{(r-D_0)\Delta t}, \\[2ex]
pu^2 + (1-p)u^{-2} = e^{(2(r-D_0)+\sigma^2)\Delta t}.
\end{cases}
$$

If Δt is very small, this problem can be approximated by

$$\begin{cases} pu + (1-p)\,u^{-1} = 1 + (r - D_0)\Delta t, \\ pu^2 + (1-p)\,u^{-2} = 1 + \left[2(r - D_0) + \sigma^2\right]\Delta t. \end{cases}$$

a) Find u and p for both cases (suppose $u > 1$).
b) Consider a more general system:

$$\begin{cases} pu + (1-p)\,u^{-1} = 1 + (r - D_0)\Delta t + O(\Delta t^2), \\ pu^2 + (1-p)\,u^{-2} = 1 + \left[2(r - D_0) + \sigma^2\right]\Delta t + O(\Delta t^2). \end{cases}$$

Show that if u and p is determined by such a system, then we always have

$$u = e^{\sigma\sqrt{\Delta t}} + O\left(\Delta t^{3/2}\right)$$

and

$$p = \frac{1}{2}\left[1 + \frac{\sqrt{\Delta t}}{\sigma}\left(r - D_0 - \frac{1}{2}\sigma^2\right)\right] + O\left(\Delta t^{3/2}\right).$$

4. *Describe the binomial methods for solving American vanilla call/put options

5. *Show that the Cox-Ross-Rubinstein binomial method for European options almost is an explicit difference scheme for the following problem:

$$\begin{cases} \dfrac{\partial V}{\partial t} + \dfrac{1}{2}\sigma^2\dfrac{\partial^2 V}{\partial y^2} + \left(r - D_0 - \dfrac{1}{2}\sigma^2\right)\dfrac{\partial V}{\partial y} - rV = 0, \\ \qquad\qquad\qquad\qquad -\infty < y < \infty, \quad t \le T, \\ V(y,T) = \max\left(\pm(e^y - 1), 0\right), \qquad -\infty < y < \infty, \end{cases}$$

where $x = \ln S$, S being the price of the stock, and assume $E = 1$.

6. Show that the relation

$$V(S, t_i^-) = V(S - D_i(S), t_i^+)$$

becomes

$$\overline{V}(\xi, \tau_i^+) = \left[1 - D_i\left(\frac{\xi P_m}{1-\xi}\right)\frac{1-\xi}{P_m}\right]\overline{V}\left(\frac{P_m\xi - D_i\left(\dfrac{\xi P_m}{1-\xi}\right)(1-\xi)}{P_m - D_i\left(\dfrac{\xi P_m}{1-\xi}\right)(1-\xi)}, \tau_i^-\right)$$

under the transformation

$$\begin{cases} \xi = \dfrac{S}{S + P_m}, \\ \tau = T - t, \\ \overline{V}(\xi, \tau) = \dfrac{V(S,t)}{S + P_m}. \end{cases}$$

7. *Suppose that we determine the price of an American vanilla call/put option through solving the following problem:

$$
\begin{cases}
\left(\dfrac{\partial \overline{V}}{\partial t} + \mathbf{L}_\xi \overline{V}\right)\left[\overline{V}(\xi,t) - \max(\pm(2\xi-1),0)\right] = 0, \\[2mm]
\dfrac{\partial \overline{V}}{\partial t} + \mathbf{L}_\xi \overline{V} \le 0, \\[2mm]
\overline{V}(\xi,t) - \max(\pm(2\xi-1),0) \ge 0, \\[2mm]
\overline{V}(\xi,0) = \max(\pm(2\xi-1),0),
\end{cases}
$$

where

$$0 \le \xi \le 1, \quad t \ge 0$$

and

$$\mathbf{L}_\xi = \frac{1}{2}\sigma^2(\xi)\xi^2(1-\xi)^2\frac{\partial^2}{\partial\xi^2} + (r-D_0)\xi(1-\xi)\frac{\partial}{\partial\xi} - [r(1-\xi)+D_0\xi].$$

Describe a numerical method for solving this problem by using a second-order implicit scheme.

8. As we know from Section 3.3, an American average strike call option is the solution of the following linear complementarity problem:

$$
\begin{cases}
\left(\dfrac{\partial W}{\partial t} + \mathbf{L}_{a,t}W\right)\left[W(\eta,t) - \max(\alpha-\eta,0)\right] = 0, \\[2mm]
\dfrac{\partial W}{\partial t} + \mathbf{L}_{a,t}W \le 0, \\[2mm]
W(\eta,t) - \max(\alpha-\eta,0) \ge 0, \\[2mm]
W(\eta,T) = \max(\alpha-\eta,0),
\end{cases}
$$

where

$$0 \le \eta, \quad t \le T, \quad \alpha \approx 1$$

and

$$\mathbf{L}_{a,t} = \frac{1}{2}\sigma^2\eta^2\frac{\partial^2}{\partial\eta^2} + \left[(D_0-r)\eta + \frac{1-\eta}{t}\right]\frac{\partial}{\partial\eta} - D_0.$$

Convert this problem into a problem on a finite domain and design a second-order implicit method for solving this new problem.

9. Based on the partial differential equation given in (2.26), design an implicit method for the LC problem of American options with discrete dividends.

10. *Suppose scheme (6.37) is used for solving an American option problem. Design a direct method, which you think is most accurate, to find the solution at each time step.

11. *Suppose an implicit finite-difference method based on the formulation (6.9) is used for solving an American option problem. Design an iteration method similar to (6.51) to find the solution of the system at each time step.
12. *When do we choose explicit first-order scheme (6.6) and when do we use implicit second-order scheme (5.34)? Why should we choose implicit second-order scheme (5.34) if we need highly accurate results?
13. a) Find a closed-form solution of the problem:

$$
\begin{cases}
\dfrac{\partial c_u}{\partial t} + \dfrac{1}{2}\sigma^2 S^2 \dfrac{\partial^2 c_u}{\partial S^2} + (r - D_0)S\dfrac{\partial c_u}{\partial S} - r c_u = 0, & 0 \le S, \ 0 \le t \le T, \\[2mm]
c_u(S,T) = \begin{cases} \max(S - E, 0), & \text{if } \ 0 \le S < g(T), \\ 0, & \text{if } \ g(T) \le S. \end{cases}
\end{cases}
$$

b) Consider the following European barrier option problem:

$$
\begin{cases}
\dfrac{\partial V}{\partial t} + \dfrac{1}{2}\sigma^2 S^2 \dfrac{\partial^2 V}{\partial S^2} + (r - D_0)S\dfrac{\partial V}{\partial S} - rV = 0, & \\[2mm]
& f(t) \le S \le g(t), \quad 0 \le t \le T, \\[1mm]
V(S,T) = \max(S - E, 0), & f(T) \le S \le g(T), \\[1mm]
V(f(t), t) = 0, & 0 \le t \le T, \\[1mm]
V(g(t), t) = 0, & 0 \le t \le T,
\end{cases}
$$

where $S = f(t)$ and $S = g(t)$ are the locations of the lower and upper barriers with $f(t) < E$ and $g(t) > E$. Assume that we need to find the solution by numerical methods. Design a SSM for this problem based on the result given in (a). (Here the problem can be defined on a non-rectangular domain.)
14. Design a SSM for European vanilla options with discrete dividends and a constant volatility, and formulate the problem on a finite domain.
15. *Design a SSM for Bermudan options with variable volatilities and formulate the problem on a finite domain.
16. Suppose r and D_0 are constant and $\sigma = \sigma(S)$. Derive the symmetry relations for Bermudan options.
17. *Find a transformation to convert an average price call option problem

$$
\begin{cases}
\dfrac{\partial V}{\partial t} + \dfrac{1}{2}\sigma^2 S^2 \dfrac{\partial^2 V}{\partial S^2} + (r - D_0)S\dfrac{\partial V}{\partial S} + \dfrac{S}{T}\dfrac{\partial V}{\partial I} - rV = 0, & \\[2mm]
& 0 \le S, \ 0 \le I, \ t \le T, \\[1mm]
V(S, I, T) = \max(I - E, 0), & 0 \le S, \ 0 \le I,
\end{cases}
$$

where

$$I = \frac{1}{T} \int_0^t S(\tau)d\tau,$$

into the problem

$$
\begin{cases}
\dfrac{\partial f}{\partial \tau} - \dfrac{1}{2}\sigma^2 \left[\xi - \dfrac{1}{(r-D_0)T}(1 - e^{-(r-D_0)\tau})\right]^2 \dfrac{\partial^2 f}{\partial \xi^2} = 0, \\
\qquad\qquad\qquad\qquad\qquad -\infty < \xi < \infty, \quad 0 \le \tau \le T, \\
f(\xi,0) = \max(\xi,0), \qquad\qquad -\infty < \xi < \infty.
\end{cases}
$$

18. Find a closed-form solution of the problem

$$
\begin{cases}
\dfrac{\partial \tilde{f}_0}{\partial \tau} - \dfrac{\sigma^2}{2(r-D_0)^2 T^2}\left(1 - e^{-(r-D_0)\tau}\right)^2 \dfrac{\partial^2 \tilde{f}_0}{\partial \xi^2} = 0, \\
\qquad\qquad\qquad\qquad\qquad -\infty < \xi < \infty, \quad 0 \le \tau \le T, \\
\tilde{f}_0(\xi,0) = \max(\xi,0), \qquad\qquad -\infty < \xi < \infty.
\end{cases}
$$

19. Convert the problem

$$
\begin{cases}
\dfrac{\partial f_1}{\partial \tau} - \dfrac{1}{2}\sigma^2 \left[\xi - \dfrac{1}{(r-D_0)T}\left(1 - e^{-(r-D_0)\tau}\right)\right]^2 \dfrac{\partial^2 f_1}{\partial \xi^2} = \dfrac{\sigma^2 \xi\, e^{-\xi^2/4\tau_1}}{4\sqrt{\pi\tau_1}} \\
\qquad \times \left[\xi - \dfrac{2}{(r-D_0)T}\left(1 - e^{-(r-D_0)\tau}\right)\right], \quad -\infty < \xi < \infty, \quad 0 \le \tau \le T, \\
f_1(\xi,0) = 0, \qquad\qquad\qquad\qquad -\infty < \xi < \infty
\end{cases}
$$

into a problem defined on $[0,1]$ and design a second-order implicit scheme for the new problem.

20. By using the transformation

$$
\begin{cases}
\xi = \dfrac{S}{S+P_m}, \\
r = r, \\
\tau = T - t, \\
u(\xi,r,\tau) = \dfrac{B_c(S,r,t)}{S+P_m},
\end{cases}
$$

the two-factor convertible bond problem for non-dividend stocks

$$
\begin{cases}
\dfrac{\partial B_c}{\partial t} + \dfrac{1}{2}\sigma^2 S^2 \dfrac{\partial^2 B_c}{\partial S^2} + \rho\sigma Sw \dfrac{\partial^2 B_c}{\partial S \partial r} + \dfrac{1}{2}w^2 \dfrac{\partial^2 B_c}{\partial r^2} + rS\dfrac{\partial B_c}{\partial S} \\
+(u - \lambda w)\dfrac{\partial B_c}{\partial r} - rB_c + kZ = 0, \\
\qquad\qquad\qquad 0 \le S, \quad r_l \le r \le r_u, \quad 0 \le t \le T, \\
B_c(S,r,T) = \max(Z, nS), \qquad 0 \le S, \quad r_l \le r \le r_u
\end{cases}
$$

can be converted into a problem on a finite domain with a bounded final condition. The one-factor convertible zero-coupon bond problem for non-dividend stocks

$$
\begin{cases}
\dfrac{\partial b_c}{\partial t} + \dfrac{1}{2}\sigma^2 S^2 \dfrac{\partial^2 b_c}{\partial S^2} + rS\dfrac{\partial b_c}{\partial S} - rb_c = 0, & 0 \le S, \quad 0 \le t \le T, \\[2mm]
b_c(S, r, T) = \max(Z, nS), & 0 \le S
\end{cases}
$$

has the following solution:

$$
nc(S, t; Z/n) + e^{-r(T-t)} Z,
$$

where $c(S, t; Z/n)$ is the price of a call option with an exercise price Z/n. Find the partial differential equation and the final condition the difference between the two bonds should satisfy. Convert the derived problem into a problem on a finite domain and design an implicit second-order scheme for the new problem.

21. Suppose that $c(S, \sigma, t)$ and $p(S, \sigma, t)$ are solutions of the following problems

$$
\begin{cases}
\dfrac{\partial c}{\partial t} + \mathbf{L}_{S,\sigma} c = 0, & 0 \le S, \quad \sigma_l \le \sigma \le \sigma_u, \quad t \le T, \\[2mm]
c(S, \sigma, T) = \max(S - E, 0), & 0 \le S, \quad \sigma_l \le \sigma \le \sigma_u
\end{cases}
$$

and

$$
\begin{cases}
\dfrac{\partial p}{\partial t} + \mathbf{L}_{S,\sigma} p = 0, & 0 \le S, \quad \sigma_l \le \sigma \le \sigma_u, \quad t \le T, \\[2mm]
p(S, \sigma, T) = \max(E - S, 0), & 0 \le S, \quad \sigma_l \le \sigma \le \sigma_u,
\end{cases}
$$

where $\mathbf{L}_{S,\sigma}$ is an operator defined by

$$
\mathbf{L}_{S,\sigma} = \dfrac{1}{2}\sigma^2 S^2 \dfrac{\partial^2}{\partial S^2} + \rho \sigma Sq \dfrac{\partial^2}{\partial S \partial \sigma} + \dfrac{1}{2}q^2 \dfrac{\partial^2}{\partial \sigma^2} + (r - D_0)S\dfrac{\partial}{\partial S} + (p - \lambda q)\dfrac{\partial}{\partial \sigma} - r.
$$

Show that the following put–call parity relation

$$
c(S, \sigma, t) - p(S, \sigma, t) = Se^{-D_0(T-t)} - Ee^{-r(T-t)}
$$

holds by the superposition principle. (Hint: Let u denote $c(S, \sigma, t) - p(S, \sigma, t)$. Show that u is the solution of the problem

$$
\begin{cases}
\dfrac{\partial u}{\partial t} + \mathbf{L}_{S,\sigma} u = 0, & 0 \le S, \quad \sigma_l \le \sigma \le \sigma_u, \quad t \le T, \\[2mm]
u(S, \sigma, T) = S - E, & 0 \le S, \quad \sigma_l \le \sigma \le \sigma_u
\end{cases}
$$

and that $Se^{-D_0(T-t)} - Ee^{-r(T-t)}$ is also the solution of this problem.)

22. *Convert the following double moving barrier call option problem

$$
\begin{cases}
\dfrac{\partial V}{\partial t} + \dfrac{1}{2}S^2\sigma^2\dfrac{\partial^2 V}{\partial S^2} + S(r - D_0)\dfrac{\partial V}{\partial S} - rV = 0, \\[2ex]
\qquad\qquad\qquad\qquad\qquad f(t) \le S \le g(t), \quad 0 \le t \le T, \\[1ex]
V(S,T) = \max(S - E, 0), \qquad f(T) \le S \le g(T), \\[1ex]
V(f(t), t) = 0, \qquad\qquad\quad 0 \le t \le T, \\[1ex]
V(g(t), t) = g(t) - E, \qquad\quad 0 \le t \le T
\end{cases}
$$

into a problem that has a smooth solution and design an implicit pseudo-spectral method for the new problem.

23. For the new problem obtained in Problem 20, design an implicit pseudo-spectral method.

Projects

General Requirements

A) *Submit a code or codes in C or C^{++} that will work on a computer the instructor can get access to. At the beginning of the code, write down the name of the student and indicate on which computer it works and the procedure to make it work.*

B) *Each code should use an input file to specify all the problem parameters and the computational parameters for each computation and an output file to store all the results. In an output file, the name of the problem, all the problem parameters, and the computational parameters should be given, so that one can know what the results are and how they were obtained. The input file should be submitted with the code.*

C) *If not specified, for each case two results are required. For the first result, a 20×12 mesh should be used. (In this case, the error of the solution might be quite large.) For the second result, the accuracy required is 0.01, and the mesh used should be as coarse as possible.*

D) *Submit results in form of tables or figures. When a result is given, always provide the problem parameters and the computational parameters.*

1. **Explicit method (6.3).**   Suppose that σ, r are constants and the dividends are given discretely or continuously. Write a code for European, Bermudan, and American calls and puts.

   • For American call and put options, give the results for the case: $S = 100$, $E = 100$, $T = 0.75$, $r = 0.1$, $D_0 = 0.05$, and $\sigma = 0.3$.

- For Bermudan call and put options, give the results for the case: $S = 100$, $E = 100$, $T = 1$, $r = 0.05$, $D_0 = 0.1$, $\sigma = 0.2$, and $K = 4$ (see Subsection 6.3.3).
- For European call and put options, give the results for the cases: $S = 100$, $E = 95, 100, 105$, $T = 1$, $r = 0.1$, $\sigma = 0.2$, and two dividend payments of $1.25 paid at $t = 2$ months and $t = 8$ months. $D(S)$ is defined by

$$D(S) = \begin{cases} S & \text{if } S \le d, \\ d & \text{if } S > d, \end{cases}$$

  where d is the dividend payment.
- Taking the European call option with $E = 100$, $T = 1$, $r = 0.1$, $D_0 = 0.05$, $\sigma = 0.2$ as an example, show that the explicit method (6.3) is unstable if $\Delta\tau$ is too large. For this problem, only one example is required. Plot the S-c curve with $t = 0$.

2. **Binomial methods (using both scheme (6.16)–(6.18) and scheme (6.17)–(6.19)).** Suppose that σ, r, D_0 are constants. Write a code for European, Bermudan, and American calls and puts. For this problem, instead of the result on a 20×12 mesh, a result with $\Delta t = T/12$ is required.
- For European call and put options, give the results for the cases: $S = 100$, $E = 95, 100, 105$, $T = 1$, $r = 0.1$, $D_0 = 0.025$, and $\sigma = 0.2$.
- For Bermudan call and put options, give the results for the case: $S = 100$, $E = 100$, $T = 1$, $r = 0.05$, $D_0 = 0.1$, $\sigma = 0.2$, and $K = 4$.
- For American call and put options, give the results for the case: $S = 100$, $E = 100$, $T = 0.75$, $r = 0.1$, $D_0 = 0.05$, and $\sigma = 0.3$.

3. **Implicit method (6.37) (solving the corresponding system by direct methods).** Suppose that σ, r, and D_0 are constants. Write a code for European, Bermudan, and American calls and puts.
- For European call and put options, give the results for the cases: $S = 100$, $E = 95, 100, 105$, $T = 1$, $r = 0.1$, $D_0 = 0.025$, and $\sigma = 0.2$.
- For Bermudan call and put options, give the results for the case: $S = 100$, $E = 100$, $T = 1$, $r = 0.05$, $D_0 = 0.1$, $\sigma = 0.2$, and $K = 4$.
- For American call and put options, give the results for the case: $S = 100$, $E = 100$, $T = 0.75$, $r = 0.1$, $D_0 = 0.05$, and $\sigma = 0.3$.

4. **Singularity-separating implicit method with scheme (6.37).** Suppose that σ, r are constants and the dividends are given discretely or continuously. Write a code for Bermudan calls and puts with continuous dividends and a code for European vanilla calls and puts with discrete dividends. Calculate the difference between the value of the option and the closed-form solution of a corresponding European vanilla option numerically. In order to calculate the price of a Bermudan put, Compute a corresponding call first and then obtain the value of the Bermudan put by using the symmetry relation.

- For Bermudan call options, give the results for the case: $S = 100$, $E = 100$, $T = 1$, $r = 0.05$, $D_0 = 0.1$, $\sigma = 0.2$, and $K = 4$. For Bermudan put options, give the results for the case: $S = 100$, $E = 100$, $T = 1$, $r = 0.1$, $D_0 = 0.05$, $\sigma = 0.2$, and $K = 12$.

- For European call and put options, give the results for the cases: $S = 100$, $E = 95, 100, 105$, $T = 1$, $r = 0.1$, $\sigma = 0.2$, and two dividend payments of \$1.25 paid at $t = 2$ months and $t = 8$ months. $D(S)$ is defined by

$$D(S) = \begin{cases} S & \text{if } S \le d, \\ d & \text{if } S > d, \end{cases}$$

where d is the dividend payment.

7

Free-Boundary Problems

As we know, a problem of pricing an American-style derivative can be for-mulated as a linear complementarity problem, and for most cases, it can also be written as a free-boundary problem. In Chapter 6, we have discussed how to solve a linear complementarity problem. Here, we study how to solve a free-boundary problem numerically. Many derivative security problems have a final condition with discontinuous derivatives at some point. In this case, their solutions are not very smooth in the domain near this point, and their numerical solutions will have relatively large error. In Chapter 6, we have suggested to deal with this problem in the following way: instead of calcu-lating the price of the derivative security, a difference between the price and an expression with the same or almost the same weak singularity is solved numerically. Because the difference is smooth, the error of numerical solution will be smaller. This method can still be used for free-boundary problems. For them there is another problem. On one side of the free boundary, the price of an American-style derivative satisfies a partial differential equation, and on the other side, it is equal to a given function. Because of this, the second derivative of the price is usually discontinuous on the free boundary. If we can follow the free boundary and use the partial differential equation only on the domain where the equation holds, then we can have less error. Hence, in Section 7.1 we not only discuss how to separate the weak singularity caused by the discontinuous first derivative at expiry but also describe how to convert a free-boundary problem into a problem defined on a rectangular domain so that we can easily use the partial differential equation only on the domain where the equation holds. The method described in Section 7.1 is referred to as the singularity-separating method (SSM) for free-boundary problems. The next two sections are devoted to discussing how to solve this problem using implicit schemes and pseudo-spectral methods for one-dimensional and two-dimensional cases. There, we also give some results on American vanilla, barrier, Asian, and lookback options, two-factor American vanilla options, and two-factor convertible bonds.

7.1 SSM for Free-Boundary Problems

7.1.1 One-Dimensional Cases

From Chapters 2, 3, and 4, we know that there are many American-style derivatives. Their major features are the same, but there are some differences among them. In this subsection, first taking an American vanilla call option as an example, we give the details of the singularity-separating method for free-boundary problems. Then, we briefly point out what modifications are needed in order to apply the method to other American-style derivatives.

From Section 2.6, we know that on the domain $[0, S_f(t)] \times [0, T]$, the price of an American call option, $C(S, t)$, is the solution of the free-boundary problem

$$
\begin{cases}
\dfrac{\partial C}{\partial t} + \dfrac{1}{2}\sigma^2 S^2 \dfrac{\partial^2 C}{\partial S^2} + (r - D_0)S\dfrac{\partial C}{\partial S} - rC = 0, \\[2mm]
\hspace{4cm} 0 \le S \le S_f(t), \quad 0 \le t \le T, \\[2mm]
C(S, T) = \max(S - E, 0), \hspace{1cm} 0 \le S \le S_f(T), \\[2mm]
C(S_f(t), t) = S_f(t) - E, \hspace{1cm} 0 \le t \le T, \\[2mm]
\dfrac{\partial C}{\partial S}(S_f(t), t) = 1, \hspace{1.5cm} 0 \le t \le T, \\[2mm]
S_f(T) = \max(E, rE/D_0);
\end{cases}
\tag{7.1}
$$

whereas on the domain $(S_f(t), \infty) \times [0, T]$, $C(S, t) = S - E$. Here, we assume $D_0 \ne 0$. Therefore, as long as we have the solution of the free-boundary problem, we can determine $C(S, t)$ for any $S \ge 0$ and any $t \in [0, T]$. The function $C(S, T) = \max(S - E, 0)$ has a discontinuous derivative at $S = E$. Therefore, $C(S, t)$ is not very smooth in the region where $S \approx E$ and $t \approx T$. Because the second derivative of $C(S, T)$ at $S = E$ goes to infinity, the truncation error of numerical methods near $S = E$ and $t = T$ is relatively large. In order to avoid such a relatively large error, we first find the numerical result of the difference between the prices of the American call option and the European call option, and then add the difference and the price of the European call option together to get the price of the American call option. Similar to those cases given in Subsection 6.3, the function representing the difference is very smooth, so numerical solution can be obtained efficiently.

Now we give the details of the method. Let $c(S, t)$ represent the price of the European call option, whose closed-form expression is given by (2.41). As we know, $c(S, t)$ is the solution of the problem

$$
\begin{cases}
\dfrac{\partial c}{\partial t} + \dfrac{1}{2}\sigma^2 S^2 \dfrac{\partial^2 c}{\partial S^2} + (r - D_0)S\dfrac{\partial c}{\partial S} - rc = 0, \quad 0 \le S, \quad 0 \le t \le T, \\[2mm]
c(S, T) = \max(S - E, 0), \hspace{4.3cm} 0 \le S.
\end{cases}
$$

Define

$$\overline{C}(S,t) = C(S,t) - c(S,t)$$

on the domain $[0, S_f(t)] \times [0,T]$. Both $C(S,T)$ and $c(S,T)$ are equal to $\max(S - E, 0)$, so $\overline{C}(S,T) = 0$. The functions $C(S,t)$ and $c(S,t)$ satisfy the same linear homogeneous partial differential equation, so the difference between them does the same. At the free boundary $S = S_f(t)$, we have

$$\overline{C}(S_f(t),t) = C(S_f(t),t) - c(S_f(t),t) = S_f(t) - E - c(S_f(t),t)$$

and

$$\frac{\partial \overline{C}}{\partial S}(S_f(t),t) = \frac{\partial C}{\partial S}(S_f(t),t) - \frac{\partial c}{\partial S}(S_f(t),t) = 1 - \frac{\partial c}{\partial S}(S_f(t),t).$$

Therefore, $\overline{C}(S,t)$ is the solution of the following free-boundary problem

$$\begin{cases} \dfrac{\partial \overline{C}}{\partial t} + \dfrac{1}{2}\sigma^2 S^2 \dfrac{\partial^2 \overline{C}}{\partial S^2} + (r - D_0)S\dfrac{\partial \overline{C}}{\partial S} - r\overline{C} = 0, & 0 \le S \le S_f(t) \\ & 0 \le t \le T, \\ \overline{C}(S,T) = 0, & 0 \le S \le S_f(T), \\ \overline{C}(S_f(t),t) = S_f(t) - E - c(S_f(t),t), & 0 \le t \le T, \\ \dfrac{\partial \overline{C}}{\partial S}(S_f(t),t) = 1 - \dfrac{\partial c}{\partial S}(S_f(t),t), & 0 \le t \le T, \\ S_f(T) = \max(E, rE/D_0). \end{cases} \qquad (7.2)$$

In the problem above, we need to determine $\overline{C}(S,t)$ on a non-rectangular domain, and one of its boundaries, $S = S_f(t)$, is also unknown.

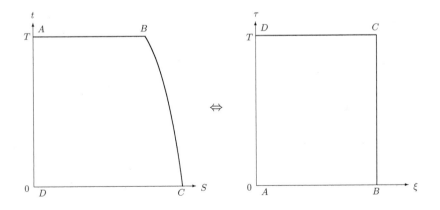

Fig. 7.1. Transforming a non-rectangular domain to a rectangular domain

In order to make discretization of the boundary conditions on the free boundary simple and convert the final-boundary value problem into an initial-boundary value problem, we introduce a new coordinate system $\{\xi, \tau\}$ through a transformation defined by

$$
\begin{cases}
\xi = \dfrac{S}{S_f(t)}, \\[2mm]
\tau = T - t.
\end{cases}
$$

This transformation converts the four boundaries of the domain of the problem (7.2), $S = 0$, $S = S_f(t)$, $t = T$, and $t = 0$, into $\xi = 0$, $\xi = 1$, $\tau = 0$, and $\tau = T$, respectively (see Fig. 7.1). Now the problem is defined on a rectangular domain, and the value of the solution at $\tau = 0$ is given, that is, the problem now is an initial-boundary value problem on a rectangular domain.

Let

$$
s_f(\tau) = \frac{1}{E} S_f(T - \tau)
$$

and

$$
u(\xi, \tau) = \frac{1}{E}\overline{C}(S, t) = \frac{1}{E}\overline{C}\left(\xi E s_f(\tau),\, T - \tau\right),
$$

that is,

$$
S_f(t) = E s_f(T - t)
$$

and

$$
\overline{C}(S, t) = E u\left(\frac{S}{E s_f(T - t)},\, T - t\right).
$$

Since

$$
\frac{\partial \overline{C}}{\partial t} = E\left[\frac{\xi}{s_f(\tau)}\frac{ds_f(\tau)}{d\tau}\frac{\partial u}{\partial \xi} - \frac{\partial u}{\partial \tau}\right],
$$

$$
\frac{\partial \overline{C}}{\partial S} = \frac{\partial u}{\partial \xi}\frac{1}{s_f(\tau)},
$$

$$
\frac{\partial^2 \overline{C}}{\partial S^2} = \frac{1}{E}\frac{\partial^2 u}{\partial \xi^2}\left[\frac{1}{s_f(\tau)}\right]^2,
$$

the problem (7.2) can be rewritten as

$$
\begin{cases}
\dfrac{\partial u}{\partial \tau} = k_2\xi^2\dfrac{\partial^2 u}{\partial \xi^2} + \left(k_1 + \dfrac{1}{s_f}\dfrac{ds_f}{d\tau}\right)\xi\dfrac{\partial u}{\partial \xi} - k_0 u, & 0 \le \xi \le 1, \\[2mm]
& 0 \le \tau \le T, \\[2mm]
u(\xi, 0) = 0, & 0 \le \xi \le 1, \\[2mm]
u(1, \tau) = g\left(s_f(\tau), \tau\right), & 0 \le \tau \le T, \\[2mm]
\dfrac{\partial u}{\partial \xi}(1, \tau) = h\left(s_f(\tau), \tau\right), & 0 \le \tau \le T, \\[2mm]
s_f(0) = \max(1, r/D_0),
\end{cases}
\tag{7.3}
$$

where $k_0 = r$, $k_1 = r - D_0$, $k_2 = \sigma^2/2$,

$$g\left(s_f(\tau), \tau\right) = s_f(\tau) - 1 - \frac{1}{E}c\left(Es_f(\tau),\, T - \tau\right)$$

and

$$h\left(s_f(\tau), \tau\right) = s_f(\tau)\left[1 - \frac{\partial c\left(Es_f(\tau),\, T - \tau\right)}{\partial S}\right].$$

The differential equation in (7.3) is a partial differential equation for u and can be understood as an ordinary differential equation for $s_f(\tau)$. This problem is a combination of an initial-boundary value problem for $u(\xi, \tau)$ on the domain $[0, 1] \times [0, T]$ and an initial value problem for $s_f(\tau)$ on the interval $[0, T]$. It can be solved using explicit schemes, implicit schemes, or pseudo-spectral methods. After we obtain $u(\xi, \tau)$, we can get the price of the American call option on the domain $[0, S_f(t)] \times [0, T]$ by

$$C(S, t) = Eu\left(\frac{S}{Es_f(T - t)}, T - t\right) + c(S, t).$$

The method described here is referred to as the singularity-separating method for American call options. The solution of the original American call option satisfies different equations in the two regions divided by the free boundary $S = S_f(t)$, and its solution has a discontinuous second derivative – a type of weak singularity – on the free boundary. In this method, the position of the free boundary is tracked accurately, so that we can use the different equations in each region exactly. Because the solution in the domain $(S_f(t), \infty) \times [0, T]$ is given by a known function, we only need to determine the solution in the region $[0, S_f(t)] \times [0, T]$. In this region, the second derivative near the free boundary is continuous, so the solution we want to get numerically is smoother than the original solution. Here, we also suggest to compute the difference between the American call option and the European call option numerically in the domain $[0, S_f(t)] \times [0, T]$, instead of directly computing the price of the American call option numerically. The difference is very smooth in the domain $[0, S_f(t)] \times [0, T]$, which make the truncation error smaller. Therefore, in the method described above, we use some techniques such that the solution we need to get numerically is much smoother than the original solution, which makes numerical methods more efficient. We refer to this as singularity-separating as we did in Subsection 6.3, because the solution becomes smoother than the original one after some singularities have been "separated." Here, the singularity that has been "separated" is the discontinuity of the derivatives of the solution, which is weak. The idea of the method was originally developed for dealing with shock problems in fluid mechanics (see [88]) and the Stefan problem (see [75]), the solutions of which had, for most of the cases, stronger discontinuities than we have here. It might be more precise if we use 'weak-singularity-separating' instead of singularity-separating. However, for simplicity we just keep the name of the method.

As pointed in Subsection 2.6.3, between American call and put options there exists the put–call symmetry relations. Using these relations, pricing a put option can be reduced to pricing a call option. There, the symmetry relations have been derived when American option problems are formulated as linear complementarity problems. Here, let us derive this conclusion when the problems are written as free-boundary problems. Let $P(S,t)$ stand for the price of an American put option. $P(S,t)$ should satisfy (2.57) on the domain $[S_f(t), \infty) \times [0, T]$ and equal $E - S$ on the domain $[0, S_f(t)) \times [0, T]$. Let

$$
\begin{cases}
\eta = \dfrac{E^2}{S}, \\[2mm]
u(\eta, t) = \dfrac{EP(S,t)}{S},
\end{cases}
$$

then it is easy to see that $u(\eta, t)$ is the solution of the free-boundary problem

$$
\begin{cases}
\dfrac{\partial u}{\partial t} + \dfrac{1}{2}\sigma^2 \eta^2 \dfrac{\partial^2 u}{\partial \eta^2} + (D_0 - r)\eta \dfrac{\partial u}{\partial \eta} - D_0 u = 0, & 0 \le \eta \le \eta_f(t), \\[2mm]
 & 0 \le t \le T, \\[2mm]
u(\eta, T) = \max(\eta - E, 0), & 0 \le \eta \le \eta_f(T), \\[2mm]
u\left(\eta_f(t), t\right) = \eta_f(t) - E, & 0 \le t \le T, \\[2mm]
\dfrac{\partial u}{\partial \eta}\left(\eta_f(t), t\right) = 1, & 0 \le t \le T, \\[2mm]
\eta_f(T) = \max\left(E, D_0 E/r\right)
\end{cases}
\tag{7.4}
$$

on the domain $[0, \eta_f(t)] \times [0, T]$; whereas on the domain $(\eta_f(t), \infty) \times [0, T]$, $u(\eta, t) = \eta - E$. As we can see, if the parameter r and the parameter D_0 in (7.1) exchange their positions, then the problem (7.1) almost becomes problem (7.4), except for the state variable. Therefore, $P(S,t)$ can be determined in the following way. First, understanding D_0 as r and r as D_0, we solve the problem (7.1) with the state variable η, instead of S, and get $u(\eta, t)$. Then, $P(S,t)$ is obtained by

$$
P(S,t) = \frac{S}{E}u\left(\frac{E^2}{S}, t\right).
$$

That is, we find $P(S,t)$ by using one of the symmetry relations.

It is not always reasonable to assume the volatility to be a constant. If the volatility is thought as a function of S, namely, $\sigma = \sigma(S)$, then the formulation (7.1) is still true after changing σ to $\sigma(S)$. Is the formulation (7.2) still true? The answer is no because in this case we do not have analytic solutions for European option. However, we can define

$$
\overline{C}(S,t) = C(S,t) - c_E(S,t; \sigma(E))
$$

on the domain $[0, S_f(t)] \times [0, T]$, where $c_E(S, t; \sigma(E))$ denotes the price of the European call option with $\sigma = \sigma(E)$. In this case, $\overline{C}(S, t)$ does not satisfy the Black–Scholes equation. Instead, it satisfies the following nonhomogeneous equation:

$$\frac{\partial \overline{C}}{\partial t} + \frac{1}{2}\sigma^2(S)S^2\frac{\partial^2 \overline{C}}{\partial S^2} + (r - D_0)S\frac{\partial \overline{C}}{\partial S} - r\overline{C} = f(S, t), \qquad (7.5)$$

where $f(S, t)$ is given by (6.58) in Subsection 6.3.2. For this case, the formulation is almost the same as (7.2) except that the partial differential equation in (7.2) should be replaced by (7.5). Therefore, the singularity-separating method still works for American options with variable volatilities because the singularity is weakened.

The same idea still works for American barrier, Asian, and lookback options. In order to remove the weak singularity at $S = E$ and $t = T$, we can use the solutions of vanilla European options for American barrier, Asian, and lookback options. However, it will be better to compute numerically the differences between American and European barrier options and between American and European lookback options because the differences are smaller in these cases. Just like the vanilla option case, the partial differential equation the differences satisfy in these cases is still the partial differential equation in (7.2). For European Asian options, explicit solutions have not been found, and the partial differential equation for Asian options is different from vanilla options. Thus, when we apply the SSM, the resulting equation for Asian options differs slightly from barrier and lookback options. For average strike options with $\alpha = 1$, the singularity-separating method will still work, and the difference will be a solution of a nonhomogeneous partial differential equation problem with a weaker singularity. It is not difficult to derive the problem in this case, and we leave this as a problem for the reader.

7.1.2 Two-Dimensional Cases

Two-factor options. In the above, we have discussed the formulation of American options if the volatility is a constant or a function of S. Now let us look at the case both the price of asset and the volatility of the asset price are random variables. As we have done in Subsection 6.3.6, we call such an option a two-factor option. Here, we discuss how to formulate the American two-factor vanilla call option as a free boundary problem if $D_0 \neq 0$.

We still assume the asset price S and the stochastic volatility σ to follow (6.77) and require (6.78)–(6.79) or (6.80)–(6.81) to hold.

Consider an American two-factor vanilla call option problem and let its value be $C(S, \sigma, t)$. As an American call option, it satisfies the condition:

$$C(S, \sigma, t) \geq \max(S - E, 0).$$

Because the European two-factor call options satisfy (6.84), the value of the two-factor vanilla American call option is the solution of the following linear complementarity problem:

$$
\begin{cases}
\left(\dfrac{\partial C}{\partial t} + \mathbf{L}_{\mathbf{s},\sigma}C\right)(C - G_c) = 0, \\[2mm]
\dfrac{\partial C}{\partial t} + \mathbf{L}_{\mathbf{s},\sigma}C \le 0, \\[2mm]
C(S,\sigma,t) - G_c(S,t) \ge 0, \\[2mm]
C(S,\sigma,T) = G_c(S,T),
\end{cases}
\tag{7.6}
$$

where $0 \le S$, $\sigma_l \le \sigma \le \sigma_u$, $t \le T$, $\mathbf{L}_{\mathbf{s},\sigma}$ is given by (6.83):

$$
\mathbf{L}_{\mathbf{s},\sigma} = \frac{1}{2}\sigma^2 S^2 \frac{\partial^2}{\partial S^2} + \rho\sigma S q \frac{\partial^2}{\partial S \partial\sigma} + \frac{1}{2}q^2 \frac{\partial^2}{\partial\sigma^2}
$$
$$
+ (r - D_0)S \frac{\partial}{\partial S} + (p - \lambda q)\frac{\partial}{\partial\sigma} - r,
$$

and

$$
G_c(S,t) = \max(S - E, 0).
$$

Consider the case $D_0 > 0$. Because

$$
\frac{\partial G_c}{\partial t} + \mathbf{L}_{\mathbf{s},\sigma}G_c < 0 \quad \text{for} \quad S > \max(E, rE/D_0)
$$

and

$$
\frac{\partial G_c}{\partial t} + \mathbf{L}_{\mathbf{s},\sigma}G_c \ge 0 \quad \text{for} \quad S \le \max(E, rE/D_0),
$$

there exists a free boundary $S = S_f(\sigma, t)$ starting from the straight line $S = \max(E, rE/D_0)$ at $t = T$ in the (S,σ,t)-space, and the entire domain is divided into two regions by the free boundary. On the domain $(S_f(\sigma,t), \infty) \times [\sigma_l, \sigma_u] \times [0,T]$, $C(S,\sigma,t) = \max(S - E, 0)$; whereas on $[0, S_f(\sigma,t)] \times [\sigma_l, \sigma_u] \times [0,T]$, $C(S,\sigma,t)$ is the solution of the following free-boundary problem:

$$
\begin{cases}
\dfrac{\partial C}{\partial t} + \mathbf{L}_{\mathbf{s},\sigma}C = 0, & 0 \le S \le S_f(\sigma,t), \\[1mm]
& \sigma_l \le \sigma \le \sigma_u, \quad 0 \le t \le T, \\[2mm]
C(S,\sigma,T) = \max(S - E, 0), & 0 \le S \le S_f(\sigma,T), \\[1mm]
& \sigma_l \le \sigma \le \sigma_u, \\[2mm]
C(S_f(\sigma,t),\sigma,t) = S_f(\sigma,t) - E, & \sigma_l \le \sigma \le \sigma_u, \quad 0 \le t \le T, \\[2mm]
\dfrac{\partial C(S_f(\sigma,t),\sigma,t)}{\partial S} = 1, & \sigma_l \le \sigma \le \sigma_u, \quad 0 \le t \le T, \\[2mm]
S_f(\sigma,T) = \max(E, rE/D_0), & \sigma_l \le \sigma \le \sigma_u.
\end{cases}
\tag{7.7}
$$

Just like the European two-factor option case, we let

$$\overline{C}(S,\sigma,t) = C(S,\sigma,t) - c_1(S,\sigma,t) \tag{7.8}$$

on the domain $[0, S_f(\sigma,t)] \times [\sigma_l, \sigma_u] \times [0, T]$. Here, $c_1(S,\sigma,t)$ is the same as the function $c(S,t)$ given by (2.41) in Subsection 2.4.4, namely, the price of the vanilla European call option when σ is a constant. Thus, the difference $\overline{C}$ is the solution of the following free-boundary problem:

$$
\begin{cases}
\dfrac{\partial \overline{C}}{\partial t} + \mathbf{L}_{s,\sigma}\overline{C} = f(S,\sigma,t), & 0 \le S \le S_f(\sigma,t), \quad \sigma_l \le \sigma \le \sigma_u, \\[2mm]
& 0 \le t \le T, \\[2mm]
\overline{C}(S,\sigma,T) = 0, & 0 \le S \le S_f(\sigma,T), \quad \sigma_l \le \sigma \le \sigma_u, \\[2mm]
\overline{C}\left(S_f(\sigma,t),\sigma,t\right) = S_f(\sigma,t) - E - c_1\left(S_f(\sigma,t),\sigma,t\right), & \sigma_l \le \sigma \le \sigma_u, \\[2mm]
& 0 \le t \le T, \\[2mm]
\dfrac{\partial \overline{C}\left(S_f(\sigma,t),\sigma,t\right)}{\partial S} = 1 - \dfrac{\partial c_1\left(S_f(\sigma,t),\sigma,t\right)}{\partial S}, & \sigma_l \le \sigma \le \sigma_u, \\[2mm]
& 0 \le t \le T, \\[2mm]
S_f(\sigma,T) = \max(E, rE/D_0), & \sigma_l \le \sigma \le \sigma_u,
\end{cases}
\tag{7.9}
$$

where

$$f(S,\sigma,t) = -\rho\sigma Sq\frac{\partial^2 c_1}{\partial S \partial\sigma} - \frac{1}{2}q^2\frac{\partial^2 c_1}{\partial\sigma^2} - (p - \lambda q)\frac{\partial c_1}{\partial\sigma},$$

$\dfrac{\partial c_1}{\partial\sigma}, \dfrac{\partial^2 c_1}{\partial\sigma^2}$, and $\dfrac{\partial^2 c}{\partial S\partial\sigma}$ being given by (6.87).

As we see from (7.7) and (7.9), the derivative of $C(S,\sigma,t)$ with respect to S is discontinuous at the point $t = T$ and $S = E$, and the derivative of $\overline{C}(S,\sigma,t)$ with respect to S at $t = T$ is identically equal to zero. It is expected that $\overline{C}(S,\sigma,t)$ is smoother than $C(S,\sigma,t)$ even though in this case the singularity only becomes weaker but is not completely removed because of the term $\dfrac{\partial^2 c_1}{\partial S\partial\sigma}$ in $f(S,\sigma,t)$. Therefore, when a numerical method is used, the truncation error for the problem (7.9) will be smaller than the problem (7.7). This is why we consider the formulation (7.9) instead of (7.7).

The free-boundary problem (7.9) is defined on the domain $[0, S_f(\sigma,t)] \times [\sigma_l, \sigma_u] \times [0, T]$ and the free boundary $S_f(\sigma,t)$ is a moving and unknown boundary. In order to make the discretization simple, we introduce the following transformation

$$
\begin{cases}
\xi = \dfrac{S}{S_f(\sigma,t)}, \\[3mm]
\sigma = \sigma, \\[2mm]
\tau = T - t.
\end{cases}
\tag{7.10}
$$

This transformation maps the domain $[0, S_f(\sigma, t)] \times [\sigma_l, \sigma_u] \times [0, T]$ in the (S, σ, t)-space onto a new domain $[0, 1] \times [\sigma_l, \sigma_u] \times [0, T]$ in the (ξ, σ, τ)-space and the moving boundary onto a plane under the new coordinate system. In the (ξ, σ, τ)-space, it is easy to construct numerical methods to solve the problem. Define

$$s_f(\sigma, \tau) = S_f(\sigma, t) = S_f(\sigma, T - \tau)$$

and

$$u(\xi, \sigma, \tau) = \overline{C}(S, \sigma, t) = \overline{C}\left(\xi s_f(\sigma, \tau), \sigma, T - \tau\right).$$

Among the derivatives of $\overline{C}$ and u, there are the following relations:

$$\frac{\partial \overline{C}}{\partial t} = \frac{\partial u}{\partial \xi}\frac{\partial \xi}{\partial t} + \frac{\partial u}{\partial \tau}\frac{\partial \tau}{\partial t} = \frac{\xi}{s_f}\frac{\partial s_f}{\partial \tau}\frac{\partial u}{\partial \xi} - \frac{\partial u}{\partial \tau},$$

$$\frac{\partial \overline{C}}{\partial S} = \frac{\partial u}{\partial \xi}\frac{\partial \xi}{\partial S} = \frac{1}{s_f}\frac{\partial u}{\partial \xi},$$

$$\frac{\partial \overline{C}}{\partial \sigma} = \frac{\partial u}{\partial \xi}\frac{\partial \xi}{\partial \sigma} + \frac{\partial u}{\partial \sigma} = -\left(\frac{\xi}{s_f}\frac{\partial s_f}{\partial \sigma}\frac{\partial u}{\partial \xi} - \frac{\partial u}{\partial \sigma}\right),$$

$$\frac{\partial^2 \overline{C}}{\partial S^2} = \frac{1}{s_f^2}\frac{\partial^2 u}{\partial \xi^2},$$

$$\frac{\partial^2 \overline{C}}{\partial S \partial \sigma} = \frac{\partial}{\partial \sigma}\left(\frac{1}{s_f}\frac{\partial u}{\partial \xi}\right) = -\frac{1}{s_f^2}\frac{\partial s_f}{\partial \sigma}\frac{\partial u}{\partial \xi} + \frac{1}{s_f}\left(\frac{\partial^2 u}{\partial \xi^2}\frac{\partial \xi}{\partial \sigma} + \frac{\partial^2 u}{\partial \xi \partial \sigma}\right)$$

$$= -\frac{1}{s_f^2}\frac{\partial s_f}{\partial \sigma}\frac{\partial u}{\partial \xi} - \frac{\xi}{s_f^2}\frac{\partial s_f}{\partial \sigma}\frac{\partial^2 u}{\partial \xi^2} + \frac{1}{s_f}\frac{\partial^2 u}{\partial \xi \partial \sigma},$$

$$\frac{\partial^2 \overline{C}}{\partial \sigma^2} = -\left[\frac{\partial}{\partial \sigma}\left(\frac{\xi}{s_f}\frac{\partial s_f}{\partial \sigma}\right)\frac{\partial u}{\partial \xi} + \frac{\xi}{s_f}\frac{\partial s_f}{\partial \sigma}\left(\frac{\partial^2 u}{\partial \xi^2}\frac{\partial \xi}{\partial \sigma} + \frac{\partial^2 u}{\partial \xi \partial \sigma}\right)\right.$$
$$\left. - \left(\frac{\partial^2 u}{\partial \sigma \partial \xi}\frac{\partial \xi}{\partial \sigma} + \frac{\partial^2 u}{\partial \sigma^2}\right)\right]$$
$$= \left\{\left(\frac{\xi}{s_f}\right)^2\left(\frac{\partial s_f}{\partial \sigma}\right)^2\frac{\partial^2 u}{\partial \xi^2} - 2\frac{\xi}{s_f}\frac{\partial s_f}{\partial \sigma}\frac{\partial^2 u}{\partial \xi \partial \sigma} + \frac{\partial^2 u}{\partial \sigma^2}\right.$$
$$\left. + \left[2\frac{\xi}{s_f^2}\left(\frac{\partial s_f}{\partial \sigma}\right)^2 - \frac{\xi}{s_f}\frac{\partial^2 s_f}{\partial \sigma^2}\right]\frac{\partial u}{\partial \xi}\right\}.$$

Substituting them into the partial differential equation in (7.9) yields

$$\frac{\partial u}{\partial \tau} = a_1\xi^2\frac{\partial^2 u}{\partial \xi^2} + a_2\xi q\frac{\partial^2 u}{\partial \xi \partial \sigma} + a_3 q^2\frac{\partial^2 u}{\partial \sigma^2} + a_4\xi\frac{\partial u}{\partial \xi} + a_5\frac{\partial u}{\partial \sigma} + a_6 u + a_7,$$

where

$$a_1 = \frac{1}{2}\sigma^2 - \frac{\rho\sigma q}{s_f}\frac{\partial s_f}{\partial \sigma} + \frac{1}{2}\left(\frac{q}{s_f}\frac{\partial s_f}{\partial \sigma}\right)^2,$$

$$a_2 = \rho\sigma - \frac{q}{s_f}\frac{\partial s_f}{\partial \sigma},$$

$$a_3 = \frac{1}{2},$$

$$a_4 = \frac{1}{s_f}\frac{\partial s_f}{\partial \tau} + r - D_0 - (\rho\sigma q + p - \lambda q)\frac{1}{s_f}\frac{\partial s_f}{\partial \sigma}$$

$$+ \left(\frac{q}{s_f}\frac{\partial s_f}{\partial \sigma}\right)^2 - \frac{1}{2}q^2\frac{1}{s_f}\frac{\partial^2 s_f}{\partial \sigma^2},$$

$$a_5 = p - \lambda q,$$

$$a_6 = -r,$$

$$a_7 = -f(S, \sigma, t) = -f(\xi s_f(\sigma, \tau), \sigma, T - \tau).$$

Therefore, noticing

$$\begin{cases} c_1(S_f, \sigma, t) = S_f e^{-D_0(T-t)} N(d_1) - E e^{-r(T-t)} N(d_2), \\ \dfrac{\partial c_1(S_f, \sigma, t)}{\partial S} = e^{-D_0(T-t)} N(d_1), \end{cases}$$

we can rewrite (7.9) as

$$\begin{cases} \dfrac{\partial u}{\partial \tau} = a_1\xi^2\dfrac{\partial^2 u}{\partial \xi^2} + a_2\xi q\dfrac{\partial^2 u}{\partial \xi \partial \sigma} + a_3 q^2\dfrac{\partial^2 u}{\partial \sigma^2} + a_4\xi\dfrac{\partial u}{\partial \xi} + a_5\dfrac{\partial u}{\partial \sigma} \\[2mm] \qquad + a_6 u + a_7, \qquad 0 \le \xi \le 1, \quad \sigma_l \le \sigma \le \sigma_u, \quad 0 \le \tau \le T, \\[2mm] u(\xi, \sigma, 0) = 0, \qquad\qquad\quad 0 \le \xi \le 1, \quad \sigma_l \le \sigma \le \sigma_u, \\[2mm] u(1, \sigma, \tau) = s_f(\sigma, \tau)\left[1 - e^{-D_0\tau}N(d_1)\right] \\[2mm] \qquad - E(1 - e^{-r\tau}N(d_2)), \qquad \sigma_l \le \sigma \le \sigma_u, \quad 0 \le \tau \le T, \\[2mm] \dfrac{\partial u(1, \sigma, \tau)}{\partial \xi} = s_f(\sigma, \tau)\left[1 - e^{-D_0\tau}N(d_1)\right], \qquad \sigma_l \le \sigma \le \sigma_u, \\[2mm] \qquad\qquad\qquad\qquad\qquad\qquad\qquad\qquad 0 \le \tau \le T, \\[2mm] s_f(\sigma, 0) = \max(E, rE/D_0), \qquad\qquad\qquad \sigma_l \le \sigma \le \sigma_u, \end{cases} \qquad (7.11)$$

where

$$d_1 = \left[\ln\frac{s_f e^{-D_0\tau}}{E e^{-r\tau}} + \frac{1}{2}\sigma^2\tau\right] \bigg/ (\sigma\sqrt{\tau}) \quad \text{and} \quad d_2 = d_1 - \sigma\sqrt{\tau}.$$

Once we have the solution of (7.11), $u(\xi, \sigma, \tau)$, we can get the value of the original American call option by

$$C(S, \sigma, t) = \overline{C}(S, \sigma, t) + c_1(S, \sigma, t)$$

$$= u\left(\frac{S}{s_f(\sigma, T-t)}, \sigma, T-t\right) + c_1(S, \sigma, t). \qquad (7.12)$$

This method is called the singularity-separating method for American two-factor call options.

For two-factor vanilla American put options, the linear complementarity problem is

$$
\begin{cases}
\left(\dfrac{\partial P}{\partial t} + \mathbf{L}_{\mathrm{s},\sigma} P \right) (P - G_p) = 0, \\[2ex]
\dfrac{\partial P}{\partial t} + \mathbf{L}_{\mathrm{s},\sigma} P \leq 0, \\[2ex]
P(S, \sigma, t) - G_p(S, t) \geq 0, \\[2ex]
P(S, \sigma, T) = G_p(S, T),
\end{cases}
$$

where $0 \leq S$, $\sigma_l \leq \sigma \leq \sigma_u$, $t \leq T$ and

$$
G_p(S, t) = \max(E - S, 0).
$$

Introducing the transformation

$$
\begin{cases}
\eta = \dfrac{E^2}{S}, \\[2ex]
\sigma = \sigma, \\[1ex]
t = t, \\[1ex]
u(\eta, \sigma, t) = \dfrac{E P(S, \sigma, t)}{S}
\end{cases}
\tag{7.13}
$$

and noticing the following relations

$$
\frac{\partial \eta}{\partial S} = -\frac{E^2}{S^2}, \qquad \frac{\partial P}{\partial t} = \frac{S}{E} \frac{\partial u}{\partial t},
$$

$$
\frac{\partial P}{\partial S} = \frac{u}{E} - \frac{E}{S} \frac{\partial u}{\partial \eta}, \qquad \frac{\partial P}{\partial \sigma} = \frac{S}{E} \frac{\partial u}{\partial \sigma},
$$

$$
\frac{\partial^2 P}{\partial S^2} = \frac{E^3}{S^3} \frac{\partial^2 u}{\partial \eta^2}, \qquad \frac{\partial^2 P}{\partial S \partial \sigma} = \frac{1}{E} \frac{\partial u}{\partial \sigma} - \frac{E}{S} \frac{\partial^2 u}{\partial \eta \partial \sigma}, \qquad \frac{\partial^2 P}{\partial \sigma^2} = \frac{S}{E} \frac{\partial^2 u}{\partial \sigma^2},
$$

we can convert the linear complementarity problem above into another linear complementarity problem

$$
\begin{cases}
\left(\dfrac{\partial u}{\partial t} + \mathbf{L}_{\eta,\sigma} u \right) (u - G_u) = 0, \\[2ex]
\dfrac{\partial u}{\partial t} + \mathbf{L}_{\eta,\sigma} u \leq 0, \\[2ex]
u(\eta, \sigma, t) - G_u(\eta, t) \geq 0, \\[2ex]
u(\eta, \sigma, T) = G_u(\eta, T),
\end{cases}
$$

where $0 \leq \eta$, $\sigma_l \leq \sigma \leq \sigma_u$, $t \leq T$,

$$G_u(\eta, t) = \max(\eta - E, 0)$$

and

$$\mathbf{L}_{\eta,\sigma} = \frac{1}{2}\sigma^2\eta^2\frac{\partial^2}{\partial\eta^2} - \rho\sigma q\eta\frac{\partial^2}{\partial\eta\partial\sigma} + \frac{1}{2}q^2\frac{\partial^2}{\partial\sigma^2}$$
$$+ (D_0 - r)\eta\frac{\partial}{\partial\eta} + [p - (\lambda - \rho\sigma)q]\frac{\partial}{\partial\sigma} - D_0.$$

This problem has the same form as the problem (7.6). The only difference is that r and D_0 are switched, and ρ and λ in (7.6) are replaced by $-\rho$ and $\lambda - \rho\sigma$ here. Therefore, a put problem can be written as a call problem.

Let $C(S, \sigma, t; a, b, c, d)$ and $P(S, \sigma, t; a, b, c, d)$ denote the prices of American call and put options and $S_{cf}(\sigma, t; a, b, c, d)$ and $S_{pf}(\sigma, t; a, b, c, d)$ be their optimal exercise prices. Here, a, b, c, and d are parameters (or parameter functions) for the risk-free interest rate r, dividend yield rate D_0, correlation coefficient ρ, and market price of volatility risk λ, respectively. Then, what we have described above can be written as a relation between the American two-factor vanilla put and call options:

$$\begin{cases} P(S, \sigma, t; a, b, c, d) = \dfrac{S}{E}C\left(\dfrac{E^2}{S}, \sigma, t; b, a, -c, d - c\sigma\right), \\ S_{pf}(\sigma, t; a, b, c, d) = E^2/S_{cf}(\sigma, t; b, a, -c, d - c\sigma). \end{cases}$$

If we let

$$\eta = E^2/S, \quad \bar{c} = -c \quad \text{and} \quad \bar{d} = d - c\sigma = d + \bar{c}\sigma,$$

then the first relation above can be written as

$$P\left(\frac{E^2}{\eta}, \sigma, t; a, b, -\bar{c}, \bar{d} - \bar{c}\sigma\right) = \frac{E}{\eta}C\left(\eta, \sigma, t; b, a, \bar{c}, \bar{d}\right)$$

or

$$C(S, \sigma, t; a, b, c, d) = \frac{S}{E}P\left(\frac{E^2}{S}, \sigma, t; b, a, -c, d - c\sigma\right).$$

The second relation can be written in a symmetric form

$$S_{pf}(\sigma, t; a, b, c, d) \times S_{cf}(\sigma, t; b, a, -c, d - c\sigma) = E^2.$$

Therefore, we can have the following relations:

$$\begin{cases} P(S, \sigma, t; a, b, c, d) = \dfrac{S}{E}C\left(\dfrac{E^2}{S}, \sigma, t; b, a, -c, d - c\sigma\right), \\ C(S, \sigma, t; a, b, c, d) = \dfrac{S}{E}P\left(\dfrac{E^2}{S}, \sigma, t; b, a, -c, d - c\sigma\right), \\ S_{pf}(\sigma, t; a, b, c, d) \times S_{cf}(\sigma, t; b, a, -c, d - c\sigma) = E^2, \end{cases} \tag{7.14}$$

which in this book are referred to as the call–put symmetry relations between American two-factor vanilla call and put options. Thus, if we have a code for one type of option, call or put, then in order to calculate another type of option, we only need to make a little change.

The free-boundary problem for a call option is defined on a finite domain and that for a put option is on an infinite domain. Consequently, it will be natural to write a code for call options and calculate a put option as a call option.

Two-factor convertible bonds. Another example of American-style derivatives depending on two random variables is two-factor convertible bonds. Let $B_c(S, r, t)$ be the price of such a bond. As was pointed out in Section 4.7, the computational domain of a two-factor convertible bond problem can be divided into two parts. On the domain $(S_f(r, t), \infty) \times [r_l, r_u] \times [0, T]$,

$$B_c(S, r, t) = \max(Z, nS);$$

whereas on the domain $[0, S_f(r, t)] \times [r_l, r_u] \times [0, T]$, $B_c(S, r, t)$ is the solution of the free-boundary problem:

$$
\begin{cases}
\dfrac{\partial B_c}{\partial t} + \mathbf{L_{s,r}} B_c + kZ = 0, & 0 \le S \le S_f(r, t), \quad r_l \le r \le r_u, \\
& 0 \le t \le T, \\
B_c(S, r, T) = \max(Z, nS), & 0 \le S \le S_f(r, T), \quad r_l \le r \le r_u, \\
B_c(S_f(r, t), r, t) = n S_f(r, t), & r_l \le r \le r_u, \quad 0 \le t \le T, \\
\dfrac{\partial B_c}{\partial S}(S_f(r, t), r, t) = n, & r_l \le r \le r_u, \quad 0 \le t \le T, \\
S_f(r, T) = \max\left(\dfrac{Z}{n}, \dfrac{kZ}{D_0 n}\right), & r_l \le r \le r_u.
\end{cases}
\tag{7.15}
$$

where

$$
\mathbf{L_{s,r}} = \frac{1}{2}\sigma^2 S^2 \frac{\partial^2}{\partial S^2} + \rho\sigma S w \frac{\partial^2}{\partial S \partial r} + \frac{1}{2} w^2 \frac{\partial^2}{\partial r^2} + (r - D_o)S\frac{\partial}{\partial S} + (u - \lambda w)\frac{\partial}{\partial r} - r.
$$

Let $b_c(S, r, t)$ be the solution of the problem

$$
\begin{cases}
\dfrac{\partial b_c}{\partial t} + \dfrac{1}{2}\sigma^2 S^2 \dfrac{\partial^2 b_c}{\partial S^2} + (r - D_0)S\dfrac{\partial b_c}{\partial S} - rb_c = 0, & 0 \le S, \\
& 0 \le t \le T, \\
b_c(S, T) = \max(Z, nS) = n\max(S - Z/n, 0) + Z, & 0 \le S,
\end{cases}
\tag{7.16}
$$

where σ, r, and D_0 are constants. This problem has the following solution:

$$b_c(S, r, t) = nc(S, t; Z/n) + e^{-r(T-t)} Z,$$

where $c(S, t; Z/n)$ is the price of a European call option with an exercise price $E = Z/n$. Define

$$\overline{B}_c(S, r, t) = B_c(S, r, t) - b_c(S, r, t). \tag{7.17}$$

For $\overline{B}_c(S, r, t)$, the free boundary problem is

$$
\begin{cases}
\dfrac{\partial \overline{B}_c}{\partial t} + \mathbf{L}_{\mathbf{s},\mathbf{r}} \overline{B}_c + kZ = f(S, r, t), \quad 0 \le S \le S_f(r, t), \\[2ex]
\qquad\qquad\qquad\qquad\qquad\qquad\quad r_l \le r \le r_u, \quad 0 \le t \le T, \\[2ex]
\overline{B}_c(S, r, T) = 0, \qquad\qquad\quad 0 \le S \le S_f(r, T), \quad r_l \le r \le r_u, \\[2ex]
\overline{B}_c\left(S_f(r, t), r, t\right) = nS_f(r, t) - b_c\left(S_f(r, t), r, t\right), \\[2ex]
\qquad\qquad\qquad\qquad\qquad\qquad\quad r_l \le r \le r_u, \quad 0 \le t \le T, \\[2ex]
\dfrac{\partial \overline{B}_c}{\partial S}\left(S_f(r, t), r, t\right) = n - \dfrac{\partial b_c\left(S_f(r, t), r, t\right)}{\partial S}, \\[2ex]
\qquad\qquad\qquad\qquad\qquad\qquad\quad r_l \le r \le r_u, \quad 0 \le t \le T, \\[2ex]
S_f(r, T) = \max\left(\dfrac{Z}{n}, \dfrac{kZ}{D_0 n}\right), \qquad r_l \le r \le r_u,
\end{cases}
\tag{7.18}
$$

where

$$f(S, r, t) = -\rho\sigma Sw \frac{\partial^2 b_c}{\partial S \partial r} - \frac{1}{2} w^2 \frac{\partial^2 b_c}{\partial r^2} - (u - \lambda w) \frac{\partial b_c}{\partial r}.$$

In order to make the discretization easy, we introduce the following transformation

$$
\begin{cases}
\xi = \dfrac{S}{S_f(r, t)}, \\[2ex]
\bar{r} = \dfrac{r - r_l}{r_u - r_l}, \\[2ex]
\tau = T - t.
\end{cases}
\tag{7.19}
$$

This transformation maps the domain

$$[0, S_f(r, t)] \times [r_l, r_u] \times [0, T]$$

in the (S, r, t)-space onto the domain

$$[0, 1] \times [0, 1] \times [0, T]$$

in the $(\xi, \bar{r}, \tau)$-space. We also introduce two new variables u and s_f defined by

$$\begin{cases} u(\xi, \bar{r}, \tau) = \dfrac{\overline{B}_c(S, r, t)}{Z}, \\[3mm] s_f(\bar{r}, \tau) = \dfrac{S_f(r, t)}{Z/n} \end{cases} \qquad (7.20)$$

and let

$$v(\xi, \bar{r}, \tau) = b_c(S, r, t)/Z.$$

For v we have

$$\begin{aligned} v(\xi, \bar{r}, \tau) &= nc(S, t; Z/n)/Z + e^{-r(T-t)} \\ &= (nS/Z)e^{-D_0(T-t)}N(d_1) - e^{-r(T-t)}N(d_2) + e^{-r(T-t)} \\ &= \xi s_f(\bar{r}, \tau)e^{-D_0\tau}N(d_1) + e^{-r\tau}N(-d_2), \end{aligned}$$

where

$$\begin{aligned} d_1 &= \left[\ln \frac{Se^{(r-D_0)(T-t)}}{Z/n} + \frac{1}{2}\sigma^2(T-t) \right] \Big/ \left(\sigma\sqrt{T-t} \right) \\ &= \left[\ln\left(\xi s_f(\bar{r}, \tau)e^{(r-D_0)\tau} \right) + \frac{1}{2}\sigma^2\tau \right] \Big/ \left(\sigma\sqrt{\tau} \right), \\ d_2 &= d_1 - \sigma\sqrt{\tau}. \end{aligned}$$

Thus, v can be expressed as a function of $\xi s_f(\bar{r}, \tau)$ and τ. Because

$$\overline{B}_c(S, r, t) = Zu(\xi, \bar{r}, \tau) = Zu\left(\frac{nS}{Zs_f\left(\dfrac{r - r_l}{r_u - r_l}, T - t \right)}, \frac{r - r_l}{r_u - r_l}, T - t \right),$$

we have

$$\frac{\partial \overline{B}_c}{\partial t} = Z\left(-\frac{\partial u}{\partial \tau} + \frac{\partial u}{\partial \xi}\frac{\xi}{s_f}\frac{\partial s_f}{\partial \tau} \right),$$

$$\frac{\partial \overline{B}_c}{\partial S} = \frac{\partial u}{\partial \xi}\frac{n}{s_f},$$

$$\frac{\partial \overline{B}_c}{\partial r} = Z\left(-\frac{\partial u}{\partial \xi}\frac{\xi}{s_f}\frac{\partial s_f}{\partial \bar{r}} + \frac{\partial u}{\partial \bar{r}} \right)\frac{1}{r_u - r_l},$$

$$\frac{\partial^2 \overline{B}_c}{\partial S^2} = \frac{1}{Z}\frac{\partial^2 u}{\partial \xi^2}\left(\frac{n}{s_f} \right)^2,$$

$$\frac{\partial^2 \overline{B}_c}{\partial S\partial r} = \left(-\frac{\partial^2 u}{\partial \xi^2}\frac{n\xi}{s_f^2}\frac{\partial s_f}{\partial \bar{r}} + \frac{\partial^2 u}{\partial \xi\partial \bar{r}}\frac{n}{s_f} - \frac{\partial u}{\partial \xi}\frac{n}{s_f^2}\frac{\partial s_f}{\partial \bar{r}} \right)\frac{1}{r_u - r_l},$$

$$\frac{\partial^2 \overline{B}_c}{\partial r^2} = Z\left\{ \frac{\partial^2 u}{\partial \xi^2}\left(\frac{\xi}{s_f}\frac{\partial s_f}{\partial \bar{r}} \right)^2 - 2\frac{\partial^2 u}{\partial \xi\partial \bar{r}}\frac{\xi}{s_f}\frac{\partial s_f}{\partial \bar{r}} \right.$$

$$+\frac{\partial u}{\partial \xi}\left[2\frac{\xi}{s_f^2}\left(\frac{\partial s_f}{\partial \bar{r}}\right)^2 - \frac{\xi}{s_f}\frac{\partial^2 s_f}{\partial \bar{r}^2}\right] + \frac{\partial^2 u}{\partial \bar{r}^2}\right\}\left(\frac{1}{r_u - r_l}\right)^2.$$

Substituting these expressions into the problem (7.18) yields

$$\begin{cases} \dfrac{\partial u}{\partial \bar{\tau}} = \mathbf{L}_{\xi,\bar{r}}u + a_7, & 0 \le \xi \le 1, \quad 0 \le \bar{r} \le 1, \quad 0 \le \tau \le T, \\[2mm] u(\xi, \bar{r}, 0) = 0, & 0 \le \xi \le 1, \quad 0 \le \bar{r} \le 1, \\[2mm] u(1, \bar{r}, \tau) = s_f(\bar{r}, \tau) - v(1, \bar{r}, \tau), & 0 \le \bar{r} \le 1, \quad 0 \le \tau \le T, \\[2mm] \dfrac{\partial u}{\partial \xi}(1, \bar{r}, \tau) = s_f(\bar{r}, \tau) - \dfrac{\partial v}{\partial \xi}(1, \bar{r}, \tau), & 0 \le \bar{r} \le 1, \quad 0 \le \tau \le T, \\[2mm] s_f(\bar{r}, 0) = \max\left(1, k/D_0\right), & 0 \le \bar{r} \le 1, \end{cases}$$ (7.21)

where

$$\mathbf{L}_{\xi,\bar{r}} = a_1\xi^2\frac{\partial^2}{\partial \xi^2} + a_2\xi w\frac{\partial^2}{\partial \xi \partial \bar{r}} + a_3 w^2\frac{\partial^2}{\partial \bar{r}^2} + \left(a_4 + \frac{1}{s_f}\frac{\partial s_f}{\partial \tau}\right)\xi\frac{\partial}{\partial \xi}$$

$$+a_5\frac{\partial}{\partial \bar{r}} + a_6,$$

$$a_1 = \frac{1}{2}\sigma^2 - \rho\sigma w\frac{1}{s_f(r_u - r_l)}\frac{\partial s_f}{\partial \bar{r}} + \frac{1}{2}w^2\left[\frac{1}{s_f(r_u - r_l)}\frac{\partial s_f}{\partial \bar{r}}\right]^2,$$

$$a_2 = \frac{1}{r_u - r_l}\left[\rho\sigma - \frac{w}{s_f(r_u - r_l)}\frac{\partial s_f}{\partial \bar{r}}\right],$$

$$a_3 = \frac{1}{2(r_u - r_l)^2},$$

$$a_4 = r - D_0 - \frac{1}{s_f(r_u - r_l)}\frac{\partial s_f}{\partial \bar{r}}(\rho\sigma w + u - \lambda w)$$

$$+\frac{1}{2}w^2\left\{2\left[\frac{1}{s_f(r_u - r_l)}\frac{\partial s_f}{\partial \bar{r}}\right]^2 - \frac{1}{s_f(r_u - r_l)^2}\frac{\partial^2 s_f}{\partial \bar{r}^2}\right\},$$

$$a_5 = \frac{u - \lambda w}{r_u - r_l},$$

$$a_6 = -r,$$

$$a_7 = k + \rho\sigma S w\frac{\partial^2 v}{\partial S \partial r} + \frac{1}{2}w^2\frac{\partial^2 v}{\partial r^2} + (u - \lambda w)\frac{\partial v}{\partial r}.$$

We will refer to this method as the singularity-separating method for two-factor convertible bonds.

In the problem (7.21), Z and n are not involved. That is, the solution of the problem, $u(\xi, \bar{r}, \tau)$ and $s_f(\bar{r}, \tau)$, does not depend on Z or n. The problem (7.21) is called the problem for a standard convertible bond.

If the asset price S, the asset price volatility σ and the interest rate r are all considered as random variables, then we have American three-factor option problems and three-factor convertible bond problems. It is not difficult to generalize the method here to such three-dimensional problems.

7.2 Implicit Finite-Difference Methods

7.2.1 Solution of One-Dimensional Problems

The problem (7.3) can be solved by different numerical methods, for example, explicit finite-difference methods, implicit finite-difference methods, pseudo-spectral methods, and so forth. In this book, we only discuss the implicit finite-difference methods and the pseudo-spectral methods. In this subsection, we discuss how to use implicit finite-difference methods to solve free-boundary problem (7.3).

As we have pointed out, the problem we are going to solve is defined on $[0, 1] \times [0, T]$ on the (ξ, τ)-plane. For simplicity, we assume that we still use the equidistant mesh given by (6.2). Let u_m^n stand for the value of u at the points $\xi = \xi_m \equiv m\Delta\xi$ and $\tau = \tau^n \equiv n\Delta\tau$, and s_f^n represent the value of s_f at $\tau = \tau^n$. At time $t = 0$, the function u and s_f are known, i.e., $u_m^0, m = 0, 1, \cdots, M$ and s_f^0 are known. We need to find $u_m^n, m = 0, 1, \cdots, M$ and $s_f^n, n = 1, 2, \cdots, N$.

The partial differential equation in (7.3) can be discretized by

$$
\frac{u_m^{n+1} - u_m^n}{\Delta\tau}
$$
$$
= \frac{1}{2} \left[k_2 m^2 \left(u_{m+1}^{n+1} - 2u_m^{n+1} + u_{m-1}^{n+1} \right) + \frac{k_1 m}{2} \left(u_{m+1}^{n+1} - u_{m-1}^{n+1} \right) - k_0 u_m^{n+1} \right]
$$
$$
+ \frac{1}{2} \left[k_2 m^2 \left(u_{m+1}^n - 2u_m^n + u_{m-1}^n \right) + \frac{k_1 m}{2} \left(u_{m+1}^n - u_{m-1}^n \right) - k_0 u_m^n \right]
$$
$$
+ \frac{s_f^{n+1} - s_f^n}{\left(s_f^{n+1} + s_f^n \right)\Delta\tau} \left[\frac{m}{2} \left(u_{m+1}^{n+1} - u_{m-1}^{n+1} \right) + \frac{m}{2} \left(u_{m+1}^n - u_{m-1}^n \right) \right]
$$

$$(7.22)$$

at $m = 0, 1, 2, \cdots, M - 1$, for $n = 0, 1, \cdots, N - 1$. Here, in all coefficients, $\xi = m\Delta\xi$ and $\tau = (n + 1/2)\Delta\tau$, so from Section 5.1, we know that the scheme has a truncation error of $O(\Delta\tau^2, \Delta\xi^2)$. At $m = 0$, the equation actually becomes

$$
\frac{u_0^{n+1} - u_0^n}{\Delta\tau} = \frac{-k_0}{2} \left(u_0^{n+1} + u_0^n \right),
$$

therefore, u_{-1}^n and u_{-1}^{n+1} do not appear in the equations. The boundary conditions at $\xi = 1$ in (7.3) can be replaced by

$$
u_M^{n+1} = g(s_f^{n+1}, \tau^{n+1}),
$$

$$(7.23)$$

and

$$\frac{3u_M^{n+1} - 4u_{M-1}^{n+1} + u_{M-2}^{n+1}}{2\Delta\xi} = h\left(s_f^{n+1}, \tau^{n+1}\right). \tag{7.24}$$

Here, the condition (7.23) is exact, and the truncation error of (7.24) is $O(\Delta\xi^2)$ because the first derivative is approximated by a one-sided second-order difference scheme. In the system (7.22)–(7.24), if u_m^n, $m = 0, 1, \cdots, M$ and s_f^n are given, then there are $M + 2$ unknowns: u_m^{n+1}, $m = 0, 1, \cdots, M$ and s_f^{n+1}. The number of equations in the system is also $M + 2$. Therefore, we can determine u_m^{n+1}, $m = 0, 1, \cdots, M$ and s_f^{n+1} from this system. From the initial conditions in (7.3), the second and the fifth equations, we can obtain

$$u_m^0 = 0, \quad m = 0, 1, \cdots, M$$

and

$$s_f^0 = \max(1, r/D_0).$$

Consequently, starting from $n = 0$, we can find the solution at τ^{n+1} from the solution at τ^n successively.

However, the system is a nonlinear one, so we cannot find the solution directly. In order to find the solution of the system, we use iteration methods. For example, (7.22)–(7.24) can be written as

$$\frac{u_m^{(j)} - u_m^n}{\Delta\tau}$$

$$= \frac{1}{2}\left[k_2 m^2\left(u_{m+1}^{(j)} - 2u_m^{(j)} + u_{m-1}^{(j)}\right) + \frac{k_1 m}{2}\left(u_{m+1}^{(j)} - u_{m-1}^{(j)}\right) - k_0 u_m^{(j)}\right]$$

$$+ \frac{1}{2}\left[k_2 m^2\left(u_{m+1}^n - 2u_m^n + u_{m-1}^n\right) + \frac{k_1 m}{2}\left(u_{m+1}^n - u_{m-1}^n\right) - k_0 u_m^n\right]$$

$$+ \frac{s_f^{(j)} - s_f^n}{\left(s_f^{(j-1)} + s_f^n\right)\Delta\tau}\left[\frac{m}{2}\left(u_{m+1}^{(j-1)} - u_{m-1}^{(j-1)}\right) + \frac{m}{2}\left(u_{m+1}^n - u_{m-1}^n\right)\right],$$

$$m = 0, 1, \cdots, M - 1, \tag{7.25}$$

$$u_M^{(j)} = g\left(s_f^{(j)}, \tau^{n+1}\right), \tag{7.26}$$

and

$$\frac{3u_M^{(j)} - 4u_{M-1}^{(j)} + u_{M-2}^{(j)}}{2\Delta\xi} = h\left(s_f^{(j)}, \tau^{n+1}\right), \tag{7.27}$$

where $u_m^{(j)}, s_f^{(j)}$ are the j-th iteration values of u_m^{n+1}, s_f^{n+1} respectively. In order to start an iteration, we set $u_m^{(0)} = u_m^n, m = 0, 1, \cdots, M$ and $s_f^{(0)} = s_f^n$. The system consisting of (7.25)–(7.27) is linear for $u_m^{(j)}, m = 0, 1, \cdots, M$, and nonlinear for $s_f^{(j)}$. This system can be solved by a modified LU decomposition method described below.

The equations (7.25) can be rewritten as

$$-\frac{1}{2}\left(k_2 m^2 + \frac{k_1 m}{2}\right)\Delta\tau u_{m+1}^{(j)} + \left[1 + \left(k_2 m^2 + \frac{k_0}{2}\right)\Delta\tau\right]u_m^{(j)}$$

$$-\frac{1}{2}\left(k_2 m^2 - \frac{k_1 m}{2}\right)\Delta\tau u_{m-1}^{(j)}$$

$$-\frac{1}{\left(s_f^{(j-1)} + s_f^n\right)}\left[\frac{m}{2}\left(u_{m+1}^{(j-1)} - u_{m-1}^{(j-1)}\right) + \frac{m}{2}\left(u_{m+1}^n - u_{m-1}^n\right)\right]s_f^{(j)}$$

$$=\frac{1}{2}\left(k_2 m^2 + \frac{k_1 m}{2}\right)\Delta\tau u_{m+1}^n + \left[1 - \left(k_2 m^2 + \frac{k_0}{2}\right)\Delta\tau\right]u_m^n$$

$$+\frac{1}{2}\left(k_2 m^2 - \frac{k_1 m}{2}\right)\Delta\tau u_{m-1}^n$$

$$-\frac{1}{\left(s_f^{(j-1)} + s_f^n\right)}\left[\frac{m}{2}\left(u_{m+1}^{(j-1)} - u_{m-1}^{(j-1)}\right) + \frac{m}{2}\left(u_{m+1}^n - u_{m-1}^n\right)\right]s_f^n,$$

$$m = 0, 1, \cdots, M - 1. \tag{7.28}$$

When $m = 0$, the equation simply becomes:

$$\left(1 + \frac{k_0}{2}\Delta\tau\right)u_0^{(j)} = \left(1 - \frac{k_0}{2}\Delta\tau\right)u_0^n.$$

Thus, no iteration for u_0^{n+1} is needed, and

$$u_0^{n+1} = \frac{1 - \frac{k_0}{2}\Delta\tau}{1 + \frac{k_0}{2}\Delta\tau}u_0^n.$$

Furthermore, noticing $u_0^0 = 0$, we have $u_0^{n+1} = 0, n = 0, 1, \cdots, N - 1$. Therefore, u_0^n can be understood as a given quantity, i.e., for each iteration, there are $M + 1$ unknowns: $u_m^{(j)}, m = 1, 2, \cdots, M$, and $s_f^{(j)}$. The $M + 1$ unknowns satisfy a system in the following form:

$$\begin{cases} b_1 u_1^{(j)} + c_1 u_2^{(j)} + e_1 s_f^{(j)} = f_1, \\ a_m u_{m-1}^{(j)} + b_m u_m^{(j)} + c_m u_{m+1}^{(j)} + e_m s_f^{(j)} = f_m, \ m=2,3,\cdots,M-1, \\ u_M^{(j)} = g\left(s_f^{(j)}, \tau^{n+1}\right), \\ d_M u_{M-2}^{(j)} + a_M u_{M-1}^{(j)} + b_M u_M^{(j)} = h\left(s_f^{(j)}, \tau^{n+1}\right). \end{cases} \tag{7.29}$$

The top $M - 1$ equations of this system are linear equations for $u_m^{(j)}, m = 1, 2, \cdots, M$ and $s_f^{(j)}$. Let us rewrite the first equation as

$$u_1^{(j)} = \alpha_1 u_2^{(j)} + \beta_1 s_f^{(j)} + \gamma_1,$$

where
$$\alpha_1 = -c_1/b_1, \quad \beta_1 = -e_1/b_1, \quad \text{and} \quad \gamma_1 = f_1/b_1.$$

Suppose we have a relation in the form
$$u_{m-1}^{(j)} = \alpha_{m-1} u_m^{(j)} + \beta_{m-1} s_f^{(j)} + \gamma_{m-1}.$$

Substituting this relation into the second equation in (7.29) and solving the equation for $u_m^{(j)}$, we have
$$u_m^{(j)} = \alpha_m u_{m+1}^{(j)} + \beta_m s_f^{(j)} + \gamma_m,$$

where
$$\alpha_m = \frac{-c_m}{b_m + a_m \alpha_{m-1}}, \quad \beta_m = -\frac{e_m + a_m \beta_{m-1}}{b_m + a_m \alpha_{m-1}}, \quad \text{and} \quad \gamma_m = \frac{f_m - a_m \gamma_{m-1}}{b_m + a_m \alpha_{m-1}}.$$

This procedure can be done for $m = 2, 3, \cdots, M-1$ successively. Therefore, the first and second equations in (7.29) are equivalent to the following relation
$$u_m^{(j)} = \alpha_m u_{m+1}^{(j)} + \beta_m s_f^{(j)} + \gamma_m, \quad m = 1, 2, \cdots, M-1, \tag{7.30}$$

where
$$\begin{cases} \alpha_m = \dfrac{-c_m}{b_m + a_m \alpha_{m-1}}, \\[2mm] \beta_m = -\dfrac{e_m + a_m \beta_{m-1}}{b_m + a_m \alpha_{m-1}}, \\[2mm] \gamma_m = \dfrac{f_m - a_m \gamma_{m-1}}{b_m + a_m \alpha_{m-1}}. \end{cases} \tag{7.31}$$

Here, we define $a_1 = 0$. Using the two relations in (7.30) with $m = M-2$ and $M-1$, we can eliminate $u_{M-2}^{(j)}$ and $u_{M-1}^{(j)}$ in the last equation of (7.29) and obtain
$$d_M \left[\alpha_{M-2} \alpha_{M-1} u_M^{(j)} + (\alpha_{M-2} \beta_{M-1} + \beta_{M-2}) s_f^{(j)} + \alpha_{M-2} \gamma_{M-1} + \gamma_{M-2} \right]$$
$$+ a_M \left(\alpha_{M-1} u_M^{(j)} + \beta_{M-1} s_f^{(j)} + \gamma_{M-1} \right) + b_M u_M^{(j)}$$
$$= h \left(s_f^{(j)}, \tau^{n+1} \right).$$

Substituting the third equation in (7.29) into this equation yields
$$[(d_M \alpha_{M-2} + a_M) \alpha_{M-1} + b_M] \, g \left(s_f^{(j)}, \tau^{n+1} \right)$$
$$+ [d_M (\alpha_{M-2} \beta_{M-1} + \beta_{M-2}) + a_M \beta_{M-1}] \, s_f^{(j)}$$
$$+ d_M (\alpha_{M-2} \gamma_{M-1} + \gamma_{M-2}) + a_M \gamma_{M-1}$$
$$= h \left(s_f^{(j)}, \tau^{n+1} \right).$$

This is an equation for $s_f^{(j)}$, and we can use the secant method to get its solution. In order to start the secant method, we need two approximate values of $s_f^{(j)}$. For $s_f^{(1)}$, we can take s_f^n and $s_f^n + \varepsilon$ as the two initial values. Here, ε is a proper positive number because $s_f(t)$ is an increasing function in τ for an American call option. For $s_f^{(j)}$, $j = 2, 3, \cdots$, we can take the last two iteration values of $s_f^{(j-1)}$ as the two initial iteration values for $s_f^{(j)}$.

After $s_f^{(j)}$ is found, we can obtain $u_M^{(j)}$ from the third equation in (7.29) and $u_m^{(j)}$ from (7.30), $m = M-1, M-2, \cdots, 1$, successively. From (7.28), we know that a_m, b_m and c_m do not depend on $u_m^{(j-1)}$ and $s_f^{(j-1)}$. Thus, a_m, b_m, and c_m remain unchanged during the iteration. Furthermore, from the expression of α_m in (7.31), we know that α_m and $b_m + a_m \alpha_{m-1}$ also remain unchanged. f_m in (7.29) is a sum of two parts:

$$\frac{1}{2}\left(k_2 m^2 + \frac{k_1 m}{2}\right)\Delta\tau u_{m+1}^n + \left[1 - \left(k_2 m^2 + \frac{k_0}{2}\right)\Delta\tau\right]u_m^n$$
$$+ \frac{1}{2}\left(k_2 m^2 - \frac{k_1 m}{2}\right)\Delta\tau u_{m-1}^n$$

and

$$\frac{1}{\left(s_f^{(j-1)} + s_f^n\right)}\left[\frac{m}{2}\left(u_{m+1}^{(j-1)} - u_{m-1}^{(j-1)}\right) + \frac{m}{2}\left(u_{m+1}^n - u_{m-1}^n\right)\right]s_f^n.$$

The first part also does not depend on $u_m^{(j-1)}$ and $s_f^{(j-1)}$. In order to make the computation efficient, all these unchanged quantities during the iteration should be computed once and stored for future use.

The iteration (7.25)–(7.27) will give a second-order accuracy if two iterations are performed. In fact, $u_m^{(1)}$ and $s_f^{(1)}$ are solutions of a first-order scheme, and $u_m^{(2)}$ and $s_f^{(2)}$ are solutions of an improved Euler method in the τ-direction, which gives second-order accuracy in the τ-direction (see any book on numerical methods for ordinary differential equations). This scheme is always second order in the ξ-direction, so the results have an accuracy of $O\left(\Delta\xi^2, \Delta\tau^2\right)$.

The way of solving the system (7.22)–(7.24) is not unique. If s_f^{n+1} is given, then the system consisting of (7.22) and (7.23) is a system with $M+1$ linear equations and $M+1$ unknowns $u_m^{(n+1)}$, $m = 0, 1, \cdots, M$. Therefore, this system determines the dependence of u_m^{n+1} on s_f^{n+1}, i.e., the functions $u_m^{n+1}(s_f^{n+1})$, $m = 0, 1, \cdots, M$. Substituting the three functions $u_{M-2}^{n+1}(s_f^{n+1})$, $u_{M-1}^{n+1}(s_f^{n+1})$, $u_M^{n+1}(s_f^{n+1})$ into (7.24), we have an equation for s_f^{n+1}:

$$f(s_f^{n+1}) \equiv \frac{3u_M^{n+1}(s_f^{n+1}) - 4u_{M-1}^{n+1}(s_f^{n+1}) + u_{M-2}^{n+1}(s_f^{n+1})}{2\Delta\xi} - h\left(s_f^{n+1}, \tau^{n+1}\right) = 0.$$

$$(7.32)$$

This equation can be solved by the secant method. When using the secant method, we need to evaluate $f(s_f^{n+1})$ for a given s_f^{n+1}. This can be done as follows. Let s_f^{n+1} in (7.22) and (7.23) take the given value, then solve the linear system consisting of (7.22) and (7.23) by the LU decomposition method described in Subsection 5.2.1. Substituting the value of $u_M^{n+1}, u_{M-1}^{n+1}, u_{M-2}^{n+1}$ into (7.32) yields the value $f(s_f^{n+1})$. As long as we have $f(s_f^{n+1})$ for two different s_f^{n+1}, we can start the iteration. When $f(s_f^{n+1})$ is very close to zero for some given s_f^{n+1}, we obtain the solution for s_f^{n+1}, and the solution of the linear system corresponding to this s_f^{n+1} gives the values for $u_m^{n+1}, m = 0, 1, \cdots, M$. This is another way to solve the system (7.22)–(7.24).

Wu and Kwok (see [74]) suggested a similar scheme to (7.22)–(7.24). The main difference is that they computed the option price directly.

7.2.2 Solution of Greeks

In practice, we usually need to know not only the price of the derivative security but also the sensitivities of the price to the parameters, i.e., the derivatives of the price with respect to parameters. As mentioned in Section 2.7, these derivatives are usually denoted by Greeks on the market. For example, $\frac{\partial V}{\partial S}, \frac{\partial^2 V}{\partial S^2}, \frac{\partial V}{\partial t}, \frac{\partial V}{\partial \sigma}, \frac{\partial V}{\partial r}$ are usually called Delta (Δ), Gamma (Γ), Theta (Θ), Vega ($\mathcal{V}$), and Rho (ρ), respectively. When we know the price of the derivative security for all S and for all $t \in [0, T]$, it is easy to get Delta, Gamma, and Theta. Here, we discuss how to get the other Greeks.

Let $V(S, t; \sigma, r, D_0)$ be the price of a derivative security. Here, we explicitly indicate that V depends on σ, r, and D_0. Thus, the sensitivities of the option price to them can be described by $\mathcal{V} = \frac{\partial V}{\partial \sigma}, \rho = \frac{\partial V}{\partial r}$, and $\rho_d = \frac{\partial V}{\partial D_0}$. In order to get $\frac{\partial V}{\partial \sigma}$, we can have $V(S, t; \sigma_1, r, D_0)$ and $V(S, t; \sigma_1 + \Delta\sigma, r, D_0)$, then get $\frac{\partial V}{\partial \sigma}$ for a σ near σ_1 by

$$\frac{V(S, t; \sigma_1 + \Delta\sigma, r, D_0) - V(S, t; \sigma_1, r, D_0)}{\Delta\sigma}.$$

We also can solve the problem derived in Section 2.7 to get $\frac{\partial V}{\partial \sigma}$.

Let us take $\frac{\partial C}{\partial \sigma}$ as an example to explain how to get such a Greek. Set $\overline{C}(S, t) = C(S, t) - c(S, t)$ and suppose $\overline{C}(S, t)$ and $S_f(t)$ have been obtained. Instead of $\frac{\partial C}{\partial \sigma}$, let us discuss how to obtain $\frac{\partial \overline{C}}{\partial \sigma}$, which will be denoted by $\overline{C}_\sigma$ in this subsection. As pointed out in Section 2.7, $\frac{\partial C}{\partial \sigma}$ is the solution of (2.68). Thus, $\overline{C}_\sigma$ should satisfy

$$\begin{cases} \dfrac{\partial \overline{C}_\sigma}{\partial t} + \dfrac{1}{2}\sigma^2 S^2 \dfrac{\partial^2 \overline{C}_\sigma}{\partial S^2} + (r - D_0)S\dfrac{\partial \overline{C}_\sigma}{\partial S} - r\overline{C}_\sigma + \sigma S^2 \dfrac{\partial^2 \overline{C}}{\partial S^2} = 0, \\[2mm] \qquad\qquad\qquad\qquad\qquad 0 \le S \le S_f(t), \quad 0 \le t \le T, \\[2mm] \overline{C}_\sigma(S,T) = 0, \qquad\qquad 0 \le S \le S_f(T), \\[2mm] \overline{C}_\sigma(S_f(t),t) = -\dfrac{\partial c\,(S,t)}{\partial \sigma}, \qquad 0 \le t \le T. \end{cases}$$

This is a problem with a known moving boundary. By using the transformation

$$\begin{cases} \xi = \dfrac{S}{S_f(t)}, \\[3mm] \tau = T - t \end{cases}$$

and letting

$$s_f(\tau) = \frac{1}{E}S_f(T - \tau)$$

and

$$W(\xi,\tau) = \frac{1}{E}\overline{C}_\sigma(S,t) = \frac{1}{E}\overline{C}_\sigma\left(\xi E s_f(\tau), T - \tau\right),$$

the problem above can be written as an initial-boundary value problem on a rectangular domain:

$$\begin{cases} \dfrac{\partial W}{\partial \tau} = k_2 \xi^2 \dfrac{\partial^2 W}{\partial \xi^2} + \left(k_1 + \dfrac{1}{s_f}\dfrac{ds_f}{d\tau}\right)\xi\dfrac{\partial W}{\partial \xi} - k_0 W + \sigma\xi^2\dfrac{\partial^2 u}{\partial \xi^2}, \\[2mm] \qquad\qquad\qquad\qquad 0 \le \xi \le 1, \quad 0 \le \tau \le T, \\[2mm] W(\xi,0) = 0, \qquad\qquad 0 \le \xi \le 1, \\[2mm] W(1,\tau) = -\dfrac{1}{E}\dfrac{\partial c\,(Es_f(\tau), T - \tau)}{\partial \sigma}, \qquad 0 \le \tau \le T, \end{cases} \tag{7.33}$$

where $u(\xi,\tau)$ and $s_f(\tau)$ are the solution of (7.3), and $c(S,t)$ is the price of the European call given in Subsection 2.4.4. The equation in (7.33) can be discretized by

$$\frac{W_m^{n+1} - W_m^n}{\Delta\tau}$$
$$= \frac{1}{2}k_2 m^2 \left(W_{m+1}^{n+1} - 2W_m^{n+1} + W_{m-1}^{n+1} + W_{m+1}^n - 2W_m^n + W_{m-1}^n\right)$$
$$+ \frac{1}{2}\left\{\left[\frac{k_1}{2} + \frac{s_f^{n+1} - s_f^n}{\left(s_f^{n+1} + s_f^n\right)\Delta\tau}\right] m\left(W_{m+1}^{n+1} - W_{m-1}^{n+1}\right) - k_0 W_m^{n+1}\right.$$
$$\left. + \left[\frac{k_1}{2} + \frac{s_f^{n+1} - s_f^n}{\left(s_f^{n+1} + s_f^n\right)\Delta\tau}\right] m\left(W_{m+1}^n - W_{m-1}^n\right) - k_0 W_m^n\right\}$$

$$+ \left(d_m^{n+1} + d_m^n\right)/2,$$
$$m = 0, 1, \cdots, M - 1, \qquad (7.34)$$

where

$$d = \sigma\xi^2 \frac{\partial^2 u}{\partial\xi^2}.$$

The boundary condition in (7.33) can be written as

$$W_M^{n+1} = -\frac{1}{E} \frac{\partial c\left(Es_f(\tau^{n+1}), T - \tau^{n+1}\right)}{\partial\sigma}. \qquad (7.35)$$

The system (7.34) and (7.35) is a linear system for W_m^{n+1}, $m = 0, 1, \cdots, M$ and we can get W_m^{n+1} by the LU decomposition method if W_m^n, $m = 0, 1, \cdots$, M, and s_f^n, s_f^{n+1}, $\frac{\partial^2 u_m^n}{\partial\xi^2}$, and $\frac{\partial^2 u_m^{n+1}}{\partial\xi^2}$ are given. As soon as we obtain W, $\frac{\partial C}{\partial\sigma}$ can be found by

$$\frac{\partial C}{\partial\sigma}(S,t) = EW\left(\frac{S}{Es_f(T-t)}, T-t\right) + \frac{\partial c}{\partial\sigma}(S,t).$$

When u and s_f are obtained, we need to solve an initial-boundary value problem in order to get $\frac{\partial C}{\partial\sigma}$ if the method above is adopted. If we obtain $\frac{\partial C}{\partial\sigma}$ by using

$$\frac{V(S,t;\sigma_1 + \Delta\sigma, r, D_0) - V(S,t;\sigma_1, r, D_0)}{\Delta\sigma},$$

then we need to solve another free-boundary problem in order to have $V(S,t;\sigma_1 + \Delta\sigma, r, D_0)$ when $V(S,t;\sigma_1, r, D_0)$ has been found. The amount of work to solve a free-boundary problem by the method described in Subsection 7.2.1 is more than twice of the amount of the work to solve an initial-boundary value problem by the method given here. This is why we formulate a problem for $\overline{C}_\sigma$ and obtain $\frac{\partial C}{\partial\sigma}$ by solving (7.33).

7.2.3 Numerical Results of Vanilla Options and Comparison

In this subsection, we will discuss some issues on the efficiency of the numerical method described in Subsection 7.2.1 and the performance of the method combined with the extrapolation technique. Here, a method combined with the extrapolation technique means that the computation is first done on a mesh by the method, then reduce the mesh sizes in the both directions by a factor of $1/2$ (or other numbers) and do the computation on the second mesh again, and finally get the results by the formula (5.51) in Section 5.5 (or other similar formulae). The method in Subsection 7.2.1 is an implicit finite-difference version of the SSM and, for simplicity, is referred to as the SSM in this subsection. Here, we also compare the results obtained by the

SSM and the combination of the SSM and the extrapolation technique with the results by other methods for two options. Finally, through the shape of the free boundaries, we point out that adopting nonuniform time steps can make the method more accurate.

Table 7.1. Parameters

Interest rates r	$0.05 \sim 0.20$ with $\Delta r = 0.025$
Volatilities σ	$0.1 \sim 0.5$ with $\Delta \sigma = 0.1$
Dividend yields D_0	$0.00 \sim 0.15$ with $\Delta D_0 = 0.025$
Expiries T	3 days, 15 days, 1$\sim$ 12 months with $\Delta T = 1$ month

Table 7.2. American call options with $r = 0.1$ and $T = 1$ year

$D_0 \backslash \sigma$	0.1	0.2	0.3	0.4	0.5
0.000	—	—	—	—	—
0.025	—	—	—	—	—
0.050	—	12×6	12×6	12×6	12×6
0.075	16×8	12×8	12×8	12×8	12×8
0.100	28×14	18×10	16×10	14×8	14×8
0.125	44×16	30×12	24×10	18×8	14×8
0.150	48×18	32×12	26×10	20×8	16×8

The SSM combined with the extrapolation technique has been tested for American vanilla call and put options with various parameters. The parameters tested are given in Table 7.1. Consider the standard American call problem, i.e., the problem with $E = 1$. Suppose $r = 0.1$, $T = 1$, and require the maximum error of C for $S \in [0.9, 1.1]$ to be less than or equal to 10^{-4}. Table 7.2 lists the numbers of mesh intervals needed for different D_0 and σ in order to get such results. There, $M \times N$ means that for the second mesh, M subintervals in the ξ-direction and N time-steps in the τ-direction are taken. In Table 7.2, '—' means that for this set of parameters, and for $S \in [0.9, 1.1]$, the difference between the American call option and the European call option is less than or only a slightly greater than 10^{-4}, so no numerical method is needed. From here, we know that if the method described in Subsection 7.2.1 is used, then a coarse mesh is enough for obtaining a result with error about 10^{-4} for $S \in [0.9, 1.1]$.

As pointed out in Chapter 2, using the symmetry relations, we can have the value of an American put option from an American call option with interchanging the interest rate and dividend yield. However, we can also solve the put option problem directly. In Table 7.3, we list the numbers of mesh intervals needed in order to have an accuracy of about 10^{-4} for $S \in [0.9, 1.1]$

Table 7.3. American put option with $r = 0.05$ and $T = 1$ year

$D_0 \backslash \sigma$	0.1	0.2	0.3	0.4	0.5
0.000	40×12	24×8	18×6	14×4	12×4
0.025	36×12	26×8	16×4	14×4	12×4
0.050	32×10	22×6	16×4	12×4	12×4
0.075	—	22×6	16×4	12×4	12×4
0.100	—	—	16×4	12×4	12×4
0.125	—	—	—	12×4	12×4
0.150	—	—	—	—	12×4

and $r = 0.05$. Thus, for both American call and put options, only a coarse mesh is needed in order to get the accuracy usually needed. From the price of the call option with $r = 0.1$ and $D_0 = 0.05$, we can have the value of the put option with $r = 0.05$ and $D_0 = 0.1$. From Tables 7.2 and 7.3, we know that in order to get the price of the put option with $r = 0.05$, $D_0 = 0.1$, and $\sigma = 0.3$, we can take a 16×4 mesh if we solve a put problem directly or we can take a 12×6 mesh if we solve a corresponding call problem and get the solution using the symmetry relations. For these two meshes, the CPU times needed are very close, so we can choose either way. However, if we already have a code to compute American call option prices, then using the second way would be a better choice since only very little code needs to be added.

Table 7.4. Optimal prices for American call options

$(\sigma = 0.2, T = 1 \text{ and } E = 100)$

$D_0 \backslash r$	0.050	0.075	0.100	0.125	0.150
0.050	141.540893	170.943495	223.764096	277.831844	331.285054
0.075	128.372144	137.454215	155.027353	186.574326	222.166283
0.100	122.069175	127.037558	134.599182	147.295598	168.445693
0.125	118.119037	121.403431	125.903014	132.417054	142.448401
0.150	115.346132	117.723481	120.800277	124.918028	130.659131

With this method, it is not difficult to get results with a high accuracy. In Table 7.4, the optimal price for American call options with various r and D_0 are listed. Analysis shows these results to be exact to at least seven digits (see [86]).

Now let us discuss the convergence rate of the SSM. Let $r = 0.1, \sigma = 0.2, D_0 = 0.05, T = 1$ year, and $S = E = 100$. In order to study the convergence rate, we have to know the exact solution. We do not have the exact solution, but we can get a solution with a very high accuracy and obtain the first few digits of the exact solution. For the parameters given above, our computation shows the exact call option price $C = 9.94092345 \cdots$ and the exact

Table 7.5. American call option

$(r = 0.1, \sigma = 0.2, D_0 = 0.05, T = 1$ year, $S = E = 100,$
and the exact value $= 9.94092345\cdots)$

Meshes	Without extrapolation			With extrapolation		
	Results	Errors	CPU(sec.)	Results	|Errors|	CPU(sec.)
32×2	9.941663	−0.000739	.00025	9.940902	0.000021	.00045
64×4	9.941097	−0.000174	.00083	9.940908	0.000015	0.0012
128×8	9.940962	−0.000038	0.0027	9.940917	0.000006	0.0038
256×16	9.940932	−0.000009	0.0099	9.9409225	0.000001	0.0125

Table 7.6. American put options

$(r = 0.1, \sigma = 0.2, D_0 = 0.05, T = 1$ year, $S = E = 100,$
and the exact value $= 5.92827717\cdots)$

Meshes	Without extrapolation			With extrapolation		
	Results	Errors	CPU(sec.)	Results	|Errors|	CPU(sec.)
12×4	5.968338	−0.040060	.00035	5.925575	0.002702	.00065
24×8	5.937883	−0.009606	.00084	5.927732	0.000545	0.0014
48×16	5.930477	−0.002200	0.0025	5.928008	0.000269	0.0035
96×32	5.928819	−0.000542	0.0078	5.928266	0.000011	0.0108
192×64	5.928409	−0.000132	0.0300	5.928272	0.000005	0.0387
384×128	5.928310	−0.000033	0.1200	5.9282767	.0000005	0.1400

put option price $P = 5.92827717\cdots$. As long as we have such a solution, we can find the error of any solution up to the eighth decimal. In Table 7.5, the results without using the extrapolation technique for 4 meshes and the errors up to the sixth decimal are listed on the second and third columns from the left. When the numbers of intervals in the both directions is doubled, the error is reduced by a factor about $1/4$. This means that the error is $O\left(\Delta\xi^2, \Delta\tau^2\right)$. Therefore, it has a second-order convergence rate. In Table 7.6, the results and errors for the put option are given. From there, we see that the convergence rate is also second order for the put option.

A method with a high convergence rate has a better performance if the mesh size is small enough. However, if the mesh size is not small enough, it might not be true. For a fixed mesh, the computational amount of work is different for different methods. Thus, from a practical point of view, a method should be judged by its performance. Therefore, we also list the CPU time needed to perform such a computation on a Space Ultra 10 computer for each mesh in Tables 7.5 and 7.6.

Using these data on errors and CPU times in Tables 7.5 and 7.6, the data given for PEFDII, Binomial, PSOR, and PIFDII in Chapter 6, the graphs of $\log_{10}$(CPU time in sec.) versus $\log_{10}$(error) for call and put options are plotted in Figs. 7.2 and 7.3, respectively. On these two figures, the lower the point, the better the performance because a lower point means that for a fixed

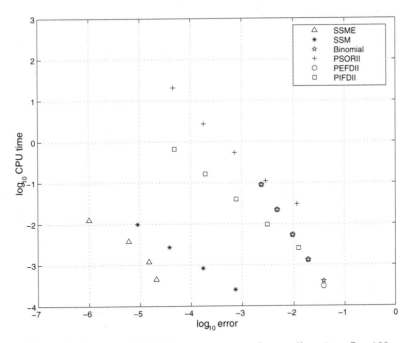

Fig. 7.2. Graphs of CPU time versus error for a call option, $S = 100$

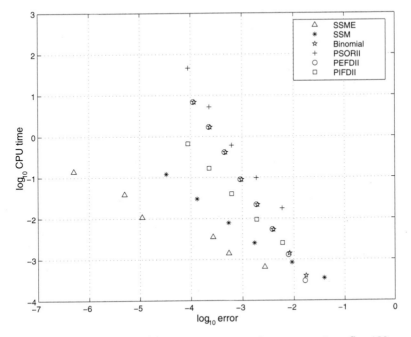

Fig. 7.3. Graphs of CPU time versus error for a put option, $S = 100$

error, it needs less CPU time. From there, we can see that the singularity-separating method (SSM) has the best performance for these two cases if the error required is less than 10^{-2}. Moreover, the higher the accuracy required, the greater the advantage of the SSM.

If the SSM is combined with the extrapolation technique, then the performance is even better. In order to explain this, the results, errors, and CPU times when the SSM is combined with the extrapolation technique are listed in the right three columns of Tables 7.5 and 7.6, and the corresponding graphs of $\log_{10}(\text{CPU time in sec.})$ versus $\log_{10}(\text{error})$ are also plotted in Figs. 7.2 and 7.3. There, SSME stands for the singularity-separating method with the extrapolation technique. From here, we can see that the extrapolation technique is very useful. At the beginning of this subsection, we showed that for various parameters, the SSM with the extrapolation technique could give very good results on quite coarse meshes. This is because due to the error function being quite smooth, the extrapolation technique is always helpful when combined with SSM.

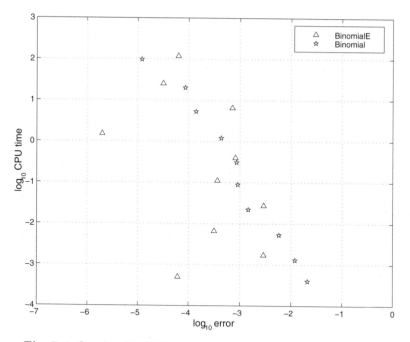

Fig. 7.4. Graphs of CPU time versus error for a call option, $S = 110$

Here, we would like to point out that the extrapolation technique is not always helpful. Let us find out if the performance is improved when the binomial method is combined with the extrapolation technique. Consider a call option with $r = 0.1$, $\sigma = 0.2$, $D_0 = 0.05$, $T = 1$, and $E = 100$. In Fig. 7.4

Table 7.7. American call option (binomial method)

$(r = 0.1, \sigma = 0.2, D_0 = 0.05, T = 1$ year, $E = 100$, $S = 110$, and the exact value $= 16.8016638\cdots)$

Numbers of	Without extrapolation			With extrapolation		
time steps	Results	Errors	CPU(sec.)	Results	|Errors|	CPU(sec.)
50	16.822670	−0.021006	0.0004	16.801602	0.000062	0.0005
100	16.813618	−0.011954	0.0013	16.804566	0.002902	0.0017
200	16.807482	−0.005818	0.0053	16.801346	0.000318	0.0066
400	16.803114	−0.001450	0.0220	16.798746	0.002918	0.0273
800	16.802573	−0.000909	0.0880	16.802032	0.000370	0.1110
1600	16.802526	−0.000862	0.3100	16.802479	0.000817	0.3980
3200	16.802096	−0.000432	1.2000	16.801666	0.000002	1.5100
6400	16.801525	+0.000139	5.2700	16.800953	0.000710	6.4700
12800	16.801578	+0.000086	20.100	16.801632	0.000032	25.370
25600	16.801652	−0.000012	97.600	16.801727	0.000063	117.70

for $S = 110$ we plot the graphs of $\log_{10}$(CPU time in sec.) versus $\log_{10}$(error) for the binomial method with and without extrapolation. There, 'Binomial' and 'BinomialE' mean the binomial method and the binomial method with extrapolation technique. From there, we can see that on some meshes, the extrapolation technique improves the results, but on other meshes, it makes the results worse. In order to have some details about why this happens, the data of the errors and the CPU times are listed for the two cases in Table 7.7. As a first-order method, the error should be reduced by a factor about $1/2$ when the number of time steps is doubled. Because the error function is not smooth due to the non-smoothness of the solution, from the table we see that from one mesh size to another, the error before extrapolation does not always show such a property and sometimes the sign of the error even changes. Thus, when the extrapolation technique is used, the error increases for some cases if the sign is unchanged and always increases if the sign changes. This phenomena occurs even if the mesh size is very small. Therefore, the extrapolation technique is not always helpful for the binomial method. However, Broadie and Detemple in [13] suggested an improved binomial method called the binomial Black and Scholes method (BBS). Examples show that the error of BBS decreases and does not change its sign when the mesh size decreases. As long as it is true, the extrapolation technique is helpful for the BBS method.

Finally, in this subsection we give two graphs on the location of the free boundaries. In Figs. 7.5 and 7.6, the location of the free boundaries is plotted for three call options and three put options, respectively. There, $E = 100, \sigma = 0.24$, and $t = 0 \sim 10$. The other parameters for the three call options are $(r = 0, D_0 = 0.06)$, $(r = 0.06, D_0 = 0.06)$, and $(r = 0.06, D_0 = 0.03)$, and for the three put options they are $(r = 0.06, D_0 = 0)$, $(r = 0.06, D_0 = 0.06)$, and $(r = 0.03, D_0 = 0.06)$. For all the cases, the location of the free boundary moves quite fast at $t \approx T$. Therefore, the time step at $t \approx T$ should be smaller

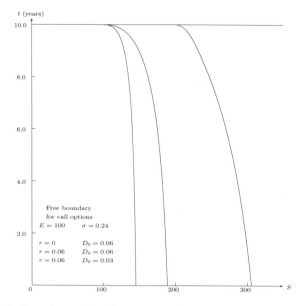

Fig. 7.5. Locations of free boundaries of call options in the (S, t)-plane
(The parameters for these curves from the left to the right are
$(r=0, D_0=0.06)$, $(r=0.06, D_0=0.06)$, and $(r=0.06, D_0=0.03)$))

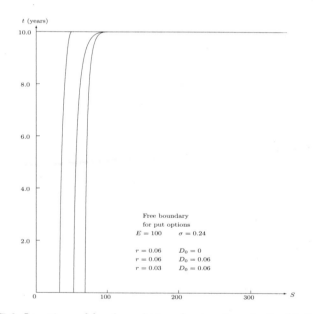

Fig. 7.6. Locations of free boundaries of put options in the (S, t)-plane.
(The parameters for these curves from the right to the left are
$(r=0.06, D_0=0)$, $(r=0.06, D_0=0.06)$, and $(r=0.03, D_0=0.06)$))

than the time step at $t << T$. In order to make computation more efficient, the time step used for all the numerical results in this subsection is not constant. When we need to find the solution for $\tau \in [0, T]$ and the total number of time step is N, then τ^n is determined by the formula

$$\tau^n = \frac{n^2}{N^2}T, \quad n = 0, 1, \cdots, N$$

and from τ^n to τ^{n+1}, the time step is $\tau^{n+1} - \tau^n$.

7.2.4 Solution and Numerical Results of Exotic Options

After making a slight change, the implicit finite-difference method described in Subsection 7.2.1 still can be used for computing free-boundary problems for American-style barrier, Asian, and lookback options. Here, we first show some results for American barrier and lookback options. Then, we discuss some modifications we have used when we compute the prices of American-style Asian options and give some results on Asian options.

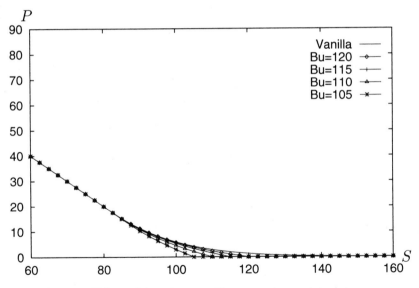

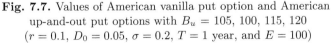

Fig. 7.7. Values of American vanilla put option and American up-and-out put options with $B_u = 105, 100, 115, 120$ ($r = 0.1$, $D_0 = 0.05$, $\sigma = 0.2$, $T = 1$ year, and $E = 100$)

For $S = 60 \sim 160$, the prices of American down-and-out call options with $B_l = 80, 85, 90, 95$ and the price of American down-and-out call option with $B_l = 0$ – the price of the American vanilla call option – have been shown

Table 7.8. American down-and-out call option

$(r = 0.1, \sigma = 0.20, D_0 = 0.05, T = 1, \text{ and } E = 100)$

S	Vanilla	$B_l = 80$	$B_l = 85$	$B_l = 90$	$B_l = 95$
80	1.769	0	0	0	0
85	3.057	2.181	0	0	0
90	4.843	4.418	3.165	0	0
95	7.145	6.943	6.242	4.251	0
100	9.941	9.846	9.464	8.243	5.361
105	13.182	13.138	12.934	12.202	10.292
110	16.802	16.782	16.674	16.244	15.005
115	20.728	20.719	20.663	20.415	19.626
120	24.893	24.889	24.861	24.720	24.226

in Fig. 3.1. There, the parameters are $r = 0.1$, $D_0 = 0.05$, $\sigma = 0.2$, $T = 1$ year, and $E = 100$. Here, for $S = 60 \sim 160$ and for the same parameters, the prices of American up-and-out put options with $B_u = 105, 110, 115, 120$ and the price of American up-and-out put option with $B_u = \infty$ – the price of the American vanilla put option – are represented in Fig. 7.7. From these curves, we see again that the price of a barrier option is less than a vanilla option. The reason is still that the holder of a barrier option has less rights than a holder of a vanilla option. In Subsection 3.2.3, we have pointed out that for call options, the higher the lower barrier B_l, the less the rights and the cheaper the option. Here, we give some data to show how big the difference between the barrier options and the vanilla options is. In Table 7.8, the prices of the American down-and-out and vanilla call options for $S = 80, 85, 90, 95$, 100, 105, 110, 115, and 120 are listed. From the data, we can see that the difference is significant for most of the cases.

In Section 3.4, for an American lookback strike call option, the values $W(\eta, t)$ as functions of η for $t = 0, 0.2, 0.4, 0.6, 0.8$ are shown in Fig. 3.6. Here, for an American lookback strike put option, similar curves are represented in Fig. 7.8. From this figure, we know that $W(\eta, t) = V(S, H, t)/S$ is an increasing function in $\eta = H/S$. That is, if S is fixed, then $V(S, H, t)$ is an increasing function in H. This is because the payoff $\max(H - S, 0)$ increases for $S \leq H$ as H increases. The highest price up to time t is of course greater than or equal to the price at time t. Thus, $\eta = H/S$ must be greater than or equal to 1. Consequently, $W(\eta, t)$ is defined only for $\eta \geq 1$ and for a fixed t, the price of the option has a minimum at $\eta = 1$. In Fig. 7.8, we can observe this being true and the value of $W(\eta, t)$ at $\eta = 1$ and $t = 0$ being about 0.16. This means that the minimum price at $t = 0$ is about 16% (the actual value is 16.37%) of S. From the last subsection, we know that the value of the vanilla put option with $S = E$ is 5.93% of S. Hence, the price of an American lookback strike put option is much higher than the price of an American vanilla put option. The reason is that the holder of an American lookback strike put option can sell a stock at any time t for the maximum price during the time interval $[0, t]$, whereas a holder of an American vanilla put option can sell a stock at any time t for the price at time t that is always less than or equal to the maximum price during the time interval $[0, t]$.

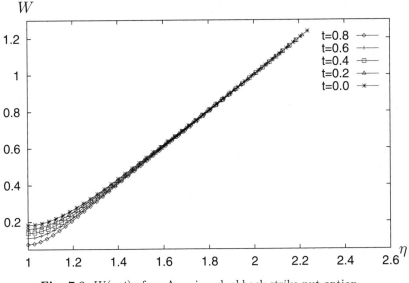

Fig. 7.8. $W(\eta, t)$ of an American lookback strike put option
($r = 0.05$, $D_0 = 0.1$, and $\sigma = 0.2$)

In Fig. 7.9, the location of the free boundary of the American lookback strike put option is given. In Fig. 3.7, a similar result for a call is represented. In Subsection 2.6.1, it has been shown that the locations of free boundaries for vanilla options are monotone functions in t. In fact, this is also true for American lookback strike options. Figures 3.7 and 7.9 show this fact. In Subsection 2.6.1, we also have pointed out that the monotonicities of the free boundary $\eta_f(t)$ and of the price with respect to t are related. This reflects that $W(\eta, t)$ should be monotone functions of t for any fixed η. Figures 3.6 and 7.8 show this feature.

For details on how to compute American barrier and lookback options by using SSM, see [17] and [83].

Now let us look at average options. In Fig. 7.10, the line with $*$ gives the solution $W(\eta, 0)$ for an American average strike call option by the singularity-separating method with the implicit finite-difference method (SSMIMP), similar to that described in Subsection 7.2.1. The result of a put option with the same parameters is given in Fig. 7.11 also by a curve with $*$. These two curves are almost horizontal straight lines except near one of the boundaries because

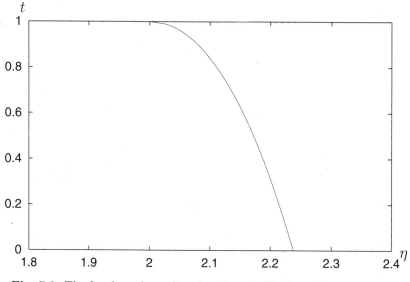

Fig. 7.9. The free boundary of an American lookback strike put option
$(r = 0.05,\ D_0 = 0.1,\ \text{and}\ \sigma = 0.2)$

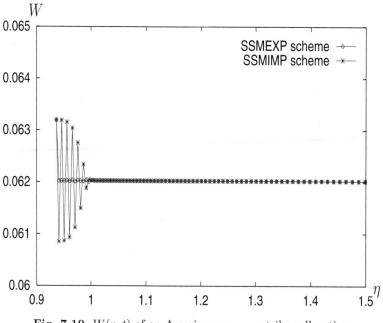

Fig. 7.10. $W(\eta, t)$ of an American average strike call option
$(r = 0.1,\ D_0 = 0.1,\ \sigma = 0.2,\ \text{and}\ t = 0)$

there is a term $(1 - \eta)/t$ in the partial differential equation of Asian options. Actually, this boundary is the free boundary. Near the free boundary, the exact solution changes very rapidly, and the numerical solution has oscillations. At time $t = 0$, the average price of the stock is always equal to the price of the stock, so we actually only need the value of $W(\eta, t)$ at $\eta = A/S = 1$, which is the level of the horizontal straight line. Therefore, we can still have a good result for $W(1, t)$ by finding the level of the horizontal straight line. However, in order to get rid of the oscillations and make the entire result nicer, we use the following scheme to approximate the first and second derivatives with respect to ξ in the partial differential equation.

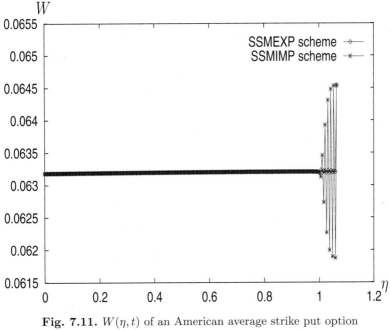

Fig. 7.11. $W(\eta, t)$ of an American average strike put option
$(r = 0.1,\ D_0 = 0.1,\ \sigma = 0.2,\ \text{and}\ t = 0)$

Let us consider the equation:

$$a_i^{n+1/2} \frac{\partial^2 U}{\partial \xi^2} + b_i^{n+1/2} \frac{\partial U}{\partial \xi} + c_i^{n+1/2} U = 0.$$

Its characteristic equation is

$$a_i^{n+1/2} \lambda^2 + b_i^{n+1/2} \lambda + c_i^{n+1/2} = 0. \tag{7.36}$$

When $a_i^{n+1/2} > 0$ and $c_i^{n+1/2} < 0$, it has two distinct real roots:

$$\begin{cases} \lambda_{1,i} = \dfrac{-b_i^{n+1/2} + \sqrt{\left(b_i^{n+1/2}\right)^2 - 4a_i^{n+1/2} c_i^{n+1/2}}}{2a_i^{n+1/2}}, \\[3ex] \lambda_{2,i} = \dfrac{-b_i^{n+1/2} - \sqrt{\left(b_i^{n+1/2}\right)^2 - 4a_i^{n+1/2} c_i^{n+1/2}}}{2a_i^{n+1/2}}. \end{cases}$$

Let

$$\varphi(\xi) = e^{\lambda_{1,i}(\xi - \xi_i)}, \quad \psi(\xi) = e^{\lambda_{2,i}(\xi - \xi_i)} \tag{7.37}$$

be the local basis functions. Then, on a subinterval $[\xi_{i-1}, \xi_{i+1}]$ near ξ_i, a function $W(\xi, \tau^{n+1/2})$ can be approximated by

$$\alpha_i \varphi(\xi) + \beta_i \psi(\xi) + \gamma_i, \tag{7.38}$$

where α_i, β_i, and γ_i are determined by the following conditions:

$$\begin{cases} \alpha_i \varphi(\xi_{i-1}) + \beta_i \psi(\xi_{i-1}) + \gamma_i = W_{i-1}^{n+1/2}, \\[1ex] \alpha_i \varphi(\xi_i) + \beta_i \psi(\xi_i) + \gamma_i = W_i^{n+1/2}, \\[1ex] \alpha_i \varphi(\xi_{i+1}) + \beta_i \psi(\xi_{i+1}) + \gamma_i = W_{i+1}^{n+1/2}. \end{cases}$$

From these conditions, we have

$$\begin{cases} \alpha_i = \alpha_{1,i} W_{i-1}^{n+1/2} + \alpha_{2,i} W_i^{n+1/2} + \alpha_{3,i} W_{i+1}^{n+1/2}, \\[1ex] \beta_i = \beta_{1,i} W_{i-1}^{n+1/2} + \beta_{2,i} W_i^{n+1/2} + \beta_{3,i} W_{i+1}^{n+1/2}, \\[1ex] \gamma_i = W_i^{n+1/2} - \alpha_i - \beta_i, \end{cases} \tag{7.39}$$

where

$$\begin{aligned} \alpha_{1,i} &= [\psi(\xi_{i+1}) - \psi(\xi_i)] / G_i, \\ \alpha_{2,i} &= [\psi(\xi_{i-1}) - \psi(\xi_{i+1})] / G_i, \\ \alpha_{3,i} &= [\psi(\xi_i) - \psi(\xi_{i-1})] / G_i, \\ \beta_{1,i} &= [\varphi(\xi_i) - \varphi(\xi_{i+1})] / G_i, \\ \beta_{2,i} &= [\varphi(\xi_{i+1}) - \varphi(\xi_{i-1})] / G_i, \\ \beta_{3,i} &= [\varphi(\xi_{i-1}) - \varphi(\xi_i)] / G_i, \\ G_i &= [\varphi(\xi_{i-1}) - \varphi(\xi_i)] [\psi(\xi_{i+1}) - \psi(\xi_i)] \\ &\quad - [\varphi(\xi_{i+1}) - \varphi(\xi_i)] [\psi(\xi_{i-1}) - \psi(\xi_i)]. \end{aligned}$$

If $b_i^{n+1/2}$ is very large, then the exponential function $\psi(\xi)$ changes very rapidly. Therefore, even if $W(\xi, \tau^{n+1/2})$ changes very rapidly, as long as its behavior is close to an exponential function, (7.38) can still give a very good approximation not only for the function itself but also for its derivatives. Differentiating (7.38) with respect to ξ yields

$$\frac{\partial W}{\partial \xi} \approx \alpha_i \lambda_{1,i} \varphi(\xi) + \beta_i \lambda_{2,i} \psi(\xi), \qquad \frac{\partial^2 W}{\partial \xi^2} \approx \alpha_i \lambda_{1,i}^2 \varphi(\xi) + \beta_i \lambda_{2,i}^2 \psi(\xi). \quad (7.40)$$

Therefore, we can have the following approximation:

$$
\begin{aligned}
a_i^{n+1/2} &\frac{\partial^2 W_i^{n+1/2}}{\partial \xi^2} + b_i^{n+1/2} \frac{\partial W_i^{n+1/2}}{\partial \xi} + c_i^{n+1/2} W_i^{n+1/2} \\
&= a_i^{n+1/2} \left[\alpha_i \lambda_{1,i}^2 \varphi(\xi_i) + \beta_i \lambda_{2,i}^2 \psi(\xi_i) \right] \\
&\quad + b_i^{n+1/2} \left[\alpha_i \lambda_{1,i} \varphi(\xi_i) + \beta_i \lambda_{2,i} \psi(\xi_i) \right] + c_i^{n+1/2} W_i^{n+1/2} \\
&= -c_i^{n+1/2} \left[\alpha_i + \beta_i - W_i^{n+1/2} \right] \\
&= -c_i^{n+1/2} \left[(\alpha_{1,i} + \beta_{1,i}) W_{i-1}^{n+1/2} + (\alpha_{2,i} + \beta_{2,i}) W_i^{n+1/2} \right. \\
&\qquad\qquad \left. + (\alpha_{3,i} + \beta_{3,i}) W_{i+1}^{n+1/2} - W_i^{n+1/2} \right],
\end{aligned}
$$

where we have used the facts that $\lambda_{1,i}$ and $\lambda_{2,i}$ are roots of (7.36) and that the expressions of α_i and β_i are given by (7.39).

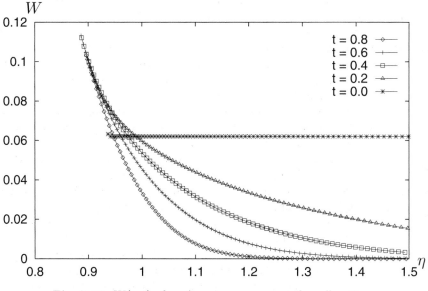

Fig. 7.12. $W(\eta, t)$ of an American average strike call option
$(r = 0.1, D_0 = 0.1, \sigma = 0.2, \text{ and } \alpha = 1)$

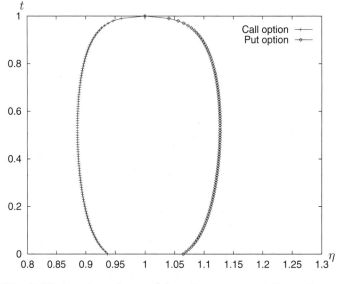

Fig. 7.13. Free boundaries of American average strike options
$(r = 0.1,\ D_0 = 0.1,\ \sigma = 0.2,\ \text{and}\ \alpha = 1)$

If the first and second derivatives with respect to ξ in the partial differential equation for an Asian option are not discretized by central schemes but by (7.40), then we have a new scheme, which is called the exponential scheme, and the method is referred to as the singularity-separating method with an exponential scheme (see [16]) and abbreviated as SSMEXP. The results of the exponential scheme are also given in Figs. 7.10 and 7.11 by the curves with ◇, which have no oscillations. From these curves, we see that this scheme improves the results. Therefore, in order to get the price of Asian options, we use this scheme. In Fig. 7.12, for an American average strike call option, the values of $W(\eta, t)$ as functions of η for $t = 0, 0.2, 0.4, 0.6, 0.8$ are given. The price of the option is $V(S, A, t) = AW(S/A, t)$. Because $A = S$ at $t = 0$, in order to find the value of the option at $S = \$100$ and $t = 0$, we need to find $\$100W(1, 0)$. From Fig. 7.12, we see that it is a little higher than $\$100 \times 0.06 = \$6.00$ (from the data we have it is $\$6.20$). In Fig. 3.3, the values of the American average strike put option with the same parameters are represented. From there, we see that the price for an American average strike put option with the same parameters at $t = 0$ is also a little higher than $\$100 \times 0.06 = \$6.00$ (from the data we have it is $\$6.32$). Thus, the difference between the call and put prices for the average options is much smaller than for the vanilla options. In Fig. 7.13, the free boundaries of the average strike call and put options are given, which shows that the locations of free boundaries are not monotone functions in t for the average strike options. This indicates that $W(\eta, t)$ is not a monotone function of t for a fixed η, which can be seen in Figs. 3.3 and 7.12.

7.2.5 Solution of Two-Dimensional Problems

In this subsection, we will discuss how to price two-factor vanilla American call options numerically. Here, "two-factor" means that both S and σ are random variables. If D_0 is not equal to zero, then pricing two-factor vanilla American options involves solving two-dimensional free-boundary problems. In what follows, we will give some details on implicit finite-difference methods for two-dimensional free-boundary problems. For the American call, the corresponding free-boundary problem is given by (7.7) or (7.9). Those problems can be converted into a problem on a rectangular domain, for example, (7.9) can be converted into (7.11). Therefore, determining the price on the domain $[0, S_f(\sigma, t)] \times [\sigma_l, \sigma_u] \times [0, T]$ can be reduced to solving (7.11) on a rectangular domain $[0, 1] \times [\sigma_l, \sigma_u] \times [0, T]$ in the (ξ, σ, τ)-space.

We use equidistant grid points on the rectangular domain. Let $\Delta\xi = 1/M$, $\Delta\sigma = (\sigma_u - \sigma_l)/I$, and $\Delta\tau = T/N$ be the mesh sizes in the ξ-, σ-, and τ-directions, respectively, where M, I and N are positive integers. We thus have $M + 1$, $I + 1$, and $N + 1$ nodes in the ξ-, σ-, and τ-directions, respectively. The $M + 1$ nodes in the ξ-direction are $\xi_m = m\Delta\xi$, $m = 0, 1, \cdots, M$, the $I + 1$ nodes in the σ-direction are $\sigma_i = \sigma_l + i\Delta\sigma$, $i = 0, 1, 2, \cdots, I$, and the $N + 1$ nodes in the τ-direction are $\tau^n = n\Delta\tau$, $n = 0, 1, \cdots, N$. In what follows, we also define $\tau^{n+1/2} = (n + 1/2)\Delta\tau$. Let $u_{m,i}^n$ stand for the approximate value of u at $\xi = \xi_m$, $\sigma = \sigma_i$, and $\tau = \tau^n$ and $s_{f,i}^n$ denote the approximate value of s_f at $\sigma = \sigma_i$, and $\tau = \tau^n$.

If $\xi \neq 1$, $\sigma \neq \sigma_l$, and $\sigma \neq \sigma_u$, then at a point $(\xi_m, \sigma_i, \tau^{n+1/2})$, the partial differential equation in (7.11) can be discretized by the following second-order approximation:

$$
\frac{u_{m,i}^{n+1} - u_{m,i}^n}{\Delta\tau}
$$

$$
= \frac{a_1 m^2}{2} \left(u_{m+1,i}^{n+1} - 2u_{m,i}^{n+1} + u_{m-1,i}^{n+1} + u_{m+1,i}^n - 2u_{m,i}^n + u_{m-1,i}^n \right)
$$

$$
+ \frac{a_2 qm}{8\Delta\sigma} \left(u_{m+1,i+1}^{n+1} - u_{m+1,i-1}^{n+1} - u_{m-1,i+1}^{n+1} + u_{m-1,i-1}^{n+1} \right. \tag{7.41}
$$

$$
\left. + u_{m+1,i+1}^n - u_{m+1,i-1}^n - u_{m-1,i+1}^n + u_{m-1,i-1}^n \right)
$$

$$
+ \frac{a_3 q^2}{2\Delta\sigma^2} \left(u_{m,i+1}^{n+1} - 2u_{m,i}^{n+1} + u_{m,i-1}^{n+1} + u_{m,i+1}^n - 2u_{m,i}^n + u_{m,i-1}^n \right)
$$

$$
+ \frac{a_4 m}{4} \left(u_{m+1,i}^{n+1} - u_{m-1,i}^{n+1} + u_{m+1,i}^n - u_{m-1,i}^n \right)
$$

$$
+ \frac{a_5}{4\Delta\sigma} \left(u_{m,i+1}^{n+1} - u_{m,i-1}^{n+1} + u_{m,i+1}^n - u_{m,i-1}^n \right)
$$

$$
+ \frac{a_6}{2} \left(u_{m,i}^{n+1} + u_{m,i}^n \right) + a_7,
$$

$$
m = 0, 1, \cdots, M - 1, \quad i = 1, 2, \cdots, I - 1.
$$

Here, q and all the coefficients a_1–a_7 should be evaluated at ξ_m, σ_i and $\tau^{n+1/2}$ in order to guarantee second-order accuracy. For a_1–a_7, the expressions are

$$a_{1,m,i}^{n+1/2} = \frac{1}{2}(\sigma_l + i\Delta\sigma)^2$$

$$- \frac{\rho_{m,i}^{n+1/2}(\sigma_l + i\Delta\sigma)q_{m,i}^{n+1/2}}{2\Delta\sigma} \frac{(s_{f,i+1}^{n+1} - s_{f,i-1}^{n+1} + s_{f,i+1}^{n} - s_{f,i-1}^{n})}{(s_{f,i}^{n+1} + s_{f,i}^{n})}$$

$$+ \frac{1}{2}\left[\frac{q_{m,i}^{n+1/2}}{2\Delta\sigma} \frac{(s_{f,i+1}^{n+1} - s_{f,i-1}^{n+1} + s_{f,i+1}^{n} - s_{f,i-1}^{n})}{(s_{f,i}^{n+1} + s_{f,i}^{n})} \right]^2,$$

$$a_{2,m,i}^{n+1/2} = \rho_{m,i}^{n+1/2}(\sigma_l + i\Delta\sigma) - \frac{q_{m,i}^{n+1/2}}{2\Delta\sigma} \frac{(s_{f,i+1}^{n+1} - s_{f,i-1}^{n+1} + s_{f,i+1}^{n} - s_{f,i-1}^{n})}{(s_{f,i}^{n+1} + s_{f,i}^{n})},$$

$$a_{3,m,i}^{n+1/2} = \frac{1}{2},$$

$$a_{4,m,i}^{n+1/2} = \frac{2}{(s_{f,i}^{n+1} + s_{f,i}^{n})} \frac{s_{f,i}^{n+1} - s_{f,i}^{n}}{\Delta\tau} + r - D_0$$

$$- \frac{[\rho_{m,i}^{n+1/2}(\sigma_l + i\Delta\sigma)q_{m,i}^{n+1/2} + p_{m,i}^{n+1/2} - \lambda_{m,i}^{n+1/2}q_{m,i}^{n+1/2}]}{2\Delta\sigma}$$

$$\times \frac{(s_{f,i+1}^{n+1} - s_{f,i-1}^{n+1} + s_{f,i+1}^{n} - s_{f,i-1}^{n})}{(s_{f,i}^{n+1} + s_{f,i}^{n})}$$

$$+ \left[\frac{q_{m,i}^{n+1/2}}{2\Delta\sigma} \frac{(s_{f,i+1}^{n+1} - s_{f,i-1}^{n+1} + s_{f,i+1}^{n} - s_{f,i-1}^{n})}{(s_{f,i}^{n+1} + s_{f,i}^{n})} \right]^2$$

$$- \frac{1}{2}\left[\frac{q_{m,i}^{n+1/2}}{\Delta\sigma} \right]^2 \frac{s_{f,i+1}^{n+1} - 2s_{f,i}^{n+1} + s_{f,i-1}^{n+1} + s_{f,i+1}^{n} - 2s_{f,i}^{n} + s_{f,i-1}^{n}}{s_{f,i}^{n+1} + s_{f,i}^{n}},$$

$$a_{5,m,i}^{n+1/2} = p_{m,i}^{n+1/2} - \lambda_{m,i}^{n+1/2}q_{m,i}^{n+1/2},$$

$$a_{6,m,i}^{n+1/2} = -r,$$

$$a_{7,m,i}^{n+1/2} = -f(m\Delta\xi(s_{f,i}^{n+1} + s_{f,i}^{n})/2, \ \sigma_l + i\Delta\sigma, \ T - (n + 1/2)\Delta\tau).$$

At the boundaries $\sigma = \sigma_l$ and $\sigma = \sigma_u$, due to $q = 0$, the partial differential equation in (7.11) becomes

$$\frac{\partial u}{\partial \tau} = a_1 \xi^2 \frac{\partial^2 u}{\partial \xi^2} + a_4 \xi \frac{\partial u}{\partial \xi} + a_5 \frac{\partial u}{\partial \sigma} + a_6 u + a_7.$$

Just like the European case, this equation possesses hyperbolic properties in the σ-direction. Hence, we can approximate the partial differential equation in (7.11) at the boundaries $\sigma = \sigma_l$ and $\sigma = \sigma_u$ by

$$\frac{u_{m,0}^{n+1} - u_{m,0}^{n}}{\Delta \tau}$$

$$= \frac{a_1 m^2}{2} (u_{m+1,0}^{n+1} - 2u_{m,0}^{n+1} + u_{m-1,0}^{n+1} + u_{m+1,0}^{n} - 2u_{m,0}^{n} + u_{m-1,0}^{n})$$

$$+ \frac{a_4 m}{4} (u_{m+1,0}^{n+1} - u_{m-1,0}^{n+1} + u_{m+1,0}^{n} - u_{m-1,0}^{n}) \tag{7.42}$$

$$+ \frac{a_5}{4\Delta\sigma} (-u_{m,2}^{n+1} + 4u_{m,1}^{n+1} - 3u_{m,0}^{n+1} - u_{m,2}^{n} + 4u_{m,1}^{n} - 3u_{m,0}^{n})$$

$$+ \frac{a_6}{2} (u_{m,0}^{n+1} + u_{m,0}^{n}) + a_7,$$

$$m = 0, 1, \cdots, M-1$$

and

$$\frac{u_{m,I}^{n+1} - u_{m,I}^{n}}{\Delta \tau}$$

$$= \frac{a_1 m^2}{2} (u_{m+1,I}^{n+1} - 2u_{m,I}^{n+1} + u_{m-1,I}^{n+1} + u_{m+1,I}^{n} - 2u_{m,I}^{n} + u_{m-1,I}^{n})$$

$$+ \frac{a_4 m}{4} (u_{m+1,I}^{n+1} - u_{m-1,I}^{n+1} + u_{m+1,I}^{n} - u_{m-1,I}^{n}) \tag{7.43}$$

$$+ \frac{a_5}{4\Delta\sigma} (3u_{m,I}^{n+1} - 4u_{m,I-1}^{n+1} + u_{m,I-2}^{n+1}$$

$$+ 3u_{m,I}^{n} - 4u_{m,I-1}^{n} + u_{m,I-2}^{n})$$

$$+ \frac{a_6}{2} (u_{m,I}^{n+1} + u_{m,I}^{n}) + a_7,$$

$$m = 0, 1, \cdots, M-1$$

respectively. Here, $\frac{\partial u}{\partial \sigma}$ is discretized by a one-sided second-order scheme in order for all the node points involved to be in the computational domain. Here, a_1 and a_4–a_7 are also evaluated at ξ_m, σ_i and $\tau^{n+1/2}$. The formulae for a_1 and a_4–a_7 are almost the same as those given above, except that the partial derivative $\frac{\partial s_f}{\partial \sigma}$ is discretized in the same way as $\frac{\partial u}{\partial \sigma}$. That is, $\frac{\partial s_f}{\partial \sigma}$ in (7.42) is approximated by

$$\frac{-s_{f,m,2}^{n+1} + 4s_{f,m,1}^{n+1} - 3s_{f,m,0}^{n+1} - s_{f,m,2}^{n} + 4s_{f,m,1}^{n} - 3s_{f,m,0}^{n}}{4\Delta\sigma}$$

and in (7.43) by

$$\frac{3s_{f,m,I}^{n+1} - 4s_{f,m,I-1}^{n+1} + s_{f,m,I-2}^{n+1} + 3s_{f,m,I}^{n} - 4s_{f,m,I-1}^{n} + s_{f,m,I-2}^{n}}{4\Delta\sigma}.$$

From the expression for a_4, we see that because $q = 0$ at $\sigma = \sigma_l$ and $\sigma = \sigma_u$, we do not need one-sided second-order finite-difference schemes for $\frac{\partial^2 s_f}{\partial \sigma^2}$.

Noticing that the coefficients of $\frac{\partial^2 u}{\partial \xi^2}, \frac{\partial u}{\partial \xi}$ in (7.11) at $\xi = 0$ are zero, $u_{-1,i}^{n}$ does not appear in the equations (7.41)–(7.43) with $m = 0$.

At $\xi = 1$, there are two boundary conditions in (7.11). One can be written as

$$u_{M,i}^{n+1} = g(s_{f,i}^{n+1}, \tau^{n+1}), \quad i = 0, 1, 2, \cdots, I, \tag{7.44}$$

where

$$g(s_f, \tau) = s_f \left[1 - e^{-D_0 \tau} N(d_1) \right] - E \left[1 - e^{-r\tau} N(d_2) \right].$$

The other can be approximated by

$$3u_{M,i}^{n+1} - 4u_{M-1,i}^{n+1} + u_{M-2,i}^{n+1} = 2\Delta\xi h(s_{f,i}^{n+1}, \tau^{n+1}), \quad i = 0, 1, \cdots, I,$$

or

$$3g(s_{f,i}^{n+1}, \tau^{n+1}) - 4u_{M-1,i}^{n+1} + u_{M-2,i}^{n+1} = 2\Delta\xi h(s_{f,i}^{n+1}, \tau^{n+1}),$$
$$i = 0, 1, \cdots, I, \tag{7.45}$$

where

$$h(s_f, \tau) = s_f \left[1 - e^{-D_0 \tau} N(d_1) \right].$$

At $\tau = 0$, from

$$u(\xi, \sigma, 0) = 0 \quad \text{and} \quad s_f(\sigma, 0) = \max(E, rE/D_0),$$

we have

$$\begin{cases} u_{m,i}^0 = 0, & m = 0, 1, \cdots, M, \ i = 0, 1, \cdots, I, \\ s_{f,i}^0 = \max(E, rE/D_0), & i = 0, 1, \cdots, I. \end{cases} \tag{7.46}$$

For a fixed n, the system (7.41)–(7.45) consists of $(M+2)(I+1)$ equations. If $u_{m,i}^n$, $m = 0, 1, \cdots, M$, $i = 0, 1, \cdots, I$ and $s_{f,i}^n$, $i = 0, 1, \cdots, I$ are known, then in the system there are $(M+2)(I+1)$ unknowns, namely, $u_{m,i}^{n+1}$, $m = 0, 1, \cdots, M$, $i = 0, 1, \cdots, I$ and $s_{f,i}^{n+1}$, $i = 0, 1, \cdots, I$, and these unknowns can be obtained from solving the system. Because (7.46) gives $u_{m,i}^0$ for all m and i and $s_{f,i}^0$ for all i, we can have $u_{m,i}^{n+1}$, $i = 0, 1, \cdots, I$, $m = 0, 1, \cdots, M$ and $s_{f,i}^{n+1}$, $i = 0, 1, \cdots, I$ for $n = 0, 1, \cdots, N-1$ successively.

There are many ways to solve the above nonlinear system. If $s_{f,i}^{n+1}$, $i = 0, 1, \cdots, I$ are given, then the system consisting of (7.41)–(7.44) is a linear system for $u_{m,i}^{n+1}$, $m = 0, 1, \cdots, M$ and $i = 0, 1, \cdots, I$. One way to solve the system is as follows. Guessing $s_{f,i}^{n+1}$, $i = 0, 1, \cdots, I$ and solving the system (7.41)–(7.44), we get all the approximate $u_{m,i}^{n+1}$, $m = 0, 1, \cdots, M$, and $i = 0, 1, \cdots, I$. Then check if (7.45) holds. If it does, we get our solution; if not, we determine new $s_{f,i}^{n+1}$, $i = 0, 1, \cdots, I$, in the following way.

For a fixed i and the approximate $u_{M-1,i}^{n+1}$ and $u_{M-2,i}^{n+1}$, (7.45) is a nonlinear equation for $s_{f,i}^{n+1}$. Hence, we can determine the new value of $s_{f,i}^{n+1}$ by Newton's method based on (7.45):

$$s_{f,i}^{(k+1)} = s_{f,i}^{(k)} - \frac{\theta(s_{f,i}^{(k)})}{\theta'(s_{f,i}^{(k)})},$$

where $s_{f,i}^{(k)}$ is the k-th iterative value of $s_{f,i}^{n+1}$ and

$$\theta(s_{f,i}, \tau^{n+1}) = 3g(s_{f,i}, \tau^{n+1}) - 4u_{M-1,i}^{n+1} + u_{M-2,i}^{n+1} - 2\Delta\xi h(s_{f,i}, \tau^{n+1}),$$

$$\theta'(s_{f,i}, \tau^{n+1}) = 3\frac{\partial g}{\partial s_{f,i}}(s_{f,i}, \tau^{n+1}) - 2\Delta\xi\frac{\partial h}{\partial s_{f,i}}(s_{f,i}, \tau^{n+1})$$

$$= (3 - 2\Delta\xi)\left[1 - e^{-D_0\tau^{n+1}}N(d_1)\right] + \frac{2\Delta\xi}{\sigma\sqrt{2\pi\tau^{n+1}}}e^{-D_0\tau^{n+1}-d_1^2/2}$$

with $d_1 = \dfrac{\ln(s_{f,i}/E) + (r - D_0 + \sigma^2/2)\tau^{n+1}}{\sigma\sqrt{\tau^{n+1}}}.$

7.2.6 Numerical Results of Two-Factor Options

Now let us show some results obtained by the numerical method above. We use the following two stochastic volatility models:

$$d\sigma = a(b-\sigma)dt + c\frac{1 - \left(1 - 2\dfrac{\sigma - \sigma_l}{\sigma_u - \sigma_l}\right)^2}{1 - 0.975\left(1 - 2\dfrac{\sigma - \sigma_l}{\sigma_u - \sigma_l}\right)^2}\sigma dX_2, \quad \sigma_l \le \sigma \le \sigma_u \quad (7.47)$$

and

$$d\sigma = a(b-\sigma)dt + c\left[\frac{(\sigma - \sigma_l)(\sigma_u - \sigma)}{(\sigma_u - \sigma_l)^2}\right]^{1/2}\sigma dX_2, \quad \sigma_l \le \sigma \le \sigma_u, \quad (7.48)$$

where a, b, and c are positive parameters. The models (7.47) and (7.48) are referred to as Model I and Model II, respectively, in what follows. Both models are in the form of (6.77). There is only a little difference between them. In Model I, $\dfrac{\partial q(\sigma, t)}{\partial \sigma}$ is bounded on $[\sigma_l, \sigma_u]$, and the reversion conditions are reduced to (6.80) and (6.81). Clearly, $q(\sigma_l) = q(\sigma_u) = 0$, so the equality conditions in (6.80) and (6.81) holds. In this case, the inequality conditions are $a(b - \sigma_l) \ge 0$ and $a(b - \sigma_u) \le 0$, which can be combined into

$$\sigma_l \le b \le \sigma_u. \quad (7.49)$$

Consequently, when (7.49) holds, Model I satisfies the reversion conditions. For Model II, the equality conditions of (6.78) and (6.79) always hold, and the inequality conditions become

$$\begin{cases} p(\sigma_l, t) - q(\sigma_l, t)\dfrac{\partial q(\sigma_l, t)}{\partial \sigma} = a(b - \sigma_l) - 0.5c^2\sigma_l^2/(\sigma_u - \sigma_l) \ge 0, \\[2mm] p(\sigma_u, t) - q(\sigma_u, t)\dfrac{\partial q(\sigma_u, t)}{\partial \sigma} = a(b - \sigma_u) + 0.5c^2\sigma_u^2/(\sigma_u - \sigma_l) \le 0. \end{cases} \quad (7.50)$$

Therefore, in order for Model II to satisfy the reversion conditions (6.78) and (6.79), we require that (7.50) hold. In the following examples, we take $\sigma_l = 0.05$ and $\sigma_u = 0.8$.

Example 1. Here, we calculate a one-year American call option with Model I. We choose $a = 0.1$, $b = 0.06$, $c = 0.12$, $\rho = 0.2$, and $\lambda = 0$. We take 20 grid points in the ξ-direction and 20 grid points in the σ-direction and 40 time steps in the τ-direction, namely, the mesh is $20 \times 20 \times 40$. The other parameters are $E = 50$, $r = 0.1$, and $D_0 = 0.05$.

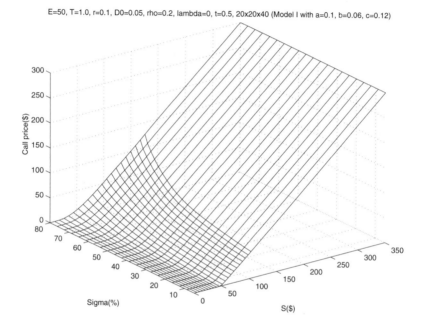

E=50, T=1.0, r=0.1, D0=0.05, rho=0.2, lambda=0, t=0.5, 20x20x40 (Model I with a=0.1, b=0.06, c=0.12)

Fig. 7.14. The American call price ($t = 0.5$, $T = 1.0$, $\rho = 0.2$, and $\lambda = 0$)

Figures 7.14 and 7.15 show the values of the American call option with $T = 1$ at time $t = 0.5$ and $t = 0$. Because those parameters a, b, c, ρ, and λ do not depend on time, Fig. 7.14 also shows the value of an option with $T = 0.5$ at time $t = 0$. Here, the strips represent the plane $C = S - E$, the solution for $S > S_f(\sigma, t)$, and the meshed surface shows the solution for $S \leq S_f(\sigma, t)$. In Figs. 7.16 and 7.17, the difference $\overline{C}$ is shown for $t = 0.5$ and 0, respectively. There, only the solution of the free-boundary problem has been shown. As we know, the derivative of C with respect to S at $t = T$ is discontinuous at $S = E$. Comparing Figs. 7.14 and 7.15, we see that the value of C becomes smoother as t decreases. However, we know from Fig. 7.15 that even at $t = 0$, for smaller σ, C still changes rapidly with respect to S near $S = E$. The difference $\overline{C}$ at $t = T$ is identically equal to zero and remains smooth as t

decreases, which can be seen from Figs. 7.16 and 7.17. Because $\overline{C}$ is much smoother than C, we can have much better numerical results if we use the partial differential equation for $\overline{C}$ instead of C when we do the computation.

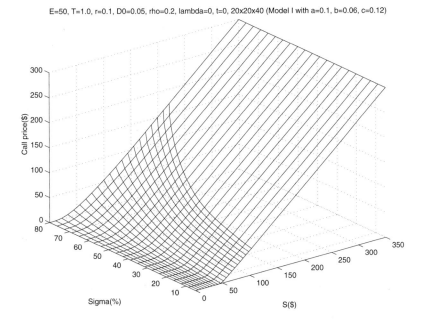

E=50, T=1.0, r=0.1, D0=0.05, rho=0.2, lambda=0, t=0, 20x20x40 (Model I with a=0.1, b=0.06, c=0.12)

Fig. 7.15. The American call price ($t = 0.0$, $T = 1.0$, $\rho = 0.2$, and $\lambda = 0$)

Example 2. In this example, we calculate a three-year American call option for Model II. All the other parameters, except ρ and λ, are the same as those in Example 1. In this case

$$a(b - \sigma_l) - 0.5c^2\sigma_l^2/(\sigma_u - \sigma_l) = 0.000976 > 0$$

and

$$a(b - \sigma_u) + 0.5c^2\sigma_u^2/(\sigma_u - \sigma_l) = -0.067856 < 0.$$

Thus, (7.50) holds, and no boundary condition is needed in order to determine the price.

First, we take $\rho = 0.2$ and $\lambda = 0$ and do the computation on different meshes. In Table 7.9, u_1 is the numerical solution using a mesh of $10 \times 10 \times 20$, u_2 is the value using a mesh of $20 \times 20 \times 40$, and u_3 is the value using a mesh of $40 \times 40 \times 80$. There, we also give results when the extrapolation technique is used. U_1^* is the extrapolation value obtained by

$$U_1^* = \frac{1}{3}(4u_3 - u_2)$$

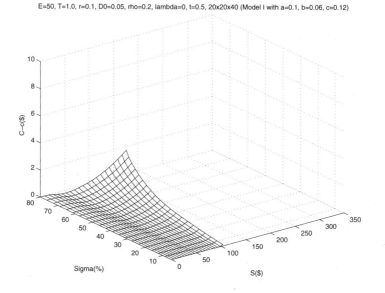

E=50, T=1.0, r=0.1, D0=0.05, rho=0.2, lambda=0, t=0.5, 20x20x40 (Model I with a=0.1, b=0.06, c=0.12)

Fig. 7.16. The difference function $\overline{C} = C - c$
$(t = 0.5,\ T = 1.0,\ \rho = 0.2,\ \text{and}\ \lambda = 0)$

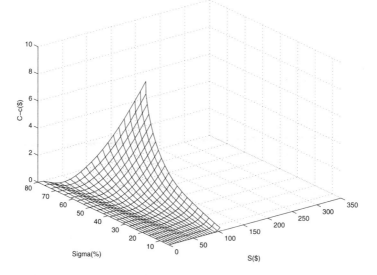

E=50, T=1.0, r=0.1, D0=0.05, rho=0.2, lambda=0, t=0, 20x20x40 (Model I with a=0.1, b=0.06, c=0.12)

Fig. 7.17. The difference function $\overline{C} = C - c$
$(t = 0.0,\ T = 1.0,\ \rho = 0.2,\ \text{and}\ \lambda = 0)$

Table 7.9. Numerical solutions with extrapolation

($E = 50$, $T = 3.0$, $r = 0.1$, $D_0 = 0.05$,
$a = 0.1$, $b = 0.06$, $c = 0.12$, $\rho = 0.2$, and $\lambda = 0$)

σ	S	u_1	u_2	u_3	U_1^*	U_2^*
0.125	45	3.93255	3.93246	3.93148	3.93115	3.93097
0.125	50	7.01873	7.02191	7.02241	7.02257	7.02253
0.125	55	10.7219	10.7224	10.7225	10.7225	10.7226
0.200	45	5.58000	5.57170	5.57137	5.57126	5.57160
0.200	50	8.49808	8.49254	8.49207	8.49191	8.49209
0.200	55	11.8781	11.8756	11.8742	11.8737	11.8736
0.350	45	8.93697	8.92810	8.92576	8.92498	8.92496
0.350	50	11.8615	11.8610	11.8607	11.8606	11.8605
0.350	55	15.1021	15.0953	15.0925	15.0916	15.0914

Table 7.10. Comparison of the results with various λ

($E = 50$, $T = 3.0$, $r = 0.1$, $D_0 = 0.05$, $a = 0.1$, $b = 0.06$, $c = 0.12$, and $\rho = 0.2$)

σ	S	$\lambda = -1.0$	$\lambda = -0.5$	$\lambda = 0$	$\lambda = 0.5$	$\lambda = 1.0$	One-factor
0.125	45	4.1031	4.0170	3.9310	3.8449	3.7589	4.1789
0.125	50	7.1681	7.0953	7.0225	6.9497	6.8770	7.2432
0.125	55	10.825	10.774	10.723	10.672	10.620	10.885
0.200	45	5.9394	5.7555	5.5716	5.3877	5.2037	6.1134
0.200	50	8.8482	8.6701	8.4921	8.3141	8.1360	9.0216
0.200	55	12.193	12.033	11.874	11.714	11.554	12.361
0.350	45	9.6937	9.3085	8.9250	8.5417	8.1576	9.9913
0.350	50	12.647	12.254	11.861	11.467	11.072	12.964
0.350	55	15.876	15.485	15.091	14.701	14.309	16.209

and U_2^* is the extrapolation value generated by

$$U_2^* = \frac{1}{21}(32u_3 - 12u_2 + u_1).$$

From the table, we see that the errors of u_1 are on the second decimal place, those of u_2 and u_3 are on the third decimal place, and for the extrapolation values U_1^* and U_2^*, they are on the fourth decimal place. This shows that the extrapolation technique increases accuracy.

Then, we take $\rho = 0.2$ and try different λ to see how the results vary. The mesh used is $40 \times 40 \times 80$. In Table 7.10, we compare the values of the options with different parameters λ. The columns with $\lambda = -1$, -0.5, 0, 0.5, 1.0 at the top contain the values of the options when $\lambda = -1$, -0.5, 0, 0.5, 1.0, respectively. For this case, the smaller the λ, the higher the call option price. The difference among the results for $\lambda \in [-1, 1]$ is about 10%–20%. This shows that we can calibrate the model to some extent even if we choose a constant λ. We also list the values of the one-factor model with a constant volatility.

From Table 7.10, we see that the one-factor model overprices the American call options.

Table 7.11. Comparison of the results with various ρ

$(E = 50, T = 3.0, r = 0.1, D_0 = 0.05, a = 0.1, b = 0.06, c = 0.12, \text{ and } \lambda = 0)$

σ	S	$\rho = 0.4$	$\rho = 0.2$	$\rho = 0$	$\rho = -0.2$	$\rho = -0.4$	One-factor
0.125	45	3.9290	3.9310	3.9329	3.9349	3.9369	4.1789
0.125	50	7.0127	7.0225	7.0323	7.0422	7.0520	7.2432
0.125	55	10.711	10.723	10.735	10.747	10.759	10.885
0.200	45	5.5735	5.5716	5.5697	5.5678	5.5659	6.1134
0.200	50	8.4815	8.4921	8.5027	8.5134	8.5240	9.0216
0.200	55	11.854	11.874	11.893	11.913	11.933	12.361
0.350	45	8.9460	8.9250	8.9038	8.8825	8.8616	9.9913
0.350	50	11.866	11.861	11.855	11.849	11.844	12.964
0.350	55	15.083	15.091	15.100	15.109	15.118	16.209

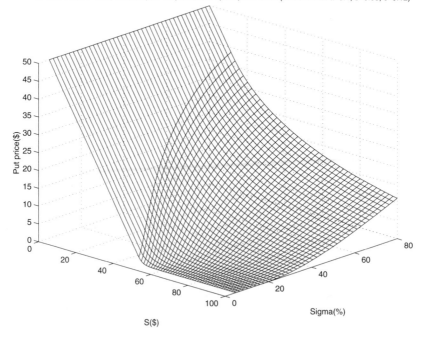

E =50, T=3.0, r=0.06, D0=0.03, rho = 0, lambda = 0, t= 0, 40x40x80 (Model II with a=0.1, b=0.06, c=0.12)

Fig. 7.18. The American put value at $t = 0.0$
$(E = 50, T = 3.0, \rho = 0, \text{ and } \lambda = 0)$

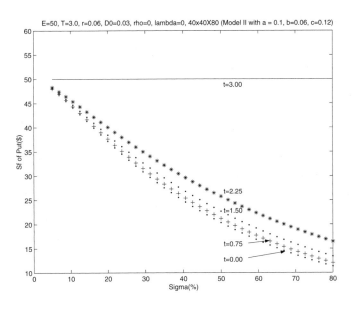

Fig. 7.19. The free boundaries of a put option for different times
$(E = 50, T = 3.0, \rho = 0, \text{ and } \lambda = 0)$

In Table 7.11, we compare the values of the options with a different correlation factor ρ and $\lambda = 0$, while the other parameters are kept unchanged. The notation is similar to Table 7.10. The results show that the option price varies a little when the correlation factor changes. Here, we again see that one factor model overprices the American call option.

An American two-factor put option problem can also be reduced to solving a free-boundary problem. However, the free-boundary problem is defined on an infinite domain. As we have pointed out, a vanilla two-factor put option can be converted into a vanilla two-factor call option with the same parameters except for r, D_0, ρ, and λ. Therefore, as long as we have a code for call options, we can also obtain the price of any put option.

Example 3. We want to have the price of a put option with $r = 0.06$, $D_0 = 0.03$, $\rho = 0$, and $\lambda = 0$ for Model II. The other parameters are the same as those in Example 2. We use a mesh $40 \times 40 \times 80$. In order to do this, we can first calculate a call option with $r = 0.03$, $D_0 = 0.06$, $\rho = 0$, and $\lambda = 0$. Then, using (7.14), we can have the price and the optimal exercise price of the put option. In Figures 7.18 and 7.19, the price of the put option at $t = 0$ for $S \in [0, 100]$ and the optimal exercise price at $t = 0, 0.75, 1.5, 2.25,$ and 3.0 are shown.

For more results and details on two-factor options, see [51] and [85]. Finally, we point out that the models given here are assumptions. In order to use such a computation in practice, the models should be found from the market data.

7.3 Pseudo-Spectral Methods

7.3.1 The Description of the Pseudo-Spectral Methods for Two-Factor Convertible Bonds

A free-boundary problem can also be solved by pseudo-spectral methods. If the solution is smooth, then the pseudo-spectral method as a high-order difference method may be more efficient. Thus, when we compute $\overline{C} = C - c$, the pseudo-spectral method might be another good tool. Also, a parabolic operator always smoothes the solution. Thus, even if the initial value is not smooth, the solution becomes smooth after a while. The life span of a convertible bond is quite long. If the time to the expiry is more than two years, the solution is already quite smooth. Thus, if expiry is not soon, then the solution of a convertible bond is quite smooth and for that time, a pseudo-spectral method might be a good choice. In the last section, we already took the American call option as an example to give the details of the implicit difference methods. In this section, we will describe the details of the pseudo-spectral method for the two-factor convertible bond problem.

In Subsection 7.1.2, a two-factor convertible bond problem with $D_0 > 0$ was reduced to (7.15) or (7.21). Suppose that we do not take the difference and want to solve $V(s, r, t)$ directly, that is, we solve the problem (7.15). Using the transformation (7.19) and defining $u(\xi, \bar{r}, \tau)$ and $s_f(\bar{r}, \tau)$ by in (7.20), we can rewrite (7.15) as the following problem on $u(\xi, \bar{r}, \tau)$ and $s_f(\bar{r}, \tau)$:

$$
\begin{cases}
\dfrac{\partial u}{\partial \tau} = \mathbf{L}_{\xi,\bar{r}} u + a_7, & 0 \le \xi \le 1, \quad 0 \le \bar{r} \le 1, \quad 0 \le \tau \le T, \\[2mm]
u(\xi, \bar{r}, 0) = \max\left(1, \xi s_f(\bar{r}, 0)\right), & 0 \le \xi \le 1, \quad 0 \le \bar{r} \le 1, \\[2mm]
u(1, \bar{r}, \tau) = s_f(\bar{r}, \tau), & 0 \le \bar{r} \le 1, \quad 0 \le \tau \le T, \\[2mm]
\dfrac{\partial u}{\partial \xi}(1, \bar{r}, \tau) = s_f(\bar{r}, \tau), & 0 \le \bar{r} \le 1, \quad 0 \le \tau \le T, \\[2mm]
s_f(\bar{r}, 0) = \max\left(1, k/D_0\right), & 0 \le \bar{r} \le 1,
\end{cases}
\tag{7.51}
$$

where $\mathbf{L}_{\xi,\bar{r}}$ is the same as given in (7.21):

$$
\mathbf{L}_{\xi,\bar{r}} = a_1 \xi^2 \frac{\partial^2}{\partial \xi^2} + a_2 \xi w \frac{\partial^2}{\partial \xi \partial \bar{r}} + a_3 w^2 \frac{\partial^2}{\partial \bar{r}^2} + \left(a_4 + \frac{1}{s_f} \frac{\partial s_f}{\partial \tau} \right) \xi \frac{\partial}{\partial \xi}
$$
$$
+ a_5 \frac{\partial}{\partial \bar{r}} + a_6
$$

and

$$
a_7 = k.
$$

Therefore, finding the value of a convertible bond is now reduced to solving a problem on a rectangular domain. Suppose that we take $M+1$ nodes in the ξ-direction: $\xi_0, \xi_1, \cdots, \xi_M$, $L+1$ nodes in the $\bar{r}$-direction: $\bar{r}_0, \bar{r}_1, \cdots, \bar{r}_L$, and $N+1$

nodes in the τ-direction: $\tau^0, \tau^1, \cdots, \tau^N$, where $\xi_0 = 0$, $\xi_M = 1$, $\bar{r}_0 = 0$, $\bar{r}_L = 1$, $\tau^0 = 0$, and $\tau^N = T$. Furthermore, we assume the nodes in the τ-direction to be equidistant with $\Delta\tau = T/N$. Let $u^n_{m,l}$ denote $u(\xi_m, \bar{r}_l, \tau^n)$ and $s^n_{f,l}$ stand for $s_f(\bar{r}_l, \tau^n)$. For a fixed n, we need to determine $u^n_{m,l}, m = 0, 1, \cdots, M$ and $l = 0, 1, \cdots, L$, and $s^n_{f,l}, l = 0, 1, \cdots, L$. In what follows, let $\left\{ u^n_{m,l} \right\}$ and $\left\{ s^n_{f,l} \right\}$ denote the sets

$$\left\{ u^n_{m,l}, m = 0, 1, \cdots, M \text{ and } l = 0, 1, \cdots, L \right\}$$

and

$$\left\{ s^n_{f,l}, l = 0, 1, \cdots, L \right\}$$

respectively. For $n = 0$, $\left\{ u^n_{m,l} \right\}$ and $\left\{ s^n_{f,l} \right\}$ are determined by the initial conditions of the problem, which gives

$$u^0_{m,l} = \max\left(1, \xi_m s^0_{f,l}\right), \quad m = 0, 1, \cdots, M, \quad l = 0, 1, \cdots, L,$$
$$s^0_{f,l} = \max\left(1, k/D_0\right), \quad l = 0, 1, \cdots, L.$$

What we need to do is to find $\left\{ u^n_{m,l} \right\}$ and $\left\{ s^n_{f,l} \right\}$ for $n = 1, 2, \cdots, N$.

According to Subsection 5.1.2, we may assume the solution on the domain $[0, 1] \times [0, 1]$ to be polynomials in each direction. Under such an assumption, for a fixed n, $\dfrac{\partial u}{\partial \xi}, \dfrac{\partial u}{\partial \bar{r}}, \dfrac{\partial^2 u}{\partial \xi^2}, \dfrac{\partial^2 u}{\partial \xi \partial \bar{r}}$, and $\dfrac{\partial^2 u}{\partial \bar{r}^2}$ at any point are linear combinations of $u^n_{m,l}, m = 0, 1, \cdots, M$ and $l = 0, 1. \cdots, L$, and $\dfrac{\partial s_f}{\partial \bar{r}}$ at any $\bar{r}$ is a linear combination of $s^n_{f,l}, l = 0, 1. \cdots, L$. Therefore, the partial differential equation and the boundary conditions in (7.51) can be discretized into algebraic equations, and solving the equations yields $\left\{ u^n_{m,l} \right\}$ and $\left\{ s^n_{f,l} \right\}$.

Now we describe the details. Suppose that for a fixed pair of n and l, $u^n_{m,l}, m = 0, 1, \cdots, M$ are known. According to these values, we can establish a polynomial in ξ with degree M. From this polynomial, we can determine $\dfrac{\partial u}{\partial \xi}, \dfrac{\partial^2 u}{\partial \xi^2}$ at any point for $\bar{r} = \bar{r}_l$ and $\tau = \tau^n$. If ξ_m is defined as follows:

$$\xi_m = \frac{1}{2}\left(1 - \cos\frac{m\pi}{M}\right), \quad m = 0, 1, \cdots, M,$$

then

$$\frac{\partial u}{\partial \xi}(\xi_m, \bar{r}_l, \tau^n) = \sum_{i=0}^{M} D_{\xi,m,i} u(\xi_i, \bar{r}_l, \tau^n),$$

$$\frac{\partial^2 u}{\partial \xi^2}(\xi_m, \bar{r}_l, \tau^n) = \sum_{i=0}^{M} D_{\xi\xi,m,i} u(\xi_i, \bar{r}_l, \tau^n)$$

and according to Subsection 5.1.2,

$$D_{\xi,m,i} = \begin{cases} \dfrac{c_m(-1)^{m+i}}{c_i(\xi_m - \xi_i)}, & m \neq i, \\[2ex] -\dfrac{2M^2 + 1}{3}, & m = i = 0, \\[2ex] \dfrac{1 - 2\xi_i}{4\xi_i(1 - \xi_i)}, & m = i = 1, 2, \cdots, M - 1, \\[2ex] \dfrac{2M^2 + 1}{3}, & m = i = M, \end{cases}$$

where $c_0 = c_M = 2$ and $c_i = 1$, $i = 1, 2, \cdots, M - 1$, and

$$D_{\xi\xi,m,i} = \sum_{k=0}^{M} D_{\xi,m,k} D_{\xi,k,i}.$$

For brevity, we define

$$\mathbf{D}_{\xi,\mathbf{m}} u_{m,l}^n = \sum_{i=0}^{M} D_{\xi,m,i} u(\xi_i, \bar{r}_l, \tau^n),$$

$$\mathbf{D}_{\xi\xi,\mathbf{m}} u_{m,l}^n = \sum_{i=0}^{M} D_{\xi\xi,m,i} u(\xi_i, \bar{r}_l, \tau^n)$$

and write the two approximations in difference operator form:

$$\frac{\partial u}{\partial \xi}(\xi_m, \bar{r}_l, \tau^n) = \mathbf{D}_{\xi,\mathbf{m}} u_{m,l}^n,$$

$$\frac{\partial^2 u}{\partial \xi^2}(\xi_m, \bar{r}_l, \tau^n) = \mathbf{D}_{\xi\xi,\mathbf{m}} u_{m,l}^n$$

Similarly, if $\bar{r}_l$ is defined by

$$\bar{r}_l = \frac{1}{2}\left(1 - \cos\frac{l\pi}{L}\right), \quad l = 0, 1, \cdots, L,$$

then

$$\frac{\partial u}{\partial \bar{r}}(\xi_m, \bar{r}_l, \tau^n) = \mathbf{D}_{\bar{r},\mathbf{l}} u_{m,l}^n,$$

$$\frac{\partial^2 u}{\partial \bar{r}^2}(\xi_m, \bar{r}_l, \tau^n) = \mathbf{D}_{\bar{r}\bar{r},\mathbf{l}} u_{m,l}^n,$$

and

$$\frac{\partial^2 u}{\partial \xi \partial \bar{r}}(\xi_m, \bar{r}_l, \tau^n) = \mathbf{D}_{\bar{r},\mathbf{l}} \mathbf{D}_{\xi,\mathbf{m}} u_{m,l}^n.$$

These difference operators are defined by

$$\mathbf{D}_{\bar{r},1} u_{m,l}^n = \sum_{j=0}^{L} D_{\bar{r},l,j} u(\xi_m, \bar{r}_j, \tau^n),$$

$$\mathbf{D}_{\bar{r}\bar{r},1} u_{m,l}^n = \sum_{j=0}^{L} D_{\bar{r}\bar{r},l,j} u(\xi_m, \bar{r}_j, \tau^n),$$

$$\mathbf{D}_{\bar{r},1} \mathbf{D}_{\xi,m} u_{m,l}^n = \sum_{j=0}^{L} D_{\bar{r},l,j} \sum_{i=0}^{M} D_{\xi,m,i} u(\xi_i, \bar{r}_j, \tau^n),$$

where

$$D_{\bar{r},l,j} = \begin{cases} \dfrac{c_l (-1)^{l+j}}{c_j (\bar{r}_l - \bar{r}_j)}, & l \neq j, \\[2mm] -\dfrac{2L^2 + 1}{3}, & l = j = 0, \\[2mm] \dfrac{1 - 2\bar{r}_j}{4\bar{r}_j (1 - \bar{r}_j)}, & l = j = 1, 2, \cdots, L - 1, \\[2mm] \dfrac{2L^2 + 1}{3}, & l = j = L, \end{cases}$$

with $c_0 = c_L = 2$ and $c_j = 1$, $j = 1, 2, \cdots, L - 1$, and

$$D_{\bar{r}\bar{r},l,j} = \sum_{k=0}^{L} D_{\bar{r},l,k} D_{\bar{r},k,j}.$$

In the τ-direction, we can approximate $\dfrac{\partial u}{\partial \tau}$ and $\dfrac{\partial s_f}{\partial \tau}$ by central differences:

$$\frac{\partial u}{\partial \tau}(\xi_m, \bar{r}_l, \tau^{n+1/2}) = \frac{u(\xi_m, \bar{r}_l, \tau^{n+1}) - u(\xi_m, \bar{r}_l, \tau^n)}{\Delta \tau},$$

$$\frac{\partial s_f}{\partial \tau}(\bar{r}_l, \tau^{n+1/2}) = \frac{s_f(\bar{r}_l, \tau^{n+1}) - s_f(\bar{r}_l, \tau^n)}{\Delta \tau}.$$

Therefore, the first equation in the problem (7.51) can be approximated by

$$\frac{u_{m,l}^{n+1} - u_{m,l}^n}{\Delta \tau}$$

$$= \frac{1}{2} \mathbf{L}_{m,l}^{n+1/2} \left(u_{m,l}^{n+1} + u_{m,l}^n \right) \tag{7.52}$$

$$+ \left(\frac{1}{s_{f,l}^{n+1} + s_{f,l}^n} \frac{s_{f,l}^{n+1} - s_{f,l}^n}{\Delta \tau} \right) \xi_m \mathbf{D}_{\xi,m} \left(u_{m,l}^{n+1} + u_{m,l}^n \right) + a_{7,m,l}^{n+1/2},$$

$$m = 0, 1, \cdots, M - 1, \quad l = 0, 1, \cdots, L.$$

Here, the operator $\mathbf{L}_{m,l}^{n+1/2}$ and the scalar $a_{7,m,l}^{n+1/2}$ are defined by

$$\mathbf{L}_{m,l}^{n+1/2} = \frac{1}{2} \left(\mathbf{L}_{m,l}^{n+1} + \mathbf{L}_{m,l}^{n} \right),$$

and

$$a_{7,m,l}^{n+1/2} = \frac{1}{2} \left(a_{7,m,l}^{n+1} + a_{7,m,l}^{n} \right),$$

where

$$\mathbf{L}_{m,l}^{n} = a_{1,m,l}^{n} \xi_m^2 \mathbf{D}_{\xi\xi,m} + a_{2,m,l}^{n} \xi_m w_{m,l}^{n} \mathbf{D}_{\bar{r},1} \mathbf{D}_{\xi,m} + a_{3,m,l}^{n} \left(w_{m,l}^{n} \right)^2 \mathbf{D}_{\bar{r}\bar{r},1}$$
$$+ a_{4,m,l}^{n} \xi_m \mathbf{D}_{\xi,m} + a_{5,m,l}^{n} \mathbf{D}_{\bar{r},1} + a_{6,m,l}^{n},$$
$$a_{i,m,l}^{n} = a_i(\xi_m, \bar{r}_l, \tau^n), \quad i = 1, 2, \cdots, 7,$$
$$w_{m,l}^{n} = w(\xi_m, \bar{r}_l, \tau^n),$$

and the derivatives $\dfrac{\partial s_f}{\partial \bar{r}}, \dfrac{\partial^2 s_f}{\partial \bar{r}^2}$ appearing in $a_1, a_2,$ and a_4 are approximated by

$$\frac{\partial s_f}{\partial \bar{r}}(\bar{r}_l, \tau^n) = \sum_{j=0}^{L} D_{\bar{r},l,j} s_f(\bar{r}_j, \tau^n),$$

$$\frac{\partial^2 s_f}{\partial \bar{r}^2}(\bar{r}_l, \tau^n) = \sum_{j=0}^{L} D_{\bar{r}\bar{r},l,j} s_f(\bar{r}_j, \tau^n).$$

The boundary conditions, the third and fourth relations in (7.51), can be discretized as follows:

$$u_{M,l}^{n+1} = s_{f,l}^{n+1}, \qquad l = 0, 1, \cdots, L, \tag{7.53}$$

$$\mathbf{D}_{\xi,M} u_{M,l}^{n+1} = s_{f,l}^{n+1}, \quad l = 0, 1, \cdots, L. \tag{7.54}$$

The system (7.52)–(7.54) has a truncation error of $O(\Delta \tau^2)$ in the τ-direction and is an M-th order scheme in the ξ-direction and an L-th order scheme in the $\bar{r}$-direction.

In the system (7.52)–(7.54), there are $(M+2)(L+1)$ equations. When $\left\{ u_{m,l}^{n} \right\}$ and $\left\{ s_{f,l}^{n} \right\}$ are given, the unknowns are $u_{m,l}^{n+1}$, $m = 0, 1, \cdots, M$, $l = 0, 1, \cdots, L$, $s_{f,l}^{n+1}$, $l = 0, 1, \cdots, L$, the total of which is also $(M+2)(L+1)$. Therefore, it is a closed system. Unfortunately, it is a nonlinear system, and we have to use iteration. Let $u_{m,l}^{(k)}, s_{f,l}^{(k)}$ represent the k-th iteration value of $u_{m,l}^{n+1}, s_{f,l}^{n+1}$, and we rewrite (7.52) in the form

$$u_{m,l}^{(k)} - \frac{\Delta \tau}{2} \bar{\mathbf{L}}_{m,l}^{(k-1)} u_{m,l}^{(k)} - \frac{s_{f,l}^{(k)}}{s_{f,l}^{(k-1)} + s_{f,l}^{n}} \xi_m \mathbf{D}_{\xi,m} \left(u_{m,l}^{(k-1)} + u_{m,l}^{n} \right)$$

$$= u_{m,l}^{n} + \frac{\Delta \tau}{2} \bar{\mathbf{L}}_{m,l}^{(k-1)} u_{m,l}^{n} - \frac{s_{f,l}^{n}}{s_{f,l}^{(k-1)} + s_{f,l}^{n}} \xi_m \mathbf{D}_{\xi,m} \left(u_{m,l}^{(k-1)} + u_{m,l}^{n} \right) + \Delta \tau a_{7,m,l},$$

$$m = 0, 1, \cdots, M-1, \quad l = 0, 1, \cdots, L, \tag{7.55}$$

where

$$\bar{\mathbf{L}}_{m,l}^{(k-1)} = \frac{1}{2}\left(\mathbf{L}_{m,l}^{(k-1)} + \mathbf{L}_{m,l}^{n}\right).$$

(7.53) and (7.54) can be written as

$$u_{M,l}^{(k)} = s_{f,l}^{(k)}, \qquad l = 0, 1, \cdots, L, \tag{7.56}$$

$$\mathbf{D}_{\xi,M}u_{M,l}^{(k)} = s_{f,l}^{(k)}, \qquad l = 0, 1, \cdots, L. \tag{7.57}$$

The system (7.55)–(7.57) is a linear one for $u_{m,l}^{(k)}$, $m = 0, 1, \cdots, M$, $l = 0, 1, \cdots, L$ and $s_{f,l}^{(k)}$, $l = 0, 1, \cdots, L$. It can be solved by a direct or iteration method. We can let $u_{m,l}^{(0)} = u_{m,l}^{n}$, $m = 0, 1, \cdots, M$, $l = 0, 1, \cdots, L$ and $s_{f,l}^{(0)} = s_{f,l}^{n}$, $l = 0, 1, \cdots, L$. When $\left\{u_{m,l}^{(k-1)}\right\}$ and $\left\{s_{f,l}^{(k-1)}\right\}$ are known, we can find $\left\{u_{m,l}^{(k)}\right\}$ and $\left\{s_{f,l}^{(k)}\right\}$ by solving the system (7.55)–(7.57). When all $u_{m,l}^{(k)} - u_{m,l}^{(k-1)}$ and $s_{f,l}^{(k)} - s_{f,l}^{(k-1)}$ become very small, we can stop the iteration. Just like the case of one-dimensional finite-difference methods, we can stop at $k = 2$, and the result should be second-order accurate in the τ-direction. This is because $\left\{u_{m,l}^{(1)}\right\}$ and $\left\{s_{f,l}^{(1)}\right\}$ can be understood as a result of a first-order scheme in τ. The results $\left\{u_{m,l}^{(2)}\right\}$ and $\left\{s_{f,l}^{(2)}\right\}$ actually are the results of a scheme in which the improved Euler method is used in the τ-direction. Therefore, if $\left\{u_{m,l}^{n}\right\}$ and $\left\{s_{f,l}^{n}\right\}$ are given, we can obtain $\left\{u_{m,l}^{n+1}\right\}$ and $\left\{s_{f,l}^{n+1}\right\}$ by solving the system (7.55)–(7.57). Because $\left\{u_{m,l}^{0}\right\}$ and $\left\{s_{f,l}^{0}\right\}$ are given by the initial conditions, we can repeat the procedure described above for $n = 0, 1, \cdots, N - 1$, and finally get $\left\{u_{m,l}^{N}\right\}$ and $\left\{s_{f,l}^{N}\right\}$.

As long as we find $\left\{u_{m,l}^{N}\right\}$ and $\left\{s_{f,l}^{N}\right\}$, for any S, r we can have the price of the convertible bond at $t = 0$ in the following way. If

$$S > Zs_f\left(\frac{r - r_l}{r_u - r_l}, T\right)\Big/n,$$

then

$$V = \max(Z, nS);$$

while

$$S < Zs_f\left(\frac{r - r_l}{r_u - r_l}, T\right)\Big/n,$$

then

$$V(S, r, 0) = Zu\left(\frac{nS}{Zs_f\left(\dfrac{r - r_l}{r_u - r_l}, T\right)}, \frac{r - r_l}{r_u - r_l}, T\right).$$

Usually, $\dfrac{r - r_l}{r_u - r_l} \neq \bar{r}_l$ for any l and $\dfrac{nS}{Zs_f\left(\dfrac{r - r_l}{r_u - r_l}, T\right)} \neq \xi_m$ for any m. In

order to find $V(S, r, 0)$, we therefore need to use interpolation.

When $t \approx T$ and $S \approx \max\left(\dfrac{Z}{n}, \dfrac{KZ}{D_0 n}\right)$, the solution in the S-direction is
not smooth. In order to overcome this problem, we need to solve (7.21) instead
of (7.51). The method for (7.21) is almost the same as the method for (7.51).
The only difference is the boundary conditions and a_7. In this case, $a_{7,m,l}$ in
(7.55) should be replaced by

$$\frac{1}{2}\left(a_{7,m,l}^{(k-1)} + a_{7,m,l}^n\right)$$

because a_7 involves the location of the free boundary. Here, $a_{7,m,l}^{(k-1)}$ is the
$(k-1)$-th iteration value of $a_{7,m,l}^{n+1}$. If we still want to solve (7.51), then at $t \approx$
T, using the finite-difference methods or using the pseudo-spectral methods
in the r-direction and using the finite-difference methods in the S-direction
might be better. The reader can find the details about how to solve the two-
factor convertible bond problems using the implicit finite-difference method
in [87] and using the mixture of the pseudo-spectral methods and the finite-
difference methods in [67]. In what follows, for brevity, we will refer to the
mixture of the pseudo-spectral method and the finite-difference method as the
pseudo-spectral method because in the entire computation, the main method
is the pseudo-spectral method. It is clear that this problem can also be solved
as a linear complementarity problem using an explicit or an implicit finite-
difference scheme.

7.3.2 Numerical Results of Two-Factor Convertible Bonds

Here, we show some numerical results of a two-factor convertible bond by
the pseudo-spectral method and compare the results by the pseudo-spectral
method with the results obtained by the finite-difference method, by the projected explicit and projected implicit finite-difference methods.

The interest rate model we adopted for the example is based on the model
used by Brennan and Schwartz (see [11]) and Druskin et al. (see [24]) even
though in practice in order to get the interest rate model, we should solve an
inverse problem by using the data on the market. Their model is

$$dr = u(r, t)dt + w(r, t)dX_2, \quad 0 \leq r,$$

where

$$\begin{cases} u(r, t) = -0.13r + 0.008 + \lambda(r, t)w(r, t), \\ w(r, t) = \sqrt{0.26r}. \end{cases}$$

We made the following modifications. We assume

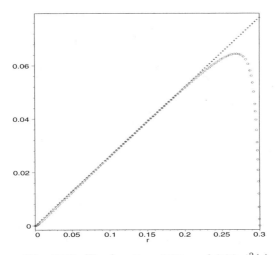

Fig. 7.20. The functions $0.26r$ and $0.26r\phi^2(r)$

$$0 \leq r \leq 0.3$$

and instead of $0.26r$, use

$$0.26r\phi^2(r),$$

where

$$\phi(r) = \frac{1 - (1 - 2r/0.3)^2}{1 - 0.975(1 - 2r/0.3)^2}.$$

Thus, our model for the example here is

$$dr = u(r,t)dt + w(r,t)dX_2, \quad 0 \leq r \leq 0.3,$$

where

$$\begin{cases} u(r,t) = -0.13r + 0.008 + \lambda(r,t)w(r,t), \\ w(r,t) = \sqrt{0.26r}\phi(r). \end{cases}$$

The functions $0.26r$ and $0.26r\phi^2(r)$ are shown in Fig. 7.20, and we can see that for $r \in [0, 0.2]$, the difference is very small. Because $\phi(0) = \phi(0.3) = 0$ and $d\phi(r)/dr$ is bounded on $[0, 0.3]$, we have

$$w(0,t)\frac{\partial w(0,t)}{\partial r} = w(0.3,t)\frac{\partial w(0.3,t)}{\partial r} = 0.$$

Therefore, the reversion conditions can be written as:

$$\begin{cases} u(0,t) \geq 0, \\ w(0,t) = 0 \end{cases}$$

and

$$\begin{cases} u(0.3, t) \leq 0, \\ w(0.3, t) = 0. \end{cases}$$

Because of $u(0, t) = 0.008 > 0$ and $u(0.3, t) = -0.13 \times 0.3 + 0.008 = -0.031 < 0$, we do not need any boundary conditions at $r = 0$ and $r = 0.3$.

We still assume the volatility and the dividend yield of the underlying stock to be

$$\sigma(S, t) = 0.20 \quad \text{and} \quad D_0 = 0.05$$

respectively, and the correlation of the two random variables dX_1 and dX_2 to be

$$\rho(S, r, t)dt = -0.01dt.$$

Let us consider a standard convertible bond with $k = 0.06$ and $T = 30$. First, we give the result obtained by the pseudo-spectral methods. Concretely, for $\tau \in [0, 2]$, in the r-direction the pseudo-spectral method described in Section 6.4 is adopted, and in the S-direction the implicit finite-difference method discussed in Subsection 7.2.1 is used, and we take $M = 60$, $L = 10$; for $\tau \in [2, 30]$ in both directions, the pseudo-spectral method is used and $M = L = 10$. In the τ-direction, a nonuniform time step is used and $N = 50$. In Fig. 7.21, the values of the two-factor convertible bond at $t = 1$ month, 6 months, 1 year, 5 years, 10 years, and 30 years are plotted. In Fig. 7.22, the location curves of the free boundary at various times are given.

Besides the method mentioned in this section, the implicit finite-difference method similar to the method in Subsection 7.2.5, the projected explicit finite-difference method and the projected implicit finite-difference method have been used to compute the same problem on various meshes. For the implicit finite-difference method, the value of the convertible bond at $r = 0.05, S = 1, t = 30$ years on a very fine mesh is $1.3116835\cdots$ [1] and these eight digits are unchanged as the mesh size further decreases. Therefore, this value is accurate to at least seven digits. After we have a highly accurate result, we can obtain the first few digits of the error of the results on different meshes. For each computation, we also record the CPU time. Thus, for each error, we can have the corresponding CPU time. Figure 7.23 is a $\log_{10}$(error) versus $\log_{10}$(CPU time in sec.) graph, and each point in the figure represents a performance of the method. Because the ranges of errors and CPU times are very large, we adopt $\log_{10}$(Error) and $\log_{10}$(CPU time in sec.) as variables. There, a "$\times$" represents the performance of the projected explicit-finite difference method, which is referred to as PEFD in the figure. A "$\circ$" indicates the performance of the projected implicit finite-difference method. The successive over relaxation method is used to get the solution. Therefore, this method is referred to as PSOR in the figure. A "$+$" stands for the performance of the implicit finite-difference method. In order to get the solution of the nonlinear algebraic equations, the alternating-direction iteration method is used (see [67]). In the figure, it is referred to as FDMI. In the figure, a "Δ" represents the

---

[1] When this figure was obtained, the function $\phi(r)$ used was $\left[4r(0.3 - r)/0.3^2\right]^{1/8}$.

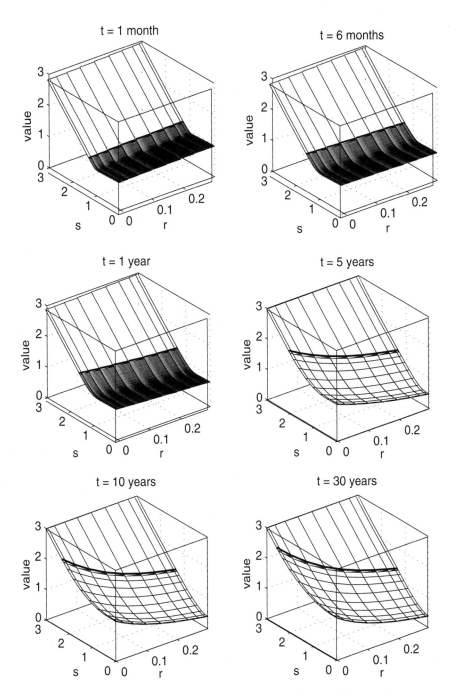

Fig. 7.21. Prices of a two-factor convertible bond at six different times

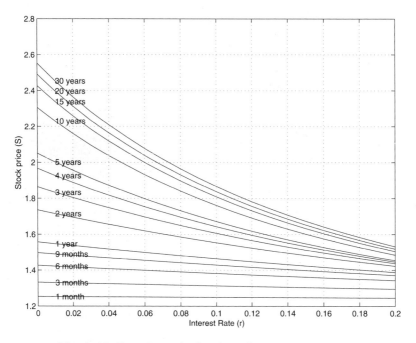

Fig. 7.22. Locations of a free boundary at various times

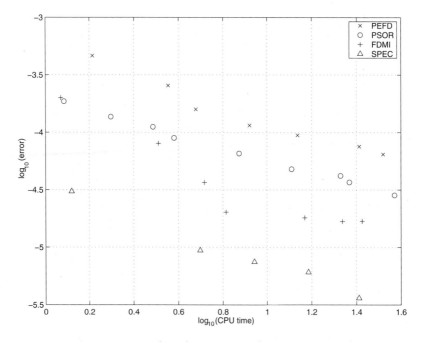

Fig. 7.23. $\log_{10}$(error) versus $\log_{10}$(CPU time in sec.)

performance of the pseudo-spectral method, which is referred to as SPEC there. Clearly, the lower the point, the better the performance. From Fig. 7.23, we see that the pseudo-spectral method has the best performance for this example.

Problems

1. Consider the following free-boundary problem that is related to American lookback strike put options:

$$
\begin{cases}
\dfrac{\partial W}{\partial t} + \dfrac{1}{2}\sigma^2\eta^2\dfrac{\partial^2 W}{\partial\eta^2} + (D_0 - r)\,\eta\dfrac{\partial W}{\partial\eta} - D_0 W = 0, \\[2mm]
\hspace{4cm} 1 \le \eta \le \eta_f(t), \quad 0 \le t \le T, \\[2mm]
W(\eta, T) = \max(\eta - \beta, 0), \quad 1 \le \eta \le \eta_f(T), \\[2mm]
\dfrac{\partial W}{\partial\eta}(1, t) = 0, \hspace{3cm} 0 \le t \le T, \\[2mm]
W(\eta_f, t) = \eta_f - \beta, \hspace{2.3cm} 0 \le t \le T, \\[2mm]
\dfrac{\partial W}{\partial\eta}(\eta_f, t) = 1, \hspace{2.8cm} 0 \le t \le T, \\[2mm]
\eta_f(T) = \beta\max(1, D_0/r).
\end{cases}
$$

Convert this problem into a problem defined on a rectangular domain and whose solution has a continuous derivative everywhere (assume $1 < \beta < \beta D_0/r$).

2. Consider the following free-boundary problem that is related to American average strike call options:

$$
\begin{cases}
\dfrac{\partial W}{\partial t} + \dfrac{1}{2}\sigma^2\eta^2\dfrac{\partial^2 W}{\partial\eta^2} + \left[(D_0 - r)\eta + \dfrac{1-\eta}{t}\right]\dfrac{\partial W}{\partial\eta} - D_0 W = 0, \\[2mm]
\hspace{5cm} \eta_f(t) \le \eta, \quad t \le T, \\[2mm]
W(\eta, T) = \max(1 - \eta, 0), \hspace{1.6cm} \eta_f(T) \le \eta, \\[2mm]
W(\eta_f, t) = 1 - \eta_f, \hspace{2.5cm} t \le T, \\[2mm]
\dfrac{\partial W}{\partial\eta}(\eta_f, t) = -1, \hspace{2.5cm} t \le T, \\[2mm]
\eta_f(T) = \min\left(1, \dfrac{1 + D_0 T}{1 + rT}\right).
\end{cases}
$$

Convert this problem into a problem defined on a rectangular domain and whose solution has a singularity weaker than the singularity here for $t \ge 0$.

3. *Let $C(S, \sigma, t; a, b, c, d)$ and $P(S, \sigma, t; a, b, c, d)$ denote the prices of American two-factor call and put options and $S_{cf}(\sigma, t; a, b, c, d)$ and $S_{pf}(\sigma, t; a, b, c, d)$ be their optimal exercise prices. Here, a, b, c, and d are parameters (or parameter functions) for the risk-free interest rate r, dividend yield rate D_0, correlation coefficient ρ, and market price of volatility risk λ, respectively. Show that between American two-factor put and call options there is the following put–call symmetry relation:

$$
\begin{cases}
P(S, \sigma, t; a, b, c, d) = \dfrac{S}{E} C\left(\dfrac{E^2}{S}, \sigma, t; b, a, -c, d - c\sigma\right), \\[2mm]
C(S, \sigma, t; a, b, c, d) = \dfrac{S}{E} P\left(\dfrac{E^2}{S}, \sigma, t; b, a, -c, d - c\sigma\right), \\[2mm]
S_{pf}(\sigma, t; a, b, c, d) \times S_{cf}(\sigma, t; b, a, -c, d - c\sigma) = E^2.
\end{cases}
$$

4. Consider the following free-boundary problem that is related to American call options:

$$
\begin{cases}
\dfrac{\partial V}{\partial \tau} - \dfrac{1}{2}\sigma^2 \xi^2 (1 - \xi)^2 \dfrac{\partial^2 V}{\partial \xi^2} - (r - D_0)\xi(1 - \xi)\dfrac{\partial V}{\partial \xi} \\[2mm]
\quad + [r(1 - \xi) + D_0\xi]V = 0, \quad 0 \le \xi < \xi_f(\tau), \quad 0 \le \tau, \\[2mm]
V(\xi, 0) = \max(2\xi - 1, 0), \quad 0 \le \xi < \xi_f(0), \\[2mm]
V\left(\xi_f(\tau), \tau\right) = 2\xi_f(\tau) - 1, \quad 0 \le \tau, \\[2mm]
\dfrac{\partial V}{\partial \xi}\left(\xi_f(\tau), \tau\right) = 2, \quad\quad\quad\quad 0 \le \tau, \\[2mm]
\xi_f(0) = \max\left(\dfrac{1}{2}, \dfrac{r}{r + D_0}\right).
\end{cases}
$$

a) Convert this problem into a problem defined on a rectangular domain and whose solution has a continuous derivative everywhere.
b) Design a second-order implicit method to solve the new problem.

5. *Consider the nonlinear system consisting of the following equations

$$
\frac{u_m^{n+1} - u_m^n}{\Delta \tau}
$$

$$
= \frac{1}{2}\left[k_2 m^2 \left(u_{m+1}^{n+1} - 2u_m^{n+1} + u_{m-1}^{n+1}\right) + \frac{k_1 m}{2}\left(u_{m+1}^{n+1} - u_{m-1}^{n+1}\right) - k_0 u_m^{n+1} \right]
$$

$$
+ \frac{1}{2}\left[k_2 m^2 \left(u_{m+1}^n - 2u_m^n + u_{m-1}^n\right) + \frac{k_1 m}{2}\left(u_{m+1}^n - u_{m-1}^n\right) - k_0 u_m^n \right]
$$

$$
+ \frac{s_f^{n+1} - s_f^n}{\left(s_f^{n+1} + s_f^n\right)\Delta \tau}\left[\frac{m}{2}\left(u_{m+1}^{n+1} - u_{m-1}^{n+1}\right) + \frac{m}{2}\left(u_{m+1}^n - u_{m-1}^n\right) \right]
$$

$$
m = 0, 1, 2, \cdots, M - 1,
$$

and

$$u_M^{n+1} = g(s_f^{n+1}, \tau^{n+1}),$$

$$\frac{3u_M^{n+1} - 4u_{M-1}^{n+1} + u_{M-2}^{n+1}}{2\Delta\xi} = h\left(s_f^{n+1}, \tau^{n+1}\right),$$

where u_m^n are known, τ^{n+1} is given, k_0, k_1, and k_2 are constants, and $g(s, \tau)$ and $h(s, \tau)$ are given functions. Discuss how to solve this system, provide at least two methods that you think are simple and effective, and give the details for one of the methods.

6. *Is the extrapolation technique always helpful and why?

7. *Design an exponential scheme to approximate

$$a(\xi)\frac{\partial^2 U}{\partial \xi^2} + b(\xi)\frac{\partial U}{\partial \xi} + c(\xi)U,$$

where $a(\xi) \geq 0$ and $c(\xi) \leq 0$.

8. Assume σ to be a random variable satisfying

$$d\sigma = p(\sigma, t)dt + q(\sigma, t)dX,$$

where dX is a Wiener process. In this case, evaluating American call options can be reduced to solving the following free-boundary problem:

$$
\begin{cases}
\dfrac{\partial C}{\partial t} + \mathbf{L}_{\mathbf{s},\sigma}C = 0, & 0 \leq S \leq S_f(\sigma, t), \\[2mm]
& \sigma_l \leq \sigma \leq \sigma_u, \quad 0 \leq t \leq T, \\[2mm]
C(S, \sigma, T) = \max(S - E, 0), & 0 \leq S \leq S_f(\sigma, T), \\[2mm]
& \sigma_l \leq \sigma \leq \sigma_u, \\[2mm]
C\left(S_f(\sigma, t), \sigma, t\right) = S_f(\sigma, t) - E, & \sigma_l \leq \sigma \leq \sigma_u, \quad 0 \leq t \leq T, \\[2mm]
\dfrac{\partial C\left(S_f(\sigma, t), \sigma, t\right)}{\partial S} = 1, & \sigma_l \leq \sigma \leq \sigma_u, \quad 0 \leq t \leq T, \\[2mm]
S_f(\sigma, T) = \max(E, rE/D_0), & \sigma_l \leq \sigma \leq \sigma_u,
\end{cases}
$$

where

$$\mathbf{L}_{\mathbf{s},\sigma} = \frac{1}{2}\sigma^2 S^2 \frac{\partial^2}{\partial S^2} + \rho\sigma S q \frac{\partial^2}{\partial S \partial \sigma} + \frac{1}{2}q^2 \frac{\partial^2}{\partial \sigma^2}$$
$$+ (r - D_0)S\frac{\partial}{\partial S} + (p - \lambda q)\frac{\partial}{\partial \sigma} - r.$$

a) *Convert this problem into a problem defined on a rectangular domain and whose solution has a singularity weaker than the singularity here.

b) *Design a second-order implicit method to solve the new problem.

c) Design a pseudo-spectral method to solve the new problem.

9. Consider the following free-boundary problem related to one-factor convertible bonds:

$$
\begin{cases}
\dfrac{\partial B_c}{\partial t} + \dfrac{1}{2}\sigma^2 S^2 \dfrac{\partial^2 B_c}{\partial S^2} + (r - D_0)S\dfrac{\partial B_c}{\partial S} - rB_c = 0, \\[2mm]
\qquad\qquad\qquad\qquad 0 \le S \le S_f(t),\ \ 0 \le t \le T, \\[2mm]
B_c(S,T) = \max(Z, nS), \qquad 0 \le S \le S_f(T), \\[2mm]
B_c(S_f(t),t) = nS_f(t), \qquad 0 \le t \le T, \\[2mm]
\dfrac{\partial B_c}{\partial S}(S_f(t),t) = n, \qquad 0 \le t \le T, \\[2mm]
S_f(T) = \dfrac{Z}{n}.
\end{cases}
$$

a) Convert this problem into a problem defined on a rectangular domain and whose solution has a continuous derivative everywhere.

b) Design a pseudo-spectral method to solve the new problem.

10. *Consider the nonlinear system consisting of the following equations:

$$
\frac{u_{m,l}^{n+1} - u_{m,l}^n}{\Delta\tau}
$$

$$
= \frac{1}{2}L_{m,l}^{n+1/2}\left(u_{m,l}^{n+1} + u_{m,l}^n\right)
$$

$$
+ \left(\frac{1}{s_{f,l}^{n+1} + s_{f,l}^n}\frac{s_{f,l}^{n+1} - s_{f,l}^n}{\Delta\tau}\right)\xi_m D_{\xi,m}\left(u_{m,l}^{n+1} + u_{m,l}^n\right) + a_{7,m,l}^{n+1/2},
$$

$$
m = 0, 1, \cdots, M - 1, \quad l = 0, 1, \cdots, L,
$$

$$
u_{M,l}^{n+1} = s_{f,l}^{n+1}, \qquad l = 0, 1, \cdots, L,
$$

and

$$
D_{\xi,M}u_{M,l}^{n+1} = s_{f,l}^{n+1}, \qquad l = 0, 1, \cdots, L.
$$

Here, $u_{m,l}^n, m = 0, 1, \cdots, M, \ l = 0, 1, \cdots, L$ and $s_{f,l}^n, l = 0, 1, \cdots, L$ are given and $u_{m,l}^{n+1}, m = 0, 1, \cdots, M, \ l = 0, 1, \cdots, L$ and $s_{f,l}^{n+1}, l = 0, 1, \cdots, L$ are unknown. In the system, $D_{\xi,m}$ is a difference operator with variable coefficients, $L_{m,l}^{n+1/2}$ is another difference operator whose coefficients depend on $s_{f,l}^{n+1}, l = 0, 1, \cdots, L$, and $a_{7,m,l}$ is a function depending on $s_{f,l}^{n+1}$, $l = 0, 1, \cdots, L$. Discuss how to solve this system and give an outline of a method that you think is simple and effective.

Projects

General Requirements

A) *Submit a code or codes in C or C++ that will work on a computer the instructor can get access to. At the beginning of the code, write down the name of the student and indicate on which computer it works and the procedure to make it work.*

B) *Each code should use an input file to specify all the problem parameters and the computational parameters for each computation and an output file to store all the results. In an output file, the name of the student, all the problem parameters, and the computational parameters should be given, so that one can know what the results are and how they were obtained. The input file should be submitted with the code.*

C) *If not specified, for each case two results are required. For the first result, a 100×50 mesh should be used. For the second result, the accuracy required is 0.001, and the mesh used should be as coarse as possible.*

D) *Submit results in form of tables. When a result is given, always provide the problem parameters and the computational parameters.*

1. **Implicit scheme (7.22)–(7.24).**   Suppose σ, r, D_0 are constant. Write a code performing implicit singularity-separating method for American calls and puts. In the code, a result of an American call option should be obtained by the implicit scheme (7.22)–(7.24), whereas a result of an American put option should be obtained through solving a corresponding call problem numerically and then using the symmetry relation.

- For American call and put options, give the results for the case: $S = 100$, $E = 100$, $T = 1$, $r = 0.1$, $D_0 = 0.05$, $\sigma = 0.2$.
- For American call and put options, give the results for the case: $S = 100$, $E = 100$, $T = 1$, $r = 0.05$, $D_0 = 0.1$, $\sigma = 0.2$.
- For American call and put options, find the results with an accuracy of 0.00001 under the help of the extrapolation technique. The problem parameters are $S = 100$, $E = 100$, $T = 1.00$, $r = 0.1$, $D_0 = 0.05$, and $\sigma = 0.2$.

8

Interest Rate Modeling

As pointed out in Section 2.9, when the spot interest rate is considered as a random variable, there is an unknown function $\lambda(r,t)$, called the market price of risk, in the governing equation. Before using the governing equation for evaluating an interest rate derivative, we have to find this function (or make some assumptions on it). This function cannot be obtained by statistics directly from the market data. In Section 4.4, the inverse problem on the market price of risk was formulated. This problem can be solved by numerical methods. However, if the problem is formulated in another way, then the inverse problem may be solved more efficiently. Therefore, in Section 8.1, we first discuss another formulation of the inverse problem and then we give numerical methods for both formulations and show some numerical examples. Then, numerical methods for one-factor interest rate derivatives are described, and some numerical results are shown in Section 8.2. Because interest rate derivative problems are so complicated, for many cases, use of multi-factor models is necessary. In the last section, we study how to price interest rate derivatives using the three-factor model and the market data.

8.1 Inverse Problems

8.1.1 Another Formulation of the Inverse Problem

As seen in Section 4.4, in order to match the bond equation with the market data, we need to find $\lambda(t)$ such that the solution $V(r,t;T^*)$ of (4.47) at $r = r^*$ and $t = 0$ is equal to today's price of the bond with maturity T^*. There, we also briefly discussed how to solve this inverse problem. Here, we reformulate the inverse problem in Section 4.4 (see [78]). This formulation may make the numerical solution easy and efficient. Let us derive this formulation. The problem (4.47) can be rewritten as follows:

$$\begin{cases} \dfrac{\partial V}{\partial t} = -\mathbf{L}_r V, & r_l \leq r \leq r_u, \quad 0 \leq t \leq T^* \leq T^*_{max}, \\[2mm] V(r, T^*; T^*) = 1, \ r_l \leq r \leq r_u. \end{cases}$$

Here, we have used the relation

$$\frac{1}{2} w^2 \frac{\partial^2 V}{\partial r^2} + (u - \lambda w) \frac{\partial V}{\partial r} - rV$$

$$= \frac{\partial}{\partial r} \left(\frac{1}{2} w^2 \frac{\partial V}{\partial r} \right) + \left[u - \left(\lambda + \frac{\partial w}{\partial r} \right) w \right] \frac{\partial V}{\partial r} - rV$$

and adopted the following notation

$$\mathbf{L}_r = \frac{\partial}{\partial r} \left(\frac{1}{2} w^2 \frac{\partial}{\partial r} \right) + \left[u - \left(\lambda + \frac{\partial w}{\partial r} \right) w \right] \frac{\partial}{\partial r} - r. \tag{8.1}$$

Let us define

$$\mathbf{L}_r^* = \frac{\partial}{\partial r} \left(\frac{1}{2} w^2 \frac{\partial}{\partial r} \right) - \frac{\partial}{\partial r} \left\{ \left[u - \left(\lambda + \frac{\partial w}{\partial r} \right) w \right] \right\} - r \tag{8.2}$$

and $U(r, t)$ satisfy the following equation:

$$\frac{\partial U}{\partial t} = \mathbf{L}_r^* U, \qquad r_l \leq r \leq r_u, \qquad 0 \leq t \leq T^*_{max}.$$

Because

$$\frac{1}{2} \int_{r_l}^{r_u} \frac{\partial}{\partial r} \left(w^2 \frac{\partial V}{\partial r} \right) U \, dr$$

$$= \frac{1}{2} \left[\left. \left(w^2 \frac{\partial V}{\partial r} U \right) \right|_{r_l}^{r_u} - \int_{r_l}^{r_u} w^2 \frac{\partial V}{\partial r} \frac{\partial U}{\partial r} \, dr \right]$$

$$= \frac{1}{2} \left[\left. \left(w^2 \frac{\partial V}{\partial r} U \right) \right|_{r_l}^{r_u} - \left. \left(w^2 \frac{\partial U}{\partial r} V \right) \right|_{r_l}^{r_u} + \int_{r_l}^{r_u} \frac{\partial}{\partial r} \left(w^2 \frac{\partial U}{\partial r} \right) V \, dr \right]$$

and

$$\int_{r_l}^{r_u} \left[u - \left(\lambda + \frac{\partial w}{\partial r} \right) w \right] \frac{\partial V}{\partial r} U \, dr$$

$$= \left. \left\{ \left[u - \left(\lambda + \frac{\partial w}{\partial r} \right) w \right] UV \right\} \right|_{r_l}^{r_u} - \int_{r_l}^{r_u} \frac{\partial}{\partial r} \left\{ \left[u - \left(\lambda + \frac{\partial w}{\partial r} \right) w \right] U \right\} V \, dr,$$

we have

$$\int_{r_l}^{r_u} \mathbf{L}_r V \, U \, dr = \int_{r_l}^{r_u} \mathbf{L}_r^* U \, V \, dr + \frac{1}{2} \left[\left. \left(w^2 \frac{\partial V}{\partial r} U \right) \right|_{r_l}^{r_u} - \left. \left(w^2 \frac{\partial U}{\partial r} V \right) \right|_{r_l}^{r_u} \right]$$

$$+ \left. \left\{ \left[u - \left(\lambda + \frac{\partial w}{\partial r} \right) w \right] UV \right\} \right|_{r_l}^{r_u}.$$

Consequently, $\mathbf{L}_r^*$ is called the adjoint operator to $\mathbf{L}_r$. Because $w(r_l, t) = w(r_u, t) = 0$ when (4.45) and (4.46) hold, we arrive at

$$\int_{r_l}^{r_u} \mathbf{L}_r V U \, dr = \int_{r_l}^{r_u} \mathbf{L}_r^* U \, V \, dr + \left\{ \left[u - \left(\lambda + \frac{\partial w}{\partial r} \right) w \right] U V \right\} \Big|_{r_l}^{r_u}. \qquad (8.3)$$

For simplicity, let us consider the case:

$$\begin{cases} u(r_l, t) - w(r_l, t) \dfrac{\partial w(r_l, t)}{\partial r} > 0, \\[2mm] u(r_u, t) - w(r_u, t) \dfrac{\partial w(r_u, t)}{\partial r} < 0. \end{cases} \qquad (8.4)$$

It is clear that when $u(r_l, t)$ and $u(r_u, t)$ are bounded and (8.4) holds, $u(r_l, t) - w(r_l, t) \dfrac{\partial w(r_l, t)}{\partial r}$ and $u(r_u, t) - w(r_u, t) \dfrac{\partial w(r_u, t)}{\partial r}$ must also be bounded even if $\dfrac{\partial w(r_l, t)}{\partial r}$ or $\dfrac{\partial w(r_u, t)}{\partial r}$ is unbounded because of $\dfrac{\partial w^2(r_l, t)}{\partial r} \geq 0$ and $\dfrac{\partial w^2(r_u, t)}{\partial r} \leq 0.$[1] In this case, in order for $\dfrac{\partial U}{\partial t} = \mathbf{L}_r^* U$ to have a unique solution, two boundary conditions are needed in addition to an initial condition. Therefore, we may add two boundary conditions on $U(r, t)$. Let us choose

$$U(r_l, t) = U(r_u, t) = 0, \quad 0 \leq t \leq T_{max}^*.$$

Under this choice, (8.3) becomes

$$\int_{r_l}^{r_u} \mathbf{L}_r V U \, dr = \int_{r_l}^{r_u} \mathbf{L}_r^* U \, V \, dr. \qquad (8.5)$$

Therefore,

$$\frac{d}{dt} \int_{r_l}^{r_u} V U \, dr = \int_{r_l}^{r_u} \frac{\partial (V U)}{\partial t} \, dr = \int_{r_l}^{r_u} \left(\frac{\partial V}{\partial t} U + V \frac{\partial U}{\partial t} \right) dr$$

$$= \int_{r_l}^{r_u} (-\mathbf{L}_r V U + \mathbf{L}_r^* U \, V) \, dr = 0,$$

from which, we further have

$$\int_{r_l}^{r_u} V(r, 0; T^*) U(r, 0) \, dr = \int_{r_l}^{r_u} V(r, T^*; T^*) U(r, T^*) \, dr.$$

Suppose we choose

$$U(r, 0) = \delta(r - r^*), \qquad r_l \leq r \leq r_u.$$

Then, noticing $V(r, T^*; T^*) = 1$, we arrive at

---

[1] This is because $w^2(r, t) \geq 0$ on $[r_l, r_u]$.

$$V(r^*, 0; T^*) = \int_{r_l}^{r_u} U(r, T^*)\, dr. \tag{8.6}$$

Consequently, for any function $\lambda(t)$, if $U(r, t)$ is the solution of the problem

$$\begin{cases} \dfrac{\partial U}{\partial t} = \mathbf{L}_r^* U, & r_l \le r \le r_u, \quad \le t \le T_{max}^*, \\[2mm] U(r, 0) = \delta(r - r^*), & r_l \le r \le r_u, \\[2mm] U(r_l, t) = U(r_u, t) = 0, & 0 \le t \le T_{max}^*, \end{cases} \tag{8.7}$$

and $V(r, t; T^*)$ is the solution of (4.47), then condition (8.6) holds.

Let today's time be $t = 0$, today's spot interest rate be r^*, and the prices of zero-coupon bonds with a face value $Z = 1$ and with various maturities T^* be $\overline{V}(T^*)$. Assume $\lambda(t)$ to be such a function that the solution $U(r, T^*)$ of (8.7) satisfies (8.6) with $V(r^*, 0; T^*) = \overline{V}(T^*)$. Then, the solution $V(r, t; T^*)$ of (4.47) at $r = r^*$ and $t = 0$ gives today's price of the zero-coupon bond with maturity T^* on the market. Consequently, matching $\lambda(t)$ with the zero-coupon bond price curve can be reduced to finding $\lambda(t)$ such that $U(r, T^*)$ satisfies (8.6) with $V(r^*, 0; T^*) = \overline{V}(T^*)$.

From (8.6), we can derive another equivalent relation that can also be used to determine $\lambda(t)$. Differentiating (8.6) with respect to T^* yields

$$\frac{\partial V(r^*, 0; T^*)}{\partial T^*} = \int_{r_l}^{r_u} \frac{\partial U(r, T^*)}{\partial T^*}\, dr = \int_{r_l}^{r_u} \mathbf{L}_r^* U(r, T^*)\, dr$$

$$= \int_{r_l}^{r_u} \left\{ \frac{\partial}{\partial r}\left(\frac{1}{2} w^2 \frac{\partial U}{\partial r} \right) - \frac{\partial}{\partial r}\left[\left(u - \left(\lambda + \frac{\partial w}{\partial r} \right) w \right) U \right] - rU \right\} dr$$

$$= -\int_{r_l}^{r_u} rU(r, T^*)\, dr$$

because $w(r_l, T^*) = w(r_u, T^*) = U(r_l, T^*) = U(r_u, T^*) = 0$. From this relation we can further have

$$\frac{\partial^2 V(r^*, 0; T^*)}{\partial T^{*2}} = -\int_{r_l}^{r_u} r \frac{\partial U(r, T^*)}{\partial T^*}\, dr = -\int_{r_l}^{r_u} r\mathbf{L}_r^* U(r, T^*)\, dr$$

$$= -\int_{r_l}^{r_u} \left(r \frac{\partial}{\partial r}\left(\frac{1}{2} w^2 \frac{\partial U}{\partial r} \right) - r \frac{\partial}{\partial r}\left\{ \left[u - \left(\lambda + \frac{\partial w}{\partial r} \right) w \right] U \right\} - r^2 U \right) dr$$

$$= -\left[\frac{r}{2} w^2 \frac{\partial U}{\partial r} \right]\Big|_{r_l}^{r_u} + \int_{r_l}^{r_u} \frac{1}{2} w^2 \frac{\partial U}{\partial r}\, dr + \left\{ r \left[u - \left(\lambda + \frac{\partial w}{\partial r} \right) w \right] U \right\}\Big|_{r_l}^{r_u}$$

$$- \int_{r_l}^{r_u} \left[u - \left(\lambda + \frac{\partial w}{\partial r} \right) w \right] U\, dr + \int_{r_l}^{r_u} r^2 U\, dr$$

$$= \int_{r_l}^{r_u} \left(\frac{1}{2} w^2 \frac{\partial U}{\partial r} + \frac{\partial w}{\partial r} wU \right) dr + \lambda \int_{r_l}^{r_u} wU\, dr + \int_{r_l}^{r_u} (r^2 - u)U\, dr$$

$$= \frac{1}{2} \int_{r_l}^{r_u} \frac{\partial (w^2 U)}{\partial r} dr + \lambda \int_{r_l}^{r_u} wU dr + \int_{r_l}^{r_u} (r^2 - u)U dr$$

$$= \lambda \int_{r_l}^{r_u} wU dr + \int_{r_l}^{r_u} (r^2 - u)U dr.$$

Consequently, λ satisfies the equation:

$$\lambda(T^*) \int_{r_l}^{r_u} wU(r, T^*) dr + \int_{r_l}^{r_u} (r^2 - u)U(r, T^*) dr = \frac{\partial^2 V(r^*, 0; T^*)}{\partial T^{*2}}. \quad (8.8)$$

Hence, instead of finding $\lambda(t)$ such that (8.6) holds, we may also find $\lambda(t)$ such that (8.8) is satisfied.

Now we discuss how to find $\lambda(t)$ from condition (8.8). From Section 4.4, the value of $\lambda(t)$ for $t \in [0, T^*]$ is determined by the portion of the zero-coupon bond price curve on $[0, T^*]$. Suppose we already have the solution of (8.7) and the value of $\lambda(t)$ for $t \in [0, T^* - \epsilon]$, ϵ being a small positive number. In order to find the value of $\lambda(t)$ for $t \in (T^* - \epsilon, T^*]$, we need to guess the value of $\lambda(t)$ for $t \in (T^* - \epsilon, T^*]$ and continue to solve the problem (8.7) from $T^* - \epsilon$ to T^* and check the condition (8.8) at any time in $(T^* - \epsilon, T^*]$. As soon as (8.8) holds, we have the value of $\lambda(t)$ on $(T^* - \epsilon, T^*]$. Such a procedure is performed from a very small T^*, gradually increasing, to $T^* = T^*_{max}$, and $\lambda(t)$ can be found for $t \in [0, T^*_{max}]$. This procedure is easy and fast, compared with the procedure of determining $\lambda(t)$ by solving the problem (4.47).

The initial-boundary value problem (8.7) is well-posed because (8.4) holds. If

$$\begin{cases} u(r_l, t) - w(r_l, t) \dfrac{\partial w(r_l, t)}{\partial r} = 0, \\[4mm] u(r_u, t) - w(r_u, t) \dfrac{\partial w(r_u, t)}{\partial r} = 0, \end{cases} \quad (8.9)$$

then from (8.3), we can still obtain (8.5) without specifying the values for $U(r_l, t)$ and $U(r_u, t)$. In this case, instead of (8.7), $U(r, t)$ is the solution of the following well-posed initial value problem

$$\begin{cases} \dfrac{\partial U}{\partial t} = \mathbf{L}^*_r U, & r_l \le r \le r_u, \quad 0 \le t \le T^*_{max}, \\[4mm] U(r, 0) = \delta(r - r^*), & r_l \le r \le r_u, \end{cases} \quad (8.10)$$

and we can still derive (8.6) and (8.8) from (8.5). For more complicated cases, the following treatment can be used. At any point on the lower boundary $r = r_l$, when

$$u(r_l, t) - w(r_l, t) \frac{\partial w(r_l, t)}{\partial r} > 0,$$

we choose $U(r_l, t) = 0$; whereas

$$u(r_l, t) - w(r_l, t) \frac{\partial w(r_l, t)}{\partial r} = 0,$$

we do not specify any value for $U(r_l, t)$ as a boundary condition. For the upper boundary, the situation is similar. Under such a treatment, (8.6) and (8.8) still hold.

8.1.2 Numerical Methods for the Inverse Problem

Again, let $\overline{V}(T^*)$ denote today's zero-coupon bond curve for bonds with a face value $Z = 1$. Suppose that the values of K zero-coupon bonds with maturities $T_1^*, T_2^*, \cdots, T_K^*$ are $V_{T_1^*}, V_{T_2^*}, \cdots, V_{T_K^*}$, which can be obtained from the market. Assume $T_K^* = T_{max}^*$ and $0 < T_1^* < \cdots < T_K^*$. Let today's time be T_0^* and $T_0^* = 0$. Clearly, $\overline{V}(T_0^*) = 1$ and $\dfrac{\partial \overline{V}(T_0^*)}{\partial T^*} = -r^*$, where r^* is today's spot interest rate. Based on the data, we can generate a zero-coupon bond price curve $\overline{V}(T^*)$ on $[0, T_{max}^*]$ by the cubic spline interpolation described in Subsection 5.1.1. Because $\dfrac{\partial \overline{V}(T_0^*)}{\partial T^*} = -r^*$, at the left end we require this condition instead of assuming $\dfrac{\partial^2 \overline{V}(T_0^*)}{\partial T^{*2}} = 0$. At the right end, we assume the function $\overline{V}(T^*)$ to be a polynomial of degree two on $[T_{K-1}^*, T_K^*]$ instead of assuming $\dfrac{\partial^2 \overline{V}(T_M^*)}{\partial T^{*2}} = 0$. Using the method described in Subsection 5.1.1 for the modified case, we can determine these polynomials on all the subintervals $[T_k^*, T_{k+1}^*]$, $k = 0, 1, \cdots, K - 1$. As soon as we have the zero-coupon bond curve, we can determine $\lambda(t)$ by solving inverse problems.

First, let us discuss how to solve the inverse problem (4.47). When $\lambda(t)$ is given on $[0, T^*]$, the partial differential equation can be discretized by (5.37). Hence, for any T^*, as long as $\lambda(t)$ is given on $[0, T^*]$, we can calculate $V(r, 0; T^*)$ from $V(r, T^*; T^*)$. Assume that we have obtained $\lambda(t)$ on $[0, T^* - \Delta t]$ from the value $\overline{V}(t)$ on $[0, T^* - \Delta t]$. We guess $\lambda(T^*)$, assume $\lambda(t)$ to be a linear function on $[T^* - \Delta t, T^*]$, and solve (4.47) from $t = T^*$ to $t = 0$. Check if $V(r^*, 0; T^*) = \overline{V}(T^*)$. If it is true, we find $\lambda(t)$ on $[T^* - \Delta t, T^*]$; if not, we adjust $\lambda(T^*)$ until we find a value $\lambda(T^*)$ such that $V(r^*, 0; T^*) = \overline{V}(T^*)$. This procedure can start from $T^* = \Delta t$ and continue successively until $T^* = T_{max}^*$. At $T^* = \Delta t$, if only $\lambda(\Delta t)$ is given, we cannot define a linear function on $[0, \Delta t]$. From (8.8), we see that $\lambda(0)$ can be determined by

$$\lambda(0) = \frac{\dfrac{\partial^2 \overline{V}(0)}{\partial T^{*2}} - r^{*2} + u(r^*, 0)}{w(r^*, 0)}. \tag{8.11}$$

Using this relation and the value $\lambda(\Delta t)$, we can define a linear function on $[0, \Delta t]$.

Now let us discuss how to solve (8.7). For the domain $[r_l, r_u] \times [0, T_{max}^*]$, we take the following partition: $r_m = r_l + m\Delta r$, $m = 0, 1, \cdots, M$, $t^n = n\Delta t$, $n = 0, 1, \cdots, N$, where $\Delta r = (r_u - r_l)/M$ and $\Delta t = T_{max}^*/N$, M, N being integers. Let U_m^n and λ^n be the approximate values of $U(r_m, t^n)$ and $\lambda(t^n)$, and $\overline{V}^n$ denote $\overline{V}(t^n)$. We also represent U_m^n, $m = 0, 1, \cdots, M$ by $\{U_m^n\}$. On this partition, (8.7) and (8.8) can be discretized as follows.

Because the initial condition in (8.7) is a Dirac delta function, we discretize the partial differential equation there by the following "conservative" scheme:

$$
\begin{aligned}
\frac{U_m^{n+1} - U_m^n}{\Delta t} \\
= \frac{1}{4\Delta r} &\left[\left(\bar{w}_{m+1/2}^{n+1/2} \right)^2 \left(\frac{U_{m+1}^{n+1} - U_m^{n+1}}{\Delta r} + \frac{U_{m+1}^n - U_m^n}{\Delta r} \right) \right. \\
&\left. - \left(\bar{w}_{m-1/2}^{n+1/2} \right)^2 \left(\frac{U_m^{n+1} - U_{m-1}^{n+1}}{\Delta r} + \frac{U_m^n - U_{m-1}^n}{\Delta r} \right) \right] \\
- &\left[\bar{u}_{m+1/2}^{n+1/2} - \left(\bar{\lambda}^{n+1/2} + \frac{w_{m+1}^{n+1/2} - w_m^{n+1/2}}{\Delta r} \right) \bar{w}_{m+1/2}^{n+1/2} \right] \\
&\times \frac{U_{m+1}^{n+1} + U_m^{n+1} + U_{m+1}^n + U_m^n}{4\Delta r} \\
+ &\left[\bar{u}_{m-1/2}^{n+1/2} - \left(\bar{\lambda}^{n+1/2} + \frac{w_m^{n+1/2} - w_{m-1}^{n+1/2}}{\Delta r} \right) \bar{w}_{m-1/2}^{n+1/2} \right] \\
&\times \frac{U_m^{n+1} + U_{m-1}^{n+1} + U_m^n + U_{m-1}^n}{4\Delta r} \\
- &\frac{r_m}{2} (U_m^{n+1} + U_m^n), \\
&\quad m = 1, 2, \cdots, M - 1,
\end{aligned}
\tag{8.12}
$$

where $\bar{w}_{m+1/2}^{n+1/2} = (w_{m+1}^{n+1/2} + w_m^{n+1/2})/2$, $\bar{u}_{m+1/2}^{n+1/2} = (u_{m+1}^{n+1/2} + u_m^{n+1/2})/2$, and $\bar{\lambda}^{n+1/2} = (\lambda^{n+1} + \lambda^n)/2$. This system can be rewritten as

$$
a_m U_{m-1}^{n+1} + b_m U_m^{n+1} + c_m U_{m+1}^{n+1} = -a_m U_{m-1}^n + (2 - b_m) U_m^n - c_m U_{m+1}^n,
$$
$$
m = 1, 2, \cdots, M - 1,
$$

where

$$
\begin{aligned}
a_m = &\frac{-\Delta t}{4\Delta r^2} \left(\bar{w}_{m-1/2}^{n+1/2} \right)^2 \\
&- \frac{\Delta t}{4\Delta r} \left[\bar{u}_{m-1/2}^{n+1/2} - \left(\bar{\lambda}^{n+1/2} + \frac{w_m^{n+1/2} - w_{m-1}^{n+1/2}}{\Delta r} \right) \bar{w}_{m-1/2}^{n+1/2} \right], \\
b_m = &1 + \frac{\Delta t r_m}{2} + \frac{\Delta t}{4\Delta r^2} \left[\left(\bar{w}_{m+1/2}^{n+1/2} \right)^2 + \left(\bar{w}_{m-1/2}^{n+1/2} \right)^2 \right] \\
&+ \frac{\Delta t}{4\Delta r} \left[\bar{u}_{m+1/2}^{n+1/2} - \left(\bar{\lambda}^{n+1/2} + \frac{w_{m+1}^{n+1/2} - w_m^{n+1/2}}{\Delta r} \right) \bar{w}_{m+1/2}^{n+1/2} \right]
\end{aligned}
$$

$$-\frac{\Delta t}{4\Delta r}\left[\bar{u}_{m-1/2}^{n+1/2} - \left(\bar{\lambda}^{n+1/2} + \frac{w_m^{n+1/2} - w_{m-1}^{n+1/2}}{\Delta r}\right)\bar{w}_{m-1/2}^{n+1/2}\right],$$

$$c_m = \frac{-\Delta t}{4\Delta r^2}\left(\bar{w}_{m+1/2}^{n+1/2}\right)^2$$

$$+\frac{\Delta t}{4\Delta r}\left[\bar{u}_{m+1/2}^{n+1/2} - \left(\bar{\lambda}^{n+1/2} + \frac{w_{m+1}^{n+1/2} - w_m^{n+1/2}}{\Delta r}\right)\bar{w}_{m+1/2}^{n+1/2}\right].$$

The initial condition in (8.7) can be approximated by

$$U_m^0 = \begin{cases} \dfrac{1}{\Delta r}\left[1 - \dfrac{r^*}{\Delta r} + \text{int}\left(\dfrac{r^*}{\Delta r}\right)\right], & m = \text{int}\left(\dfrac{r^*}{\Delta r}\right), \\[2ex] \dfrac{1}{\Delta r}\left[\dfrac{r^*}{\Delta r} - \text{int}\left(\dfrac{r^*}{\Delta r}\right)\right], & m = \text{int}\left(\dfrac{r^*}{\Delta r}\right) + 1, \\[2ex] 0, & \text{otherwise}, \end{cases} \qquad (8.13)$$

where int(x) is the integer part of the number x, and we assume $r^* \in [r_l + \Delta r, r_u - \Delta r]$.

By the trapezoidal rule (see Subsection 5.1.3), the condition (8.8) can be approximated by

$$\lambda^{n+1}\Delta r\left(\frac{1}{2}w_0^{n+1}U_0^{n+1} + \sum_{m=1}^{M-1}w_m^{n+1}U_m^{n+1} + \frac{1}{2}w_M^{n+1}u_M^{n+1}\right)$$

$$+\frac{\Delta r}{2}\left[(r_0^{n+1})^2 - u_0^{n+1}\right]U_0^{n+1} + \Delta r\sum_{m=1}^{M-1}\left[(r_m^{n+1})^2 - u_m^{n+1}\right]U_m^{n+1} \qquad (8.14)$$

$$+\frac{\Delta r}{2}\left[(r_M^{n+1})^2 - u_M^{n+1}\right]U_M^{n+1} = \frac{\partial^2\overline{V}(t^{n+1})}{\partial T^{*2}}.$$

From (8.13) and (8.11), we can have $\{U_m^0\}$ and λ^0. Therefore, we can have the following procedure for $n = 0, 1, \cdots, N - 1$ successively. Suppose we already have $\{U_m^n\}$ and λ^n. Guessing λ^{n+1}, we can obtain $\{U_m^{n+1}\}$ by solving the system consisting of (8.12) and the boundary conditions in (8.7) at $t = t^{n+1}$. Then, we check if (8.14) holds. If not, we need to find a new guess by solving λ^{n+1} from (8.14) or by other iteration methods, and obtain new $\{U_m^{n+1}\}$ and check again; if it is, we find the value λ^{n+1}. When this procedure is done for $n = 0, 1, \cdots, N - 1$ successively, we find the values for λ^n, $n = 1, 2, \cdots, N$. Another condition that can be used to determine λ^{n+1} is (8.6). The advantage of using (8.6) is to let the value of the zero-coupon bonds be exactly equal to the data from the market. In this case, we have to design an iteration method to find the next iterative value of λ^{n+1}. It is clear that if (8.7) needs to be replaced by (8.10), the procedure above is almost the same.

For the method based on (4.47), in order to do one iteration to determine λ^{n+1}, we need to integrate the partial differential equation $n + 1$ times from

t^{n+1} to t^0. For the method based on (8.7), in order to do the same thing, we need to integrate the partial differential equation only once from t^n to t^{n+1}. Therefore, we pay more attention to the method based on (8.7). The only complication is that the computation based on (8.7) involves the Dirac delta function. This requires us to use more grid points in the r-direction. In order for a function $\lambda(t)$ to be used in practice, we have to check whether or not the computed zero-coupon bond values are matched with the real market data well enough. If the formulation (4.47) is adopted, then such a condition is used directly when $\lambda(t)$ is determined. Thus, no further check is needed for this case. However, when the formulation (8.7) is used, theoretically the computed zero-coupon bond values should be consistent with the real market data if (8.6) or (8.8) holds. Because there exists numerical error, this fact will be true only if the the numerical error is controlled.

8.1.3 Numerical Results on Market Prices of Risk

Here, we give two examples on numerical results of inverse problems. Because the method of solving (8.7) is faster, we only give the results obtained by this method. As an example, we take the following spot interest rate model:

$$dr = (r^{**} - r)dt + r(0.2 - r)dX, \quad r_l = 0 \le r \le r_u = 0.2,$$

where r^{**} is a constant between r_l and r_u, and $r^{**} = 0.05345$ in these examples given here. This model satisfies conditions (4.45) and (4.46), so these partial differential equation problems we are going to solve are well-posed.

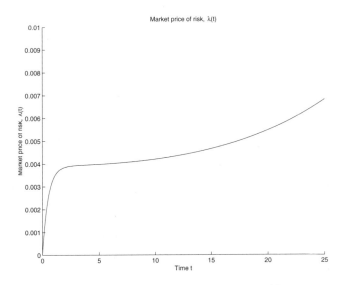

Fig. 8.1. The market price of risk $\lambda(t)$

Table 8.1. Comparison between given and computed bond prices

($V_{b,g}$ denotes given bond prices and V_b stands for computed bond prices)

T^*	0.5	1	2	3	5	7	10	15	20	25
$V_{b,g}$	97.36	94.80	89.86	85.18	76.55	68.79	58.60	44.85	34.335	26.283
V_b	97.36	94.80	89.86	85.18	76.55	68.79	58.60	44.85	34.333	26.279

Example 1. Suppose today's bond prices are given by the exponential function $100e^{-0.05345T^*}$. According to this function, we can use the method given in the last subsection to find the market price of risk $\lambda(t)$. In Fig. 8.1, the function $\lambda(t)$ is shown. As soon as we have the market price of risk, we can compute the bond price by solving the bond equation. In Table 8.1, we list both the numerical results and the values from the given function. From the table, we see that the difference is on the third decimal place, which means that the inverse problem has been solved quite accurately. In order to do this computation, a 1000×1000 mesh was used.

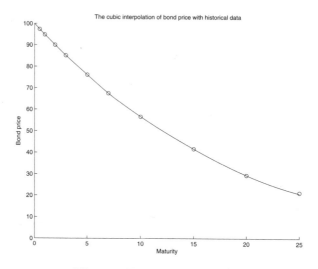

Fig. 8.2. Prices of today's bonds

Example 2. From the market, we obtained the data on the spot interest rate and the zero-coupon bond prices with maturities 0.5, 1, 2, 3, 5, 7, 10, 15, 20, and 25 years on November 30, 1995. Using the data, we generate a bond price function by the cubic spline interpolation described in Subsection 5.1.1. In Fig. 8.2, the data are given by "o," and the function is shown by a solid curve. Using the bond price function, we find the market price of risk, which is shown in Fig. 8.3. This function is not as smooth as the market price of risk given in Fig. 8.1. From (8.8), we see that λ is closely related to the second

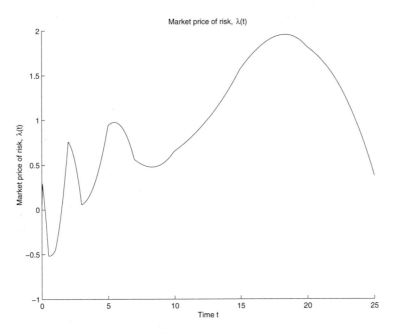

Fig. 8.3. The market price of risk λ

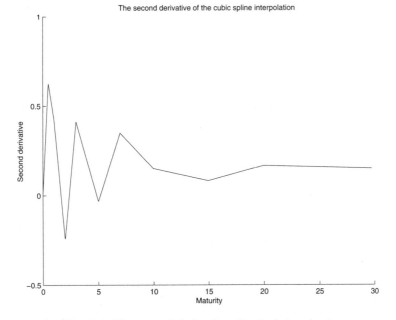

Fig. 8.4. The second derivative of today's bond prices

Table 8.2. Comparison between market and computed bond prices

($V_{b,m}$ represents market bond prices and V_b stands for computed bond prices)

T^*	0.5	1	2	3	5	7	10	15	20	25
$V_{b,m}$	97.35	94.83	90.01	85.16	76.18	67.62	56.72	41.76	29.49	21.00
V_b	97.35	94.83	90.01	85.16	76.18	67.62	56.73	41.77	29.50	21.02

derivative of today's bond curve. For this case, the second derivative of bond prices is not smooth (see Fig. 8.4), so the market price of risk has the shape shown in Fig. 8.3. Using the market price of risk, we can compute the bond price by solving the bond equation. In Table 8.2, both the computed bond prices and the bond prices on the market are listed. Their difference is also very small, which means that the inverse problem has been solved successfully even if the market data are used.

8.2 Numerical Results of One-Factor Models

In order to price interest rate derivatives, the market price of risk for the spot interest rate and today's spot interest rate r^* must be given. In this section, the market price of risk is given numerically and is based on the data from November 30, 1995. Today's spot rate is assumed to take the value of the spot interest rate on that day, namely, $r^* = 0.05345$. Also, we suppose today's time t to be zero.

First, let us briefly discuss how to price bond options. Suppose that we need to find today's price of a T-year option with an exercise price E on a $(T_b - T)$-year bond that has a face value $Z = 1$ and a coupon rate k. Let $V_b(r, t; T_b)$ and $V(r, t)$ be the prices of the bond and the option, respectively. What we need to find is $V(r^*, 0)$. In order to do this, we first need to find $V_b(r, T; T_b)$ for $r \in [r_l, r_u]$ by solving the problem (4.48):

$$
\begin{cases}
\dfrac{\partial V_b}{\partial t} + \dfrac{1}{2}w^2\dfrac{\partial^2 V_b}{\partial r^2} + (u - \lambda w)\dfrac{\partial V_b}{\partial r} - rV_b + k = 0, & r_l \le r \le r_u, \quad t \le T_b, \\
V_b(r, T_b; T_b) = 1, & r_l \le r \le r_u
\end{cases}
$$

from $t = T_b$ to $t = T < T_b$. Based on the function $V_b(r, T; T_b)$, we then obtain $V(r, 0)$ by solving the problem (4.49):

$$
\begin{cases}
\dfrac{\partial V}{\partial t} + \dfrac{1}{2}w^2\dfrac{\partial^2 V}{\partial r^2} + (u - \lambda w)\dfrac{\partial V}{\partial r} - rV = 0, & r_l \le r \le r_u, \quad t \le T, \\
V(r, T) = \max\left(V_b(r, T; T_b) - E, 0\right), & r_l \le r \le r_u
\end{cases}
$$

from $t = T$ to $t = 0$. When the market price of risk is given numerically, the problems (4.48) and (4.49) have to be solved numerically, and the scheme

(5.37) or a modified (5.31) can be adopted. The modified scheme (5.31) is a scheme that is the same as (5.31) for any interior point and the same as (5.37) for the boundary points. In Table 8.3, the numerical results of the prices on 40 bond options are listed. There, the exercise price E is equal to 0.95 and 1, and the bond pays coupons continuously with a coupon rate $k = 0.055$. The expiries of the options are 0.25, 0.5, 0.75, and 1 year and the life spans of bonds are 0.5, 1, 2, 3, and 5 years.

Table 8.3. Prices of bond options with $E = 0.95, 1$ and $k = 0.055$

E	$T\backslash T_b - T$	0.5	1	2	3	5
0.95	0.25	0.0502	0.0516	0.0536	0.0530	0.0515
0.95	0.50	0.0499	0.0514	0.0525	0.0519	0.0498
0.95	0.75	0.0495	0.0509	0.0512	0.0507	0.0478
0.95	1.00	0.0489	0.0500	0.0497	0.0494	0.0457
1.00	0.25	0.0011	0.0024	0.0044	0.0039	0.0026
1.00	0.50	0.0014	0.0029	0.0041	0.0037	0.0022
1.00	0.75	0.0017	0.0031	0.0036	0.0034	0.0016
1.00	1.00	0.0017	0.0029	0.0030	0.0029	0.0010

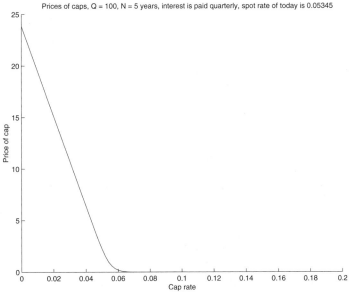

Fig. 8.5. Prices of caps

Pricing a cap is done in the following way. Consider a N-year cap and suppose that money is paid quarterly. As pointed out in Section 4.5, the cap is a sum of $4N - 1$ caplets in this case and the maturities of the bonds associated with the $4N - 1$ caplets are $t_k = k/4$, $k = 2, 3, \cdots, 4N$. Let us call the bond with maturity t_k the k-th bond, and its value is denoted by $V_{bk}(r, t)$. In order to have the value of the k-th bond, we solve the problem (4.63):

$$
\begin{cases}
\dfrac{\partial V_{bk}}{\partial t} + \dfrac{1}{2}w^2 \dfrac{\partial^2 V_{bk}}{\partial r^2} + (u - \lambda w)\dfrac{\partial V_{bk}}{\partial r} - rV_{bk} = 0, & r_l \leq r \leq r_u, \\
 & t_{k-1} \leq t \leq t_k, \\
V_{bk}(r, t_k) = (1 + r_c/4)\, Q, & r_l \leq r \leq r_u.
\end{cases}
$$

After we have all the values of the bonds, we can obtain the total value of the $4N - 1$ caplets by solving the problem (4.64):

$$
\begin{cases}
\dfrac{\partial V_c}{\partial t} + \dfrac{1}{2}w^2 \dfrac{\partial^2 V_c}{\partial r^2} + (u - \lambda w)\dfrac{\partial V_c}{\partial r} - rV_c \\
\qquad + \displaystyle\sum_{k=2}^{4N} \max\left(Q - V_{bk}(r, t_{k-1}), 0\right) \delta(t - t_{k-1}) = 0, \\
\qquad\qquad\qquad\qquad r_l \leq r \leq r_u, \quad t^* \leq t \leq t_{4N-1}, \\
V_c(r, t_{4N-1}) = 0, \qquad r_l \leq r \leq r_u.
\end{cases}
$$

The value $V_c(r^*, t^*)$ gives the premium of the cap.

The way to solve (4.63) and (4.64) numerically is similar to the way to solve (4.48), namely, by using the scheme (5.37) or modified (5.31). The only difference is that in (4.64) there exist the Dirac delta functions. In this case, the treatment of the Dirac delta function is simple: after $V_c(r, t_{k-1}^+)$ is obtained, we should let $V_c(r, t_{k-1}^-) = V_c(r, t_{k-1}^+) + \max\left(Q - V_{bk}(r, t_{k-1}), 0\right)$ and then continue the computation by using the scheme (5.37) or modified (5.31). We take $Q = 100$, $N = 5$ years, and $r_c = 0, 0.002, 0.004, \cdots, 0.2$ and find these values of caps numerically.[2] The values of caps as a function of r_c are plotted in Fig. 8.5. The curve resembles a price curve of a put option, that is, the price is a decreasing function of r_c and changes rapidly near $r_c = r^* = 0.05345$.

The way to price floors is similar to the way to price caps. For the floor rate $r_f \in [0, 0.2]$, the floor prices are shown in Fig. 8.6. Their parameters are the same as the caps. The floor resembles a call option, that is, the price is an increasing function of r_f and changes rapidly near $r_f = r^* = 0.05345$. As soon as we have the prices of a cap with a cap rate r_c and a floor with a floor rate r_f, the difference between them is the price of a collar for the pair of r_c and r_f. If the price of a cap is equal to the price of a floor, then the price of the collar with this pair of r_c and r_f is zero. In Table 8.4, eight such pairs of

---

[2]When these values of caps were computed, a cap was defined as a sum of $4N$ caplets. For those results on floors, the situation is similar.

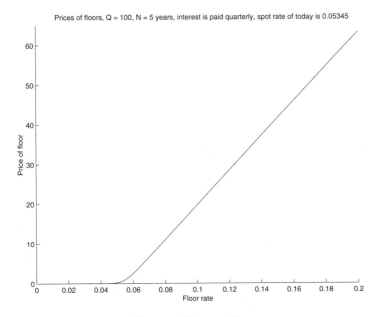

Fig. 8.6. Prices of floors

Table 8.4. Pairs of caps and floors with the same values

$(Q = 100, N = 5$ years, and the interest is paid quarterly)

r_c	r_f	Prices of caps or floors
0.05502	0.05466	0.8
0.05557	0.05422	0.7
0.05618	0.05370	0.6
0.05687	0.05310	0.5
0.05768	0.05242	0.4
0.05868	0.05160	0.3
0.05999	0.05055	0.2
0.06204	0.04899	0.1

r_c and r_f are listed. That is, on November 30, 1995, the price of a collar with one of these pairs of r_c and r_f should be zero. In Table 8.4, the corresponding prices of caps and floors are also shown.

Now let us discuss how to price swaps and swaptions – options on swaps. Let $V_s(r, t; r_s, T)$ be the value of an N-year swap with a swap rate r_s at time t when it is initiated at time T, $t \geq T$. Here, the notation is a little different from the notation used in Chapter 4: the time of the swap being initiated, T, is explicitly given in the notation as a parameter.

In order to find such a function at time $t = t^* \geq T$, we can first solve numerically the problem (4.60):

$$\begin{cases} \dfrac{\partial V_{s1}}{\partial t} + \dfrac{1}{2}w^2\dfrac{\partial^2 V_{s1}}{\partial r^2} + (u - \lambda w)\dfrac{\partial V_{s1}}{\partial r} - rV_{s1} + \sum_{k=1}^{2N} \delta(t - T - k/2) = 0, \\[4mm] \qquad\qquad\qquad\qquad\qquad r_l \le r \le r_u, \quad t \le T + N, \\[2mm] V_{s1}(r, T + N) = 0, \qquad\qquad r_l \le r \le r_u \end{cases}$$

and the problem (4.61):

$$\begin{cases} \dfrac{\partial V_{s2}}{\partial t} + \dfrac{1}{2}w^2\dfrac{\partial^2 V_{s2}}{\partial r^2} + (u - \lambda w)\dfrac{\partial V_{s2}}{\partial r} - rV_{s2} = 0, \\[4mm] \qquad\qquad\qquad\qquad\qquad r_l \le r \le r_u, \quad t \le T + N, \\[2mm] V_{s2}(r, T + N) = 1, \qquad\qquad r_l \le r \le r_u \end{cases}$$

from $t = T + N$ to $t = t^*$ and obtain $V_{s1}(r, t^*; T)$ and $V_{s2}(r, t^*; T)$, where instead of $V_{s1}(r, t^*)$ and $V_{s2}(r, t^*)$, we use $V_{s1}(r, t^*; T)$ and $V_{s2}(r, t^*; T)$ in order to indicate their dependence on the parameter T in (4.60) and (4.61). Then, the value of a swap at time $t = t^*$ is given by the expression (4.59):

$$V_s(r, t^*; r_s, T) = Q[1 - r_s V_{s1}(r, t^*; T)/2 - V_{s2}(r, t^*; T)],$$

where Q is the notional principal. The way to solve (4.60) and (4.61) numerically is the same as the way to solve (4.64) and (4.63), respectively. Thus, there is no difficulty to find their solutions.

Suppose that today $t = 0$ and $r = r^*$ and that the swap is initiated at $t = 0$. Then, the value of the swap today is given by

$$V_s(r^*, 0; r_s, 0) = Q[1 - r_s V_{s1}(r^*, 0; 0)/2 - V_{s2}(r^*, 0; 0)].$$

As we know, the value of a swap should be zero when it is initiated. From this condition, we know that the fixed swap rate in the contract should be

$$r_s(N) = \frac{1 - V_{s2}(r^*, 0; 0)}{V_{s1}(r^*, 0; 0)/2}. \tag{8.15}$$

As pointed out in Subsection 4.5.2, the swap rate should also be determined by (4.55). Based on the data on November 30, 1995, we can find $r_s(N)$ for $N = 2, 3, 5, 10$ by using (4.55) from the market data or by using (8.15) after getting $\lambda(t)$ and solving the bond equation. In Table 8.5, both results are given. They are very close, so we know that $\lambda(t)$ is determined accurately.

Because the expression (4.59) for $V_s(r, t; r_s, t)$ with $t = T$

$$V_s(r, T; r_s, T) = Q[1 - r_s V_{s1}(r, T; T)/2 - V_{s2}(r, T; T)]$$

gives the value of a swap when it is initiated for any $r \in [r_l, r_u]$ and any r_s, as soon as we have such an expression, we can find the values of European and American swaptions. Suppose that we want to calculate the price of a

Table 8.5. Swap rates on November 30, 1995

N	2	3	5	10
Market	0.05336	0.05425	0.05509	0.05728
Computed	0.05335	0.05423	0.05506	0.05712

T-year option on N-year swap with a notional principal Q and an exercise swap rate r_{se}. Let $V_{so}(r, t)$ represent the value of the swaption. $V_{so}(r, t)$ can be determined as follows. We first solve (4.60) and (4.61) from $T + N$ to T and find $V_{s1}(r, T; T)$ and $V_{s2}(r, T; T)$. Then, using (4.59) with $t = T$, we have the value of the swap with $r_s = r_{se}$ when it is initiated at time T:

$$V_s(r, T; r_{se}, T) = Q\left[1 - r_{se} V_{s1}(r, T; T)/2 - V_{s2}(r, T; T)\right].$$

As an option, the value of the swaption must be positive. Thus

$$V_{so}(r, T) = \max\left(V_s(r, T; r_{se}, T), 0\right).$$

Now we can solve (4.62):

$$\begin{cases} \dfrac{\partial V_{so}}{\partial t} + \dfrac{1}{2} w^2 \dfrac{\partial^2 V_{so}}{\partial r^2} + (u - \lambda w)\dfrac{\partial V_{so}}{\partial r} - r V_{so} = 0, \; r_l \leq r \leq r_u, \quad t \leq T, \\ V_{so}(r, T) = \max\left(V_s(r, T; r_{se}, T), 0\right), \qquad\qquad r_l \leq r \leq r_u \end{cases}$$

from $t = T$ to $t = 0$. $V_{so}(r^*, 0)$ gives today's value of the European swaption.

For an American swaption, its value $V_{so}(r, t)$ at any time $t \in [0, T]$ must be greater than or equal to $\max\left(V_s(r, t; r_{se}, t), 0\right)$:

$$V_{so}(r, t) \geq \max\left(V_s(r, t; r_{se}, t), 0\right), \tag{8.16}$$

where

$$V_s(r, t; r_{se}, t) = Q\left[1 - r_{se} V_{s1}(r, t; t)/2 - V_{s2}(r, t; t)\right].$$

Therefore, in order to obtain $V_{so}(r, t)$, we need to solve the following linear complementarity problem:

$$\begin{cases} \left[\dfrac{\partial V_{so}}{\partial t} + \dfrac{1}{2} w^2 \dfrac{\partial^2 V_{so}}{\partial r^2} + (u - \lambda w)\dfrac{\partial V_{so}}{\partial r} - r V_{so}\right] \\ \times \left[V_{so}(r, t) - \max\left(V_s(r, t; r_{se}, t), 0\right)\right] = 0, \\ \dfrac{\partial V_{so}}{\partial t} + \dfrac{1}{2} w^2 \dfrac{\partial^2 V_{so}}{\partial r^2} + (u - \lambda w)\dfrac{\partial V_{so}}{\partial r} - r V_{so} \leq 0, \\ V_{so}(r, t) - \max\left(V_s(r, t; r_{se}, t), 0\right) \geq 0, \\ V_{so}(r, T) = \max\left(V_s(r, T; r_{se}, T), 0\right), \end{cases} \tag{8.17}$$

where $t \in [0, T]$ and $r \in [r_l, r_u]$. Because $V_s(r, t; r_{se}, t)$ are not given functions, we have to solve (4.60) and (4.61) with $T = t$ from $t + N$ to t when $V_{so}(r, t)$ for time t needs to be determined. Of course, this problem can also be formulated as a free-boundary problem. The reader is asked to write down the free-boundary problem for this case as an exercise.

Table 8.6. Prices of European and American swaptions with $Q = 100$

(For each N, take the value of computed r_s given in Table 8.5
as r_{se} of the options on the swap)

	$T \backslash N$	2	3	5	10
European	0.5	0.167	0.196	0.269	0.278
	1	0.276	0.288	0.499	0.490
	2	0.492	0.548	1.083	1.021
American	0.5	0.213	0.248	0.331	0.342
	1	0.450	0.474	0.731	0.722
	2	0.678	0.753	1.338	1.273

The problems (4.62) and (8.17) can be solved by the scheme (5.37) or modified (5.31). In Table 8.6, we list some numerical results on European and American swaptions. The exercise swap rates r_{se} are 0.05335, 0.05423, 0.05506, 0.05712 for $N = 2, 3, 5, 10$, respectively. The other parameters are given in the table.

8.3 Pricing Derivatives with Multi-Factor Models

8.3.1 Determining Models from the Market Data

In Section 4.6, a three-factor interest rate model was proposed. In this section, we will discuss implicit finite-difference methods for the three-factor interest rate derivative problems and some other related problems. In order to use that model to price an interest rate derivative, we need to know how to find the payoff of the derivative and to determine those coefficients in the partial differential equation (4.82). In this subsection, we will discuss these two problems, and the next subsection is devoted to implicit finite-difference methods.

Suppose we want to price a half-year option on five-year swaps with an exercise swap rate r_{se}. Assume the day we want to price the swaption (the option on swaps) to be denoted as $t = 0$. Thus, according to the notation given in Subsection 4.5.2, $T = 0.5$ and $N = 5$.

First, let us discuss how to determine the final value. On the market, the prices of 3-month, 6-month, 1-year, 2-year, 3-year and 5-year zero-coupon bonds are given every day. Set $T_1^* = 0.25, T_2^* = 0.5, T_3^* = 1, T_4^* = 2, T_5^* = 3$, and $T_6^* = 5$, let Z_i denote the price of the bond with maturity T_i^*, and define

$S_i = Z_i/T_i^*$, $i = 1, 2, \cdots, 6$. Suppose we have these values on a period of L days and let $S_{i,l}$ stand for the value of S_i at the l-th day, $l = 1, 2, \cdots, L$. By b_i^2 and $b_i b_j \rho_{i,j}$, we denote the variance of S_i and the covariance between S_i and S_j, respectively. From statistics, we know that b_i^2 and $\rho_{i,j}$ can be estimated by

$$
\begin{aligned}
b_i^2 &= \frac{1}{L-1} \sum_{l=1}^{L} \left(S_{i,l} - \frac{1}{L} \sum_{l=1}^{L} S_{i,l} \right)^2 \\
&= \frac{1}{L-1} \left[\sum_{l=1}^{L} (S_{i,l})^2 - \frac{1}{L} \left(\sum_{l=1}^{L} S_{i,l} \right)^2 \right]
\end{aligned}
$$

and

$$
\begin{aligned}
\rho_{ij} &= \frac{\sum_{l=1}^{L} \left(S_{i,l} - \frac{1}{L} \sum_{l=1}^{L} S_{i,l} \right) \left(S_{j,l} - \frac{1}{L} \sum_{l=1}^{L} S_{j,l} \right)}{\sqrt{\left[\sum_{l=1}^{L} \left(S_{i,l} - \frac{1}{L} \sum_{l=1}^{L} S_{i,l} \right)^2 \times \sum_{l=1}^{L} \left(S_{j,l} - \frac{1}{L} \sum_{l=1}^{L} S_{j,l} \right)^2 \right]}} \\
&= \frac{\sum_{l=1}^{L} (S_{i,l} S_{j,l}) - \frac{1}{L} \left(\sum_{l=1}^{L} S_{i,l} \times \sum_{l=1}^{L} S_{j,l} \right)}{\sqrt{\left[\sum_{l=1}^{L} (S_{i,l})^2 - \frac{1}{L} \left(\sum_{l=1}^{L} S_{i,l} \right)^2 \right] \left[\sum_{l=1}^{L} (S_{j,l})^2 - \frac{1}{L} \left(\sum_{l=1}^{L} S_{j,l} \right)^2 \right]}}.
\end{aligned}
$$

Using the data for the period from January 4, 1982, to February 25, 2002, we obtain

$$
\begin{aligned}
\mathbf{B} &= \begin{bmatrix}
b_1^2 & b_1 b_2 \rho_{1,2} & \cdots & b_1 b_6 \rho_{1,6} \\
b_1 b_2 \rho_{1,2} & b_2^2 & \cdots & b_2 b_6 \rho_{2,6} \\
\vdots & \vdots & \ddots & \vdots \\
b_1 b_6 \rho_{1,6} & b_2 b_6 \rho_{2,6} & \cdots & b_6^2
\end{bmatrix} \\
&= 10^{-3} \begin{bmatrix}
0.4644 & 0.4224 & 0.2993 & 0.4758 & 0.4637 & 0.3776 \\
0.4224 & 0.4109 & 0.3014 & 0.4413 & 0.4392 & 0.3724 \\
0.2993 & 0.3014 & 0.2289 & 0.3145 & 0.3161 & 0.2766 \\
0.4758 & 0.4413 & 0.3145 & 0.4916 & 0.4818 & 0.3956 \\
0.4637 & 0.4392 & 0.3161 & 0.4818 & 0.4760 & 0.3952 \\
0.3776 & 0.3724 & 0.2766 & 0.3956 & 0.3952 & 0.3392
\end{bmatrix}.
\end{aligned}
$$

By the **QR** method given in Subsection 5.2.4 or other methods, we can find the eigenvalues and the unit eigenvectors of $\mathbf{B}$. As soon as we have them, $\mathbf{B}$ can be rewritten as

$$
\mathbf{B} = \mathbf{A}^T \mathbf{C} \mathbf{A},
$$

where

$$A = \begin{bmatrix} 0.4366 & 0.4151 & 0.3011 & 0.4533 & 0.4479 & 0.3745 \\ -0.5426 & 0.2650 & 0.5706 & -0.3546 & -0.0918 & 0.4190 \\ -0.5871 & 0.2779 & -0.5143 & 0.1231 & 0.5461 & -0.0121 \\ -0.3980 & -0.4305 & 0.3912 & 0.6808 & 0.0016 & -0.1994 \\ 0.1082 & -0.4366 & 0.2864 & -0.4337 & 0.7019 & -0.1869 \\ -0.0031 & -0.5516 & -0.2902 & 0.0448 & 0.0113 & 0.7806 \end{bmatrix}$$

and

$$C = 10^{-3} \times \text{diag}\,(2.366, 0.04109, 0.003240,$$
$$3.953 \times 10^{-4}, 1.996 \times 10^{-4}, 4.498 \times 10^{-5}).$$

Because the last three components of C are very small compared with the first three components, the six random variables, $S_1, S_2, \cdots, S_6$, almost depend on only three variables. Because

$$\begin{vmatrix} a_{1,1} & a_{1,4} & a_{1,6} \\ a_{4,1} & a_{4,4} & a_{4,6} \\ a_{6,1} & a_{6,4} & a_{6,6} \end{vmatrix} = \begin{vmatrix} 0.4366 & 0.4533 & 0.3745 \\ -0.3980 & 0.6808 & -0.1994 \\ -0.0031 & 0.0448 & 0.7806 \end{vmatrix} \approx 0.3711 \neq 0,$$

we can choose S_1, S_4, and S_6 as the three independent components, which will be denoted by S_{i_1}, S_{i_2}, and S_{i_3} in what follows. From Subsection 4.6.2, we know that the values of S_i, $i \neq i_1, i_2$, and i_3, are uniquely determined by (4.67) for a given set of S_{i_1}, S_{i_2}, and S_{i_3} when A is found and $S_i^*, i = 1, 2, \cdots, 6$, are specified. Based on the six values of $S_1, S_2, \cdots, S_6$, a zero-coupon bond curve with a maximum maturity $T_{\max}^* = 5$ can be found by using the cubic spline interpolation. Assume that for the period $t \in [0, T] = [0, 0.5]$, S_i^* are constants, for example, are equal to the values of zero-coupon bonds at $t = 0$. Thus, the possible zero-coupon bond curves for any $t \in [0, T]$ are the same, i.e.,

$$\bar{Z}\,(T^*; Z_{i_1}, Z_{i_2}, Z_{i_3}, t) = \bar{Z}\,(T^*; Z_{i_1}, Z_{i_2}, Z_{i_3}, 0).$$

Here in order to indicate the dependence of the zero-coupon bond curves on $Z_{i_1}, Z_{i_2}, Z_{i_3}$, instead of $\bar{Z}\,(T^*; t)$, we use $\bar{Z}\,(T^*; Z_{i_1}, Z_{i_2}, Z_{i_3}, t)$. As soon as we have a zero-coupon bond curve, using (4.54) with $r_s = r_{se}$:

$$Q \left[1 - \frac{r_{se}}{2} \sum_{k=1}^{2N} Z\,(T; T + k/2) - Z\,(T; T + N) \right],$$

we can determine the value of a swap with an exercise rate r_{se}. Here, Q is the notional principal and $Z(T; T + k/2) = \bar{Z}(k/2; Z_{i_1}, Z_{i_2}, Z_{i_3}, T) = \bar{Z}(k/2; Z_{i_1}, Z_{i_2}, Z_{i_3}, 0)$. Therefore, the final value of a swaption is

$$Q \max \left(1 - \frac{r_{se}}{2} \sum_{k=1}^{2N} \bar{Z}\,(k/2; Z_{i_1}, Z_{i_2}, Z_{i_3}, 0) - \bar{Z}\,(N; Z_{i_1}, Z_{i_2}, Z_{i_3}, 0), 0 \right).$$

$$(8.18)$$

Before discussing how to determine the coefficients in the partial differential equation, we would like to give some information about how these zero-coupon bond curves generated above are close to the real zero-coupon bond curves. Suppose that one day, the prices of zero-coupon bonds are

$$\begin{cases} Z_1 = 0.9811, \ Z_2 = 0.9559, \ Z_3 = 0.9047, \\ Z_4 = 0.7979, \ Z_5 = 0.7068 \quad \text{and} \quad Z_6 = 0.5475, \end{cases} \quad (8.19)$$

which correspond to the following interest rates:

$$\begin{cases} r_1 = 0.0776, \ r_2 = 0.0923, \ r_3 = 0.1027, \\ r_4 = 0.1161, \ r_5 = 0.1191 \quad \text{and} \quad r_6 = 0.1242. \end{cases}$$

Here, r_i is associated with Z_i by the following expression:

$$Z_i = (1 + r_i/2)^{-2N_i},$$

where N_i is the maturity of the i-th zero-coupon bond. From this set of data, we can determine a class of zero-coupon bond curves with $Z_{i_1}, Z_{i_2}, Z_{i_3}$ as parameters. For any day in the period from January 4, 1982, to February 25, 2002, we take the values of $Z_{i_1}, Z_{i_2}, Z_{i_3}$ as input and find a zero-coupon bond curve from the class. From the zero-coupon bond curve, we obtain the values of Z_i, $i \neq i_1, i_2$, and i_3, and the differences between the values determined from the curve and the values from the original market data. We do this for every day. The average value of the differences divided by $(1 - Z_i)$, $i \neq i_1, i_2$, and i_3, is 0.005. The same thing to the swap rate and to the value of the swaption on a 5-year swap with $r_{se} = 0.1225$ is also done. The maximum difference between the swap rates from the market curve and the model curve is 0.0004 (4 basis points), and the average difference is 0.00008 (0.8 basis points). The average error of the swaption value is 0.02 if the notional principal is 100. Therefore, we may conclude that these zero-coupon bond curves reflect the market situation.

Now let us discuss how to determine the coefficients in the partial differential equation. Suppose that derivative securities depend on $Z_{i_1}, Z_{i_2}, Z_{i_3}$, and t. Let

$$\begin{cases} \xi_1 = \dfrac{Z_{i_1} - Z_{i_1,l}}{1 - Z_{i_1,l}}, \\[2ex] \xi_2 = \dfrac{Z_{i_2} - Z_{i_2,l}}{Z_{i_1} - Z_{i_2,l}}, \\[2ex] \xi_3 = \dfrac{Z_{i_3} - Z_{i_3,l}}{Z_{i_2} - Z_{i_3,l}}, \end{cases} \quad (8.20)$$

where $Z_{i_1,l}, Z_{i_2,l}$, and $Z_{i_3,l}$ are minimums of $Z_{i_1}, Z_{i_2}, Z_{i_3}$ and we set $Z_{i_1,l} = 0.9597$, $Z_{i_2,l} = 0.7209$, and $Z_{i_3,l} = 0.4332$, which are a little less than the observed minimums 0.9634, 0.7463, and 0.4847, respectively. From Subsection

4.6.3, we know that the value of a derivative security, $V(\xi_1, \xi_2, \xi_3, t)$, satisfies (4.82), where coefficients depends on $r, \tilde{\sigma}_1, \tilde{\sigma}_2, \tilde{\sigma}_3, \tilde{\rho}_{1,2}, \tilde{\rho}_{1,3}, \tilde{\rho}_{2,3}$ besides ξ_1, ξ_2, and ξ_3. Therefore, in order to use that equation, we have to know $r, \tilde{\sigma}_1, \tilde{\sigma}_2, \tilde{\sigma}_3, \tilde{\rho}_{1,2}, \tilde{\rho}_{1,3}$, and $\tilde{\rho}_{2,3}$. It is clear that r can be determined by the slope of zero-coupon bond curves at the left end, i.e.,

$$r(\xi_1, \xi_2, \xi_3, t) = -\frac{\partial \bar{Z}}{\partial T^*}(0; Z_{i_1}, Z_{i_2}, Z_{i_3}, 0), \tag{8.21}$$

where

$$\begin{cases} Z_{i_1} = Z_{i_1,l} + \xi_1(1 - Z_{i_1,l}), \\ Z_{i_2} = Z_{i_2,l} + \xi_2[Z_{i_1,l} + \xi_1(1 - Z_{i_1,l}) - Z_{i_2,l}], \\ Z_{i_3} = Z_{i_3,l} + \xi_3\{Z_{i_2,l} + \xi_2[Z_{i_1,l} + \xi_1(1 - Z_{i_1,l}) - Z_{i_2,l}] - Z_{i_3,l}\}. \end{cases} \tag{8.22}$$

As we know, for $\tilde{\sigma}_1, \tilde{\sigma}_2, \tilde{\sigma}_3$ we need to require (4.84):

$$\begin{cases} \tilde{\sigma}_1(0, \xi_2, \xi_3, t) = \tilde{\sigma}_1(1, \xi_2, \xi_3, t) = 0, \\ \tilde{\sigma}_2(\xi_1, 0, \xi_3, t) = \tilde{\sigma}_2(\xi_1, 1, \xi_3, t) = 0, \\ \tilde{\sigma}_3(\xi_1, \xi_2, 0, t) = \tilde{\sigma}_3(\xi_1, \xi_2, 1, t) = 0. \end{cases}$$

Let us assume $\tilde{\sigma}_i$ to be in the form

$$\tilde{\sigma}_i(\xi_1, \xi_2, \xi_3, t) = \tilde{\sigma}_i(\xi_i) = \tilde{\sigma}_{i,0}\frac{1 - (1 - 2\xi_i)^2}{1 - p_i(1 - 2\xi_i)^2}, \quad i = 1, 2, 3, \tag{8.23}$$

where $\tilde{\sigma}_{i,0}$ and p_i are positive constants, and $p_i \in (0, 1)$. It is clear that in this case, condition (4.84) is fulfilled. On each day, we have the values of $Z_{i_1}, Z_{i_2}, Z_{i_3}$. Because ξ_1, ξ_2, ξ_3 are defined by (8.20), we can also have the values of ξ_1, ξ_2, ξ_3 every day. Therefore, we can find $\tilde{\sigma}_i(\xi_i)$ from the data on the market using the method described in Subsection 5.6.2 with

$$g(\xi_i) = \frac{1 - (1 - 2\xi_i)^2}{1 - p_i(1 - 2\xi_i)^2} \quad \text{and} \quad N = 0.$$

For $\tilde{\rho}_{1,2}, \tilde{\rho}_{1,3}$, and $\tilde{\rho}_{2,3}$, there is no requirement. We assume that they are constant and that the value can also be obtained using the method described in Subsection 5.6.2.

Taking $p_1 = p_2 = p_3 = 0.8$ and using the data on the market for the period between January 4, 1982, and February 25, 2002, we obtain

$$\tilde{\sigma}_{1,0} = 0.09733, \quad \tilde{\sigma}_{2,0} = 0.08622, \quad \tilde{\sigma}_{3,0} = 0.08148$$

and

$$\tilde{\rho}_{1,2} = 0.5682, \quad \tilde{\rho}_{1,3} = 0.4996, \quad \tilde{\rho}_{2,3} = 0.8585.$$

8.3.2 Numerical Methods and Results

From Section 4.6, we know that for a European swaption, $V(\xi_1, \xi_2, \xi_3, t)$ satisfies (4.82):

$$\begin{cases} \dfrac{\partial V}{\partial t} + \mathbf{L}_{3\xi} V = 0 \quad \text{on } \tilde{\Omega} \times [0, T], \\[2mm] V(\xi_1, \xi_2, \xi_3, T) = V_T\left(Z_{i_1}(\xi_1), Z_{i_2}(\xi_1, \xi_2), Z_{i_3}(\xi_1, \xi_2, \xi_3)\right) \quad \text{on } \tilde{\Omega}, \end{cases}$$

where $\tilde{\Omega}$ is the domain $[0,1] \times [0,1] \times [0,1]$ in the (ξ_1, ξ_2, ξ_3)-space, $\mathbf{L}_{3\xi}$ is defined by

$$\begin{aligned} \mathbf{L}_{3\xi} = {} & \frac{1}{2}\tilde{\sigma}_1^2 \frac{\partial^2}{\partial \xi_1^2} + \frac{1}{2}\tilde{\sigma}_2^2 \frac{\partial^2}{\partial \xi_2^2} + \frac{1}{2}\tilde{\sigma}_3^2 \frac{\partial^2}{\partial \xi_3^2} \\ & + \tilde{\sigma}_1 \tilde{\sigma}_2 \tilde{\rho}_{1,2} \frac{\partial^2}{\partial \xi_1 \partial \xi_2} + \tilde{\sigma}_1 \tilde{\sigma}_3 \tilde{\rho}_{1,3} \frac{\partial^2}{\partial \xi_1 \partial \xi_3} + \tilde{\sigma}_2 \tilde{\sigma}_3 \tilde{\rho}_{2,3} \frac{\partial^2}{\partial \xi_2 \partial \xi_3} \\ & + b_1 \frac{\partial}{\partial \xi_1} + b_2 \frac{\partial}{\partial \xi_2} + b_3 \frac{\partial}{\partial \xi_3} - r \end{aligned}$$

and $Z_{i_1}(\xi_1)$, $Z_{i_2}(\xi_1, \xi_2)$, $Z_{i_3}(\xi_1, \xi_2, \xi_3)$ are given by (8.22). For b_1, b_2, and b_3 we have expression (4.85):

$$\begin{cases} b_1 = \dfrac{r Z_{i_1}}{1 - Z_{i_1, l}}, \\[3mm] b_2 = \dfrac{r\left(Z_{i_2} - Z_{i_1}\xi_2\right)}{Z_{i_1} - Z_{i_2, l}} - \dfrac{\tilde{\sigma}_1 \tilde{\sigma}_2 \tilde{\rho}_{1,2}\left(1 - Z_{i_1, l}\right)}{Z_{i_1} - Z_{i_2, l}}, \\[3mm] b_3 = \dfrac{r\left(Z_{i_3} - Z_{i_2}\xi_3\right)}{Z_{i_2} - Z_{i_3, l}} - \dfrac{\tilde{\sigma}_1 \tilde{\sigma}_3 \tilde{\rho}_{1,3}\xi_2\left(1 - Z_{i_1, l}\right)}{Z_{i_2} - Z_{i_3, l}} - \dfrac{\tilde{\sigma}_2 \tilde{\sigma}_3 \tilde{\rho}_{2,3}\left(Z_{i_1} - Z_{i_2, l}\right)}{Z_{i_2} - Z_{i_3, l}} \end{cases}$$

and r is given by (8.21).

Let

$$\tau = T - t \quad \text{and} \quad \overline{V}(\xi_1, \xi_2, \xi_3, \tau) = V(\xi_1, \xi_2, \xi_3, T - \tau),$$

the above problem becomes

$$\begin{cases} \dfrac{\partial \overline{V}}{\partial \tau} = \mathbf{L}_{3\xi} \overline{V} \quad \text{on } \tilde{\Omega} \times [0, T], \\[2mm] \overline{V}(\xi_1, \xi_2, \xi_3, 0) = V_T\left(Z_{i_1}(\xi_1), Z_{i_2}(\xi_1, \xi_2), Z_{i_3}(\xi_1, \xi_2, \xi_3)\right) \quad \text{on } \tilde{\Omega}. \end{cases} \tag{8.24}$$

In the last subsection, we discussed how to determine the final value $V_T\left(Z_{i_1}(\xi_1), Z_{i_2}(\xi_1, \xi_2), Z_{i_3}(\xi_1, \xi_2, \xi_3)\right)$ for a swaption, which is given by (8.18), and find r, $\tilde{\sigma}_1$, $\tilde{\sigma}_2$, $\tilde{\sigma}_3$, $\tilde{\rho}_{1,2}$, $\tilde{\rho}_{1,3}$, $\tilde{\rho}_{2,3}$, $Z_{i_1, l}$, $Z_{i_2, l}$, and $Z_{i_3, l}$ from the market. Therefore, we have everything we need in order to solve (8.24) numerically.

Suppose that M, L, I, and N are given integers. Let $\Delta \xi_1 = 1/M$, $\Delta \xi_2 = 1/L$, $\Delta \xi_3 = 1/I$, and $\Delta \tau = T/N$ and $u_{m,l,i}^n$ be an approximate value of $\overline{V}$ at $\xi_1 = m\Delta \xi_1$, $\xi_2 = l\Delta \xi_2$, $\xi_3 = i\Delta \xi_3$, and $\tau = n\Delta \tau$. Here, $m = 0, 1, \cdots, M$, $l = 0, 1, \cdots, L$, $i = 0, 1, \cdots, I$ and $n = 0, 1, \cdots, N$.

The partial differential equation in (8.24) is discretized at $\tau = (n+1/2)\Delta \tau$, $n = 0, 1, \cdots, N - 1$. At any point, the partial derivative with respect to t is discretized by the central difference:

$$\frac{\partial \overline{V}_{m,l,i}^{n+1/2}}{\partial \tau} \approx \frac{u_{m,l,i}^{n+1} - u_{m,l,i}^n}{\Delta \tau}.$$

At any interior point, in $\tilde{\Omega}$, first- and second-order partial derivatives with respect to ξ_i are approximated by central schemes. For example,

$$\frac{\partial \overline{V}_{m,l,i}^{n+1/2}}{\partial \xi_1} \approx \frac{1}{2} \left(\frac{u_{m+1,l,i}^{n+1} - u_{m-1,l,i}^{n+1}}{2\Delta \xi_1} + \frac{u_{m+1,l,i}^n - u_{m-1,l,i}^n}{2\Delta \xi_1} \right)$$

and

$$\frac{\partial^2 \overline{V}_{m,l,i}^{n+1/2}}{\partial \xi_1^2} \approx \frac{1}{2} \left(\frac{u_{m+1,l,i}^{n+1} - 2u_{m,l,i}^{n+1} + u_{m-1,l,i}^{n+1}}{\Delta \xi_1^2} + \frac{u_{m+1,l,i}^n - 2u_{m,l,i}^n + u_{m-1,l,i}^n}{\Delta \xi_1^2} \right).$$

Mixed second-order partial derivatives are discretized by the central finite-difference for mixed partial derivatives. For example,

$$\frac{\partial^2 \overline{V}_{m,l,i}^{n+1/2}}{\partial \xi_1 \partial \xi_2} \approx \frac{1}{2} \left(\frac{u_{m+1,l+1,i}^{n+1} - u_{m+1,l-1,i}^{n+1} - u_{m-1,l+1,i}^{n+1} + u_{m-1,l-1,i}^{n+1}}{4\Delta \xi_1 \Delta \xi_2} \right.$$
$$\left. + \frac{u_{m+1,l+1,i}^n - u_{m+1,l-1,i}^n - u_{m-1,l+1,i}^n + u_{m-1,l-1,i}^n}{4\Delta \xi_1 \Delta \xi_2} \right).$$

At the boundary $\xi_1 = 0$, because $\tilde{\sigma}_1 = 0$, only $\dfrac{\partial}{\partial \xi_1}$, $\dfrac{\partial}{\partial \xi_2}$, $\dfrac{\partial}{\partial \xi_3}$, $\dfrac{\partial^2}{\partial \xi_2^2}$, $\dfrac{\partial^2}{\partial \xi_3^2}$, and $\dfrac{\partial^2}{\partial \xi_2 \partial \xi_3}$ appear in the partial differential equation. In this case, we can always deal with $\dfrac{\partial}{\partial \xi_1}$ by the second-order one-sided scheme:

$$\frac{\partial \overline{V}_{0,l,i}^{n+1/2}}{\partial \xi_1} \approx \frac{1}{2} \left(\frac{-u_{2,l,i}^{n+1} + 4u_{1,l,i}^{n+1} - 3u_{0,l,i}^{n+1}}{2\Delta \xi_1} + \frac{-u_{2,l,i}^n + 4u_{1,l,i}^n - 3u_{0,l,i}^n}{2\Delta \xi_1} \right)$$

because of $b_1 \geq 0$ at $\xi_1 = 0$. If $\xi_1 = 0$, $\xi_2 \neq 0$, $\xi_2 \neq 1$, $\xi_3 \neq 0$, and $\xi_3 \neq 1$, then $\dfrac{\partial}{\partial \xi_2}$, $\dfrac{\partial}{\partial \xi_3}$, $\dfrac{\partial^2}{\partial \xi_2^2}$, $\dfrac{\partial^2}{\partial \xi_3^2}$, and $\dfrac{\partial^2}{\partial \xi_2 \partial \xi_3}$ can still be discretized by central schemes. If $\xi_1 = 0$, $\xi_3 \neq 0$, $\xi_3 \neq 1$, and $\xi_2 = 0$ or $\xi_2 = 1$, then both $\tilde{\sigma}_1$ and $\tilde{\sigma}_2$ are equal to zero and only $\dfrac{\partial}{\partial \xi_1}$, $\dfrac{\partial}{\partial \xi_2}$, $\dfrac{\partial}{\partial \xi_3}$, and $\dfrac{\partial^2}{\partial \xi_3^2}$ are left. The treatment of

$\dfrac{\partial}{\partial \xi_1}$, $\dfrac{\partial}{\partial \xi_3}$, and $\dfrac{\partial^2}{\partial \xi_3^2}$ is unchanged, and $\dfrac{\partial}{\partial \xi_2}$ is approximated by the second-order one-sided differences. For example, at $\xi_2 = 1$, we can use the following approximation:

$$
\frac{\partial \overline{V}_{m,L,i}^{n+1/2}}{\partial \xi_2} \approx \frac{1}{2} \left(\frac{3u_{m,L,i}^{n+1} - 4u_{m,L-1,i}^{n+1} + u_{m,L-2,i}^{n+1}}{2\Delta \xi_2} \right.
$$
$$
\left. + \frac{3u_{m,L,i}^{n} - 4u_{m,L-1,i}^{n} + u_{m,L-2,i}^{n}}{2\Delta \xi_2} \right)
$$

because of $b_2 \leq 0$ at $\xi_2 = 1$. If $\xi_1 = 0$, $\xi_2 = 0$ or 1, and $\xi_3 = 0$ or 1, then $\tilde{\sigma}_1 = \tilde{\sigma}_2 = \tilde{\sigma}_3 = 0$ and only $\dfrac{\partial}{\partial \xi_1}$, $\dfrac{\partial}{\partial \xi_2}$, and $\dfrac{\partial}{\partial \xi_3}$ are left. In this case, all of them need to be dealt with by proper one-sided second-order differences.

For the other boundaries, the situations are similar. All the approximations have a second-order accuracy. In order for the truncation error of the finite-difference equations to have a second-order accuracy, all the coefficients need to take values at the point: $\xi_1 = m\Delta \xi_1$, $\xi_2 = l\Delta \xi_2$, $\xi_3 = i\Delta \xi_3$, and $\tau = (n + 1/2)\Delta \tau$, and the term $r\overline{V}$ should be approximated by

$$
\frac{1}{2} \left(r_{m,l,i}^{n+1} u_{m,l,i}^{n+1} + r_{m,l,i}^{n} u_{m,l,i}^{n} \right).
$$

The number of the finite-difference equations for the time level $\tau = (n + 1/2)\Delta \tau$ is $(M + 1) \times (L + 1) \times (I + 1)$. If all the values $u_{m,l,i}^{n}$ are known, the number of unknowns $u_{m,l,i}^{n+1}$ is equal to the number of the equations. Thus, $u_{m,l,i}^{n+1}$ can be determined by the system. It is clear that this system is linear. This system is quite large and usually solved by iteration methods, for example, by successive over relaxation described in Subsection 5.2.2 because iteration methods need less memory space and are usually more efficient than direct methods for this case. The initial condition $\overline{V}(\xi_1, \xi_2, \xi_3, 0) = V_T \left(Z_{i_1}(\xi_1), Z_{i_2}(\xi_1, \xi_2), Z_{i_3}(\xi_1, \xi_2, \xi_3) \right)$ gives $u_{m,l,i}^{0}$. Thus, the computation can start with $n = 0$ and continue for $n = 1, 2, \cdots, N - 1$ successively. Finally, we obtain $u_{m,l,i}^{N}$, the price of the derivative security at time $t = 0$.

This problem can also be solved by explicit schemes. If the partial differential equation is discretized at $\tau = n\Delta \tau$ and the time derivative is approximated by the forward finite-difference, then we have an explicit scheme. In this case, $\Delta \tau$ should be small enough so that the stability of computation is guaranteed.

For American swaptions, the value must be greater than or equal to the constraint. In the model here, the value of the constraint does not depend on t and equals to (8.18). Therefore, for American swaptions, the method needs to be modified in the following way. At each time step, we should choose the

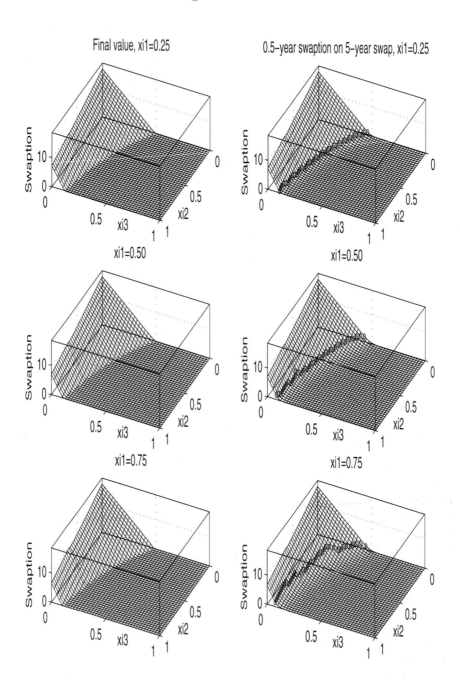

Fig. 8.7. The price of a swaption

maximum between the computed value by the PDE and the constraint (8.18) as the value of an American swaption.

Consider an American swaption with $r_{se} = 0.1225$, $T = 0.5$, and $N = 5$. We want to have the price of the swaption today. Suppose that the prices of zero-coupon bonds today are given by (8.19), then we can use the numerical methods described here to find the price of the American swaption value. Its value today for $\xi_1 = 0.25, 0.5, 0.75$, $0 \leq \xi_2 \leq 1$, and $0 \leq \xi_3 \leq 1$ is shown by the right three graphs in Fig. 8.7. There, the circles are the approximate locations of the free boundary. The final value is also plotted on the left-hand side for comparison. We can see that the derivative of the final value is discontinuous and that for the solution at $t = 0$ it is continuous. The result shown in this figure is obtained by the implicit scheme. When for $\tilde{\sigma}_i$, $i = 1, 2, 3$ and $\tilde{\rho}_{i,j}$, $i, j = 1, 2, 3$, we use more complicated expressions, the procedure of evaluating the interest rate derivatives is the same. For the details of the procedure and more results, see [77].

Problems

1. *Define
$$\mathbf{L}_r = \frac{\partial}{\partial r}\left[f_1(r,t)\frac{\partial}{\partial r}\right] - f_2(r,t)\frac{\partial}{\partial r} + f_3(r,t).$$

a) Find an operator $\mathbf{L}_r^*$ such that

$$\int_{r_l}^{r_u} \mathbf{L}_r V\, U dr = \int_{r_l}^{r_u} \mathbf{L}_r^* U\, V dr + \left[f_1\left(U\frac{\partial V}{\partial r} - V\frac{\partial U}{\partial r}\right) - f_2 V U\right]\Big|_{r_l}^{r_u}.$$

This operator is called the conjugate operator of $\mathbf{L}_r$.
b) Suppose
$$\frac{\partial V}{\partial t} = -\mathbf{L}_r V, \quad \frac{\partial U}{\partial t} = \mathbf{L}_r^* U,$$
$$f_1(r_l,t) = f_1(r_u,t) = 0, \quad f_2(r_l,t) \leq 0, \quad f_2(r_u,t) \geq 0$$

and
$$U(r_l,t) = U(r_u,t) = 0.$$

Show
$$\int_{r_l}^{r_u} U(r,t)V(r,t)dr = constant.$$

c) Let $U(r,0) = \delta(r - r^*)$ and $V(r,T^*) = 1$. Prove that there is the following relation:
$$V(r^*,0) = \int_{r_l}^{r_u} U(r,T^*)dr.$$

2. *Consider the following problem

$$
\begin{cases}
\dfrac{\partial U}{\partial t} = \dfrac{\partial}{\partial r}\left[f_1(r,t)\dfrac{\partial U}{\partial r}\right] + \dfrac{\partial}{\partial r}\left[f_2\left(r,t,\lambda(t)\sqrt{f_1}\right) U\right] + f_3(r,t)U, & \\
 & r_l \le r \le r_u, \quad 0 \le t, \\
U(r,0) = \delta(r-r^*), & r_l \le r \le r_u, \\
U(r_l,t) = 0, & 0 \le t, \\
U(r_u,t) = 0, & 0 \le t,
\end{cases}
$$

where
$$f_1(r,t) \ge 0 \quad \text{and} \quad f_1(r_l,t) = f_1(r_u,t) = 0,$$
and
$$f_2\left(r_l,t,\lambda(t)\sqrt{f_1}\right) \le 0, \quad f_2\left(r_u,t,\lambda(t)\sqrt{f_1}\right) \ge 0.$$

Here, $\lambda(t)$ is a unknown function with a known $\lambda(0)$. Design a second-order numerical method such that
$$\int_{r_l}^{r_u} U(r,t)dr = f(t) \qquad \text{for any } t \in [0, T_{max}^*],$$

where $f(t)$ is a given function.

3. *Design a numerical method for finding the market price of risk by taking the prices of today's zero-coupon bonds with various maturities as input.

4. Design an implicit second-order accurate finite-difference method based on the bond equation to solve the European bond option problem.

5. Design an explicit first-order accurate finite-difference method based on the bond equation to solve a cap problem.

6. What is the difference between the numerical methods for a cap problem and for a floor problem if the bond equation is adopted.

7. Design an implicit second-order accurate finite-difference method based on the bond equation to solve the European swaption problem.

8. What is the difference between the numerical methods for the European swaption problem and for the American swaption problem formulated as a linear complementarity problem if the bond equation is adopted.

9. Assume that the prices of American swaptions are the solutions of the following linear complementarity problem:

$$
\begin{cases}
\left[\dfrac{\partial V_{so}}{\partial t} + \dfrac{1}{2}w^2\dfrac{\partial^2 V_{so}}{\partial r^2} + (u-\lambda w)\dfrac{\partial V_{so}}{\partial r} - rV_{so}\right] \\
\quad \times \left[V_{so}(r,t) - \max\left(V_s(r;t,r_{se}),0\right)\right] = 0, \\
\dfrac{\partial V_{so}}{\partial t} + \dfrac{1}{2}w^2\dfrac{\partial^2 V_{so}}{\partial r^2} + (u-\lambda w)\dfrac{\partial V_{so}}{\partial r} - rV_{so} \le 0, \\
V_{so}(r,t) - \max\left(V_s(r;t,r_{se}),0\right) \ge 0, \\
V_{so}(r,T) = \max\left(V_s(r;T,r_{se}),0\right),
\end{cases}
$$

where $t \in [0, T]$ and $r \in [r_l, r_u]$ and $V_s(r; t, r_{se})$ is the price of the swap. Suppose that the price of the swap has been found and assume that there is only one free boundary. Formulate this problem as a free-boundary problem.

10. Design an implicit second-order accurate finite-difference method to solve the free-boundary problem of the American swaption problem based on the bond equation.

11. *Describe how to solve a European swaption problem by using the three-factor interest rate model.

12. Design a method for determining the value of a bond option by using a two-factor or three-factor interest rate model.

13. Design a method for evaluating the price of a cap by using a two-factor or three-factor interest rate model.

Projects

General Requirements

A) *Submit a code or codes in C or C++ that will work on a computer the instructor can get access to. At the beginning of the code, write down the name of the student and indicate on which computer it works and the procedure to make it work.*

B) *Each code should use an input file to specify all the problem parameters and the computational parameters for each computation and an output file to store all the results. In an output file, the name of the student, all the problem parameters, and the computational parameters should be given, so that one can know what the results are and how they were obtained. The input file should be submitted with the code.*

C) *For each case, two results are required. One result is on a 20 × 20 mesh, and the accuracy of the other result will be specified individually. (The error of the solution on a 20 × 20 mesh might be quite large.)*

D) *Submit results in form of tables. When a result is given, always provide the problem parameters and the computational parameters.*

1. **Implicit method (5.31) with modification at the boundaries for European bond options and swaptions.** Suppose

$$dr = (0.05345 - r)dt + r(0.2 - r)dX, \quad r_l = 0 \le r \le r_u = 0.2$$

and $\lambda(t)$ has been found and is given as a function in C. Also, assume that today's spot interest rate is 0.05345. Write a code to calculate European bond options and a code to calculate European swaptions.

- For European bond options, give results for the case: $E = 0.95, 1$, $k = 0.055$, $T = 0.25, 0.5$, and $T_b - T = 1, 2$. The requirement on the accuracy of the other result is 0.0001, and the mesh used should be as coarse as possible.

- For swaptions, give the results for the cases: $Q = 100$, $N = 5, 10$, $T = 0.5, 1, 2$, $r_{se} = 0.05506$ for $N = 5$, and $r_{se} = 0.05712$ for $N = 10$. The requirement on the accuracy of the other result is 0.001, and the mesh used should be as coarse as possible.

References

1. Andersen, T.G., Lund, J.: Estimating continuous-time stochastic volatility models of the short-term interest rate. Journal of Econometrics, **77**, 343–377 (1997)
2. Andreasen, J.: The pricing of discretely sampled Asian and lookback options: a change of numeraire approach. Journal of Computational Finance, **2**, 5–30 (1998)
3. Badea, L., Wang, J.: A new formulation for the valuation of American options, I: solution uniqueness. In: Park, E.-J., Lee, J.-W. (eds) Analysis and Scientific Computing, Proceeding of the 19th Daewoo Workshop in Analysis and Scientific Computing, 3–16. Kyowoosa Publishing Co., Ltd, Seoul (2000)
4. Badea, L., Wang, J.: A new formulation for the valuation of American options, II: solution existence. In Park, E.-J., Lee, J.-W. (eds) Analysis and Scientific Computing, Proceeding of the 19th Daewoo Workshop in Analysis and Scientific Computing, 17–33. Kyowoosa Publishing Co., Ltd, Seoul (2000)
5. Baxter, M., Rennie, A.: Financial Calculus, An introduction to derivative pricing. Cambridge University Press, Cambridge (1996)
6. Behboudi, R.: Existence and regularity of generalized solutions of degenerate parabolic problems via finite difference schemes. Ph.D. dissertation, University of North Carolina at Charlotte, Charlotte (2002)
7. Black, F.: The pricing of commodity contracts. Journal of Financial Economics, **3**, 167–179 (1976)
8. Black, F., Derman, E., Toy, W.: A one-factor model of interest rates and its application to treasury bond options. Financial Analysts Journal, **46**, 33–39 (1990)
9. Black, F., Karasinski, P.: Bond and option pricing when short rates are lognormal. Financial Analysts Journal, **47**, 52–59 (1991)
10. Black, F., Scholes, M.: The pricing of options and corporate liabilities. Journal of Political Economy, **81**, 637–654 (1973)
11. Brennan, M., Schwartz, E.: Analyzing convertible bonds. Journal of Financial and Quantitative Analysis, **15**, 907–929 (1980)
12. Brennan, M.J., Schwartz, E.S.: An equilibrium model of bond pricing and a test of market efficiency. Journal of Financial and Quantitative Analysis, **17**, 301–329 (1982)

504   References

13. Broadie, M., Detemple, J.: American option valuation: new bounds, approximations, and a comparison of existing methods. Review of Financial Studies, **9**, 1211-1250 (1996)
14. Brooks, R.: Multivariate contingent claims analysis with cross-currency options as an illustration. Journal of Financial Engineering, **2**, 196–218 (1993)
15. Brooks, R., Corson, J., Wales, J.D.: The pricing of index options when the underlying assets all follow a lognormal diffusion. Advances in Futures and Options Research, **7**, 65–85 (1994)
16. Chen, B.-m.: Modelling of American average strike options. Working paper, Academy of Mathematics and System Sciences, Chinese Academy of Sciences, AMSS-2000-129 (2000)
17. Chen, B.-m., Liang Z.-x.: A numerical method for American barrier options. Working paper, Academy of Mathematics and System Sciences, Chinese Academy of Sciences, AMSS-1999-066 (1999)
18. Chen, R.R., Scott, L.: Pricing interest rate options in a two-factor Cox–Ingersoll–Ross model of the term structure. Review of Financial Studies, **5**, 613–636 (1992)
19. Chesney, M., Cornwall, J., Jeanblanc-Picqué, M., Kentwell, G., Yor, M.: Parisian Pricing. Risk, **10**, 77–79 (1997)
20. Conze, A., Viswanathan: Path dependent options: the case of lookback options. Journal of Finance, **46**, 1893–1907 (1991)
21. Cox, J.C., Ross, S.A., Rubinstein, M.: Option pricing: a simplified approach. Journal of Financial Economics, **7**, 229–263 (1979)
22. Cox, J.C., Ingersoll, J.E., Ross, S.A.: A theory of the term structure of interest rates. Econometrica, **53**, 385–407 (1985)
23. Detemple, J.: American options: symmetry properties. In: Jouini, E., Cvitanic, J., Musiela, M. (eds) Option pricing, interest rates and risk management, 67–104. Cambridge University Press, Cambridge (2001)
24. Druskin, V., Knizhnerman, L., Tamarchenko, T., Kostek, S.: Krylov subspace reduction and its extensions for option pricing. Journal of Computational Finance, **1**, 63–79 (1997)
25. Duffie, D., Harrison, J.M.: Arbitrage pricing of Russian options and perpetual lookback options. Annals of Applied Probability, **3**, 641–651 (1993)
26. Duffie, D.: Dynamic Asset Pricing Theory, Second Edition. Princeton University Press, Princeton, New Jersey (1996)
27. Fong, H.G., Vasicek, O.A.: Fixed-income volatility management. Journal of Portfolio Management, **17**, 41–46 (1991)
28. Friedman, A.: Variational Principles and Free-Boundary Problems. John Wiley & Sons, Inc., New York (1982)
29. Geske, R.: The valuation of compound options. Journal of Financial Economics, **7**, 63-81 (1979)
30. Gihman, I.I., Skorohod, A.N.: Stochastic Differential Equations. Springer-Verlag, New York (1972)
31. Goldman, M.B., Sosin, H.B., Gatto, M.A.: Path dependent options: "buy at the low, sell at the high." Journal of Finance, **34**, 1111–1127 (1979)
32. Golub, G.H., Van Loan, C.F.: Matrix computations, Third Edition. Johns Hopkins University Press, Baltimore (1996)
33. Gottlieb, D., Hussaini, M.Y., Orszag, S.A.: Theory and applications of spectral methods. In: Voigt, R.G., Gottlieb, D., Hussaini, M.Y. (eds) Spectral Methods for Partial Differential Equations, 1–54. Society for Industrial and Applied Mathematics (SIAM), Philadelphia (1984)

34. Haber, R.J., Schönbucher, P.J., Wilmott, P: Pricing Parisian options. Journal of Derivatives, **6**, 71–79 (1999)

35. Heath, D., Jarrow, R., Morton, A.: Bond pricing and the term structure of interest rates: a discrete time approximation. Journal of Financial and Quantitative Analysis, **25**, 419–440 (1990)

36. Heath, D., Jarrow, R., Morton, A.: Contingent claim valuation with a random evolution of interest rates. Review of Futures Markets, **9**, 54–76 (1990)

37. Heath, D., Jarrow, J., Morton, A.: Bond pricing and the term structure of interest rates: a new methodology for contingent claims valuation. Econometrica, **60**, 77–105 (1992)

38. Ho, T.S.Y., Lee, S.B.: Term structure movements and pricing interest rate contingent claims. Journal of Finance, **41**, 1011–1029 (1986)

39. Hull, J.C.: Options, Futures and Other Derivatives, Fourth Edition. Prentice Hall, Upper Saddle River, New Jersey (2000)

40. Hull, J.C., White, A.D.: Pricing interest-rate-derivative securities. Review of Financial Studies, **3**, 573–592 (1990)

41. Huynh, C.B.: Back to baskets. Risk, **7**, 59–61 (1994)

42. Ingersoll, J.: Theory of Financial Decision Making, Rowman and Littlefield, Totowa, New Jersey (1987)

43. James, J., Webber, N.: Interest Rate Modelling. John Wiley & Sons, Ltd, England (2000)

44. Jarrow, R.A.: Modelling Fixed Income Securities and Interest Rate Options. The McGraw-Hill Companies, Inc., New York (1996)

45. Johnson, H.: Options on the maximum or the minimum of several assets. Journal of Financial and Quantitative Analysis, **22**, 277–283 (1987)

46. Karatzas, I., Shreve, S.E.: Brownian Motion and Stochastic Calculus. Springer-Verlag, New York (1991)

47. Kevorkian, J.: Partial Differential Equations: Analytical solution techniques. Chapman & Hall, London (1990)

48. Kholodnyi, V.A., Price, J.F.: Foreign Exchange Option Symmetry. World Scientific, Singapore (1998)

49. Kwok, Y.K.: Mathematical Models of Financial Derivatives. Springer-Verlag, Singapore (1998)

50. Lamberton, D., Lapeyre, B.: Introduction to Stochastic Calculus Applied to Finance. Chapman and Hall, London (1996)

51. Li, J.: Numerical solutions for American options on assets with stochastic volatilities. Ph.D. dissertation, University of North Carolina at Charlotte, Charlotte (2002)

52. Longstaff, F.A., Schwartz, E.S.: Interest rate volatility and the term structure: a two-factor general equilibrium model. Journal of Finance, **47**, 1259–1282 (1992)

53. Luo, J., Wu, X.: Numerical methods for pricing Parisian options. Working paper, Tongji University, Shanghai (2001)

54. Lyuu, Y.-D.: Financial Engineering and Computation. Cambridge University Press, Cambridge (2002)

55. Maryrabe, W.: The value of an option to exchange one asset for another. Journal of Finance, **33**, 177–186 (1978)

56. McDonald, R.L., Schroder, M.D.: A parity result for American options. Journal of Computational Finance, **1**, 5–13 (1998)

57. Merton, R.C.: Theory of rational option pricing. Bell Journal of Economics and Management Sciences, **4**, 141–183 (1973)

58. Musiela, M., Rutkowski, M.: Martingale Methods in Financial Modelling. Springer-Verlag, Berlin and Heidelberg (1997)
59. Oleĭnik, O.A., Radkevič, E.V.: Second Order Equations With Nonnegative Characteristic Form. American Mathematical Society, Providence, Rhode Island and Plenum Press, New York (1973)
60. Pelsser, A.: Efficient Methods for Valuing Interest Rate Derivatives. Springer-Verlag, London (2000)
61. Rogers, L., Shi, Z.: The value of an Asian option, Journal of Applied Probability, **32**, 1077–1088 (1995)
62. Rubinstein, M.: Options for the undecided. In: From Black–Scholes to Black Holes, New frontiers in options, 187–189. Risk Magazine Ltd, London (1992)
63. Rumsey, J.: Pricing cross-currency options. Journal of Futures Markets, **11**, 89–93 (1991)
64. Saad, Y.: Iterative Methods for Sparse Linear Systems. PWS Publishing Co., Boston (1996)
65. Smithson, C.: Multfactors options. Risk, **10**, 43–45 (1997)
66. Stulz, R.M.: Options on the minimum or the maximum of two risky assets: analysis and applications. Journal of Financial Economics, **10**, 161–185 (1982)
67. Sun, Y.: High order methods for evaluating convertible bonds. Ph.D. dissertation, University of North Carolina at Charlotte, Charlotte (1999)
68. Sun, Z.-z., Yan, N.-n., Zhu, Y-l.: Convergence of second-order difference schemes and extrapolation algorithm for degenerate parabolic equations. Working paper, University of North Carolina at Charlotte, Charlotte (2001)
69. Tavella, D., Randall, C.: Pricing Financial Instruments: The Finite Difference Method. John Wiley & Sons, Inc., New York (2000)
70. Vasicek, O.A.: An equilibrium characterization of the term structure. Journal of Financial Economics, **5**, 177–188 (1977)
71. Wilmott, P., Dewynne, J., Howison, S.: Option Pricing, Mathematical models and computation. Oxford Financial Press, Oxford (1993)
72. Wilmott, P.: Derivatives: The theory and practice of financial engineering. John Wiley & Sons, Ltd, Chichester (1998)
73. Wilmott, P.: Paul Wilmott on Quantitative Finance, John Wiley & Sons, Ltd, Chichester (2000)
74. Wu, L., Kwok, Y.-K.: A front-fixing finite difference method for the valuation of American options. Journal of Financial Engineering, **6**, 83–97 (1997)
75. Xu, Q.-s., Zhu, Y.-l.: Solution of two dimensional Stefan problem by the singularity-separating method. Journal of Computational Mathematics, **3**, 8–18 (1985)
76. Yan, J.-A.: Introduction to Martingale Methods in Option Pricing. Lin Bie Ju Centre for Mathematical Sciences, City University of Hong Kong, Hong Kong (1998)
77. Yang, C.: Evaluating interest rate derivatives by using one-factor and three-factor models. Ph.D. dissertation, University of North Carolina at Charlotte, Charlotte (2004)
78. Zhang, G.-Q, Li, P.-J.: An inverse problem of derivative security pricing. In: Hon, Y.-C., Yamamoto, M., Cheng, J., Lee, J.-Y. (eds) Recent Development in Theory & Numerics, International Conference on Inverse Problems, 411–419. World Scientific, Singapore (2003)

79. Zhang, J.: A semi-analytical method for pricing and hedging continuously sampled arithmetic average rate options. Journal of Computational Finance, **5**, 59–79 (2001)

80. Zhu, Y.-l.: Evaluation of discretely sampled Asian options by finite-difference methods. Working paper, University of North Carolina at Charlotte, Charlotte (2000)

81. Zhu, Y.-l.: Three-factor interest rate models. Communications in Mathematical Sciences, **1**, 557-573 (2003)

82. Zhu, Y.-l., Abifaker, A.: Applications of the singularity-separating method to double moving barrier options with rebates. Working paper, University of North Carolina at Charlotte, Charlotte (1999)

83. Zhu, Y.-l., Chen, B.m., Ren, H., Xu, H.: Application of the singularity-separating method to American exotic option pricing. Advances in Computational Mathematics, **19**, 147–158 (2003)

84. Zhu, Y.-l., Li, J.: Numerical solutions for two-factor American options. Working paper, University of North Carolina at Charlotte, Charlotte (2002)

85. Zhu, Y.-l., Li, J.: Multi-factor financial derivatives on finite domains. Communications in Mathematical Sciences, **1**, 343–359 (2003)

86. Zhu, Y.-l., Ren, H., Xu, H.: Improved effectiveness evaluating American options by the singularity-separating method, Working paper, University of North Carolina of Charlotte, Charlotte (1996)

87. Zhu, Y.-l., Sun, Y.: The singularity-separating method for two-factor convertible bonds. Journal of Computational Finance, **3**, 91–110 (1999)

88. Zhu, Y.-l., Zhong, X.-c., Chen, B.-m., Zhang, Z.-m.: Difference Methods for Initial-Boundary-Value Problems and Flow Around Bodies. Springer-Verlag, Heidelberg, and Science Press, Beijing (1988)

Index